Insect Phylogeny

Insect Phylogeny

WILLI HENNIG

Translated and edited by
ADRIAN C. PONT

Revisionary notes by
DIETER SCHLEE

With the collaboration of
Michael Achtelig, Martin Baehr, Ragnar Kinzelbach,
Eberhard Königsmann, Niels Kristensen, Gerhard Mickoleit,
Wolfgang Seeger, Rainer Willmann, and Peter Zwick

A Wiley-Interscience Publication

JOHN WILEY & SONS
Chichester · New York · Brisbane · Toronto

British Library Cataloguing in Publication Data:

Hennig, Willi
  Insect phylogeny.
  1. Insects—Evolution
  I. Title   II. Pont, Adrian Charles
  III. Schlee, Dieter
  595.7′03′8   OL468.7   80-40853

ISBN 0 471 27848 3

Text set in 10/12pt Linotron 202 Times, printed and bound
in Great Britain at The Pitman Press, Bath

Contents

viii

Foreword

The gods did not reveal, from the beginning,
All things to us, but in the course of time
Through seeking we may learn and know things better.
But as for certain truth, no man has known it,
Nor shall he know it, neither of the gods
Nor yet of all the things of which I speak.
For even if by chance he were to utter
The final truth, he would himself not know it:
For all is but a woven web of guesses.

(Xenophanes, translated by Karl Popper)

Hennig's *Die Stammesgeschichte der Insekten* was published in German in 1969, and has been hailed by N. P. Kristensen (1975, p. 1) as 'one of the most important milestones in the study of the phylogeny of the insects and their allies'. In view of this fundamental importance, I was all the more surprised to learn in December 1976, when the idea of the present translation was first mooted, that no English translation had ever been attempted and that the book had remained largely unknown in the English-speaking world. Professor Claude Dupuis, in a detailed review of the historical development, theoretical basis, and diffusion of Hennigian phylogenetic systematics, has also expressed surprise that no English translation of this book had been made. He wrote that the opening chapter of this work 'représente l'expression la plus complète, la plus nuancée, et peut-être la plus heureuse de la pensée de Hennig' (Dupuis, 1979, p. 10), but that it has been almost completely ignored outside Germany, much as *Grundzüge einer Theorie der phylogenetischen Systematik* was prior to the English version of 1966.

It is a matter of deep regret that Willi Hennig died prematurely on 5 November 1976 and was unable to undertake a revision of the text, as he had done for his earlier book *Grundzüge einer Theorie der phylogenetischen Systematik* (1950; revised and translated as *Phylogenetic Systematics*, 1966). It is extremely fortunate, however, that Dr Dieter Schlee, Hennig's erstwhile colleague in the Department of Phylogenetic Research in the State Museum of Natural History, Stuttgart, was willing to update the text of the book and was able to enlist the collaboration of another nine specialists to assist with this task. Furthermore, Professor Wolfgang Hennig, the author's son, has made available the annotations and corrections in Hennig's own working copy of the book. The present edition therefore has the advantage of presenting Hennig's original ideas and arguments, together with up-to-date views and synopses of more recent work. The alternative of a team of specialists completely re-writing the book is scarcely practicable at present and would hardly give the uniform and consistent overview of insect phylogeny that is one of the conspicuous merits of Hennig's book.

This edition, then, contains an unmutilated translation of Hennig's original text. At the end of each section are groups of revisionary notes and comments which are numbered consecutively throughout the book and which are clearly attributed to their authors. Most have come from Dr Schlee and his collaborators, but some are from Hennig's personal copy of the book and a few have been added by myself. It should be noted that many of Hennig's notes are no more than non-committal references to papers that he evidently intended to discuss when preparing a revised edition of the book. My own role, in addition to translating the original text, has been to correct the English of the revisionary notes organized by Dr Schlee or, in a few cases, to translate them from the German; and to translate and intercalate the notes provided by Professor Wolfgang Hennig. I have also edited the book to the extent of correcting obvious slips in the original such as spellings, omissions of dates, references, or sources of figures, revising the original bibliography, and adding certain items. I should stress, however, that in doing this I have not tampered with the text by changing the original meaning or emphasis. I have prepared a revised bibliography, author index, and subject index to include the material from the revisionary notes, have prepared the final typescript, and have seen the book through the press. I should add that all geographical localities have been modernized and, so far as possible, quoted in the form given in the *Times Atlas of the World*, Comprehensive Edition (1968). Russian names and localities have been transliterated following the Royal Geographical Society US Board (BGN/PCGN), 1959, and consequently appear in English versions rather than German (e.g. Usachev and Chernova for Ussatschov and Tshernova). New illustrations have been given Roman numerals in order to avoid changing the sequence and numbering from the original edition.

Translations come in many forms. At one extreme is the dream of every schoolboy who discovers a word-for-word 'crib' for Caesar's *Gallic Wars* or Horace's *Odes*. At the other is a translation such as Edward Fitzgerald's *Rubaiyyet of Omar Khayaam*, where the original is transmogrified into a fresh and unique work of art (even though Robert Graves condemned Fitzgerald both for his infidelity to Omar and for his lack of poetic gifts!). Fitzgerald wrote that 'in the absence of a poet who can recreate in his own language the body and soul of a foreign poet, the best translator is one who paraphrases the original work while conserving the author's spirit'. Conscious that no translation can satisfy every reader, I can only hope that those able to read the work in German will not be too critical of my attempts to convey the spirit and sense of the original without necessarily producing a verbatim translation. I hope, too, that those unfamiliar with German will find the present version of *Die Stammesgeschichte* acceptable if not congenial, bearing in mind the formidable style of the original.

Willi Hennig wrote in his Preface that mastery of all the palaeontological, zoological, and morphological facts necessary for a work such as this could be accomplished only with great difficulty by one man. In a sense the same applies to the more humble role of his translator, who must combine proficiency in the German language with a working knowledge of palaeoentomology, phylogenetic systematics, insect morphology, and classification: if accuracy has been achieved in all these fields, then it is in large measure due to my good fortune in having colleagues at the British Museum (Natural History), too numerous to mention individually, who have read critically through my translation of the sections dealing with their own special groups. To all of them I am extremely grateful, but particular thanks are due to Brian Cogan, Edmund Jarzembowski, Richard Jefferies, Klaus Sattler, and Paul Whalley for their patience and assistance with innumerable problems. A special acknowledgement must go to my wife who, in addition to the burdens of small children and an entomologist husband, has cheerfully 'lived with Hennig' for the last 3 years.

Finally, on behalf of the publishers, I should like to thank Professor Wolfgang Hennig, Dr Dieter Schlee, and the other collaborators who have made this English edition possible.

London ADRIAN C. PONT

Introduction to the English Edition

Willi Hennig's book was originally published in 1969 as *Die Stammes-geschichte der Insekten*. At that time Hennig was head of the Department of Phylogenetic Research at the State Museum of Natural History, Stuttgart, Germany.

The enormous amount of data that Hennig collected, and the analysis of so much fossil and recent material with his 'Hennigian' phylogenetic theory, have made this an outstandingly important and admirable book. It has aged far less than one might have expected in the 12 years that have elapsed since it was published.

I was very glad to hear that John Wiley and Sons were interested in publishing an English edition, as this will make the book available to a much wider public, including many for whom the language of the original was an insuperable barrier.

Additional material

It was felt that at least some additions should be made to the book to take account of progress in the fields of entomological systematics, palaeoento-mology, and phylogenetic research during the last 12 years. Dr Michael Achtelig, Dr Martin Baehr, Prof. Dr Ragnar K. Kinzelbach, Dr Eberhard Königsmann, Dr Niels Kristensen, Dr Gerhard Mickoleit, Dr Wolfgang Seeger, Dr Rainer Willmann, Dr Peter Zwick, and myself have provided the additional material. I have organized and co-ordinated the preparation of this material and have intercalated it into the original text. Adrian Pont has made the linguistic improvements and a few translations, has intercalated the notes from Willi Hennig's own copy of the book, and has edited the whole text for publication. It was neither possible nor practical to update *all* the original text, much less to re-write the entire book, and the new material should be considered for what it is: some additions and comments.

Some contributors wanted to intercalate the new material directly into Hennig's text, using italic type to distinguish it from the original, but this did not prove practicable. The method chosen will entail some inconvenience to the reader, but it does have the merit of preserving the integrity of the original text.

Unfortunately, Willi Hennig died on 5 November 1976 and was unable to participate directly in this new edition. However, his personal copy of the book contained many annotations and additional references, which were forwarded for inclusion by his son, Professor Wolfgang Hennig.

Obituaries and tributes to Hennig's achievement have been published by Ax (1977a, 1977b), Byers (1977), Kühne (1978), Schlee (1977b, 1978b), and Steffan (1978).

Hennig's recent papers

Since this book was first published in 1969, Hennig published many papers on the fossils in Cretaceous amber that I collected in the Lebanon (Hennig, 1970, 1971b, 1972b), the fossils in Baltic amber (see below), and on the phylogeny of recent insects. A complete list of his papers has recently been published (Anon., 1978). In many of these papers he modified and applied his phylogenetic theories. Special mention should be made of his controversy with the 'evolutionary systematists' (Hennig, 1974, 1975), and there are further discussions of 'Hennigian phylogenetic systematics' by Schlee (1970, 1971, 1975, 1978a) and by Hennig and Schlee (1978).

Two of his own major fields of research receive rather scant attention in this book: the Diptera, which are given a relatively brief treatment, and the amber fossils, which are hardly mentioned at all.

His knowledge of the Diptera was summarized in the *Handbuch der Zoologie* (Hennig, 1973) and in the revision of Volume 1 of *Die Larvenformen der Dipteren* (Hennig, 1948), on which he was working when he died and which will be published by the Akademie-Verlag, Berlin.

His studies of the fossils in Baltic amber were unique in their quality and quantity ratio. They were published in the journal *Stuttgarter Beiträge zur Naturkunde*†, Series A, Nos. 127, 145, 150, 153, 154, 162, 165, 166, 174, 175, 176, 185, 209, 233, and 240 (1964–1972).

Other recent literature

Two important works have been published on amber insects: Larsson (1978) has dealt with Baltic amber, particularly the collections in the Zoological Museum, Copenhagen; and Schlee and Glöckner (1978) have discussed amber from the Baltic, Dominican Republic, Lebanon, and other

---

† Published by the Staatliches Museum für Naturkunde, Schloss Rosenstein, D–7000 Stuttgart 1, West Germany.

sources, especially that in the collection of the State Museum of Natural History, Stuttgart. Schlee (1980) has just published another booklet illustrating the diversity of morphological and other characters preserved in amber fossils, accompanied by a large number of coloured photomicrographs.

Reviewing *Insect Phylogeny* and the enormous scope of the groups and structures that are dealt with, one might wish for a more detailed survey of certain groups of insects, or for more copious illustrations. *The Insects of Australia* (CSIRO, 1970, 1974) shows the type of monumental and modern collaborative work that can be achieved, with illustrations of outstandingly high quality. Recent work on insect phylogeny has been reviewed by Kristensen (1975).

Two further detailed sources of information on all groups of insects have been published by Matsuda, written mainly from the viewpoint of comparative morphology: on the insect thorax (Matsuda, 1970) and the insect abdomen (Matsuda, 1976).

Ludwigsburg, March 1980   DIETER SCHLEE

Preface to the German Edition

To describe the phylogenetic development of a group of animals is a task in which palaeontological, zoological, and morphological facts must all be taken into consideration. However, in a group of such antiquity and extreme diversity as the insects, this is something that one man can accomplish only with great difficulty. This is why no recent comprehensive account has attempted to unite palaeontological and zoological–morphological facts into one comprehensive picture from a methodologically consistent standpoint. Strictly speaking, no such attempt has been made in this book. Its title, chosen for the sake of brevity, may encourage greater expectations than the book can in fact fulfil. Its sole purpose is to see what monophyletic groups can be recognized with some certainty among the insects, using the factual material currently available, and how far these can be traced back into the geological past through the fossil record. As different authors have often put forward the most widely diverging views, I have had to examine carefully the evidence for these. In the sections on individual insect groups, I have not hesitated to refer again and again to methodological principles which are discussed in the opening chapter, though this may lead to the charge of repetition. However, I have found that very few authors succeed in applying to their own particular groups those methodological principles which have been proposed only in a generalized fashion or only with examples from other groups.

I have not hesitated to put forward my own views when they have seemed to me to be well founded, or to make decisions between the views of other authors for reasons that seem to me cogent, although I am aware that I shall often have been mistaken or shall have overlooked some fundamental point. If many experts in specialized fields of study feel obliged to point out incorrect statements or omissions, then one explicit object of this book, which aims to be a working document and not a textbook, will have been achieved.

I must particularly stress that critical debate with an author should not be seen as diminishing the value of his work. The reverse is actually true. It is not worth expending critical energy on unimportant work.

Finally, I must explain certain formal aspects.

The form of the palaeogeographical illustrations does not imply that I favour the land-bridge theory as opposed to that of continental drift. The real purpose of these maps is to show the present geographical position of fossil-insect localities. Nevertheless, I did not want to disregard the palaeogeography of each geological period, and particularly the position of localities in relation to Tethys. The maps are therefore a compromise, and the 'land-bridge' between Africa and South America, both of which are shown in their present positions, only means that the South Atlantic did not then exist.

In the phylogenetic trees statements of the numbers of recent species are often based on very crude estimates.†

I have refrained from giving any categorical rank ('order', 'suborder', etc.) to groups of higher rank. I have done this because I have found that the fundamental questions of phylogenetic systematics so often become bound up with the subsidiary question of the rank of each group, and I wanted to avoid this kind of unfruitful debate.

Instead of this, individual groups have been given a sequence of numbers from which their position in the hierarchical system can immediately be seen. The sister-group relationship can always be deduced from the last figure in the series. For example, I consider the Ectognatha (2) to be the sister-group of the Entognatha (1), the Coccina (2.2.2.2..3.2..2.2.2.1.2.) the sister-group of the Aphidina (...1.), and so on. Higher figures (for example, 1–6 in the Holometabola) are used when I am still uncertain about the exact relationships (sister-group relationships).

The list of contents is therefore a précis of my views on the classification of insects.

I have given separate accounts of the history of individual groups during the Palaeozoic and Mesozoic, because the prerequisites for future investigations are different in each case. This is discussed more fully in the last chapter[1].††

I have been greatly assisted by publication of the Arthropod volume in the Russian *Textbook of Palaeontology* (Rodendorf, 1962). Without it I would hardly have been able to make an adequate survey of the large and important Russian palaeoentomological literature. I feel greatly indebted to Professor B. B. Rodendorf and Dr A. G. Sharov for sending me this and other important papers. Although I do not know Russian, I have spent a great deal of time making literal translations of the sections that seemed to me most important in this and other Russian papers. However, I am sure that I must have missed a number of important comments.

References to important new papers have been included in the text wherever possible during the proof stage, but more detailed consideration of their results has not been possible. I recommend that papers from the last few years should be carefully studied in the original.

I am very grateful to Professor K. Günther (Berlin) who has read through the entire manuscript, and to Dr G. von Wahlert (Ludwigsburg) who has read through the section on evolutionary ecology; and especially to Professor O. Kraus (Hamburg) who has arranged for the publication of this book in the series of *Senckenberg-Bücher* and who, most unselfishly, has sacrificed an unusually large amount of time and effort in making formal revisions of the manuscript.

---

† The numbers of recent species have been updated from notes in Hennig's personal copy of the book. [Adrian Pont.]

†† Superscript numerals refer to the Additional Notes and Revisionary Notes throughout the text.

My wife has carried out most of the laborious work of preparing the subject and author indexes: these will greatly enhance the utility of the book.

Finally, I need to thank Dr Waldemar Kramer who has undertaken the publication of the book and who has met my every wish with unfailing good will and understanding.

Ludwigsburg, April 1969 WILLI HENNIG

Additional note

1. I have deliberately omitted any account of the stratigraphical importance of fossil insects. Martynova (1961a) referred to Guthörl, who discussed the stratigraphical significance of the Palaeodictyoptera from the Carboniferous of Saarbrücken. Martynova also wrote: '. . . scorpionflies of the Kuznetsk Stage bear on traces of their veins the attachment of large-sized hairs which scorpionflies of higher strata are lacking. The beetles of the Il'inskoe Stage possess a particularly rough structure of the elytra. The width and height of cells of anterior wings in Glosselytrodea alter from stratum to stratum'. [Willi Hennig.]

List of Contributors
to the English Edition

Dr Michael Achtelig
Naturwissenschaftliches Museum, Peutingerstrasse 11, D-8900 Augsburg,
West Germany (B.R.D.).
Neuropteroidea.

Dr Martin Baehr
Institut für Biologie III, Universität Tübingen, Auf der Morgenstelle 28,
D-7400 Tübingen, West Germany (B.R.D.).
Coleoptera.

Professor Dr Ragnar K. Kinzelbach
Institut für Zoologie, Johannes Gutenberg-Universität, Saarstrasse 21, Post-
fach 3980, D-6500 Mainz, West Germany (B.R.D.).
Strepsiptera.

Dr Eberhard Königsmann†
Zoologisches Museum, Museum für Naturkunde der Humboldt-Universität,
Invalidenstrasse 43, DDR 104 Berlin, East Germany (D.D.R.).
Hymenoptera.

Dr Niels Peder Kristensen
Zoologisk Museum, Universitetsparken 15, DK 2100, København, Denmark.
Amphiesmenoptera.

Dr Gerhard Mickoleit
Institut für Biologie III, Universität Tübingen, Auf der Morgenstelle 28,
D-7400 Tübingen, West Germany (B.R.D.).
Antliophora, Mecoptera.

Dr Dieter Schlee
Abteilung für Stammesgeschichtliche Forschung, Staatliches Museum für
Naturkunde Stuttgart, Arsenalplatz 3, D-7140 Ludwigsburg, West Germany
(B.R.D.).
Chapter 1 (phylogenetic methodology), chapter 2 (fossil localities, amber),
Hemiptera, Diptera and other sections of chapter 3, collation and co-
ordination of the revisionary notes.

† Unfortunately, Dr Königsmann died in November 1980, whilst this book was in press.

Dr Wolfgang Seeger
Abteilung für Stammesgeschichtliche Forschung, Staatliches Museum für Naturkunde Stuttgart, Arsenalplatz 3, D-7140 Ludwigsburg, West Germany (B.R.D.).
Psocodea, Thysanoptera.

Dr Rainer Willmann
Geologisch-Paläontologisches Institut und Museum, Universität Kiel, Olshausenstrasse 40–60, D-2300 Kiel, West Germany (B.R.D.).
Chapter 2 (fossil localities), Siphonaptera and Mecoptera.

Dr Peter Zwick
Limnologische Flussstation des Max-Planck-Instituts für Limnologie, Postfach 260, D-6407 Schlitz, West Germany (B.R.D.).
Plecoptera.

Translator and Editor
Adrian C. Pont
Department of Entomology, British Museum (Natural History), Cromwell Road, London SW7 5BD, England.

CHAPTER ONE

Methodological Introduction[2]

A. The tasks, methods, and limits of phylogenetic research[2]

As previous writers have shown, the basic unit of phylogeny is the species, at least in groups like the insects where (with few exceptions) bisexual reproduction takes place. It is only because the species exist as reproductive communities that we can talk about phylogeny rather than about evolution in general or the history of morphotypes.

No one doubts that during earlier periods of the geological past the animal kingdom was organized into species exactly as it is now, except when life first began. There can only be three different relationships between the species of an earlier period, such as the Eocene, and the present: (1) the descendants of an Eocene species may have died out before reaching the present; or (2) they may have survived, changed or unchanged, to form a single living species; or (3) new species-limits may have developed amongst these descendants so that the Eocene species has now been replaced by a group of several living species.

All recent and fossil species must have evolved from such ancestral species (stem-species) and should be regarded, either singly or in groups, as the descendants of these species. It is the fundamental task of phylogenetic research to reconstruct the phylogenetic tree of these species, at whose points of branching there would again have been single species. This task is inextricably linked to further tasks by the principle of reciprocal illumination. These are to describe and explain the changes in form, behaviour, mode of life, and geographical distribution that have taken place within species in the course of geological history.

a. Fossils as the source material for phylogenetic information

Many zoologists, and certainly most palaeontologists, believe that phylogeny can only really be described with the aid of fossils and consequently only in those groups where adequate numbers of fossils are known. Quite frankly, this view is false, although I must admit that it contains a certain element of truth. Undoubtedly it *would* be ideal if we could take a particular species from an earlier period of the geological past and describe the history of its descendants right down to the present by means of a temporally and spatially uninterrupted series of fossils. However, a little reflection shows that, even in this case, a complete phylogenetic account would be impossible. This is mainly because fossils consist only of the dead bodies of organisms, and they are seldom if ever complete. It is only the *morphology* of organisms from the geological past that can be directly studied. However, the species as the basic unit of phylogeny is not defined morphologically. Although it is true that even now species can generally be recognized only by morphological characters, there is always the possibility of undertaking experiments to investigate the limits between different species. This possibility in principle does not exist for the palaeontologist. Every entomologist knows of cases where many species can only be distinguished in the male sex by means of their genitalia. Such detailed morphological investigations are seldom possible in fossils, even in the best preserved cases such as those in amber and similar fossil resins[3].

So there is not even a starting-point for ideal phylogenetic research. Paradoxically, in fact, species limits are most uncertain in cases where particularly numerous fossils are available from one geological period[4]. A striking example is provided by the thirteen species of the hymenopterous genus *Pseudosirex* which Handlirsch described from the lithographic limestone of Solnhofen (Malm): Carpenter thought that only two different species were actually represented. Nor do we know how many of the 800 or more so-called 'cockroach' species from the Upper Carboniferous actually belong to different species, for the species names are often based only on fore-wing fragments and the limits of individual variation in Carboniferous species are completely unknown.

A further insuperable obstacle[5] to an ideal phylogenetic account is that only the dead bodies of organisms are preserved as fossils[5]. Wings or wing fragments are often all that is preserved in a fossil insect. These remains do not usually enable us to make even a partially complete description of the morphological[5] changes which have taken place in the course of phylogeny, even though conclusions extrapolated from the part to the whole can sometimes be drawn with some certainty. However, there are groups where the important features are found in other characters. For example, no one would dispute that the parthenogenetic origin of Hymenoptera males (male haploidy) and the heterogametic females of the Amphiesmenoptera (Trichoptera + Lepidoptera) are at least as remarkable and are probably even more important characteristics of these groups than features of the wing venation. Yet fossils will never give us any clue as to when and where these characteristics originated.

A third obstacle lies in the temporal and spatial deficiencies of the fossil record. Carpenter (1953, 1954a) estimated that some 13 000 species of fossil insect had been described. Many more have been described since then, and Martynova (1961a) mentioned that some 300 papers describing fossil insects had been published during the preceding decade. The figure of 15 000 is probably not too high for the number of formally described species. Compared with the figure of perhaps 750 000 described species of living insects and, in the opinion of many authors, 2–3 million actually existing, even twice this figure would not be very impressive. Its importance is diminished still further by the fact that the described fossil insects are restricted to comparatively few localities from comparatively few horizons. Furthermore, for some important periods no insects are known at all (Lower Carboniferous) and for others only a very few (Devonian, Lower Triassic, Cretaceous[6]).

Many of the descriptions[7] are almost worthless for phylogenetic research, and many of the well described species are not of very great importance. They only show that one particular large group, such as the beetles, was represented at one particular time and in one particular place by numerous species whose relationships to recent species cannot be clearly established.

It seems reasonable to expect that the temporal and spatial deficiencies of the fossil material will gradually dwindle[6] and become less obvious, but they will always be there. In principle it is impossible to make good this holomorphological[8] incompleteness[7].

b. The methodological procedure for phylogenetic analysis

In these circumstances we must ask which of the targets of complete phylogenetic research can be salvaged[9] and what methods can be used to compensate for the deficiencies of the fossil record.

The main result of these holomorphological, temporal, and spatial deficiencies[6-9] is that it is not possible to follow the natural course of phylogeny, which is to follow certain species in the geological *past* through their descendants down to the present[10]. We actually have to work back from the *present* into the past and attempt to relate the known fossils to living species. In fact, phylogenetic *research* has never been carried out in any other way, even though the *description* of phylogeny has usually been done in the opposite order by dealing with the fossils first and then moving on to the recent species. Long before the first fossils were known or were recognized for what they were, the insects were distinguished as one group of the recent animal kingdom and the beetles, butterflies, dragonflies, flies, etc., as groups within the insects. So each time a fossil was discovered the *a priori* question was to what recent animal group did it belong. It follows from the hierarchical structure of the classificatory system that *every* fossil must in fact belong to some group of the recent animal kingdom. The only problem is the relative rank within the hierarchical system occupied by the group to which it belongs. It is possible that one particular fossil can only be interpreted as an insect, another as a beetle, a third as a weevil, and a fourth as a species of the weevil genus *Curculio*. However, for theoretical reasons it is impossible for a fossil not to belong to any known group of animals. At the most it is possible that it is too fragmentary for any decision to be made about where it belongs or whether it even represents the remains of an animate organism at all.

What importance do these ideas have for phylogenetic research? I began by establishing that organic evolution only becomes phylogeny because the species exist as phylogenetic units, and because all living species or species groups are the physical descendants of single species of the past. The first task of phylogenetic research, therefore, is to reveal the genealogical relationships[11] that exist between all known species.

If we propose to carry out phylogenetic research using the fossils, we can only mean 'genealogical relationships' when we say that we intend to reveal the 'relationships' between particular fossils and recent species or species groups. Strictly speaking, however, it is impossible[12] to work out the exact genealogical relationships existing between fossils and recent species or species groups. In many cases we can be sure that fossil species or their descendants have become extinct before reaching the present. But we can never be sure that a particular group of recent species includes *all* the descendants of a particular fossil species.

Does this mean that we should not attempt phylogenetic research because it is impossible in practice[13] to be sure about the genealogical relationships existing between recent species and particular fossil species? Some writers believe this. They prefer to describe the history of morphotypes or of structures instead of pursuing phylogenetic research, which aims at the most exact representation possible of the genealogy of species. Handlirsch (1937) cannot be construed in any other way when he explains the aim of his work: 'hypothetical stem-forms should gradually be replaced by particular types whose actual former existence could be proved by the fossils'. If he had written 'stem-species' instead of 'types', he would have been describing the aims of true genealogical, phylogenetic research. However, I would have to object that in practice this goal is unattainable, because it is impossible in principle to prove[14] that a particular fossil is really the stem-species of particular recent species. Although the goal of ideal phylogenetic research can never be attained[15], we should not capitulate to the extent of giving it up altogether. On the contrary, we should ask how closely the available aids and methods enable us to approach these goals. To recapitulate:

1. The recent animal kingdom is the fixed starting point for phylogenetic research, and it is legitimate to trace its history back into the past.
2. All recent species are the physical descendants of ancient species, and so as we go back further into the past we are bound to encounter species that have given rise to more and more extensive groups of recent species.
3. The recent animal kingdom provides a scheme in which all living species are grouped together hierarchically, into larger and smaller groups.

These three points highlight the first task in the practice of phylogenetic research: each individual group in the recent animal kingdom must be examined to see whether it includes the descendants of a single stem-species, by which I mean—the vital point—*all* the recent descendants of this stem-species.

Experience has shown that various groups in the present classification can provide three different answers to this question (Fig. 1):

1. The first answer arises when a particular group seems only to include species descended from a single stem-species of the past, of which they are the sole living descendants. Most of the present insect 'orders' are groups of this kind.

No one has ever suggested, and probably never will seriously suggest, that there was once a single species whose descendants include some (and only some) of the recent fleas (e.g. the Pulicidae) together with some other recent insects (e.g. some or all of the flies or the beetles). This would compel us to go still further back into the past to find a species whose descendants include all the other fleas, and not just some of them (e.g. the Pulicidae).

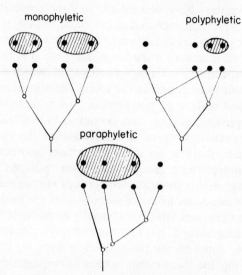

Fig. 1. Diagram to explain the concepts of 'monophyletic', 'polyphyletic', and 'paraphyletic' groups[16]. From Hennig (1965b). (See also Figs. I–III on pp. 17–18.)

Groups like the fleas, which are believed to contain all the known descendants of a single stem-species, are called *monophyletic groups*.

2. The second answer arises when a particular group almost certainly does not include all the recent descendants of a single species of the past.

In his classification of the insects, Linnaeus defined an order 'Aptera' in which he included the so-called Apterygota (primarily wingless insects), Isoptera (termites), Psocoptera (book lice), Mallophaga and Anoplura (biting and sucking lice), Siphonaptera (fleas), and certain other groups that are not insects at all. We cannot now believe that there was ever a single species in the past whose only descendants are the groups that Linnaeus included under the name Aptera.

We call such groups *polyphyletic groups*. The present insect classification contains virtually no such groups. Any that were once included, as in Linnaeus' time, were split up long ago.

3. The third answer arises when a particular group is unlikely to include all the descendants of a single species of the past. However, the descendants

of the stem-species that are excluded from the group are descended from a species of the group younger than the stem-species.

The Psocoptera (Corrodentia)[17] are an example of this kind of group. This group probably includes the descendants of a single species of the past, but the Phthiraptera (biting and sucking lice) have also descended from this stem-species. It is probable that there was a species younger than this stem-species from which the Phthiraptera as well as part of the so-called Psocoptera have descended. So if the group Psocoptera, excluding the Phthiraptera, is included in our classification, it cannot be called a monophyletic group. By definition, a monophyletic group must include *all* the recent descendants of a species of the past, and this is unlikely to be true of the Psocoptera[17]. Groups of this kind, which we call *paraphyletic groups*, are still to be found in the insect classification.

The common feature of polyphyletic and paraphyletic groups is that, unlike monophyletic groups, they do not include *all* the descendants of a single stem-species: some of the descendants are excluded from these groups. In fact, there is no sharp distinction between paraphyletic and moderately polyphyletic groups. Moreover, paraphyletic groups sometimes resemble strictly monophyletic groups very closely, especially when only a few descendants from the stem-species are excluded.

In the literature, only the two concepts monophyletic and polyphyletic are usually distinguished. It is therefore often very difficult to decide under which of these concepts the paraphyletic groups are concealed. Different authors may include them either in their monophyletic or their polyphyletic groups[18].

Interpreting the fossils involves assigning them to a particular group within the present classification (see above). To do this it is extremely important to know whether the group is monophyletic, paraphyletic, or even polyphyletic.

Before explaining why this is so important, I shall consider how to show that a group is monophyletic. I have already explained that the direct approach, which is to search for and find the stem-species from which all the living species of the group are descended, cannot be used. The only possibility, especially where no fossils are known, is to analyse the characters of recent species. The methodology of analysing characters has been described so frequently (e.g. Hennig, 1953, 1965b, 1966g, 1967a) that it does not need to be repeated here in detail. Fundamentally, it is based on the insight that all the differences and similarities between species have arisen in the course of phylogeny. When species split up, their individual characters are transmitted with or without change to their descendants. In this way a group of closely related species acquires a mosaic-like distribution of primitive and derived stages of expression of the same characters ('heterobathmy of characters'). This enables us to draw conclusions about the sequence in which the recent species have evolved from species of the past (Fig. 2A). This is the method used to determine whether the groups in our classification may be considered as monophyletic or not[20].

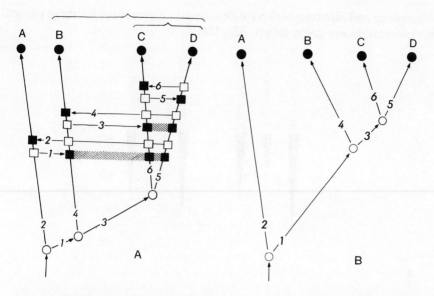

Fig. 2. A: Argumentation plan of phylogenetic systematics[19]. From Hennig (1965b). B: Branching in the phylogenetic tree taking place at different times from those shown in A, but the phylogenetic relationships between groups A–D the same as in A. See p. 11.

c. Phylogenetic trees

The results of this investigation can be shown in a diagram in which the recent species considered to be the descendants of single species of the past are joined together by lines that meet at a single point, representing the stem-species. This diagram is called a phylogenetic tree or dendrogram. The precision of the details shown in such a tree depends partly on the certainty of our results and partly on the aim of the diagram[21].

For example, Fig. 3 shows the following. We believe that there was once a species (1) whose descendants include all the recent butterflies and moths (Lepidoptera), and only these groups. We believe that the same applies to the caddis flies (Trichoptera: 2). If we go further back into the past, we shall find the species (3) from which all the recent Lepidoptera and Trichoptera, and only these two groups, have descended. Still further back, we shall find a species (4) whose descendants include the scorpion flies (Mecoptera), fleas (Siphonaptera), and flies (Diptera), as well as the recent Lepidoptera and Trichoptera. We believe that each of these three groups contains only the descendants of its own individual stem-species (6, 7, 8), and also that there must once have been a species (5) whose descendants include all the recent Mecoptera, Diptera, and Siphonaptera. However, it is not clear what happened during the interval between the periods when species 5 and species 6, 7, and 8 lived. It may well be that species 5 was followed by another species from which the Diptera and Siphonaptera, or Mecoptera and Diptera, or

8

Mecoptera and Siphonaptera have descended. The reasons for these particular conclusions are given on pp. 322–324.

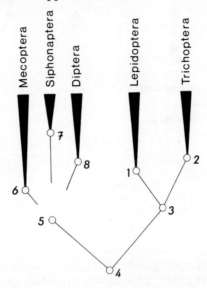

Fig. 3. Relationships amongst the orders of the Mecopteroidea[22]. The Siphonaptera probably do not belong to this group, but are considered here only for methodological reasons. See pp. 7–8. (See also Fig. IV on p. 20.)

The best method of showing such results is a phylogenetic tree, something that is completely unambiguous, clear, and trenchant. Drawing a phylogenetic tree is therefore an entirely justifiable procedure.

Unfortunately, some totally false ideas about the value or shortcomings of such phylogenetic trees have been gaining ground. For example, Tuxen (1960, 1963) has argued that the era of the phylogenetic tree is coming to an end. He has justified this with the assertion that two or three dimensions are not sufficient to show everything that has taken place in the course of evolution. This is correct, but diagrams such as Fig. 3 should not show, and do not pretend to show, everything that has taken place in the course of evolution. Organic evolution becomes phylogeny because species exist as reproductive communities and because all existing species are the descendants of earlier species. If we recognize this and wish to pursue phylogenetic research, then we have no alternative but to put forward well founded hypotheses about the genealogical affinities of recent species. These hypotheses are valid just as long as they are not refuted or replaced by better ones. They supply, furthermore, an indispensable background for understanding everything else that may have taken place in the course of phylogeny. These genealogical hypotheses can be represented most clearly and unambiguously by diagrams. The era of the phylogenetic tree will therefore most certainly not come to an end.

Tuxen's aversion to phylogenetic trees is easy to understand, for so many of them are poorly founded or are not based on any sound arguments at all. Moreover, they often do not show what they ought to show. One vital criterion for deciding this is whether the lines between the groups of species lead back to a single point. Some phylogenetic trees show individual groups issuing as side-shoots from other groups (Fig. 4). Fundamentally, these are not true genealogical trees but diagrams showing typological connections. They show, in fact, that the author admits paraphyletic groups into his classification, and these are typological rather than genealogical units.

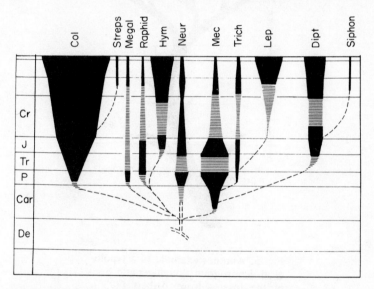

Fig. 4. Example of a typological dendrogram: relationships amongst the Holometabola according to Rodendorf (1962, Fig. 718)[23].

As well as showing genealogical or typological connections, many phylogenetic trees try to give other information about the phylogeny of each group. Most commonly this involves expressing the degree of morphological divergence between groups or their 'degree of apomorphy', that is their degree of derivation (Illies). Other authors try to show what they think they know about different types of holomorphological development, such as adaptiogenesis/aromorphosis or cladogenesis/anagenesis[24]. The phylogenetic trees of Aubert (1950) and Wille (1960) even suggest that their authors have followed the old idea of the *scala rerum* or some such belief. For they seem to be showing that there is a main direction to evolution, with different groups diverging from it at various stages (Fig. 5). On the other hand, it could be that

10

they have been influenced by too literal an interpretation of the term phylogenetic 'tree'[25].

It does not matter whether phylogenetic trees should, or even can, show anything more than the genealogical relationships between genealogical units. The vital point about the value of such a tree for phylogenetic research is that it should at least do *this*. Certain authors treat this matter so frivolously that, in one and the same work, they show phylogenetic trees giving quite different conclusions about the genealogical relationships of particular groups

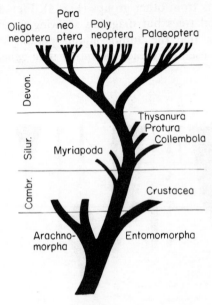

Fig. 5. Another example of a typological dendrogram: 'phylogenetic tree' of the insects from Aubert (1950). Aubert originally included typical representatives of the individual groups, but these have been omitted for the sake of simplicity.

(Martynova, 1959, Figs. 1 and 2; Rodendorf, 1962, Figs. 26 and 718[23]). Moreover, these trees differ not because they are supposed to show alternative interpretations but just because their authors have not even noticed that there is any difference. I share Tuxen's hope that the era of this type of phylogenetic tree will soon come to an end.

d. The role of fossils in the reconstruction of dendrograms

The next question is what role the fossils can play in phylogenetic research. For I have shown that 'phylogenetic trees' can be worked out without their help, and that they must actually be available before we can begin to interpret the fossils.

We may perhaps agree to differ about whether a phylogenetic tree should show anything other than the genealogical relationships of species or monophyletic species groups. Even so, the tree is always defective in *one* respect: it does not show when the events actually took place. All that it and the analysis of characters[26] can tell us is the sequence in which particular events took place. It does not allow[26] us to decide whether their actual position in the geological time scale is as in Fig. 2A, or as in Fig. 2B, or is one of many other possibilities.

This can only be done by studying the fossils themselves, and by assigning them to particular groups in the phylogenetic tree. This then gives the minimum age of each group. Fossils are assigned to particular groups by using exactly the same method that is used for grouping recent species into monophyletic groups. I have often explained that two points need to be made to confirm that several recent species comprise a monophyletic group (see Hennig, 1965b, 1966g, 1967a). Firstly, they must share certain characters considered to be the derived transformation stages of characters present in other species; and secondly, it must be improbable that these shared characters have arisen through convergence. It is then likely that these characters arose in one particular species of the past and have been transmitted to all the descendants of this species, sometimes with further transformations.

If we want to assign a fossil to a particular group, this can only be done if we can show that the two share the same derived characters. This is usually much more difficult with fossils than with recent species. In the next section I shall deal with a number of difficulties and errors that can arise in the study of fossils.

Revisionary notes

2. Several recent papers have an important bearing on the subjects discussed in this chapter, e.g. Peters (1972a), Gilyarov (1969), Kevan, Chaloner and Savile (1975), and Kristensen (1975). [Willi Hennig.]

Hennig opens this chapter with a largely theoretical discussion of so-called 'ideal phylogenetic research'. In this context 'ideal' means that we would have had to have been present throughout geological history in order to make first-hand observations and records of the division of species and origin of taxa. It is self-evident that this is a purely hypothetical and theoretical proposition for which it would be more appropriate to use the phrase 'ideal observation' rather than 'ideal research'. In fact, what we actually deal with are the theoretical and practical tasks of *reconstructing* phylogeny using indirect, circumstantial evidence. In the opening pages of the book, Hennig placed a great deal of emphasis on this unattainable 'ideal' situation. He came to some very pessimistic conclusions, unnecessarily so in my opinion, and these could well lead to serious misunderstandings. Consequently, I shall be giving below a number of additional arguments and comments, based on my experience of 10 years' daily contact with Hennig, a joint paper with him on phylogenetic principles (Hennig and Schlee, 1978), and several other papers (Schlee, 1970, 1971, 1978a). As a result of this experience I hope that I shall be able to argue from Hennig's point of view: to counter a number of attacks, and to present a much more optimistic view of the clear-cut and convincing results that can be achieved by using the methods of Hennigian phylogenetic systematics.

Some parts of this opening chapter could give the impression that phylogenetic research could or should be carried out mainly or solely at the species level. I should point out that phylogenetic reconstructions can in fact be made at all levels—genera, families, orders, etc.

Furthermore, Hennig's statements about recognizing stem-species, and grouping together recent species and certain fossil species, give a very pessimistic impression, but they should be regarded as purely hypothetical and theoretical. In fact, when a practical analysis is made using the methods of Hennigian phylogenetic reconstruction, the fundamental role is not taken by the stem-species but by the *stem-group* and its alternative, the **group*. These two kinds of group are recognizable and the method itself is practicable in fossil studies, particularly with well preserved amber fossils where an extensive set of characters is visible.

For example, a particular *group may have the synapomorphic characters x, y, z, and is thus shown to be a monophyletic group, whereas the stem-group may have only x + y in a synapomorphic state and z still in the plesiomorphic state.

Another of Hennig's opinions that needs some comment is his pessimistic view of the 'enlightening power' of fossils. Those who know his analyses of amber fossils will be astonished to read many of the statements in Chapter 1.

It is possible that he felt frustrated by the large number of unsuccessful attempts in the literature to erect classifications and draw 'phylogenetic' conclusions from fossils that consisted only of damaged wing fragments. This part of the book may have been conceived and written when he was analysing groups like the 'so-called Palaeozoic cockroaches' (pp. 203–208).

I believe that these sections of Chapter 1 must have been written some time ago, before he had begun his intensive study of amber fossils in 1963–64, after moving to Stuttgart. They were not revised when the decision to publish the book was suddenly taken and the manuscript sent to the publisher. This came about in the following way. In 1967 or 1968, when I had been doing some work on Hemiptera phylogeny, I showed him my manuscripts and the following day he brought me 30–40 manuscript pages to read which had high page numbers and did not look freshly typed. When I asked him in some astonishment if these came from a new book, he said that they were only part of a rough manuscript which he had completed some time before but that he did not know whether to publish it; at all events, I should compare his results with mine. I found that we had independently reached the same phylogenetic conclusions. Shortly after this he took the manuscript to Frankfurt, 'to ask if they wanted to publish it'. They did indeed, and there was no time to make any major revisions or additions. This is probably why the Tertiary was not included, although 'insect phylogeny' obviously took place during this period too, as Hennig himself explicitly stressed when he criticized Peus' assertion that 'nothing has taken place in the evolution of the fleas since the Eocene at least, that is to say for the last 50–60 million years' (Hennig, 1969c, p. 63). What is more, throughout this entire period Hennig was busy with his huge monograph of the anthomyiid flies for *Die Fliegen der palaearktischen Region* and with other projects.

I am sure that in any new edition of the book he would have changed several of the statements dealing with the information and conclusions to be gained from fossil studies, and these changes would have been very useful. However, as it is the publisher's intention to reprint the original text unchanged in English translation, this chapter is in some respects a historical document rather than Hennig's final view. Hennig's own intensive work on the analysis of amber fossils resulted in 15 papers between 1964 and 1969 (Anon., 1978), and in the same year that *Die Stammesgeschichte der Insekten* appeared he also published two short papers (Hennig, 1969b, 1969c) in which he emphatically showed the importance and potential of the analysis of amber fossils.

Since Hennig did not give an optimistic appraisal of the amber fossils, either Tertiary or Mesozoic, I propose to make some additions and comments of my own in this chapter, based on my experience of the positive conclusions that can be reached through study of the amber fossils (see also the additional notes 101–103 on pp. 83–85). [Dieter Schlee.]

3. This statement needs some qualification, because it is frequently possible to study the male genitalia of amber fossils in as much detail and with as much precision as if they were recent species. What Hennig probably intended to say was this: it is rarely possible to see the male genitalia of fossils in groups where the taxonomic characters are on parts that are withdrawn into the body or are concealed by other structures. [Dieter Schlee.]

4. Uncertainties in defining species limits result from the state of preservation rather than the abundance of fossils. The real reason for these difficulties is the poor state of compression fossils, whereas large numbers of 'similar' amber fossils present no problem with species limits if adequate analytical methods are employed. [Dieter Schlee.]

5. This line of argument does not apply to amber fossils, where we can be more optimistic. Far more substantial results are to be expected from the analysis of these fossils.

It is possible to make a complete description of the morphology (= external morphology, sometimes including anatomical details) of well preserved amber fossils, and then to draw various conclusions from this. Comparisons with observations and analyses of recent relatives may lead to conclusions regarding function, ethology, etc.—far more than the description of 'the remains of dead bodies'! Examples taken from birds' feathers, the 'social life' of termites, midge behaviour, etc., have been given by Schlee and Glöckner (1978) and Schlee (1980). [Dieter Schlee.]

6. So far as the Cretaceous and other periods are concerned, the situation has greatly improved as knowledge has been extended to new geological periods and geographical areas. New amber fossils have been found in the Lower Cretaceous (Lebanon), Middle Cretaceous (France), and Upper Cretaceous (Siberia, Canada). New Tertiary fossiliferous ambers have also been discovered, such as the Dominican amber of Central America. For a review of the amber fossils, see Schlee and Glöckner (1978). [Dieter Schlee.]

7. These two paragraphs are also concerned mainly with compression fossils. They would also apply to analyses of amber fossils which are poorly documented or incompletely carried out—and unfortunately these are still found all too frequently in contemporary publications. The best available technical and scientific methods for the comprehensive analysis of amber fossils have been described by Schlee (1970, 1973), Hennig and Schlee (1978), and Schlee and Glöckner (1978). If these are followed, then it is possible to attain morphological results of almost the same quality and quantity as are obtained in the course of studies of recent species with the light microscope and at the lower magnifications of the scanning electron microscope. It is possible to study structures of $0.000\,1$–$0.000\,2$ mm in diameter, even when these are arranged in several superimposed layers (which the SEM cannot do!), and to observe the entire external morphology from all angles. The potential of amber fossils brings us very close to 'holomorphology' (see note 8), in the sense of whole (complete) morphology.

However, the term holomorphology also includes the anatomy and histology (internal morphology) of all ontogenetic stages, and even behaviour. Such features are rarely visible in fossils, but when they are it can only be in amber fossils. This is because of the mode of fossilization, involving intrusion of the fluid resin, and because of the autecologically and synecologically relevant assemblages contained in each piece of amber. Examples of the range of features preserved are: wing retractor muscles and mouth-part muscles from the Lower Cretaceous (Schlee, 1970), or Tertiary cornea lenses and book lungs (Mierzejewski, 1976a, 1976b); females with extruding eggs; male and female *in copula*; several larval stages (Poinar, 1977); emerging adults and their preimaginal exuviae; ants with their transported larvae (various stages) and pupae (Dominican amber, Stuttgart amber collection; Schlee, 1980; parasitic associations, including nematode + midge and acari + midge, and phoretic associations (Schlee and Glöckner, 1978); indirect evidence of the presence of special internal glands and of specialized behaviour, provided by the modified head-capsules of termite soldiers (Schlee and Glöckner, 1978) or by the silken threads spun by pseudoscorpions (Schawaller, 1978) or direct evidence of glandular secretions (Schlee, 1980), etc. [Dieter Schlee.]

8. The term 'holomorphological' is derived from 'holomorphe', which means the complete set of morphological, anatomical, ontogenetic, and other attributes. [Dieter Schlee.]

9. It will now be evident from notes 6–8 that the deficiencies in the Coenozoic and even Mesozoic fossil record are slowly decreasing with the discovery of amber fossils from different geological periods and geographical locations. For many groups, the number of fossil records is slowly reaching an acceptable level. For example, Schlee and Glöckner (1978) recorded more than 50 orders and families for the first time from Dominican amber in the Stuttgart amber collection. Since then, this number has been considerably augmented. If adequate methods are used in the basic analysis of their comparative morphology, then we can take an optimistic view of the relevance of our phylogenetic results. [Dieter Schlee.]

10. It would be perfectly possible in theory to establish an independent taxonomy for these fossils: comparative descriptions could be made, taxa created, and 'relationships' suggested in much the same way as taxonomy was conducted in the early years of entomological investigations. In practice, this would conflict with the requirements of nomenclature, and with the desire to compare and co-ordinate fossils and recent species. Only a short step from this is needed to reach the conclusion that the most illuminating feedback of information takes place between well preserved fossils and their well studied recent relatives. The same is true, on a correspondingly reduced scale, of fossil wing impressions. [Dieter Schlee.]

11. 'Genealogical relationships' refer to the branching sequence in the phylogenetic tree (see Hennig, 1974, 1975). [Dieter Schlee.]

12. This sentence sounds extremely pessimistic, unnecessarily so in my opinion. If Hennig himself had actually believed this, he would never have written this book, nor would he have devoted so many years to the analysis of fossil insects! I think that he is dealing with a purely hypothetical notion, and that the emphasis is on (1) 'strictly speaking', and (2) 'to work out the exact'. (1) He may have intended to say that in principle it is impossible to provide *absolute* proof with any method of reconstruction that uses circumstantial evidence. However, this is not a problem specific to fossil analyses, since it will apply to any reconstruction of historical events (Schlee, 1978a), and consequently does not really deserve any special mention here. (2) He may have intended to say that in any case only a part of the complete genealogical sequence is actually available. Yet this in itself is no obstacle to genealogical investigations or to successful phylogenetic reconstructions. Hennig himself has demonstrated that a dendrogram showing phylogenetic relationships can be reconstructed even when enormous gaps exist in our knowledge of the actual genealogical sequence, and he often used the example of man, the lion, and the earthworm. [Dieter Schlee.]

13. For this sentence too, I can only repeat my comments in note 12. The phrase 'to be sure' has given rise to some unduly pessimistic statements, which are certainly not reflected in Hennig's own work on the fossils. It is by no means 'impossible in practice' to make analyses. At the most, it is 'theoretically impossible to attain complete and absolute certainty', because of the indirect nature of the proof. Or, one might say that it is impossible in practice to work out the genealogical relationships of all fossils, particularly those that are only represented by wing impressions. As a final comment, I should point out that Hennig's phylogenetic methodology makes it possible in practice to work out the genealogical relationships between well preserved fossils and their recent relatives, if these have been soundly analysed. [Dieter Schlee.]

14. Once again, see notes 12 and 13. At this point, Hennig is concentrating on the species as 'the basic unit of phylogeny' (see the opening sentence of the book, p. 1), and this leads to a statement that sounds like a serious defect in his phylogenetic methodology. However, this statement does not actually raise any difficulties for the work of practical analysis. The central part of a Hennigian analysis of fossils is the differentiation of the 'stem-group' and the '*group', which are defined by sets of synapomorphies within a well defined context. There is no problem with the stem-species, which naturally belongs to the 'stem-group'. (For further concise discussion, see section B on p. 31 or Schlee, 1970, 1971, 1978a, or Hennig and Schlee, 1978.) [Dieter Schlee.]

15. See notes 12 onwards. Hennig can only be using the concept of 'ideal phylogenetic research' in a purely hypothetical sense (see note 2), that is to say to mean the complete and simultaneous first-hand observation of the origin of taxa over millions of years. Of course, there can be no disputing such a proposition, nor is it really necessary to be influenced by it here. Moreover, it should really be called 'ideal historio*graphy*' rather than 'historical *re-search*'. There is a practicable method for making phylogenetic reconstructions which yields reliable and testable results (see Hennig's own practical papers or the references given at the end of the preceding note). [Dieter Schlee.]

16. Hennig's original diagram of a 'polyphletic' group could give rise to some misunderstandings because of the different angles of the forks. These angles play an important role in evolutionary systematics but are not an essential part of phylogenetic systematics. Hennig's use of them in this diagram should not be interpreted as meaning that he considered them to be part of his methodology. He may have used them simply as part of his general practice of illustrating his ideas with widespread or generally known examples, or with examples he knew would be within his audience's sphere of knowledge.

I should like to offer the following additional diagrams and comments.

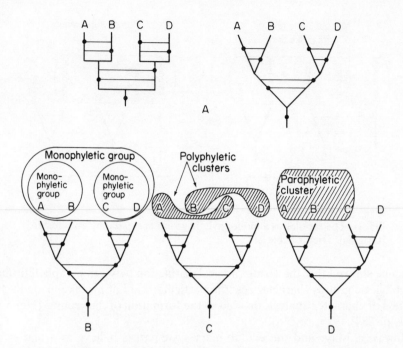

Fig. I. Additional diagrams to illustrate the concepts of monophyly (A, B), polyphyly (C), and paraphyly (D). [Original. Dieter Schlee.]

The diagrams in Fig. IA–D show different interpretations of the same cladistic sequence (shown in Fig. IA). Fig. IA shows two versions of the scheme of synapomorphies. I prefer the one on the left, but have prepared the one on the right (and Fig. IB–D) in the style of Hennig's original Fig. 1 (p. 5). Fig. IB shows Hennig's and the phylogenetic systematist's interpretation of the kinship relations. There are seven monophyletic groups: (1) A + B + C + D, (2) A + B, (3) C + D, and (4–7) A, B, C, and D individually. Fig. IC illustrates polyphyly, in which several subgroups from different lines of descent are grouped together, for example 'A + C' or 'B + D'. Such groups are the result of a relatively close 'overall similarity' produced by convergence or parallelism, such as the Linnaean 'Aptera' which Hennig discusses on p. 5 or the Apterygota (p. 100). Fig. ID illustrates paraphyly, in which 'all but one' are grouped together: in other words, a paraphyletic group includes most of the subgroups belonging to a more extensive group but the subgroup excluded is not the sister-group of all the other subgroups. Such groups are recognized (by non-Hennigians) because of the extreme divergence of the excluded subgroup and the 'relative considerable plesiomorphy' of all the other subgroups. The birds and 'reptiles' are an example of this (Fig. II).

Important note: Hennig and other phylogenetic systematists do not accept that the so-called 'polyphyletic groups' and 'paraphyletic groups' are real groups, i.e. are natural taxa. Hennig gave a number of clear definitions in

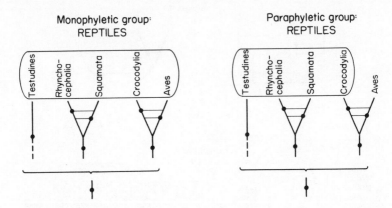

Fig. II. The reptiles as a monophyletic group or a paraphyletic cluster. [Original. Dieter Schlee.]

which he showed that the terminological distinction between paraphyletic and polyphyletic groups characterizes 'the particular kind of mistake made in the process of character analysis that led to the formation of the groups' (Hennig, 1974, p. 284; 1975, p. 248).

However, Mayr and the evolutionary systematists believe as a matter of conviction that paraphyletic groups (clusters) are real taxa, even when they are familiar with genealogical situations like those shown in Fig. IA–D. This is the essence of the controversy between Mayr (1974) and Hennig (1974, 1975).

The example in Fig. III shows how the distinction between paraphyletic 'groups' and polyphyletic 'groups' is determined by the sort of character states used for the clustering. A 'taxon' based on species agreeing in plesiomorphic character states can be called paraphyletic. A 'taxon' based on species agreeing in convergences can be called polyphyletic. See Hennig (1975, Fig. 1.)

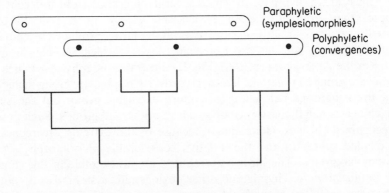

Fig. III. Paraphyly and polyphyly based on symplesiomorphies and convergences, respectively. From Hennig (1975).

Since the aim of Hennigian phylogenetic systematics is the search for monophyletic groups and the avoidance of non-monophyletic clusters, both polyphyletic and paraphyletic 'groupings' should be avoided. [Dieter Schlee.]

17. Hennig's example of the Psocoptera as a 'paraphyletic group' can no longer be used since it has now been shown that the Psocoptera are a monophyletic group: see the section by Seeger in this edition (notes 200–203 on pp. 233–234). Instead, the 'Nematocera' (Diptera) would provide a good example, and Hennig himself (1968b) has dealt explicitly with this problem. The 'Symphyta' (Hymenoptera: p. 396) would be another example, and they have been discussed in detail by Königsmann (1977). [Dieter Schlee.]

18. In a subsequent paper, Hennig (1974, 1975) discussed monophyly and the way in which polyphyletic and paraphyletic groups are formed, and also discussed several 'patent solutions' (holophyly, etc.). [Dieter Schlee.]

19. This kind of argumentation plan originated during the early days of phylogenetic systematics, and could give rise to a number of misunderstandings. For a detailed discussion of the diagrammatic presentation of the results of phylogenetic systematics, see Hennig and Schlee (1978) and Schlee (1978a). [Dieter Schlee.]

20. At this point I should like to draw attention to the evidence to be obtained from nucleic acids and proteins, and also to Crowson's view (1970, p. 164) that 'phylogenetic classification is concerned to express genetical relationships, and there is reason to suppose that these relationships would be most clearly seen if we could compare the genotypes directly'. Crowson (1970, p. 166) also referred to 'phylogenetic affinity (or genotypic resemblance, if you prefer the term)'. [Willi Hennig.]

21. With reference to the interpretation of fossils, we should be aware of the dangers inherent in reconstructing phylogenetic trees with the method that uses the minimum number of forks. This allows typology to 'get in through the back door'. [Willi Hennig.]

22. This diagram should only be used as an illustration of the accompanying text. Out of context it could give rise to a number of misunderstandings. The way in which the left-hand part of the diagram has been drawn (Mecoptera + Siphonaptera + Diptera) shows only two of the three possible relationships between these three groups. It could indicate that there is probably a sister-group relationship between the Siphonaptera and Diptera or between the Mecoptera and Diptera, but it does not show the third possibility, that the Mecoptera and Diptera could be sister-groups (see Fig. IV).

The left-hand diagram in Fig. IV shows the uncertain sister-group relationships between the subgroups (Mecoptera, Siphonaptera, Diptera), of a monophyletic group of higher rank, whilst the other three diagrams demonstrate the three possible solutions to this problem. [Dieter Schlee.]

20

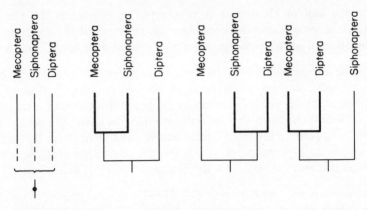

Fig. IV. Possible relationships amongst the Siphonaptera, Diptera, and Mecoptera, and a method of demonstrating the uncertainty. [Original. Dieter Schlee.]

23. Compare this with Fig. 26 in Rodendorf (1962). In his Fig. 26 the Neuropteroidea are shown as the sister-group of the Hymenoptera, whereas in his Fig. 718 (reproduced here as Fig. 4) the Raphidioptera are shown to be this sister-group. [Willi Hennig.]

24. See also Hennig's (1974, 1975) controversy with Mayr. [Dieter Schlee.]

25. For a discussion of the most appropriate and convincing type of phylogenetic tree, see Hennig and Schlee (1978) and Schlee (1978a). [Dieter Schlee.]

26. This can only refer to analysis of the characters of recent species. [Dieter Schlee.]

1. Distinguishing the constitutive and diagnostic characters of a group

In Chapter III I shall discuss the characters that show that the Lepidoptera (butterflies and moths) and Trichoptera (caddis flies) are monophyletic groups. These are the *constitutive* characters of these groups[27]. The presence of wing scales is one of the constitutive characters of the butterflies and moths (Lepidoptera). Caddis flies (Trichoptera), on the other hand, have haired wings. Wing hairs, however, are not one of the constitutive characters of the Trichoptera, because hairs are the precursors of scales, and the common ancestors of the Lepidoptera and Trichoptera must also have had haired wings. However, wing hairs are important as a *diagnostic* character of the recent Trichoptera, and are used to distinguish the group from the Lepidoptera. In deciding whether a species belongs to the Trichoptera or to the Lepidoptera, as in an identification key, we can use the wing hairs as a character of the Trichoptera[27].

However, we cannot do this with the fossils. Diagnostic characters that are adequate for the differentiation of recent groups are not valid here. A fossil that agrees with recent Trichoptera in every detail of the venation and of the wing hairs may well belong to the Trichoptera (i.e. the *Trichoptera) but may equally well belong to the stem-group of the Amphiesmenoptera (Trichoptera + Lepidoptera). We can only decide if a particular fossil really belongs to the Trichoptera if it possesses the constitutive characters of this group. These are to be found in the structure of the mouth-parts, the larvae, etc. (p. 326). However, in fossils it is usually only the wings that are preserved. Consequently, one of the commonest errors of palaeoentomology is to assign a fossil to a recent group solely on the basis of agreements in the wing, without considering whether the constitutive characters of the group are in fact found in the wing.

For example, Hinton (1958) was right to point out that many of the fossils now placed in the Mecoptera[28] according to their wings would be assigned to different groups if other structures or their larvae were known. Several other examples are discussed in Chapter III. The assignment of many Carboniferous insects to the cockroaches, and of Palaeozoic beetles to the Cupedoidea, also provide very striking examples. Rodendorf (1944) believed that fossil 'Cupedoidea' were known from the Permian onwards and that this group was 'undoubtedly' much more abundant in the geological past than it is now. However, the assignment of Permian beetles to the 'Cupedoidea' is based solely on their agreement in elytral structure. This feature is adequate as a *diagnostic* character for differentiating *recent* Cupedoidea from all other recent beetles, but other characters are used to show that the Cupedoidea are a monophyletic group (see p. 305)[29]. The constitutive characters of the Cupedoidea are not visible in Palaeozoic fossils. However, the elytral characters retained only by the recent 'Cupedoidea' must have been present in the common ancestors of all beetles. Consequently, we cannot take any agreement in elytral structure to prove that the Permian fossils belong to the 'Cupedoidea' and that certain groups of recent beetles were already in existence during the Palaeozoic. Later I shall show why it is so important to establish this.

Revisionary notes

27. 'Constitutive characters' = the autapomorphies of the taxon = the synapomorphies of its subgroups (which may be species within a genus, or families within an order). [Dieter Schlee.]

28. See the critical catalogue of fossil Mecoptera by Willmann (1978), and Willmann's contribution to the present edition (notes 346–348 on pp. 341–342). [Dieter Schlee.]

29. See the recent papers on the Cupedidae by Baehr (1975, etc.), and

Baehr's contribution to the present edition (notes 312–313 on p. 308). [Dieter Schlee.]

2. Constitutive characters and convergence (parallel development)

I have described how a particular group of recent species can be shown to be monophyletic: this is done by demonstrating the presence of a number of derived characters which indicate that the species had a single stem-species in common.

It is well known that a character may often develop independently into more or less identical derived states on several occasions. This is known as convergence or, in certain circumstances, parallel development. Consequently, the presence of a particular derived character in a group of species does not necessarily mean that the group is monophyletic.

The more numerous and peculiar the derived characters, the more certain we can be that a group is monophyletic. Females of the Trichoptera and Lepidoptera are heterogametic, which is a remarkable feature amongst insects. In itself, this is a strong indication that the two groups form a single monophyletic group, the Amphiesmenoptera, as White (1957) was right to point out. The same is true of parthenogenesis and male haploidy in the Hymenoptera. In other cases, such as the fleas (Siphonaptera), the large number of derived characters shows that the group is undoubtedly monophyletic.

However, the derived characters used as 'constitutive' characters to show that a group is monophyletic are rarely all visible in the fossils. This is particularly true of the most convincing characters, such as the cytological features of the Amphiesmenoptera and Hymenoptera mentioned above. Moreover, *individual* derived characters can often arise convergently, and this is a further source of error which we must certainly be aware of when interpreting the fossils.

Fossils of female Amphiesmenoptera (Trichoptera + Lepidoptera) can obviously never be recognized as heterogametic, but the group has another derived character in the looped anal veins. The anal veins usually reach the wing margin, but in the Amphiesmenoptera their apical sections have been lost although the veins themselves have remained connected by the cross-veins that originally joined them together. This character is found in the Permian Paramecoptera. Amongst recent insects, however, looped anal veins also occur in the Hymenoptera. This makes it even more difficult to assign fossils to recent groups. The Paramecoptera were considered by Tillyard (1919b) to be the ancestors of the Amphiesmenoptera (Trichoptera + Lepidoptera), whereas Riek (1953a) thought it possible to derive the Hymenoptera from them. There are other cases where fossils have been regarded as belonging to the 'stem-group' of several recent groups. These groups have the same character as the fossils, but for other reasons they could not have had a stem-group in common.

With the recent species it is often difficult to make assignments to particular monophyletic groups, but this can be achieved by examining a large number of characters. A similar procedure has to be followed when studying fossils, but frequently a solution will only be found when individuals in a better state of preservation are discovered. These may make it possible to decide whether the agreements should be considered as synapomorphies or convergent developments. In some cases it may even be possible to make a decision by refining a character concept that has been formulated too crudely, but more completely preserved specimens are often necessary for this too. The extinct Megasecoptera are an example of this. It is known that some of the species could flex their wings back over the abdomen like the Neoptera. Consequently, it has been suggested that they might belong to the Neoptera, or even to the Holometabola. Here it is only the genealogical aspect of this problem that interests me: are the Megasecoptera, or at least the species that could flex their wings back over the abdomen, derived from a single stem-species from which only they and the Neoptera have descended? No answer can be given to this because nothing is known about the basal articulation of the wing in the Megasecoptera. It is impossible to say whether the 'neoptery' of the Megasecoptera is the same as the neoptery of the Neoptera.

Several other questions are raised by this. At first sight these only seem to be important from the point of view of formal terminology or formal systematics, but in fact they have very real consequences.

'Neoptery' can be defined as the ability to flex the wings back over the abdomen, and all the insects with this ability can be assigned to one group, the Neoptera. They all belong to one *morphotype*, and neoptery as just defined is their essential and typologically determinant character. The question of whether this character has arisen monophyletically or polyphyletically in this group is a secondary one. If it has arisen polyphyletically, the group does not need to be divided up when it is included in a typological classification, because by definition it is not necessary for a morphotype to have arisen monophyletically. However, the genealogical problem still remains and it is concealed rather than solved by a typological classification. On the other hand, if a particular character like neoptery has arisen several times polyphyletically it does not mean that it has arisen polyphyletically every time it is found. For example, Zalesskiy (1958) apparently believed that the 'neoptery' of the Megasecoptera originated independently, and he suggested that the division of the insects into Palaeoptera and Neoptera was a 'morpho-functional' division, not a 'systematic' one. Zalesskiy's use of the word 'systematic' in this context shows that his term 'systematic classification' means a genealogical classification in which only monophyletic groups are admitted. In such a classification, the fact that the Megasecoptera have acquired their 'neoptery' independently means that they should not be placed in the Neoptera. This is because there are good reasons for believing that the neoptery of the Neoptera has arisen monophyletically and is one of the constitutive characters of this group. It is quite another matter that it has arisen elsewhere as well. On the other hand, the 'palaeoptery' of the Palaeoptera is only a diagnostic character as far as the recent species are concerned. The constitutive characters of this group, which are still rather problematic, are found in different structures such as the antennae and certain larval characters. It is unfortunate that many people are misled by the semantic meaning of a group's name, which is not necessarily based on one of its constitutive characters, and this is something that we must guard against.

It is usually difficult to find the fundamental constitutive characters of a group amongst the random features that happen to be preserved during the process of fossilization. This has often led palaeontologists to group fossils according to agreements in the characters that can actually be seen, and then to derive all those recent groups from such a fossil group whose characters will give some formal support to such a derivation. As the characters of the fossils are not numerous, this is often a very simple matter, but for precisely this reason it often produces some absurd results. In some cases it has led to the construction of an exclusively fossil 'stem-group' such as the Palaeodictyoptera, which Handlirsch (1919) described as '. . . *the* proto-insects and not one of their side-branches'; or the 'Paraplecoptera', from which most authors still try to derive the Plecoptera and the Embioptera.

In other cases, the same method has been used for assigning fossils to recent groups, such as the Blattariae or Mecoptera, and then deriving several other recent groups from these. This is a totally inappropriate method of describing phylogeny, and I shall discuss it in the section dealing with the stem-group concept.

3. Constitutive characters and additive typogenesis

I have already explained that the greater the number[30] of derived characters present in a particular group of recent species, the more soundly it can be shown that the group is monophyletic. Many groups, such as the Siphonaptera, are distinguished by a large number of such characters, and so present an impressively closed structural plan.

Obviously it is highly improbable that all these characters arose at the same time, and in fortunate cases we can actually work out the sequence in which they developed[31]. Heberer called this process 'additive typogenesis', and it presents a further problem in the interpretation of fossils.

If a group of recent species is thought to be monophyletic because it has a large number of derived characters, then it is safe to say that the full complement of these constitutive characters could only have been present in the latest common stem-species of the recent species of the group. This is because the idea that these species form a monophyletic group is based on the belief that they inherited their constitutive characters from a common stem-species. In Fig. 3, point 1 would represent the latest common stem-species of all recent Lepidoptera and point 2 the latest common stem-species of all recent Trichoptera. As the Trichoptera and Lepidoptera form a monophyletic group of higher rank, the Amphiesmenoptera, we must assume that there was an earlier stem-species (3) common to species 1 and 2. This would be the latest common stem-species of the Amphiesmenoptera. Between the epochs when species 3 and species 1 + 2 lived, there were long periods when the complexes of constitutive characters of the recent Lepidoptera (period 3 to 1) and Trichoptera (period 3 to 2) originated.

How should we delimit groups in these circumstances? There are three possibilities which I shall discuss by reference to the Trichoptera (Fig. 6).

(a) The Trichoptera could be defined to include only the species that appear to have descended from the latest common stem-species of this group (point 2 in Fig. 6). However, this immediately raises considerable practical difficulties when we attempt to assign fossils to the 'Trichoptera': it would mean that we must show that the fossils have all the constitutive characters which are present in recent Trichoptera and which provide evidence that the group is monophyletic. In practice this is impossible and we must always be content to show that only one or a few[32] of the derived characters are present. This means that if we cannot show that a fossil[32] belongs to a subgroup of the recent Trichoptera, we can never be certain that it really does belong to the Trichoptera as defined here: it could have arisen during the period between points 3 and 2, either as a member of the direct ancestral line of the 'Trichoptera' or as a side-branch from the section of the tree between points 3 and 2[33].

(b) The Trichoptera could be defined so that most weight is given to their sister-group relationship[34] with the Lepidoptera, which is founded on the

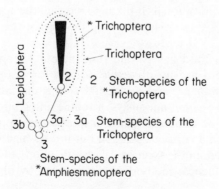

Fig. 6. Diagram to illustrate the 'stem-species' concept and the problems[35] that can arise when tracing recent monophyletic groups back into the past. See p. 30[36].

belief that there was a single stem-species (3) common only to these two groups. This would mean that when the common stem-species (3) of the Amphiesmenoptera first divided into two daughter-species, one of them (3a) gave rise to descendants that also include all the recent Trichoptera: the Trichoptera are defined to include all these descendants, including species 3a.

This has the practical advantage that we could assign to the Trichoptera all the fossils that have at least one of the constitutive characters of the group, which have obviously had to be worked out from the recent species. Nor would we need to consider whether the fossils possessed only this single constitutive group character, or whether there were others which cannot be seen because the condition of the fossil is too poor[37], or because in principle they are never preserved in fossils. Difficulties of this kind are often insoluble in practice.

In fact, the assignment of fossils to particular groups of recent species is often based on this method. For example, the assignment of Permian beetles to the Coleoptera is based solely on the presence of elytra, which is the only feature preserved in the fossils. It is not known if they actually possessed any of the other constitutive characters of the group Coleoptera: the prognathous head which is closed ventrally by a gula, the invaginated genitalia, etc.

(c) The Trichoptera could be defined both as a structural type and as a monophyletic group. These two concepts are not fundamentally irreconcilable. There are many monophyletic groups that are also impressively discrete structural types (beetles, flies, etc.). Nevertheless, the concept of morphotypes also involves the differentiation of fundamental (essential, typologically determinant) characters, and of ancillary (accidental) characters which need not necessarily be present. It could be thought that only fossils with the essential, typologically determinant, characters of the group do actually belong to the Trichoptera. This is the method usually followed in palaeontology. It is used for assigning fossils to individual groups of recent insects, and then the age of the oldest fossils is taken to indicate the age of the group itself. Carpenter's table (1953, 1954a) showing the age of the insect orders is an example of this.

However, this method has certain fundamental theoretical and practical disadvantages. There are no objective criteria for distinguishing essential, typologically determinant, and accidental characters. Sharov (1965) stated that *Archaeopteryx* is simultaneously[38] a reptile and a bird, and in a 'natural' classification could be treated either as a specialized reptile or as a primitive bird. This may seem a curious assertion but it does highlight a very real and fundamental problem in the typological classification of fossils. It is not at all clear how the 'dialectical method' recommended by Sharov could help to solve this problem.

The practical difficulties arise because it is often[39] impossible to show that the typologically determinant characters are present in particular fossils, even assuming that they can be defined objectively. The difficulties are often as great as those involved in finding the constitutive characters of a genealogical unit. It is the customary practice of palaeontologists to search for typological-ly determinant characters only among the features that can be and actually are preserved in fossils. In insects these are mainly in the wing venation. It must be obvious that this method does not solve any phylogenetic problems but merely shelves them.

Revisionary notes

30. This phrase should be modified to: 'the more numerous and the more peculiar are the derived characters . . .'. [Dieter Schlee.]

31. See Crowson's (1970, p. 118) discussion of the principle of non-congruence, and also his statements that selection only operates by changing one character rather than several characters simultaneously, in other words that some kinds of character evolve successively rather than concurrently (Crowson, 1970, pp. 210–211). [Willi Hennig.]

32. These pessimistic comments refer primarily to compression fossils. In amber fossils the (external) morphology is almost intact: it has not been subjected to compression, and is readily visible. This means that there is an extensive array of characters available for study. These form a realistic basis for obtaining substantial results, even though nothing is known about anatomical or histological features.

It goes without saying that some anatomical and other results would be welcome! However, for how many recent arthropod groups can we say that features of the internal morphology are known and are an essential part of identification, i.e. determine the position of the specimen or the group within the classificatory system? [Dieter Schlee.]

33. Reference should be made at this point to the Ephemeroptera: the Protephemeroptera (*Triplosoba*) and Permoplectoptera represent 'different layers' of the stem-group. See also note 46 on p. 37. [Willi Hennig.]

34. Note: Sister-groups are the most closely related groups that show a distribution of synapomorphies as in Fig. 2 (p. 7) and, in addition, a single synapomorphy at the base of the diagram which proves that the group as a whole (A, B, C, D) is a monophyletic unit. [Dieter Schlee.]

35. In practice the stem-*group* is more important than the stem-*species* in the analysis of fossils. See also p. 29. [Dieter Schlee.]

36. For the definition of *groups, for example the *Trichoptera, see p. 29. [Dieter Schlee.]

37. I should point out once again that in amber fossils the full complement of characters available for precise and detailed study is only slightly reduced. The chances of recognizing a fossil as belonging to the stem-group of a *group are good: see, for example, the stem-group of the Aleyrodina (Schlee, 1970). [Dieter Schlee.]

38. From the point of view of phylogenetic systematics, *Archaeopteryx* is a member of the stem-group of the birds (stem-group Aves). This is because only some of the synapomorphic/constitutive characters of the *Aves have been acquired, whilst the rest of the characters are still in a 'primitive' (symplesiomorphic) and 'reptile-like' state. [Dieter Schlee.]

39. The practical difficulties are caused by the poor state of preservation, except for the amber fossils. [Dieter Schlee.]

B. Phylogenetic research as a systematic problem. The stem-group concept in phylogenetic systematics

1. Theoretical basis of the stem-group concept

I have shown in the preceding sections that phylogenetic research is essentially a systematic problem. It operates by recognizing groups of recent species and by assigning individual fossils or groups of fossils to these groups: it shows the relationships between fossils and recent species-groups. This is pure systematics.

The principles we use in systematic work determine the way in which we should interpret the results of phylogenetic research. For example, in deciding how much weight we should attach to a statement about the age of a particular group, it is obviously important to know which of the three methods discussed in the preceding section is used for assigning the fossils to groups of recent species and whether these groups are monophyletic, paraphyletic, or polyphyletic. Determining the age of various groups of insects provides the basis for discussing further problems. For example, can the origin of various groups be associated with particular geological events, or have there been brief periods in the geological past when the rapid development of many new insect groups took place?

In the preceding section I have also discussed the various methods that can be used for assigning fossils to groups of recent insects. However, the proposition *omne vivum e vivo* is now a fundamental tenet of biology, and several things follow from this: individuals only arise as the physical descendants of older individuals; the inheritance of characters, either unchanged or considerably changed, only takes place in this way; and the procreation of offspring and the inheritance of characters takes place, at least in insects, exclusively within the reproductive communities called species. This means that phylogenetic research as a biological science is only possible if its primary aim is to discover the genealogical relationships amongst species, no matter what other important aims it may also wish to pursue.

It follows from this that establishing genealogical relationships, or hypotheses about genealogical relationships, is the only principle that can be followed in phylogenetic research when grouping species together. We can only achieve something in phylogenetic research with a group like the Trichoptera because we believe that there was a single stem-species from which all the recent Trichoptera and only these have descended. The task of phylogenetic research is to investigate the history of the Trichoptera, and if possible to trace them back to the moment of their origin. In the following discussion I shall use the Trichoptera wherever possible as an example.

There was a point in time when the Trichoptera originated: this is a theoretical conclusion drawn from the well founded conviction that there was once a stem-species whose descendants include both the Lepidoptera and the Trichoptera (point 3 in Fig. 6). This stem-species does not belong to either group and cannot be assigned to either. The two daughter-species are the first species that can belong to the line of descent of the Trichoptera (3a in Fig. 6) or Lepidoptera (3b in Fig. 6). Consequently, they can and must be assigned to one of these groups because they are genealogically related to either one or the other. For this reason I believe that the second of the three methods for assigning fossils to recent groups that I discussed in the preceding section is the most suitable one for phylogenetic research (Fig. 7).

The first method is also compatible with the aims of phylogenetic research. Its main disadvantage is that we would have to create a new name, for example the 'Prototrichoptera', for all the species that are more closely related genealogically to the recent Trichoptera than to the recent Lepidoptera but which are not descendants of the latest common stem-species of recent Trichoptera (2 in Figs. 6 and 7). Furthermore, if we restrict the name Trichoptera to the descendants of the latest common stem-species of all recent Trichoptera, we would have to use yet another name for the 'Prototrichoptera' + Trichoptera, for example the 'Trichopterodea'. This would be the only way to express the sister-group relationship within the Amphiesmenoptera. This kind of procedure would lead to unimaginable nomenclatorial complications—and in some cases has already done so! Such a situation should always be avoided, if it can be done without jeopardizing the fundamental aims of a scientific discipline.

a. *Groups

In the rest of this book I shall take the names of recent monophyletic groups (e.g. Lepidoptera, Trichoptera, Diptera) and use them to include all the fossils which are more closely related to the recent species of each group (e.g. Trichoptera) than they are to the recent species of the sister-group (in

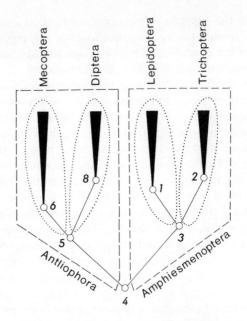

Fig. 7. Diagram to illustrate the delimitation of monophyletic groups on the basis of the sister-group principle. The orders of the Mecopteroidea are shown. See p. 29, and Figs. 3 and 6. The Siphonaptera, shown in Fig. 3 with stem-species 7, are omitted.

this case, the Lepidoptera). When I want to refer only to the more restricted monophyletic group within the Trichoptera that contains all the descendants of the latest common stem-species of all recent species, I shall call it the *Trichoptera.

I shall refer to fossils that belong to the Trichoptera, but cannot be shown to belong to the *Trichoptera, as the 'stem-group of the Trichoptera'. This is both simple and unambiguous (Fig. 6).

Such a stem-group would include the species which are in the direct line of descent of recent species (for example, of the *Trichoptera) and also the 'side-branches' of the phylogenetic tree[40]. This stem-group concept is partly typological and not purely genealogical, and as such its use in phylogenetic research cannot be completely justified. It represents a compromise with the limitations of practical palaeontology and can only be tolerated as long as it does not seriously jeopardize the fundamental aims of phylogenetic research[41].

Its use has a number of definite advantages. The age of the Trichoptera can be established as the moment when the sister-group relationship between the Trichoptera and the Lepidoptera actually originated, by the division of their common stem-species. If this procedure is correct, then every fossil that belongs to the stem-group of the Trichoptera is equally important. It does not really matter whether it belongs to the direct line of descent of the *Trichoptera or should be considered as a 'side-branch of the phylogenetic tree'. The literature is overburdened with discussions about whether, for example, the Palaeodictyoptera are the direct ancestors or a side-branch of the Pterygota or of the Palaeoptera, or whether the Protoperlaria are the direct ancestors or a side-branch of the Plecoptera, etc. It is relatively unimportant whether the fossils are the direct ancestors of recent species or are side-branches of the phylogenetic tree. The most important matter is the group (order, family, genus, etc.) to which they actually belong. In any case, the question of whether particular fossils belong to the actual line of descent of recent species-groups can only be answered when the answer is 'no', i.e. when this possibility can be *ruled out*. On the other hand, the assertion that particular fossils *must* have had recent descendants can *never* be proved.

b. Stem-groups

The name of a group of recent species can therefore also be applied to fossils that are more closely related to these species than they are to the recent species of the sister-group. Even when an unsatisfactory result is obtained by forming a stem-group from all the fossil species that can reasonably be grouped together, this use of the name will still hold good. It will always appear to be unsatisfactory when the interval between the origin of the group and the latest common stem-species of all recent species is particularly long (the period between 3 and 1 or 3 and 2 in Fig. 3), and when large numbers of fossils are known from this period. Prolonged study of the fossils will show that some are much more closely related to the recent species of the whole group than are others. The desire to reflect this in the hierarchy of the classification will then become irresistible. This can easily be done by introducing intermediate categories. There is no need to change the name of any recent group, nor do the intermediate categories have to be included in a classification covering only the recent species. As the modern animal kingdom and its classification are the fixed reference point for all phylogenetic research, the advantages of this can hardly be overestimated.

The 'stem-group' concept as defined here[42] is a genuine compromise which recognizes that the results that can be achieved by palaeontological work are limited[43]. At the same time it does not obstruct the way forward towards the real goal of phylogenetic research, and it leaves all options open[44].

Revisionary notes

40. As a further illustration of the stem-group concept, see the Tertiary amber Diptera genus *Meoneurites:* originally this was tentatively assigned to the stem-group of the Carnidae (Hennig, 1965a), but this interpretation had to be revised when the Chilean genus *Neomeoneurites* was discovered, with the result that *Meoneurites* is assigned to the stem-group of *Neomeoneurites*, and *Meoneurites* + *Neomeoneurites* are considered to be the sister-group of the rest of the Carnidae (Hennig, 1972a). [Willi Hennig.]

41. The assessment given in this sentence is a very hypothetical and theoretical one. It can only be of interest to someone who considers these stem-groups to be natural groups (taxa) that are equivalent to monophyletic groups, the recognition and definition of which is the aim of phylogenetic systematics. However, if we examine realistically the way in which *a fossil is assigned into the stem-group of a particular *group using the methodology of phylogenetic systematics, which involves weaving it into a particular topographical position within the fabric of a dendrogram constructed from synapomorphies* (Schlee, 1970, 1971, 1978a; Hennig and Schlee, 1978), then there can be no doubt that the procedure involved and the term 'stem-group' are correct. [Dieter Schlee.]

42. For the 'stem-group' considered as a 'topographical term', see note 41 above. [Dieter Schlee.]

43. The 'results are limited' because of the smaller number of characters preserved in the fossils, not because of any intrinsic defect in palaeontological work itself. [Dieter Schlee.]

44. Other methods, and even the most fantastic ones, could leave a diffuse array of options open. The 'stem-group' concept, as defined here by Hennig, provides a framework into which concrete future results can be incorporated with a *very high degree of precision*. A fossil in the stem-group provides direct evidence that the corresponding *group was probably in the process of arising (i.e. had not yet acquired its complete set of characters)—a rather important conclusion. [Dieter Schlee.]

2. Difficulties in practice

There are many difficulties in the practical application of the stem-group concept, although its theoretical basis is defined very precisely. This is partly because only a few of the available characters are preserved in the fossils and partly because of the way in which the characters have developed. This has given rise to many pointless arguments because pragmatically inclined authors are unwilling or unable to understand that theoretical clarity in the definition of a concept and practical difficulties in its application exist on two completely different levels and should not be confused. Such practical difficulties should not induce us to alter a well founded concept like that of the stem-group in order to adapt it for practical work, thereby creating an invalid compromise approach. In fact, the original theoretical form of this concept shows the goal that we are trying to reach: it is impossible to know whether, when, and how far the practical difficulties now obscuring this goal will eventually be overcome[45]. If we modify the goal so that we can cope with our present practical difficulties, there will no longer be any stimulus to search for methods that may eventually solve them.

I shall again use the Trichoptera and Lepidoptera to show the types of difficulty that may arise in the practical application of the stem-group concept (Fig. 6).

We can begin with the latest common stem-species of the *Amphiesmenoptera (Trichoptera + Lepidoptera) (point 3 in Fig. 6). We must accept that it had the two constitutive characters that all the *Amphiesmenoptera have inherited from it: heterogametic females and looped anal veins. The second character is the only one that is preserved in fossils. Fossils with this character can always be assigned to the stem-group of the Amphiesmenoptera, but it can never be proved that they actually belonged to the latest common stem-species of the *Amphiesmenoptera. Even though it cannot be proved from the fossil record, the latest common stem-species of the *Amphiesmenoptera must once have existed. If we accept that it separated into two daughter-species, it is absolutely clear in theory that one of them belonged to the stem-group of the Lepidoptera (3b in Fig. 6). However, this does not imply that these daughter-species had already acquired one or more of the constitutive characters of the *Trichoptera or *Lepidoptera. Their differentiation could have involved some quite insignificant characters that were later lost. The later descendants of the earliest stem-species of the Trichoptera and Lepidoptera may have been the first species to acquire the constitutive characters of the *Trichoptera and *Lepidoptera. We do not know the sequence in which this took place and may never be able to work it out. Moreover, we might not even be able to find these characters in the species in which they first occurred, either because they happen not to have been preserved or because in principle they can never be preserved in fossils. If we group fossils according to the characters preserved, and in practice this is all that can be done, our groups will include, for example, species from the stem-group of the Amphiesmenoptera together with species from the stem-groups of the Trichoptera and Lepidoptera (Fig. 8). Many 'stem-groups' established by palaeontologists are of this kind. They are 'invalid stem-groups', which include species from the valid stem-group of a monophyletic group together with species from the stem-groups of its monophyletic subgroups. Wherever possible, we must separate these invalid stem-groups into their component parts, the valid stem-groups. As I have already pointed out, this aim is not generally recognized, nor can it be completely attained[46].

Handlirsch (1937) accused Lameere of making the mistake, 'like many of his predecessors, of arriving at his higher groups by simple vertical division of the phylogenetic tree and of including within his modern groups everything

34

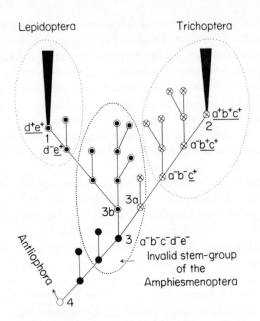

Fig. 8. Diagram to illustrate the concept
'invalid stem-group'. See p. 33.

situated in their ancestral lines, without regard for any morphological differentiation'. This statement shows Handlirsch explicitly advocating the retention of invalid stem-groups, and elsewhere he described a 'purely phylogenetic classification' as 'something impossible'. However, the task of phylogenetic research is to trace the history of 'modern groups' as far back into the past as possible: this can only be done if we assign to each 'modern group' all the fossils that belong to its 'ancestral line'. Handlirsch would have been more correct to use the term 'stem-group' rather than 'ancestral line'. To say that it is not completely possible to make such assignments in practice is one thing. To give up and to avoid giving systematic expression to the genealogical relationships, even when they are clear, is quite another.

It is now clear that Lameere's views, so strongly opposed by Handlirsch, have been generally accepted in all their essentials. Often authors expect too many results, even using the methods now available. An example of this are the attempts to trace 'modern' groups of the Auchenorrhyncha, such as the Cicadelloidea, Fulgoroidea, etc., back into the Permian or even the Carboniferous. It is fascinating to see how authors who do not support a phylogenetic classification in principle, or who even reject it, often give the most radical practical assistance in dividing up supposedly invalid stem-groups, even though there is sometimes no evident theoretical justification for doing so.

The practical and theoretical limitations in the assessment of fossils are so important for phylogenetic systematics that I shall discuss the theoretical aspects of this problem once again, without reference to any specific examples (Fig. 9).

I shall begin with three monophyletic groups, A, B, and C. Each is characterized by autapomorphic features which are also the synapomorphic characters of the included species: group A has characters d^+ and e^+, group B has character f^+, and group C has characters a^+, b^+, and c^+. I suggest that one of the characteristic (constitutive) features of C, character a^+, was acquired from the stem-species (3a) of this group, and that characters b^+ and c^+ were subsequently acquired by 'additive typogenesis'. All species with character a^+ belong to group C and some of them belong to its stem-group (circles with a cross). Unlike 3a, the other daughter-species of 3 (3b) inherited the characters of 3 unaltered and remained morphologically 'the same'. When 3b divided up, one of the daughter-species acquired character d^+, which is one of the constitutive characters of A. As in group C, character e^+ was subsequently acquired by additive typogenesis. All species with character d^+ (circle with a black dot), both fossil and recent, belong to monophyletic group A. The single constitutive character of group B (f^+) did not develop until a very late stage amongst the descendants of 3b.

In this particular case it is clear that it would be impossible to show that groups A and B form a monophyletic group, the sister-group of group C. It would not even be possible to interpret the fossils, shown as empty circles in Fig. 9[47]. They could belong to the stem-group of A + B + C or to the stem-group of any of these three groups. But this is precisely where my diagram shows that to a certain extent we can work independently of the fossils in determining the age of monophyletic groups. If we can establish that

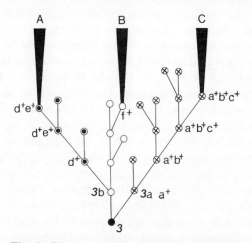

Fig. 9. Diagram to illustrate the problems that can arise from the operation of the deviation rule[47] when attempting to assign fossils to particular recent monophyletic groups. See pp. 35–36.

groups A, B, and C form a single monophyletic group, then the existence of fossils with characters a^+ and d^+, respectively, proves that at the same time the stem-group of B must also have been in existence, although nothing in the fossil record could prove this. The more precise our knowledge of the genealogical relationships of monophyletic groups, the greater the results that can be obtained with only a few fossils. If a more penetrating analysis of characters were to prove that groups A and B form a single monophyletic group, one fossil with character d^+ would prove that all three groups A, B, and C must have been in existence at this time as separate units. For the sake of argument I have not taken into account the possibility of convergence, because this belongs to a different methodological context.

In spite of this, it remains a fact that tracing the history of animal groups into the past often leads back into the limits of invalid stem-groups. Even these attempts have their limitations, since it is never possible to reach back to the actual point of origin of a group, and all that can ever be pinpointed is the moment *before which* the group must have originated. However, the genealogical relationships of monophyletic groups often enable us to work out the minimum age of a group, even though it cannot be confirmed by any fossils. We have to do this in many groups if we wish to give some sort of account of their phylogeny, because it is unlikely that any fossils will ever be found which belong to them. The following example will illustrate this.

The phylogenetic relationships of the oldest monophyletic groups of the Insecta are now well known and it is almost universally accepted that they are as shown in Fig. 21 (p. 108). The oldest known insect is *Rhyniella praecursor* Hirst and Maulik, from the Middle Devonian of Scotland[48], and its assignment to the Collembola is seldom disputed. All the phylogenetic events that led to the origin of a particular group must obviously have taken place before the epoch from which the fossils are available. Consequently, in this particular case, a single fossil collembolan from the Middle Devonian shows clearly that the primary divisions of the insects must have taken place before the Middle Devonian: these are the divisions of the Insecta into the monophyletic sister-groups Ectognatha and Entognatha, and the division of the Entognatha into Diplura, Protura, and Collembola. This conclusion is indirectly confirmed by the discovery of a dicondylic mandible (*Rhyniognatha*) from the Middle Devonian, which must belong to the Ectognatha-Dicondylia, and of a winged insect (*Eopterum*)[49] from the Upper Devonian, which must belong to the Pterygota. The discovery of *Eopterum* proves that the division of the Ectognatha into Archaeognatha, Zygentoma and Pterygota must also have taken place before the Upper Devonian. Unfortunately, it has not been possible to make a definitive assignment of *Eopterum*, for example by showing that it could belong to the Neoptera or even to the Paurometabola. If this could be done, this single fossil would enable us to draw several more far-reaching conclusions. However, *Eopterum* probably belongs to the valid stem-group of the Pterygota[49].

Amongst other things, this example shows that statements about the course of phylogeny do not depend on the discovery of fossils. On the contrary, the fossils can only be interpreted when monophyletic groups of species have first been established and their phylogenetic relationships worked out. The process of defining monophyletic groups and their genealogical relationships is called phylogenetic systematics. The example I have just given shows why I prefer this method to all other methods of biological classification.

Handlirsch and other authors, and more recently Sharov (1959 on), have recommended a system that only allows the genealogical relationships to emerge when a certain degree of morphological divergence has been achieved. This is why Sharov has not recognized the Zygentoma + Pterygota as the Dicondylia, a subgroup of the Ectognatha, which is one of the conclusions of phylogenetic systematics: he has instead continued to treat the Zygentoma + Archaeognatha as the 'Thysanura'. As a result, he has deprived himself of the deductive potential available in cases like the one I discussed above. Sharov (in Rodendorf, 1962) treated the 'Monura' as a distinct order alongside the Thysanura, Diplura, and Collembola, and in his discussion of this typological group only suggested a close relationship with the 'Thysanura'. The 'Monura' have been found in the Carboniferous and their existence in this period is of little importance from the phylogenetic point of view. However, had they been found in the Devonian or even earlier, their assignment to the 'Thysanura' would prevent us from deducing very much about the phylogeny of this group of orders. Following the principles recommended by Sharov (and others), we could even retain the old division of the Lepidoptera into Microlepidoptera and Macrolepidoptera, since the Microlepidoptera are a paraphyletic group exactly like the 'Thysanura'. It would mean virtually nothing to assign a fossil to the Microlepidoptera and hardly anything to assign one to the Macrolepidoptera. We must try to work out the phylogenetic classification in the greatest possible detail and we shall then be able to draw some far-reaching conclusions from the few available fossils, such as indirect inferences about the synchronous existence of other groups which are not known from fossils and possibly never will be.

Revisionary notes

45. For example, there may be new discoveries of amber fossils, which will provide additional morphological data, additional taxa, and additional geological periods. [Dieter Schlee.]

46. For example, Illies' (1974) concept of 'geologically old' or 'geologically young' groups and their zoogeographic implications, as well as his basic principles 5 and 6 of causal zoogeography, should be viewed in the light of this and the preceding few sentences. They can be partly explained by his confusion of *groups and groups. [Willi Hennig.]

47. The so-called deviation rule is not defined in this book, nor does it play a

38

very important role. However, it has been a source of great difficulty for those who have tried to understand and use Hennigian phylogenetic systematics. As a result, I have felt it necessary to give some discussion of it in my papers (Schlee, 1971, pp. 27–30, with a comment by Hennig himself; Schlee, 1978a).

It seems to me that there is no need to introduce the 'deviation rule' at this point. It is not the operation of the 'deviation rule' that makes it impossible to recognize and reconstruct a phylogenetic process like that shown in Fig. 9, but rather the fact that hardly any characters are visible in the various fossils available. It is a simple (and obvious) fact that no reliable conclusions can be reached if there is no concrete information available! The same confusion would arise if there was insufficient information on the derivation and the 'deviation rule' was not in operation. [Dieter Schlee.]

48. For further discussion of *Rhyniella*, which might even be a modern organism accidentally contaminating rock specimens, see Crowson (1970, pp. 65–66). [Willi Hennig.] *Rhyniella* is now considered to be Lower Devonian (see note 57 on p. 50). [Adrian Pont.]

49. This section of four sentences on *Eopterum* is no longer valid. Riek (in Hamilton, 1971a) and Rodendorf (1972) both came to the conclusion that *Eopterum* and *Eopteridium* were fragments of Crustacea and not insect wings at all (Riek in CSIRO, 1974). [Dieter Schlee.]

C. Phylogenetic research and evolutionary ecology

The view is now being expressed more and more frequently that phylogenetics must develop into evolutionary biology or ecology and become 'causal research'. The Russian author Martynov is credited with introducing this point of view into palaeoentomology: his descriptions of each new fossil are said always to have been accompanied by an attempt to interpret the evolutionary process. This principle has been closely followed in the Russian literature, and the description of the phylogenetic development of various groups is still usually accompanied by an explanation of the origin, survival, and development of each group in evolutionary biological terms. This point of view is undoubtedly a legitimate one, and without it phylogenetic research would remain incomplete as a scientific discipline. It is also correct to talk about causal research in this context.

It is possible to explain the holometaboly of the Holometabola by showing that a change has taken place in the hormonal system regulating metamorphosis, causing the corpora allata to release an inhibiting hormone. This ensures that ecdysis continues for several more stages as a purely larval ecdysis and that the hormone initiating pupal and imaginal ecdysis is only released at one particular moment. In one sense this is a 'causal explanation' of the fact of complete metamorphosis during the ontogenetic development of a single individual. Yet it reveals nothing about the reasons for the origin of holometaboly in the course of phylogeny, on the basis of its 'biological significance' in the sense of Bock and von Wahlert (1965), or for the origin of the group Holometabola. No experimental work can explain this, and any suggestion can only be in the nature of a *vaticinium ex eventu*. Such explanations attempt to harmonize theories about the effect of known evolutionary factors with what is known about the environment which probably existed at the time of the origin and development of the Holometabola. Obviously, we shall never have any detailed knowledge about this environment. Consequently, all such evolutionary biological explanations are very speculative and often sound extremely facile, not least those of the Russian authors.

So many of these evolutionary biological interpretations give an unsatisfactory impression because they ignore two fundamentally important factors:

1. It is vitally important to know the genealogical relationships of each group whose evolutionary course is to be explained 'causally'.
2. No adequate distinction is made between the origin of a group (e.g. the Diptera) and the origin of the morphological–functional type which it now represents (*Diptera). Connected with this is the lack of any clear distinction between an explanation of the origin of a group and an explanation of its development, in other words of its evolutionary success.

A few examples will illustrate the importance of these factors.

1. The origin of holometaboly will have to be explained differently if the 'Holometabola', that is to say the species exhibiting holometaboly, are regarded as a monophyletic group, or if they are considered to be polyphyletic as they were by Weber. In the first case there is only a single event to be explained. In the second case, however, several independent events must have taken place: these could all have had a different 'biological significance' and could have had different causes.

   A similar example is provided by the two 'palaeopterous' groups, Ephemeroptera and Odonata, which have survived to the present day. Different explanations for this survival will have to be given, depending on whether we believe that they form a monophyletic group with certain derived ground-plan characters in common or that there is no close relationship between them. Both views are still held.
2. It is important to distinguish between the origin of a group and the origin of the discrete morphological–functional structural type which now represents it. I suggested above that a group 'originates' at the moment when the latest common stem-species that it shared with its sister-group divides into two daughter-species. These two daughter-species must be regarded as the stem-species of two separate groups, but it follows from the

deviation rule that one of them hardly differed or did not differ at all from the common stem-species[50]. To begin with, the derived characters in which the two sister-groups now differ from each other could at the most have only been present in the stem-species of one of them. The other will have retained the mode of life and morphology of the common stem-species. However, it could have been this second group that subsequently acquired most of the derived characters and became most successful, although conservatism was the principle of its 'origin' (see Fig. 9). This is covered by 'Cope's law of the unspecialized'.

Each group is now characterized by a structural type, and the additive development of this usually took place a step at a time. We do not know the sequence of these steps, nor shall we ever know it in cases where the characters involved cannot be preserved in the fossils.

There are two sister-groups in the Mecopteroidea. The first is the Antliophora (Mecoptera + Diptera), characterized by the presence of a sperm pump[51], reduction of larval legs, and perhaps originally by predaceous adults. The second is the Amphiesmenoptera (Trichoptera + Lepidoptera), characterized by heterogametic females and a peculiarity in the wing venation, the looped anal veins. It is not known which of the derived ground-plan features that are now characteristic of these sister-groups arose first or in what sequence they originated in each group. Nor is the selective value of these characters known, either singly or in combination, although it must have determined their evolutionary success. None of the evolutionary biological statements made about these groups has taken any account of the internal characters, such as the sperm pump[51] of the Antliophora and heterogametic females of the Amphiesmenoptera, and so they can never be verified.

In many cases knowledge of the genealogical relationships may be of assistance. This is because different ideas about genealogical relationships must have a fundamental effect on the evolutionary biological interpretation, as the following example will show.

The derived characters of the beetles include the transfer of the flight function to the hind-wings (posteromotoria), the modification of the fore-wings into elytra, and the prognathous head which is associated with predaceous habits. There is no obvious functional connection between these characters, and we have no direct knowledge of the sequence in which they arose.

Weber suggested that the sister-group of the beetles is the 'Blattoidea'. If we accept this it would mean that the beetles took over the formation of elytra and the transfer of the flight function to the hind-wings from ancestors that they shared with the 'Blattoidea'. In this case these characters cannot have determined the evolutionary success of the beetles, or can only have done so when combined with further characters: prognathy and holometaboly are the acquisitions peculiar to the beetles that would have to be considered in this context.

On the other hand, if we believe that the beetles are most closely related to the Neuropteroidea, they would have inherited holometaboly and prognathy from the ancestors that they shared with other groups. The modification of the fore-wings into elytra and the transfer of the main flight function to the hind-wings would then be acquisitions peculiar to the beetles, together with other characters.

There is a further difficulty in giving evolutionary biological interpretations of the phylogenetic development of monophyletic groups of insects. This is because it is much more difficult with such small creatures to recognize their connections with particular ecological zones and niches. This is quite different from the situation in the vertebrates. The fact that certain insects are able to complete a vital part of their life-cycle in the tiniest and most ephemeral bodies of water, such as tree hollows or thin films, makes it much more difficult to distinguish between an aquatic and a terrestrial mode of life than it is with larger animals. Even phytophagy covers a much wider range of habits in the insects than it does in larger animals, because insect larvae can feed externally on the leaves, or can live in galls or mines in the stem, in the parenchyme of the leaf, in the roots or in the flowers. There are also greater possibilities for them to form specific links with particular species of plant (monophagy). In many cases the individual life-cycle consists of several stages, each with fundamentally different ecological requirements. Very little is known about these stages and their life histories, particularly about the eggs and their adaptive characters which may well have made significant contributions to the evolutionary success of a group in the way of selective advantages[52]. Even the functional significance of quite obvious imaginal characters has not been adequately studied.

It is likely that a considerable period of time may often have elapsed between the origin of a group (e.g. the Siphonaptera) and the time when the latest common stem-species of all recent species lived (i.e. the origin of the *Siphonaptera). The mode of life may have gradually changed during this period, at least in the direct ancestral line of the recent species. Nevertheless, the foundations for the origin of the present ecological–functional–morphological characters of the group must have been partly laid by the relevant adaptive characters, as preadaptations. This then connects with the matter discussed above: in what sequence, and when (during which geological periods), did the various derived characters now present in a group come into existence? It is rare that anything at all is known about this.

The Phasmatodea are now a group of arboreal leaf-feeding insects. No authentic fossils are known before the Tertiary. Gangwere (1965) suggested that the expansion of this group in the Jurassic was due to the appearance and diffusion of tree-like dicotyledons. Yet he placed the origin of the Phasmatodea back in the Upper Carboniferous. It is not clear how the species of this group actually lived and what they looked like during the Palaeozoic and early Mesozoic. Evolutionary biological explanations of the phylogeny of an extinct group are even more difficult: the Palaeodictyoptera provide a typical example of this, and will be discussed in Chapter III.

In the light of all these difficulties, I have decided not to devote much space to evolutionary biological discussions in the following account. This is not because I underestimate the value of this approach, which can also shed light on the genealogical aspects of phylogeny by the principle of reciprocal illumination, but because I am only too conscious of the harm that has often been done by basing such speculation on inadequate foundations.

One of the aims of phylogenetic studies must be to explain the actual historical process of the phylogeny of a particular group in terms of the action of known natural laws. Ideally a comparative study should be made of as many phylogenetic processes as possible. This aspect of phylogenetic research resembles research into the processes of human history because both have given rise to similar theories: those that suggest something like an immanent law directing development towards a particular goal or towards generally higher forms of organization, those that suggest a cyclical process, and those that attempt to combine both these elements. As an example of the cyclical theories I should mention Schindewolf who, like other authors, suggested that the 'life' of each group follows a course conforming to natural laws, which can be compared with the life of an individual: birth, youth, maturity, old age, death. However, one objection to this theory is that there are monophyletic groups that have existed since the Pre-Cambrian. So if it is suggested that animal groups are ineluctably condemned to die out after a certain time (A. H. Müller), this can only be true if the groups are understood not as genealogical units but as 'morphotypes', some of which are genealogically connected with later morphotypes. In this respect, this resembles the idea that at least in certain groups phylogeny follows a 'rocket scheme', that a monophyletic group produces a succession of 'radiations' and that each cluster of radiations proceeds from a single branch of the preceding one (Fig. 10)[53]. This raises several other questions that need to be studied:

1. Does the development of all animal groups, including the insects, follow such a 'rocket scheme'? There is no clear answer to this. For example, Heslop-Harrison (1952b) supported the idea that the insects have undergone explosive development, whilst in a discussant's comment on this lecture Sylvester-Bradley opposed it.

   This can only be answered by establishing the times of origin of as many monophyletic groups as possible and checking whether they are concentrated within certain periods. This is difficult to do because the time of origin of a group is often considered to be the moment when it first occurs as a discrete structural type, perhaps supported by fossils. An example of this is Carpenter's (1953, 1954a) table of the first appearance of various 'insect orders', some of which, such as the Blattariae, are purely typological concepts and cannot be compared with each other as genealogical units. However, it is obviously wrong to suggest that the oldest fossil marks the actual moment when a group originated. It is therefore usually suggested that this moment occurred during an earlier period, and it is calculated by

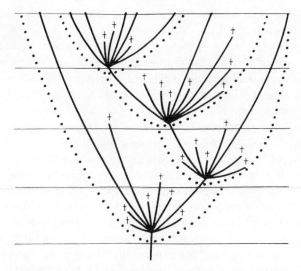

Fig. 10. A 'rocket scheme', which the development of
animal groups is supposed to follow. From Gross
(1964).

a linear extrapolation of the rate of development observed in each group.
This rate is measured by morphological change, although it is clear that in
the course of time it may not always have been constant. As a result, some
very early times of origin have been given for quite subordinate groups.
An example of this is Borkhsenius' (1958) 'phylogenetic tree' of the
scale-insects, in which many subgroups are said to have originated during
the Carboniferous!

Wille's (1960) paper shows how difficult it still must be to say what
radiations have taken place amongst the insects. He suggested three separate
'main radiations': the Apterygota, Palaeoptera, and Neoptera[54]. It is not
clear why he did not also suggest a separate radiation for the Holometabola,
to which more than three-quarters of all recent insects belong.

2. If it were proved that the phylogenetic development of a large animal
group has followed a 'rocket scheme', it would still be necessary to
establish whether the periods of explosive development are identical or
different between individual subgroups. There is no agreement about this.
For example, in a review of an American symposium on this subject,
Colbert (1953) asserted that the periods of evolutionary radiation were not
the same in different groups of animals. There were three great periods of
vertebrate expansion: Devonian, Permian–Triassic, and late Cretaceous–
Tertiary. However, the periods of invertebrate expansion are supposed to
have been the Cambrian–Ordovician, Lower Carboniferous (Mississip-
pian) and Triassic–Jurassic.

This is scarcely credible. I fail to understand why the invertebrates as a group should show such uniform periods of expansion which are yet so different from those of the vertebrates.

3. A third question is whether there is any correlation between the periods of radiation, which may have been the same or different in various groups of animals, and particular geological events. 'Organic life and the earth itself have both changed in the course of time, and this prompts the question: are these two processes connected, and if so, how? Should we believe that geological events are the main cause of organic evolution?' (Kuhn-Schnyder, 1965). Kuhn-Schnyder continued: 'The various phases of anagenesis, of higher development, are correlated with particular geological epochs. Within any particular lineage, the breakthrough to a higher level of development has taken place more or less simultaneously in different lines and in different regions'. However, even he believed that 'no intrinsic correlation between important turning-points in the post-Cambrian history of life and the phases of mountain formation or displacements of land and sea can be discerned'[55]. Similar opinions have been expressed by Brinkmann (1954) and by Dehm (1963). Dehm found it 'impossible to detect any obvious connection between the expansion and decline of life on the one hand and the external events of geological history on the other (mountain formation, marine transgressions, continental drift, etc.)'. Dehm believed that in different groups of marine animals 'the climaxes of their expansion took place in different periods, which were not the same in any two groups'. These views sometimes overlap and sometimes differ, but they all arouse the suspicion that in the final analysis they are based upon typological methods of thought. Different authors' inability to detect any correlation between the development of the earth and the development of life could be because they have only been comparing particularly striking structural types. They should instead have been comparing the times of origin of genealogical units with each other and relating them to geological events. Only this can provide the basis for realistic historical studies.

We do not yet know what insights will be gained when the times of origin of strictly monophyletic groups are actually compared with each other. Palaeontologists, who mainly deal with such questions, appear to have paid too much attention to the direct influence of tellurian or cosmic factors. This is seen most clearly in the views of Kuhn-Schnyder, whom I have just quoted and who wrote of organisms having changed or having become more highly evolved, and even more so in the ideas of Schindewolf. Schindewolf 'recognized that there were certain periods, such as the transition of the Mesozoic and Coenozoic, when several animal lineages disappeared and others appeared. He believed that these were preceded by periods of increased cosmic radiation which could have resulted in a general increase in the rate of mutation. Schindewolf thus felt obliged to put forward a *deus ex machina* solution in the absence of any terrestrial explanation' (Dehm, 1963).

However, biologists may prefer to think that geological events could influence the evolutionary process by shaping ecological–evolutionary factors such as selection, isolation, etc.

It is probably an established fact that drastic changes in the external environment are the characteristic features of certain epochs, such as the Permian and the Cretaceous (see pp. 57 and 79). These changes must have had an effect on the evolutionary process, for example by altering the selective value of particular adaptive characters, and in this way they must have acted as a kind of filter. This must have been an extremely complicated process. It is easy to imagine that a changing environment may have condemned entire morphological groups to extinction, and may have provided others with the impetus for greater development (i.e. for evolutionary radiation). However, large monophyletic groups are rarely closed ecological adaptive types. In many cases a few species will have survived the filtering process and will subsequently have undergone considerable radiation. Such species need not necessarily have been those that appeared to be new types or the nuclei of new types right from the start. Cope's law of the unspecialized, which I discussed above in the context of the deviation rule, shows that on the contrary the species that derived the greatest advantage were frequently those with no particularly specialized adaptations to restricted ecological zones. The new possibilities that faced them as entire groups began to decline towards extinction may have enabled their descendants to ensure the complete extinction of less favoured groups. These descendants must, of course, have had the necessary preadaptations for the required ecological development.

We may still be justified in believing that the evolutionary process was influenced by individual geological events even if the periods when different groups radiated were not exactly synchronous but followed each other at certain intervals. For example, the insects frequently have close links with the plant kingdom and these associations require particularly close study. It is still possible to talk about the influence of geological events on the course of evolution if only a few large groups of organisms are affected initially, whereas others, which may be directly or indirectly dependent on them, do not react until later.

These connections are very obscure at present, but they will become much clearer when we have established the precise moment of origin, as I defined this above, of as many monophyletic groups as possible. I shall be concentrating on this task in Chapter III.

The connection between palaeogeographic changes and phylogeny is a separate question. It is *a priori* probable that there must have been some influence, because of the important evolutionary factor of isolation amongst other things, but it is extraordinarily difficult to investigate because only a few regions of the world have produced any extensive fossil material. In his account of the history of the insects, Jeannel (1950)[56] laid great stress on the separate development of different lineages in Laurentia, Gondwanaland, and

46

Angaraland in the Palaeozoic, and, after a period of faunal interchange, again in the Mesozoic. However, the evidence for this view is too scanty for it to be regarded as anything more than a stimulating working hypothesis.

Revisionary notes

50. See also note 47. If one of the 'daughter-species' does not differ from the stem-species, it cannot be recognized as a *'daughter'* and most people would prefer to speak of a species (stem-species) which persists even after a daughter-species has separated off. The possibilities of reconstructing phylogenetic sequences containing stem-species and daughter-species, which may survive or become extinct, have been discussed by Schlee (1971, pp. 30–37, 'Rekonstruktionsmöglichkeit bei fortlebender Stammart'). In practice, none of these situations will actually give rise to any serious scientific problems, irrespective of whether there is any justification for the 'deviation rule', because a properly conducted analysis using the methods of Hennigian phylogenetic systematics makes it possible to recognize precisely the limits of conclusions that are warranted from the available facts.

The following is an example of this. If the only result that can be obtained from analysis so far is like that shown in Fig. VB, rather than a sister-group relationship like that in Fig. VA, then concise use of Hennig's phylogenetic methodology immediately reveals several unresolved questions:

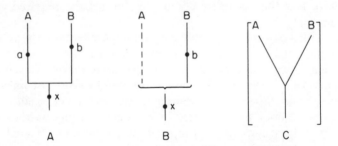

Fig. V. Correct and incorrect methods of expressing the limits of conclusions when a surviving stem-species is involved. [Original. Dieter Schlee.]

1. It is uncertain whether 'A' is a monophyletic taxon or a doubtful group (cluster).
2. It is uncertain whether 'A' as a whole is the sister-group of B; or whether part of 'A' (A_1) is the sister-group of B whilst another section of 'A' (A_2) might be more distantly related (the sister-group of $A_1 + B$).
3. The actual phylogenetic history of 'A' is uncertain: it could either belong to the stem-group, which 'continued unchanged' after B had arisen, or it could be a daughter-species which has not yet been adequately studied.

The conclusion must be that there is no solution at present, and this fact can be seen from Fig. VB. Such a diagram also shows the various solutions that are possible.

Brundin (1976) has disagreed with this by explicitly stressing that it is not necessary to have a complete set of synapomorphies or autapomorphies (Fig. VA) in order to obtain acceptable phylogenetic conclusions, at least at the species-level. He has claimed that a dendrogram like Fig. VC is an acceptable interpretation of the actual state of knowledge shown in Fig. VB. This line of argument must be rejected since Brundin has ignored the three areas of uncertainty listed above: for a refutation of Brundin, see Schlee (1978a, especially pp. 10–12). There is a particularly grave danger if Brundin's principle is used with taxa above the species-level. This has often been done in the past, and now that Brundin has invoked the name of phylogenetic systematics to absolve specialists from the necessity of making a clear distinction between their proven and unproven results, there could be an increase in the number of papers containing premature conclusions.

If there is not a complete set of synapomorphies (as in Fig. VA), then those who want to work with Hennigian phylogenetic systematics should use a diagram like Fig. VB to show the actual state of knowledge, and should avoid equating Figs. VB and VC. They should bear in mind Hennig's (1976, pp. 61–62) own admonishment: progress can only be achieved by making intensive comparative studies with specifically directed aims, and by pinpointing existing areas of weakness in the classification. [Dieter Schlee.]

51. I do not think that the sperm pumps of the Mecoptera and Diptera are homologous. See also note 338 on p. 332. [Dieter Schlee.]

52. Several papers have recently been published on egg structures and their phylogenetic significance: Cobben (1968) on the Heteroptera, Seeger (1979) on the Psocoptera and related groups, and Hinton (1980) on the Insecta generally. [Dieter Schlee.]

53. Donovan (1964, p. 269, Fig. 6) has published a 'rocket scheme' to illustrate the evolution of the ammonoids, showing extinctions at the end of the Permian and the end of the Triassic, and the final extinction of the group at the end of the Cretaceous. [Willi Hennig.]

54. Smart (1963) has published a diagram that uses the concept of 'explosive evolution' to relate orders to each other, rather than the usual phylogenetic tree. [Willi Hennig.]

55. For a consideration of the effect of oxygen consumption on the extinction of animal groups, see McAlester (1970). An absolute geological chronology and the problem of cyclical development have been discussed by Kölbel (1969). [Willi Hennig.]

56. Jeannel (1950) discussed the history of the origin and dispersal of the insects in relation to the history of the three great continental blocs ('refugia') of Laurentia, Angaraland, and Gondwanaland. The alternating connection and separation of these continents is said to explain the origin and distribution of the insect lineages.

In *Laurentia*, which had a tropical climate during the Carboniferous and Permian, the Palaeoptera and Polyneoptera are said to have arisen from more northern, Devonian ancestors. After the Carboniferous, Laurentia became progressively cooler.

On the other hand, *Gondwanaland*, which was originally close to the pole and had a correspondingly cold climate, is said to have become progressively warmer after the Carboniferous. The Paraneoptera and Holometabola are said to have arisen there, as the fauna of a cold area. Jeannel considered the 'nymphal diapause' of the Holometabola to be an adaptation to a temperate climate with cold winters.

Unlike the other two continents, *Angaraland* enjoyed a rather stable climate and for this reason became a faunal refugium.

At the end of the *Permian*, climatic assimilation took place and a retreat of the mesogaean oceans facilitated a faunal exchange. At this period, therefore, there was a mixture throughout the world of elements that had originated from Laurentia and Gondwanaland.

During the *Mesozoic*, development in isolation took place as Gondwanaland broke up.

The history of the Angaraland element did not begin until the *Cretaceous* and was restricted to Laurentia (= Holarctis). [Willi Hennig.]

CHAPTER TWO

Localities for Palaeozoic and Mesozoic Insects

A. Palaeozoic

1. Devonian
(Fig. 11)

Only three Devonian insects are known, and they come from two different localities. These are Rhynie in Scotland (Middle Devonian), which has yielded *Rhyniella* and *Rhyniognatha*, and Ukhta (Komi Republic), in the region of the Timanskiy Kryazh ridge (Upper Devonian), from which came *Eopterum*, probably representing the stem-group of the Pterygota[57]. Both localities are in the Northern Continent, which included Fennoscandia and the Canadian Shield. According to North (1931), conditions in the Devonian were such that there can be little hope of ever finding greater numbers of insect fossils.

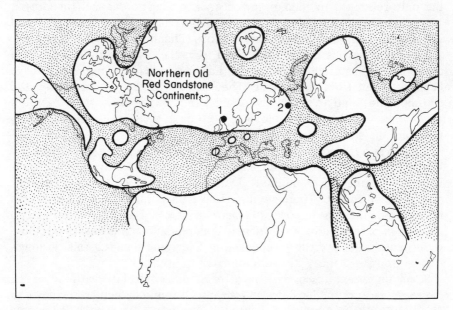

Fig. 11. Palaeogeography and insect localities in the Devonian. From the palaeogeographical maps of Joleand, Termier and Termier, Babnoff, and Kossmat. 1, Rhynie (Scotland; Middle Devonian); 2, Ukhta (Komi Republic; Upper Devonian)[58].

Revisionary notes

57. The situation is now as follows. Only two Devonian insects are known, from one locality: *Rhyniella* and *Rhyniognatha*, from Scotland. The reference to *Eopterum* from Ukhta should be deleted, as it appears not to be an insect wing. '*Eopterum* and *Eopteridium* have been found to be parts of the tail-fan of eumalacostracan Crustacea (Riek, in Hamilton, 1971a; Rodendorf, 1972)' (Riek in CSIRO, 1974, p. 28). [Dieter Schlee.]

The Rhynie Chert, in which *Rhyniella* and *Rhyniognatha* were found, has now been assigned to the Lower Devonian (Siegenian) on the basis of its spore assemblages (Richardson, 1967; Richardson, in House *et al.*, 1977, p. 76). [Adrian Pont.]

58. Locality no. 2 (Ukhta) should be deleted as it is no longer a locality for fossil insects (see note 57 above). [Dieter Schlee.]

2. Carboniferous
(Fig. 12)

Although insects must obviously have existed in the Lower Carboniferous, the only reference to their occurrence, according to Jeannel (in Grassé, 1949), is Pruvost's statement that the wing of a palaeodictyopteron was found in his presence in the Lower Carboniferous (Dinantian) of Nova Scotia.

The oldest insects actually described from the Carboniferous have been found in the Namurian. According to Laurentiaux (1952), five species have been recorded from this epoch. They belong to the genera *Erasipteron* (Protodonata); and *Stygne*[59], *Limburgina*, and *Ampeliptera*, which are probably all in the Paurometabola but, according to Sharov (1966b), partly represent the stem-group of the Pterygota.

There was a period of at least 30–40 million years between the Middle and Upper Devonian, where insect remains are generally scanty, and the lower Upper Carboniferous (Namurian). During this period the Variscan Orogeny took place in Europe. There are numerous Upper Carboniferous fossils, and these include the most important Palaeozoic insects. They have been found in deposits in paralic basins that originated along the margins of salt water and were periodically overrun by the sea; and in some limnic basins, in longitudinal troughs of the Variscan fold-belts.

Some important discoveries have been made recently in these areas. However, most of the localities and the insects found in them were known to Handlirsch (1922), who listed them individually in his review of Palaeozoic insects. Guthörl (1934) has given a tabular review of the 'age relationships of the groups of strata in the north-west European coal belt', which extended from England to Upper and Lower Silesia and included the Saar[60]. More

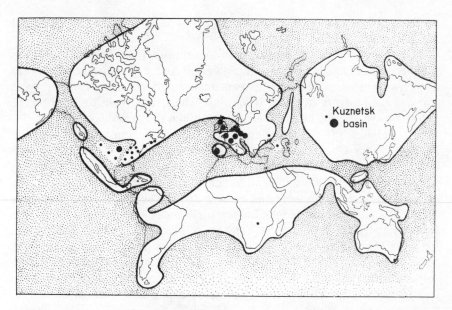

Fig. 12. Palaeogeography and insect localities in the Upper Carboniferous. From the palaeogeographical maps of Termier and Termier, Schuchert, and Bubnoff. The largest of the North American black dots: Mazon Creek (Illinois). For the other localities, see text, pp. 51–57. [The new localities mentioned in notes 59–66 have been added to Hennig's original map. Adrian Pont.]

recent monographs or comprehensive reviews are available for individual parts of the region as follows.

a. Paralic basins in Europe

Great Britain. The insects of the 'British Coal Measures' were monographically revised by Bolton (1921–22). There were some later additions in Bolton (1930, 1934) and Wallis (1939a, 1939b), and a lucid review has been given by North (1931). According to North, an upland spur separated the true fore deeps of the Variscan fold-belts in South Wales from the more extensive northern shelf-area. He also showed how this is reflected in the insects of South Wales, which resemble those of Kent and North France, whereas those from Central and North England do not. These British fossils are not of outstanding importance.

North France, Belgium, Westphalia (Ruhr valley Coal Measures). 'In the northern part of the Ardennes, the Upper Carboniferous is only preserved in a narrow trough between the Ardennes and the older Brabant uplands. As the latter dip towards the east, the Upper Carboniferous basin undergoes considerable expansion in Limburg, the Lower Rhine, and Westphalia. Its northern limit here is not known, and probably lies under the southern part of the North Sea' (Bubnoff, 1941, pp. 236–237).

There is no comprehensive review of this entire region. Pruvost (1930) has summarized the fossils from the Belgian area. About 15 years ago there were also reports in the popular newspapers about the discovery of *Protoprosbole straeleni* Laurentiaux from the Namurian, which could then claim to be the oldest known species of the Pterygota. 'Unlike the neighbouring coalfields (Saar, Belgium, Dutch Limburg, North France), those of Westphalia (= Ruhr valley Coal Measures) have only produced a few insects although their Carboniferous beds are very well developed' (Laurentiaux, 1958). Nothing has changed since then, although between 1958 and 1962 the number of known insect fossils increased from 7 to 13. The oldest fossil from this region (*Patteiskya bouckaerti* Laurentiaux) comes from Namurian B. For further details, Demoulin (1958a) and the 'critical review of the insect fauna from the Carboniferous of Lower Rhine-Westphalia' (Carboniferous of the Ruhr valley and the Erkelenz Horst) by W. Schmidt (1962) should be consulted, but Schmidt's assessment of the so-called Mecoptera needs to be treated with some caution and he uses Haupt's controversial system of wing-vein nomenclature.

Upper Silesia (Moravian Ostrawa-Karwiná). The generally accepted view is that the Upper Silesian basin is a continuation of the northern paralic corner of the Variscan fold-belts, but Bubnoff (1941, p. 239) suggested that Upper Silesia should be treated as a separate basin 'which was bounded in the north by Variscan fold-arcs (Sudeten-Polish uplands). The facies is also paralic, but the connection with the open sea must have been in the south where, in the Carpathians, the marine facies persists for longer. Slow submersion for several thousands of metres produced a thick sequence of strata'. Amongst other things, the Namurian of the Upper Silesian basin has produced the oldest known odonatan, *Erasipteron larischi* Pruvost. It is now being worked intensively and successfully by Czech palaeontologists (see Kukalová, 1961), to whom we are indebted for describing the basal articulation of a palaeo-dictyopterous wing.

b. Limnic basins in Europe

'The inland basins lie in longitudinal troughs within the fold-belts and had no connection with the sea (limnic). They formed individual basins rather than continuous channels' (Bubnoff, 1941: 234).

French Central Massif. 'Some 30 small Upper Carboniferous basins are known here', according to Bubnoff (1941, p. 242). Only a few of these have produced any fossil insects. The one really important locality, and in fact one of the most important localities for Carboniferous insects generally, is Commentry: according to Laurentiaux (1950) this locality has achieved immortality through the methodical investigations of Fayol, and it will always be associated with the name of Brongniart.

Lameere (1917) has given a more recent review of this famous fauna, which includes the celebrated giant insects with wing-spans of almost 70 cm (see also Laurentiaux and Laurentiaux-Vieira, 1960).

According to Laurentiaux (1950), Commentry is still the main source for our knowledge of insects from the Upper Carboniferous (Stephanian).

Everything else known from the French Central Massif also belongs to the Stephanian. Laurentiaux (1949, 1950) has mentioned the localities Saint-Éloy-les-Mines (Puy-de-Dôme), Saint-Étienne (Loire), Rive-de-Gier, and Avaize (both Gard). In addition, I know of La Mure (Isère: see Haudour *et al.*, 1960) and La Balorais, in Saint-Pierre-La-Cour (Mayenne: see Péneau, 1930).

None of these localities has produced anything remarkable. Handlirsch and Bolton (1925) dealt with the fossils from Commentry, and more recently some of them have been critically reviewed by Carpenter (1943–1962b)[61].

Saar–Nahe–Palatinate region. As in the basins of the French Central Massif, the Namurian stage is also missing in this region. However, according to Guthörl (1934), the stratigraphical sequence is uninterrupted from the middle Upper Carboniferous (Westphalian B) up to the lowest beds of the Lower Permian (the Rothliegende). Guthörl (1934) and Waterlot (1934a, 1934b) have given a review of the fossil insects of the Saar region which occupies a special place in the early history of palaeoentomology on account of the activities of Goldenberg between 1873 and 1877. Most of the insects found have been Palaeodictyoptera and cockroach-like species[62]. Richardson (1956) has reported that Waterlot (1934b) attributed the abundance of Palaeodictyoptera in the Saar basin to the absence or scarcity of fish. However, this suggestion was based on the premise that the larvae of the Palaeodictyoptera were aquatic, which is almost certainly incorrect.

'Erzgebirge basin of Lugau-Ölsnitz' and *'Wettin basin east of the Harz'.* These terms are taken from Bubnoff (1941, p. 242). The fossils were all listed by Handlirsch, and nothing of any note has been found there since[63].

'North Bohemian basin of Plzeň (= Pilsen)-Kladno' and *'Lower Silesian basin'.* Kukalová (1961) has published a report on the North Bohemian basin, in which the coal gas of Nýřany (Nürschan) is particularly remarkable because of its unusually well preserved fossils. In the same paper Kukalová also reported on the Lower Silesian basin.

c. Southern Europe (Portugal)

Several insects, especially cockroach-like species and Palaeodictyoptera, and even a *Metoedischia*, have been described recently from the Upper

Carboniferous of the Portô area. The localities mentioned are Douro Litoral (= Baixo-Douro = Bas-Douro), Pejão-Castelo de Paiva, Valdeão-Valongo, São Pedro da Cova, Santa Susana: see Fleury (1936), and the papers of Teixeira (1939–48) and Laurentiaux and Teixeira (1948–58). According to Fleury (1936), three basins on the western edge of the Meseta can be recognized as continental deposits of the Upper Carboniferous in Portugal, and in several respects these are reminiscent of the French limnic basins: the northern basin or Douro basin (São Pedro da Cova-Pejão zone), the central basin of Buçaco basin, and the southern basin or Sado basin (Santa Susana-Moinho da Ordem zone). Fleury gave the age of the northern and southern basins as Westphalo-Stephanian and of the central basin as Stephano-Autunian.

d. Eastern Europe

In his review of the localities of Palaeozoic insects in the Soviet Union, Rodendorf (1957b) illustrated two 'cockroach' wings, one from the upper Upper Carboniferous (Stephanian) on the River Teberda and one from the 'Carboniferous' of the Donets basin (eastern Ukraine). Both wings are apparently still undescribed and are certainly of no particular importance. I only mention them here for the sake of completeness.

e. Asia (Siberia, Angaraland)

There is no doubt that one of the most important events in the more recent history of palaeoentomology has been the discovery and working-up of fossils from the Kuznetsk basin. Before 1939, when the first fossils were found by M. F. Neuburg, only a few fortuitous Palaeozoic insects were known from Asia. According to Bubnoff (1941, p. 296), the Kuznetsk basin has all the features of a marginal or inland basin in the Asian Variscan fold-belt.

According to Rodendorf (1958), the great importance of these fossils is that they provide a complete stratigraphical sequence from the middle Upper Carboniferous (see below) to the Upper Permian. Previously, almost nothing was known from Palaeozoic Angaraland. From this sequence no fewer than 4870 individual fossils have been described, according to the comprehensive account published by Rodendorf et al. (1961). In addition, both Triassic and Jurassic strata and fossils are known from the same region.

The stratigraphical divisions are different from those customary in Europe. In the most recent work (Rodendorf et al., 1961), both the lower and the upper Balakhonka Stages have been placed in the Carboniferous, whilst earlier (Rodendorf, 1956) the division between the Carboniferous and the Permian was identified with the division between the lower and upper Balakhonka Stages. In the Russian Textbook of Palaeontology (Rodendorf, 1962) the Carboniferous has been divided into three series, the lowest of which coincides with the Namurian. No fossils are known from this series. As the lower Balakhonka Stage is equated with the Upper Carboniferous (C_3) in the work on the Kuznetsk basin, the two Balakhonka Stages most probably correspond to the Westphalian and Stephanian of western stratigraphy. The localities in the Kuznetsk basin are listed and briefly described by Rodendorf (1957b): they are situated in the region of Kemerova on the River Tom' and its tributaries (see the map in Rodendorf et al., 1961). Some 1250 fossils have been described from these Carboniferous deposits (lower and upper Balakhonka Stages), and most of them consist of so-called cockroaches. Rodendorf considered two facts to be particularly striking: the very small numbers of Palaeoptera (Rodendorf, 1958) and the high percentage of forms which have only been found here (Rodendorf, 1961a). He attributed this to the pronounced palaeogeographical isolation of Angaraland in the later Palaeozoic (Upper Carboniferous and Permian).

The most productive locality, with 1000 fossils, is in the area of the oldest deposits at Zhëlty Yar (= Gelty Yar, Zhelty Jar; lower Balakhonka Stage).

In the Kuznetsk basin, the Permian fossils (357 species) are much more important than the Carboniferous ones (see below, p. 62). Only 72 Carboniferous species have been described, 62 from the lower and 10 from the upper Balakhonka Stages.

Rodendorf mentioned the Podkamennaya Tunguska (= stony Tunguska) in the region of Krasnoyarsk as a further Carboniferous locality in Angaraland. Apparently only two 'cockroach' wings have been found, and they have not yet been described.

Laurentiaux (1947) published a short preliminary paper on an insect fauna from the coalfields (Westphalian C) of K'ai-p'ing in China (near T'ang-shan). He only mentioned some 'cockroaches' of the families Archimylacrididae and Hemimylacrididae, but these do not appear to have been worked up yet.

f. North America

I do not know of any recent clear review of the origins of the Upper Carboniferous deposits which contain insect fossils. In one respect, the situation in North America resembles that in Europe because the Appalachian range was uplifted at about the same time 'as the Variscan fold-belt of Europe, of which it can be seen as a continuation' (Bubnoff, 1941, p. 253).

As in the Variscan fold-belt of Europe, there was a paralic zone of 'littoral peat bogs' (Kossmat, 1936, p. 103) extending along the marginal depression of the Appalachians, and in Nova Scotia and New Brunswick there were extensive intra-montane troughs which apparently correspond to the limnic inland basins of Europe (Kossmat, 1936).

In addition to the Appalachian region, Bubnoff (1941, p. 253) distinguished 'two giant basins in the area of the great Mississippi plain', the Midcontinental basin, 'and these were separated at the Mississippi by a shallow north-south positive area consisting of older rocks'. Unfortunately, it is not clear into which of these two regions the fossil localities listed by Handlirsch should be placed (Cape Breton, Nova Scotia[64]; Rhode Island; Maryland; Pennsylvania; West Virginia; Alabama; Georgia; Ohio; Illinois; Indiana; Arkansas; Montana; Kansas). However, this is unlikely to have been of any significance for the phylogenetic assessment of the fossils.

The age of the Carboniferous fossils of North America corresponds roughly to that of the European ones. The oldest 'Pottsville-series' corresponds to the 'Upper Namurian to Westphalian B' according to Bubnoff (1941, p. 253). The most recent revisions of the North American Carboniferous insects are by Cockerell (1927a, Maryland), Carpenter (1934a, Kansas; 1934b, Pennsylvania; 1961b, Georgia; etc.) and Richardson (1956, Mazon Creek)[65]. Richardson has also given a general description of Mazon Creek in Illinois, a particularly rich and celebrated locality. The insects, which occur as fossils in nodules of ironstone, lived in the middle Upper Carboniferous on a plain close to the sea and not far above sea-level. The vegetation was much less dense than in a typical bituminous coal swamp. Richardson concluded from the comparatively small number of 'cockroaches' that the forest must have been drier and more open in the region of Mazon Creek 'if roaches preferred then, as they do now, a dense and moist forest habitat'. However, this conclusion is open to dispute because the Carboniferous 'cockroaches' were not true cockroaches but belonged to the stem-group of several groups of the Paurometabola[66].

Revisionary notes

59. For further discussion of *Stygne*, see Schwarzbach (1939). [Willi Hennig.]

60. Kurtén (1968) has described the rhythmic succession of marine transgressions and regressions in the coal basins, which he considered to resemble the alternation of glaciations and interglacials of the Pleistocene. He suggested that this conspicuous rhythm was probably connected with the formation of continental icefields in the southern hemisphere. [Willi Hennig.]

61. Demoulin (1958c) has given a new interpretation of *Lithoptilus* (Palaeodictyoptera) from Commentry. [Willi Hennig.]

62. Guthörl has subsequently continued to publish descriptions of new species from this area (e.g. Guthörl, 1963). [Willi Hennig.]

63. Simon (1971) has described some Protoblattoidea, Blattodea, and Protorthoptera from the Stephanian of the Halle trough. [Willi Hennig.]

64. Carpenter (1963) has revised some of the Carboniferous species described from Nova Scotia. [Willi Hennig.]

65. Both Carpenter and Richardson have continued to describe new species from the Pennsylvanian (e.g. Carpenter, 1938, 1964; Carpenter and Richardson, 1972). [Willi Hennig.]

66. The following section should be added.

g. Southern hemisphere

A single fossil insect is known from Africa (Riek, 1974a): *Hadentomoides dwykensis* Riek (Paraplecoptera: Hadentomidae), from the Dwyka Series of Rhodesia, which is probably of Carboniferous age. [Rainer Willmann.]

Riek (1973b) has also described a species 'with Palaeodictyoptera and Megasecoptera features' from the Upper Carboniferous of Tasmania (see also Riek, in CSIRO, 1974, p. 28). [Dieter Schlee.]

3. Permian
(Fig. 13)

The Permian was 'one of the most peculiar and important epochs in the development of terrestrial relief and climate. In these respects it represents the most remarkable turning-point in the geological past since the Algonkian—it was a critical epoch of the first order . . .' (Kossmat, 1936, p. 108). Schuchert and Dunbar (1945) have expressed a similar opinion. According to these authors the extensive regression of all the epeiric seas removed one of the main factors responsible for stabilizing temperatures and generating the rain-bearing winds that swept across the continental centres. Most regions of Europe and North America were predominantly arid during the Permian, and extreme variations in local climate were not uncommon. Amongst the plants, lycopods and pteridosperms were the dominant forest trees of the early Permian as of the Carboniferous, but towards the end of the epoch tougher forms with reduced leaves became dominant. 'After the Zechstein, conifers (and Cycadaceae) became the dominant elements in the vegetation' (Mägdefrau, 1959).

When Handlirsch wrote his comprehensive work on fossil insects (Handlirsch, 1906–08), only 14 fossils were known from the Permian apart from some 'cockroach-like' species. According to Carpenter (1930b), the difference between the archaic fauna of the Upper Carboniferous and the Triassic fauna is just as great as that between the Triassic fauna and the recent fauna. He therefore considered that the Permian was the geological epoch when the insects evolved most rapidly.

The discovery of some Permian faunas rich in fossils has been one of the most important events in palaeoentomology since the appearance of Handlirsch's famous work. In particular, Permian fossils are now known from North America and Australia (for a preliminary appraisal, see Tillyard, 1926a), and from the Soviet Union.

58

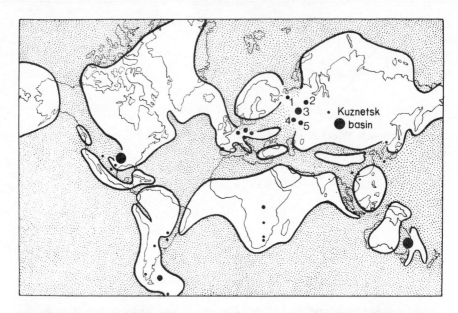

Fig. 13. Palaeogeography and insect localities in the Permian. From the palaeogeographical maps of Schuchert. 1, Arkhangelsk region; 2, Pechora basin; 3, Chekarda and other localities in the Kama basin; 4, Tikhiye Gory; 5, Kargala copper mines, Chkalov region. For the other localities, see text, pp. 58–66. [The new localities mentioned in notes 67–77 have been added to Hennig's original map. Adrian Pont.]

a. North America

'At the height of the Permian there was a marine basin in the south of the central plain and in the eastern Cordilleras; this basin, a continuation of the Upper Carboniferous basin, expanded towards the north-west during the Leonardian (= upper Rothliegende according to Shimer, 1934) and joined the Californian sea. Somewhat later (Word; = lower Zechstein according to Shimer, 1934) it extended even further northwards and joined the Arctic basin of Alaska via British Columbia and the Yukon' (Bubnoff, 1941, p. 298).

Geologically the oldest insects are from the Belle Plains Formation of Texas (Seymour-Baylor County: Carpenter, 1962a; Fulda: Carpenter, 1948b). Fossils from the Grand Canyon (Hermit Shale) are only slightly more recent, but Protodonata are the only insects known from here (see Carpenter, 1928, 1930c).

The most important localities belong to the Wellington Formation. This was an era when parts of Oklahoma, Kansas, Nebraska, and Colorado were inundated by an epicontinental sea (Lake Wellington according to Carpenter, 1945), which extended across from the Gulf of Mexico and was periodically cut off from it.

The locality of Elmo (Kansas), discovered by Sellards in 1906, has been known for longer than any other locality (see Dunbar and Tillyard, 1924). According to Carpenter (1945) Lake Elmo was a freshwater lake or lagoon that developed from an earlier swamp. There was some vegetation close at hand, and insect nymphs lived in the water itself. The fauna of this famous locality has been thoroughly worked out by Tillyard (1923–37) and Carpenter (1930–66). Drevermann (1930) has given a popular account in which he has emphasized the excellent condition of the fossils and has drawn the wings of several of them.

Tasch and Zimmermann (1959) have mentioned two insect faunas from the Lower Permian of Kansas: the 'Carlton insect bed' of Dunbar in Dickinson county, and a stratigraphically somewhat older bed in Marion, Harvey, and Sedgwick counties.

In 1939 a new locality for fossil insects was discovered by Raasch in Noble County, Oklahoma, and this also belongs to the Wellington Formation. Carpenter (1945) worked out this fauna and described Midco salt lake as a playa which was only colonized by algae and ostracods, unlike Lake Elmo. He added that there was no reason to think that insects lived in the water of the lake itself: the occasional nymphs that have been found were probably washed in by rivers, and there were not even any plants growing close to the lake. Lake Elmo and Lake Midco were almost contemporary according to Carpenter.

Tasch and Zimmermann (1959) and Zimmermann (1959) have reported the discovery of a somewhat younger insect fauna from Oklahoma and Sumner County (Kansas), but apparently none of the insects has been described[67].

Martynov (1929) pointed out that the insects from the Lower Permian of Kansas differ from the Carboniferous insects of Europe by their small size. He suggested that this fauna developed for a long time in isolation and enjoyed a more moderate climate than did the European Carboniferous fauna.

On the other hand, Tillyard (1926e) believed that the fossils from the Lower Permian of Kansas showed how the fauna reacted to the change of climate between the Upper Carboniferous and Lower Permian. The onset of a hot arid climate and the formation of extensive salt lakes signalled the end of the era of giant insects. Tillyard also believed that holometaboly (i.e. the pupal stage), which he considered to have arisen polyphyletically, was correlated with this climatic change.

b. Central and South America

Two unimportant localities in Mexico (Valle de Las Delicias, south-west of Coahuila: Carpenter and Miller, 1937) and Brazil (Carpenter, 1930a) should be mentioned as an appendix to the North American Permian localities, but I have no further details about them.

c. Western Europe

In contrast with the Carboniferous, Europe has produced few important Permian insects[68]. The species described from Germany have recently been reviewed by Staesche (1963). He listed 68 species, but most of them, including those that he assigned to the 'Fulgoroidea', are cockroach-like species and are of no phylogenetic importance. Staesche's 'Protocoleoptera' are also related to the cockroach-like species, in fact to the Dermaptera, and have nothing to do with the beetles (see also Forbes, 1928; Peyrimhoff, 1934).

At that time Europe, like all the continents of the northern hemisphere, had a 'predominantly arid climate, which intensified to produce desert conditions' (Kossmat, 1936, p. 108). The insect fossils from Central Europe are all from the Lower Permian (Rothliegende) and are associated with inland basins in the Variscan fold-belt whose limnic coal measures, as already mentioned, are so important for our knowledge of the Upper Carboniferous fauna. In the upper Rothliegende all these basins merged into a single large longitudinal trough, the 'Saar–Saale depression' (Bubnoff, 1941).

So far as I know, no insects have been described from the French part of this inland depression. From Germany Staesche has listed localities in the Saar–Nahe–Rhine region, the southern Harz (Ilfeld), the area of Halle on the Saale, and the region of Thuringia–Upper Franconia–Saxony[68], which is probably part of the Stockheim–Erzgebirge–Döhlen trough according to Stille (Bubnoff, 1941)[69].

Like the troughs mentioned above, the Boskovice trough in Moravia was also an inland depression in the Variscan fold-belt, and recently its insect fauna has been the subject of investigation. Kukalová (1965) thought that the well known disjunction between the Carboniferous and Permian insect faunas only appears to be so abrupt because virtually all the known Carboniferous species were lowland forms. Progressive desiccation during the Lower Permian induced the more derived species that had lived on xerophilous plants in the mountain ranges during the Upper Carboniferous to follow these plants down into the lowlands.

In Portugal, the Buçaco basin, the central one of the three basins situated on the western edge of the Meseta, also belongs partly to the lower Rothliegende (Autunian) according to Fleury (1936). Teixeira (1948) described a wing similar to that of *Dictyoneura* from the Autunian.

d. Eastern Europe and Angaraland

The localities of the Arkhangelsk region and the Urals are situated between Angaraland and the eastern edge of the continent that represented Europe during the Palaeozoic. As in the Upper Carboniferous, there was a marine connection here between Arctis and Tethys, and this channel slowly retreated towards the east (Bubnoff, 1941). According to Martynov (1938a), the east

European localities for Permian insects, which are listed in detail in Rodendorf's (1957b) review, are situated in two southern zones.

The localities of one zone (Martynov's 'red-coloured zone') are freshwater deposits from lakes and rivers in the Permian Urals. In some localities they contain aquatic insect larvae, but no brachiopods. Almost all these localities belong to the Kungurian and upper Artinskian Stages which, according to Bubnoff (1941), correspond to the upper Rothliegende (or lower Middle Permian). In Russian papers the age of these fossils is given as 'Lower Permian'. Several of the localities listed by Rodendorf (1957b) belong here: those in the Pechora basin, which have produced comparatively few fossils, and some of those in the Kama basin, including Chekarda which is the best known and most productive locality. Rodendorf (1939) thought it likely that the fauna of Chekarda, on the banks of the Sylva, was almost exactly contemporary with that of Kansas (Elmo). However, it is now assigned to the Kungurian Stage, which is somewhat younger than the oldest finds in Kansas. Finally, the localities in the copper mines of Kargala, in the region of Chkalov (= Orenburg), also belong to this freshwater zone.

In Martynov's second zone, the localities represent shallow marine bays and lagoons. Brachiopods have been found quite commonly. All these localities belong to the Kazanian Stage, that is to say to the most recent Permian. According to Bubnoff (1941, p. 290) the Kazanian Stage was the 'Russian Zechstein' and represented 'a new marine transgression that extended from the north as far as the Central Volga'. This zone includes some localities that have been discovered since 1926 on the Soyana River in the Arkhangelsk region, where Iva-Gora, Sheymo-Gora, and Letopola are particularly well known and productive; and also some of the localities in the Kama basin, especially those in the 'Silent Mountains' (Tikhiye Gory of Martynov, 1930, 1931) on the Kama River.

According to Martynov (1929), the Permian insect fauna of the northern European part of the Soviet Union differs considerably from the Palaeozoic fauna of central and western Europe. Three elements can be recognized.

Only about one-quarter of the species have any relationship with the late Palaeozoic insects of Europe (Upper Carboniferous and Lower Permian). They are characterized by their comparatively large size and, according to Martynov, represent the vanishing thermophilous faunal element of late Palaeozoic Europe.

A second faunal element, to which more than one-third of the species belong (according to Martynov, 1929!), consists mainly of smaller species which have more or less distinct affinities with the species of the Lower Permian of Kansas. Martynov's (1929) theory about the origin of the Permian insects in the north-eastern European part of the Soviet Union was based upon this, the most extensive faunal element. He suggested that the fauna to which they belong developed during the Upper Carboniferous in a land area situated to the north or north-west of Kansas. From here, moving to the east and south, it reached Kansas during the Lower Permian or somewhat earlier.

To the west, it crossed the north Pacific land-bridge between America and Asia to reach Angaraland, and during the Upper Permian penetrated as far as the eastern and north-eastern European part of the Soviet Union. Here it has been preserved in the fossil beds on the east coast of the channel that connected the Arctic Ocean with Tethys during the Permian. According to Martynov, the existence of a north–south channel also explains why this fauna never reached central and western Europe.

It is difficult to say how far Martynov's (1929) theory still fits the facts. In his day the Permian fauna of the Kuznetsk basin was completely unknown, and even the Permian faunas that were known were still very incompletely worked out. I do not know of any more recent discussion of this problem or of any critical assessment of his views. However, it is interesting that in one respect his ideas agree with those of Kukalová (see above): both authors considered that the Permian fauna of the regions where fossil deposits have been preserved originated in migrations from regions where it had already developed before the Permian.

In addition to these two faunal elements in the Permian of eastern Europe, Martynov (1929) distinguished a third which also includes almost one-third of the total number of species. He thought that the groups to which these species belong are still very characteristic of the north European part of the Soviet Union, but that some have affinities with the Permian fauna of Australia. He pointed out that Gondwanaland plants (*Gangamopteris* according to Bubnoff, 1941) are also known from the Upper Permian of north-east Europe and that, according to Potonié, they are supposed to have migrated during the youngest Carboniferous and during the Permian through India and into Angaraland, where they have also been found in Siberia and Mongolia.

The discoveries in the Kuznetsk basin (Rodendorf, 1958; Rodendorf *et al.*, 1961) occupy a special position because they contain a complete stratigraphical sequence from the Upper Carboniferous down to the Upper Permian, with only a few breaks. Triassic and Jurassic fossils have also been found here. I have already discussed the Carboniferous fossils. There are almost four times as many fossils from the Permian deposits as there are from the Carboniferous, 4870 as against 1252. The most productive locality is Kaltan, with 2046 individual fossils, and this belongs to the Lower Permian (Kuznetsk Stage). The most productive localities in the Upper Permian (Il'inskoe and Erunakova Stages) are Surikovo I (670) and Sokolovo II (569). Altogether 206 species have been described from the Lower Permian (Kuznetsk Stage) and 151 from the Upper Permian (Il'inskoe and Erunakova Stages). This gives a total of 357 species from the entire Permian, which is almost five times the number of species known from the Carboniferous (72 species).

According to Rodendorf (1956), the most unusual features of the Permian fauna of the Kuznetsk basin are the complete absence of the cockroach-like species that were so abundant during the Carboniferous, and the abundance of Homoptera, Mecoptera, and true beetles, all of which were completely absent during the Carboniferous. Seventy-one families of insects have been recognized from the Kuznetsk basin, and according to Rodendorf (1961a) 64 are known only from here. 'These remarkable facts can only be explained by the pronounced palaeogeographical isolation of Angaraland during the Palaeozoic. The peculiarities of the insect fauna reflect the isolation of the Angaraland fauna, which has remained virtually unknown until now, and the unusual climatic conditions to which it was subjected' (Rodendorf, 1961a). According to Rodendorf (1956), the climate of this region during the Permian was different from that prevailing in Europe and North America at the same time: he suggested that it was temperate. Somewhat at odds with Rodendorf, Martynova (1961a) has emphasized the close relationship between the insects of Angaraland and Australia.

In addition to these localities, Rodendorf (1957b) has mentioned a further locality from Angaraland at the Nizhnaya Tunguska (= lower Tunguska), in the Krasnoyarsk region of the Yenisey basin, where a palaeodictyopteron wing was found in 1879; and at Vladivostok, where an apparently unimportant fossil was found which has not been described in detail since[70].

e. Southern hemisphere

'During no earlier period does the contrast between conditions in the northern and southern hemispheres emerge so strikingly as in the Permian. The northern hemisphere had a predominantly arid climate which intensified to produce desert conditions. On the other hand the continents of the southern hemisphere laid down sediments that had essentially the same facies as those of the productive Carboniferous' (Kossmat, 1936, p. 108). In the southern hemisphere the climatic extremes followed 'a completely different direction and induced glacial conditions which are all the more striking because most of these regions are now in the subtropics or even in the tropics' (Kossmat, 1936, p. 122).

In these circumstances it could be extremely interesting to have a detailed knowledge of the Permian fauna of the continents of the southern hemisphere (Gondwanaland). In fact, however, only the Upper Permian fauna of Australia is at all well known. The localities, which have been described and mapped by Knight (1960), are in New South Wales between Belmont and Warner's Bay, in the Newcastle coal measures, a little below the Triassic (Tillyard: discussant's comment on Martynov, 1929). The two most remarkable features of the Australian Upper Permian fauna, according to Tillyard (1926b), are its great specialization and its restriction to a very few orders[71]: virtually all the fossils have been Homoptera and Holometabola ('Mecoptera', 'Neuroptera' and Coleoptera). Tillyard (1922a) also stated that almost 50% of the Belmont fossils[72] belong to the Permochoristidae and that almost all the individuals are of small to medium size, which is also true of the North American Permian faunas.

From the fact that these Upper Permian species are the oldest insects to have been found in Australia[73], Tillyard (1922a) drew the somewhat illogical conclusion that Australia was not populated by insects until long after the northern hemisphere. Furthermore, because only representatives of the 'most highly specialized' groups Hemiptera and Holometabola have been found in Australia, he drew the equally questionable conclusion that Australia lay far away from the area where the earliest insects developed. However, he did not dare suggest where the Upper Permian insects of Australia originally came from. The fact that this fauna was apparently associated with *Glossopteris* led him to conclude that it had reached Australia from other parts of Gondwanaland with the typical *Glossopteris* Flora.

Unfortunately, the fossils that have been found elsewhere in the southern hemisphere give us no insight into the insect fauna of these regions. Zeuner (1955) has described a single unimportant cockroach-like species (*Rhodesiomylacris*, family Mylacrididae) from the Upper Permian (Karroo System, Lower Beaufort Series) of Rhodesia (Madziwadzido on the Serami River, Sebungu district)[74]. The only known insects from the Permian of South Brazil (Paraná near Teixeira Soares city: stratigraphy unclear) are also 'cockroaches', such as *Phyloblatta oliveirai* Carpenter (1930a; see Oliveira, 1930). In addition, Fossa-Manzini (1941) has mentioned three undescribed insect wings from San Luis province, Argentina[75].

The fauna of the Falkland Islands could be of the greatest interest for illuminating the transantarctic connections between Australia and South America which many authors believe existed at least during the Palaeozoic. However, the only fossil to have been described from the Upper Permian (Gondwana Series) is a dragonfly, *Protagrion falklandicum* Tillyard (1928a). This is naturally of great interest for the phylogeny of the Odonata, but no dragonflies have yet been found in the Australian Palaeozoic[76]. Apart from this, a single palaeodictyopteron wing has been found in the Falkland Islands (Fossa-Manzini, 1941)[77].

A list of Permian insects has been published by Branson (1948) but this is already very much out-of-date.

Revisionary notes

67. Jeannel (in Grassé, 1949) listed two further Permian localities in North America, at Fairplay (Colorado) and Cassville (Virginia), which he dated as Autunian. [Willi Hennig.]

68. Müller (1975, 1977a, 1977b) has described some new Mylacrididae and Blattinopsidae from the lower Rothliegende (Autunian) of Thuringia, which he compared with other species from Central Europe. He has also given a detailed description of *Eugereon* (Palaeodictyoptera) which has settled many unresolved questions (Müller, 1977b). [Dieter Schlee.]

Kinzelbach (1970) has described a species of Permoplectoptera from the lower Rothliegende of Bad Kreuznach (Saar–Nahe–Palatinate region), and

Guthörl (1965) has described *Protereisma rossenrayensis* (Ephemeroptera) from the Zechstein of the Lower Rhine. [Willi Hennig.]

69. Bachmayer and Vasicek (1967) have recorded a Permian locality at Zöbing, near Krems (Austria), from which they have described some Spiloblattinidae (Blattodea). [Willi Hennig.]

70. Lin (1978a) described two Blattoidea from the Upper Permian of China (Na-yung, Kweichow province, and Fu-yüan, Yunnan province). Subsequently (Lin, 1978b), he again mentioned these species but with a slightly different specific name for one of them. [Dieter Schlee.]

71. Riek (in CSIRO, 1970, p. 170) described this fauna as follows: 'The insect fauna is curiously unbalanced as compared with that of the corresponding period in the northern hemisphere. On the one hand there are no Palaeoptera except an undescribed meganisopteron, no blattoids, and no orthopteroids other than a single species of Plecoptera. There is one species of Glosselytrodea. On the other hand there are many Homoptera, several Psocoptera, three families of Neuroptera, a few Coleoptera and Trichoptera, an abundance of Mecoptera, and a few probable ancestors of the Diptera'. [Dieter Schlee.]

72. Riek (1968b) has given a list of the Upper Permian insects from Belmont. [Willi Hennig.]

73. This is no longer true: there is an Upper Carboniferous locality in Tasmania (see note 66). [Dieter Schlee.]

74. Pruvost (1934) described the paraplecopteran *Boutakovia saalei* (Homalophlebiidae) from Zaïre (Kivu, Lukuga). Riek (1973a, 1976a) has recently recorded some 30 species from several Upper Permian localities in Natal, South Africa (Middle Beaufort Series, *Daptocephalus* zone). This fauna consists mainly of Homoptera and Paraplecoptera, and some species were assigned to genera that are otherwise known only from Australia. Riek (1976b, 1976c) has also described a few insect remains from the Lower Permian Middle Ecca of South Africa (Hammanskraal near Pretoria, Transvaal). One of these, the collembolan *Permobrya mirabilis*, was said to be surprisingly similar to recent species. [Rainer Willmann.]

75. Pinto (1972a) has listed some additional insects from the Irati Formation (Prosbolidae, Pereboriidae, Permithonidae, Cupidae, Mecoptera), including several specimens from a second outcrop at Passo de São Borja (southern Brazil, São Gabriel). Only a few of these have been described so far. The fauna is associated with *Mesosaurus brasiliensis* and *Glossopteris* leaves, but the precise age of the Irati Formation is not certain, though it is probably younger than the Lower Permian. A blattoid from Patagonia (Argentina) is probably Lower Permian in age (Pinto, 1972b). [Rainer Willmann.]

76. Apparently this is still true (see Riek, in CSIRO, 1970, 1974). [Dieter Schlee.]

77. The first Permian insect from the Antarctic was described at the time that this book was first published (Tasch and Riek, 1969). It was found near 2000 m in the Polarstar Formation of the Sentinel Mountains. It consists of the basal part of a small wing, and was assigned to the Stenoviciidae (Homoptera). A short time after this, Carpenter (1970) described *Uralonympha schopfi* from the Mount Glossopteris Formation (Ohio Range; see also Tasch, 1971). [Rainer Willmann.]

Kukalová (1966) has recorded some Protelytroptera from the Upper Permian of Australia (see Riek, in CSIRO, 1974). [Dieter Schlee.]

B. Mesozoic

1. Triassic
(Fig. 14)

The general appearance of the Mesozoic insect fauna is relatively modern in character from the Triassic onwards: this was stressed by Carpenter (1953, 1954a) and even by Handlirsch (in Schröder, 1920–21). The main reason for this is that only a few of the groups that can be recognized as being side-branches of the stem-groups of recent orders survived beyond the end of the Permian: Protoperlaria (*Tomia costalis* Martynov, from the Triassic of the Kuznetsk basin), Archimylacrididae (*Phyloblatta*, from the Rhaetic of Coimbra: Teixeira, 1947), Glosselytrodea (*Mesojurina*, from the Triassic of Issyk-Kul'), and a few others.

Most of the insect fossils are closely related to the recent species of the orders to which they belong, and some of those from the early Mesozoic can even be assigned to more restricted subgroups within their particular orders.

This conforms with what is known of the vertebrates and plants. According to Schuchert and Dunbar (1945) the main groups of Palaeozoic plants had completely or almost completely died out by the Triassic. Pteridosperms had almost disappeared and the great lycopods are known only from a few *Sigillaria* fossils from the early Triassic. *Lepidodendron* was absent and *Cordaites* was no longer abundant. The forests consisted mainly of conifers and cycads, with a ground vegetation of ferns and rush-like plants.

In striking contrast with this, it has only been possible to detect small differences in the general palaeogeographical and climatic conditions of the

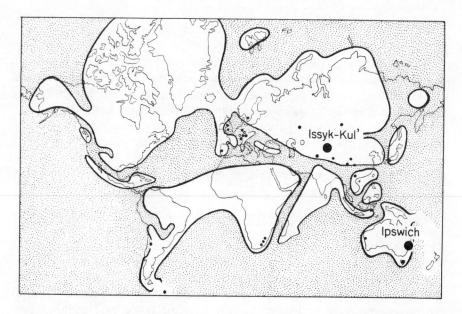

Fig. 14. Palaeogeography and insect localities in the Triassic. From the palaeogeographical maps of Termier and Termier, Schuchert, Bubnoff, and Kossmat. White circles: Lower Triassic. Black dots: Upper Triassic. For the names of the less important localities (small circles and dots), see text, pp. 68–72. [The new localities mentioned in notes 78–85 have been added to Hennig's original map. Adrian Pont.]

Permian and Triassic. In the Triassic 'the axis of the palaeogeographical picture was still the great equatorial sea of Tethys, which continued to separate the continental masses of the northern and southern hemispheres' (Bubnoff, 1949, pp. 395–396). 'North of Tethys were large continental masses, which at the most contained extensive continental lagoons; in the Old World only Fennosarmatia and Angaraland were united, whilst in the New World North America formed a similar but smaller continental nucleus'. 'In the southern hemisphere, only non-marine deposits are known, except in western South America and the adjacent parts of the Pacific, and these are similar everywhere. However, it is doubtful if a united land-mass existed: the fragmentation of Gondwanaland had already made considerable progress by the Triassic'. 'Climatologically the Triassic resembles the arid Permian'. 'Warm and arid conditions are the main features of the entire epoch throughout the northern hemisphere and extensive areas of the southern. In many places, such as the Bunter Sandstones of Europe and the United States, this led to the development of desert- or steppe-like landscapes and the formation of rich salt deposits. It was relatively warm right up into the Arctic, and the contrasts in temperature in various regions of the world could not have been very great. It did not become damp again until the end of the Keuper' (Schwarzbach, 1950).

a. Northern hemisphere

North America. Virtually nothing is known about the Triassic insects of North America. In Northrop's (1928) review, the only species to be listed are *Mormolucoides articulatus* Hitchcock from Turners Falls, Massachusetts (probably the 'larva of a sialidan neuropteran' according to Lull, 1953) and 32 other species most of which are not insects at all according to Handlirsch. In addition, Walker (1938) described five 'species' of insect galleries in fossilized trunks of *Araucarioxylon arizonicum*, from the petrified forest of Holbrook in Arizona. None of these fossils is of any value for insect phylogeny.

Europe. In his list of German Triassic insects, Kuhn (1937) mentioned only two species from the Bunter Sandstone, which belongs to the family Chaulioditidae: *Triadosialis zinkeni* Heer, found near Bremke (survey sheet Reinhausen near Göttingen), and *Chauliodites picteti* Heer, from Salzmünde (between Halle and Eisleben). Both species, and *Anasialis langei* Handlirsch (see Handlirsch, 1939), were unreservedly assigned by Laurentiaux (1953) to the Megaloptera, an assignment which was first made by Handlirsch and which is even indicated by their names. However, in the Russian *Textbook of Palaeontology* (Rodendorf, 1962), they were tentatively assigned by Sharov to the Liomopteridae (Paraplecoptera or Protorthoptera of other authors). So far as I know, the only other fossils from the Bunter Sandstone are an ephemeropterous larva (*Mesoplectopteron longipes* Handlirsch) from the Voltzia Sandstone of Sulzbach in Alsace, an odonatan (*Triadotypus guillaumei* Grauvogel & Laurentiaux) from the Vosges (Bas-Rhin: Bust), *Pseudodiptera gallica* Laurentiaux & Grauvogel, and a cockroach (Mesoblattinidae: *Billia triadis* Handlirsch from the Vosges). Almost all of these seem to be species with freshwater larvae, and they must have inhabited some of the freshwater lakes which Mägdefrau (1959) described as extending in a broad belt along the coast of the Triassic Ocean. In the region of the central Variscan zone, this German Triassic basin was separated from Tethys by the Vindelician ridge (Bubnoff, 1949).

The insect fossils from the Bunter Sandstone are of no phylogenetic significance. The customary assignment of *Triadotypus* to the Protodonata (Grauvogel & Laurentiaux, 1952) does not automatically mean that this species, which has also been found in the Upper Triassic, was more closely related to the Palaeozoic Protodonata than to the other Mesozoic and recent Odonata (see p. 143). The same may be true of *Thuringopteryx gimmii* from the middle Bunter Sandstone of Saalfeld in Thuringia (Kuhn, 1937), if it really does belong to the Protodonata as Kuhn (1937) suggested; and also of *Pseudodiptera gallica*, which may be more closely related to the Diptera than to the Palaeozoic Paratrichoptera to which it is usually assigned.

Kuhn (1938) has described *Kulmbachiellon fragile*, an unimportant 'cockroach', from the middle Bunter Sandstone of Kulmbach (Bavaria).

Fossils from the Muschelkalk are even more scarce. Kuhn (1937) has only listed two species, and no more have been discovered since. One of the species, *Reisia gelasii* Reis from Münnerstadt in Lower Franconia, is a doubtful protodonatan according to Handlirsch (1939). *Flichea lothringia* Fliche, from Lunéville in Lorraine, is a beetle of unknown affinities.

Fossils from the Keuper, especially from the Rhaetic, are rather more numerous in Europe. 'During the lower Keuper, the landscape was characterized by numerous lakes and extensive swamps whose lush vegetation led to the formation of small coal seams' and 'the appearance of the landscape and vegetation at the turn of the Rhaetic-Lias was very similar to that in the lower Keuper, although many of the plant species were fundamentally different' (Mägdefrau, 1959).

Most of the fossils are beetles, and some 20 species were listed by Kuhn (1937) and Handlirsch (1939). Some of their names suggest modern families, although it is not really possible to demonstrate that there are any actual relationships. However, there is good reason to believe that at least some of them do in fact belong to subgroups of the recent beetles. The localities are in Switzerland (Mythen, Basle), Liechtenstein (Vaduz), Germany (Bayreuth, Hildesheim, Westphalia, Gross-Leuthen, Witzenhausen, Teufelsgraben near Altdorf, Strullendorf, and Veitlahm near Kulmbach in Bavaria), southern Sweden (Höganäs and Kulla Gunnarstorp), and Britain (Cnap Twt quarry, near Bridgend, South Wales: Gardiner, 1961).

Several other species have been described in addition to these beetles: a cockroach (*Pedinoblattina stromeri* Handlirsch, from the Rhaetic of Central Franconia), a species of Planipennia [*Osmylopsychops radialis* Ellenberger, Laurentiaux and Ricour (1953) from Savoy: Dent de Villard, Vanoise], an odonatan (*Piroutetia liasina* Meunier, from Fort Mouchard: from Handlirsch, 1939), and the 'protodonatan' described from the Bunter Sandstone, *Triadotypus guillaumei* (also from the Keuper of the Dent de Villard: Laurentiaux-Vieira et al., 1953). *Osmylopsychops* and *Piroutetia* are said to belong to subgroups of the recent *Planipennia and *Odonata, respectively[78].

Away from the Central European localities, there is apparently a Rhaetic outcrop in the region of Coimbra (Peneireiro) in Portugal, but only one cockroach (*Phyloblatta* sp.) has been found there (Teixeira, 1947). As the genus *Phyloblatta* belongs to the otherwise exclusively Palaeozoic Archimylacrididae, this fossil would be extremely interesting if the identification could be confirmed.

Nakyz, in the southern Urals (Bashkirskaya Republic, Kyurgazinskiy district) should also be included amongst the European localities. *Triassomachilis* was found here, a species that appears to be more primitive than all the recent Archaeognatha.

Angaraland. As has already been mentioned, there was no marine channel in the Urals area to separate Angaraland from the Fennosarmatian continent during the Triassic as there had been during earlier periods.

Apparently no insects have been described from the three localities listed by Rodendorf (1957b) from western Siberia (Chelyabinsk).

In Central Asia the region of Fergana (Tadzhikistan) has produced two Hemiptera from two localities (Madygen and Shurab IV) which apparently belong to subgroups of the recent Hemiptera: *Tingiopsis* (Heteroptera-Tingidae according to Russian authors, Auchenorrhyncha-Cercopoidea according to Evans) and *Maguviops* (Auchenorrhyncha, 'Membracidae').

The locality Issyk-Kul', a lake in the Tien Shan Mountains (Kirgiziya), has produced some 3000 insect impressions, according to Rodendorf (1957b), and is the most productive locality for Triassic insects in the northern hemisphere. Not all of them have been worked out yet. Whilst Rodendorf gave a total of only 25 insect species from the Triassic of the Soviet Union in 1957, a few years later (Rodendorf, 1961c, 1962, 1964) he described no less than 53 species of Diptera just from this one locality. Some of them were described as surviving side-branches of the stem-group of the Diptera ('Archidiptera'), but most belong to various subgroups of the recent Diptera. The 14 Mecoptera from Issyk-Kul' also belong to various subgroups of this recent order. The discovery of a species of *Mesojurina*, a genus of the otherwise exclusively Palaeozoic Glosselytrodea, has been of particular interest. One neuropteron (*Petrushevskia*), one homopteron (*Cicadoprosbole*), one mecopteron (*Liassochorista*; allegedly Permochoristidae[79]) and one unrecognizable para-neopteron (*Mononeura*) are more difficult to assess.

The Kuznetsk basin (locality Babiy Kamen') has only produced three fossils. One (*Tomia*) is said to belong to the otherwise exclusively Palaeozoic Protoperlaria, a second (*Ademosynoides*) to an exclusively Permian family of the Coleoptera (Permosynidae), and a third (*Bittacopanorpa*) to a subgroup of the *Mecoptera[80].

Vietnam. Three cockroaches (Poroblattinidae: *Kebaona*; Mesoblattinidae: *Hongaya* and *Rhaetoblattina*) have been listed by Handlirsch from Ke Bao Island and Hon Gay in the Tonkin region of Vietnam. According to the palaeogeographical maps, these localities were on a small continental area which was situated like an island in the East Asian part of Tethys[81].

b. Southern hemisphere

By far the most important contribution to our knowledge of Triassic insects has come from Australia (Tillyard, 1926a)[82]. Compared with this, all the discoveries in the northern hemisphere as well as those in other continental areas of Gondwanaland pale into insignificance.

South America and *Antarctica* (Graham Land). A few insects have been described from the Rhaetic of Mendoza (Argentina). Two of these are 'Saltatoria': *Baiera argentina* Kurtz, which was first described as a ginkgoalean, and *Notopamphagopsis bolivari* Cabrera (1928) (for both of these, see

Zeuner, 1939). Two others were originally described as Diptera (Wieland, 1925; see also Tillyard, 1926c): one of these (*Tipulidites affinis*) has remained unrecognized, whilst the other (*Tipuloidea rhaetica*) has proved to be a homopteron (family Scytinopteridae according to Tillyard). The odonatan *Triassothemis mendozensis*, recently described from Mendoza by Carpenter (1961a), is said to be close to the Archizygoptera.

Unfortunately, the exact age of two beetle fossils (*Grahamelytron crofti* and *Ademosynoides antarctica*) from the Mount Flora Beds of Graham Land is unknown. Zeuner (1959b) has reported that plants found at the same locality are considered to be Middle Jurassic, and he argued that the beetles should be Upper Triassic to Jurassic. In his opinion they indicate that there was a palaeogeographical connection with Australia at this time. He probably based this assertion on the fact that one of them is said to belong to the genus *Ademosynoides*, several species of which have been described from the Upper Triassic of Australia. However, according to Rodendorf and Ponomarenko (in Rodendorf, 1962), the genus has also been found in the Kuznetsk basin. Unfortunately, then, we can draw just as few palaeogeographical conclusions from the Mesozoic fossils of Graham Land as we could from the Upper Permian fossils of the Falkland Islands (p. 64).

Africa. Fossils have been known for a long time from the Stormberg Series (the uppermost series of the Permo-Triassic Karroo suite) of Lesotho and Transvaal (see Haughton, 1924). One unnamed beetle, an ephemeropterous nymph (*Phthartus africanus*), a cockroach (Mesoblattinidae: *Stratiotegmen africanum*) and a 'cricket' (*Archaegryllodes stormbergensis*) have been described.

Zeuner (1961) has described three beetles (families Cupedidae, Silphidae and ?Carabidae) and a cockroach (*Rhipidoblattopsis natalensis*) from the Middle Triassic Molteno Beds, which still contain *Glossopteris* according to Bubnoff (1949), from the same Stormberg Series of Natal (Burnera Water Fall, Upper Umkomaas)[83].

Australia and *Tasmania*[84]. By far the most important and productive localities in Australia are near Brisbane in Queensland: Ipswich (Rhaetic: see Tillyard, 1923b, 1936b) and Mt. Crosby (somewhat older than Ipswich; see Tillyard, 1936b; Middle Triassic according to Dodds, 1949). Detailed information on the stratigraphy and insects of Mt. Crosby has been given by Tindale (1945).

Less productive localities are Beacon Hill, near Deewhy (not far from Manly; Middle Triassic; see Tillyard, 1925b) and Brookvale (Middle Triassic; see Riek, 1950, 1953b) in New South Wales, and Hobart in Tasmania (Riek, 1962, 1967a)[85].

The insects described from the Australian Triassic are so important and so numerous that they can only be mentioned individually in the general treatment of Mesozoic insects.

72

Revisionary notes

78. An unnamed wing has been recorded by Thompson (1967) from lower
Keuper marls at Styal, Cheshire (England). It was tentatively identified as
either Euplecoptera (Ephemeroptera) or Calvertiellidae (Palaeodictyoptera).
[Willi Hennig.]

79. Willmann (1978) has retained *Liassochorista* in the Permochoristidae.
[Dieter Schlee.]

80. Fujiyama (1973) has reviewed the Upper Triassic insect fauna of East
Asia and has described a number of species from Ominé, Honshu, Japan
(orders 'Blattodea', Paraplecoptera, Homoptera, and Coleoptera). [Willi
Hennig.]

81. Lin (1978a) has described some Blattoidea from the Upper Triassic of
China (Shensi and Szechwan provinces), and the Sinkiang-Uighur Auton-
omous Region (Tudupin); and (Lin, 1978b) two Paraplecoptera from the
Triassic of Kweichow (= Guizhou). See also note 80. [Dieter Schlee.]

82. Riek (in CSIRO, 1970, 1974) has published two recent reviews. [Dieter
Schlee.]

83. Riek (1974b) has described another twelve species and a number of
undetermined specimens from seven additional localities in the Molteno Beds
(see Anderson, 1974). The fauna includes some 'Orthoptera', Blattodea,
Homoptera, Megaloptera (*Euchauliodes distinctus*, which Riek considered to
be the oldest corydaloidean wing), a scorpion fly, and several beetles. Riek
(1976d) has recorded another Upper Triassic insect assemblage from Cape
Province (South Africa: Birds River and Mount Fletcher). [Rainer Willmann.]

84. See note 82. [Dieter Schlee.]

85. Riek (1968c) has described a locality in the Hill River area of Western
Australia which he believed to be Triassic. The only insects were a cockroach
forewing, beetle elytra, and a beetle pronotum. [Willi Hennig.]

2. Jurassic
(Figs. 15–17)

The Jurassic fossils can only be compared with the Upper Triassic fossils of
Australia in their importance for the history of Mesozoic insects.

Palaeogeographically there was at first no fundamental difference between the Jurassic and the Triassic: 'the Tethyan geosyncline was still the centre of the picture that the earth presented during the Jurassic . . . it continued to separate the continents of the northern and southern hemispheres' (Bubnoff, 1949, p. 464). 'Conditions in the Lower Lias were very similar to those of the Rhaetic and the two are difficult to distinguish'. 'The main changes in the appearance of the earth's surface did not take place until the Callovian transgression' (Bubnoff, 1949, pp. 463–464).

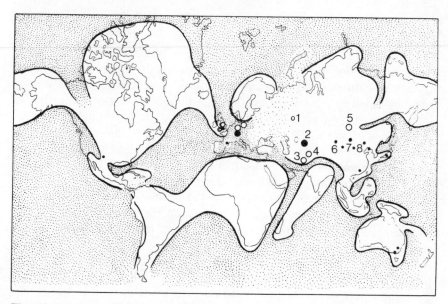

Fig. 15. Palaeogeography and insect localities in the Jurassic. White circles: Lias. Black dots: Malm. Light stippling between Europe and Angaraland: submergence during the Lias. From the palaeogeographical maps of Termier and Termier, Schuchert, Kossmat, and Bubnoff. 1, Sukhomesovo (Chelyabinsk region); 2, Karatau; 3, Shurab; 4, Kyzyl-Kiya; 5, Ust'-Baley; 6, T'u-lu fan (= Turfan); 7, Mongolia; 8, Ch'eng-te-Shih (= Jehol). For European localities, see Figs. 16 and 17. [The new localities mentioned in notes 86–91 have been added to Hennig's original map. Adrian Pont.]

Schuchert and Dunbar (1945) considered the Cycadeoideae to be the most characteristic plants, and they called the Jurassic the Age of the Cycadeoideae. No deciduous trees are known from this epoch. According to Schuchert and Dunbar (1945), the forests consisted, as before, of evergreen conifers intermixed with ginkgo trees and tree-ferns, with a ground vegetation of herbaceous ferns and reeds. Dry slopes were covered with open stands of Cycadeoideae and ferns.

Our knowledge of Jurassic insects is almost entirely limited to two of the continents in the northern hemisphere, Europe and Angaraland. According to Northrop (1928), no Jurassic insects are known from North America[86].

Jurassic fossils are also virtually unknown from the southern hemisphere. Caroll (1962) has listed two insect fossils from southern Australia[87] (Victoria: Koonwarra, near Leongatha, in South Gippsland): a plecopteran ('Nemourid') and a hymenopteron, neither of which has been described in detail.[88].

In Europe, fossil insects have mainly been found in the Lower Jurassic (Lias) and Upper Jurassic (Malm), but only rarely in the Middle Jurassic (Dogger). Angaraland localities are from the Lias and Malm.

a. Lias

'During the Lias the sea covered large parts of north-west, central, and southern Germany. It covered the area between the Bohemian landmass, including a south-west 'Vindelician' spur, in the east and south-east, and an Ardennes island in the west. Local uplifts running south-east to north-west [Harz, Thuringian Mountains (Thüringer Wald), Riesbarre] delimited local basins; one uplift in the Elbe region (Pompeckj' Swell) delimited an estuarine area with occasional marine ingressions in Scania and East Germany. These basins were connected with a basin in North France and England via the North Sea and Lorraine, where similar downwarps were separated by uplifted areas. The western margin of the sea lay on the Armorican land-mass (Brittany, Wales), and the southern margin on the Central Massif; it was occasionally connected to the Mediterranean via the Rhone basin and the Straits of Poitou' (Bubnoff, 1949, p. 429).

According to Schwarzbach (1950), climatic conditions during the Lias were rather warmer than now, but 'on the whole it was rather cooler than during the warmest periods of the geological past. In many areas the climate was humid. The rich floras in both polar regions offer particularly striking evidence that there were no great climatic contrasts'.

Europe (Fig. 16). The fossils from the upper Lias of eastern Lower Saxony (Brunswick and district: see Bode, 1953, as well as Handlirsch) and Mecklenburg (Dobbertin; worked up by Handlirsch; see Handlirsch, 1939) are almost contemporary and are particularly famous. Kuhn (1951, 1952) has described some insect fossils from northern Franconia (Bamberg district).

These all belong to the Lias (*Posidonia* shales), and in Germany this has 'the facies of a partly enclosed basin with hydrogen sulphide fermentation, analogous to the Kupferschiefer Sea or the Black Sea' (Bubnoff, 1949, p. 430).

Bode (1953) has suggested that the insects from the Brunswick area originated from the Bohemian mass or the Ardennes island, and those from Mecklenburg (Dobbertin) from the north European land-area. This would explain the dissimilarities between the fossils from these two areas.

Handlirsch (in Schröder, 1920–21) has listed two further localities: Schambelen in the Aargau (Switzerland), and Weyer (Upper Austria).

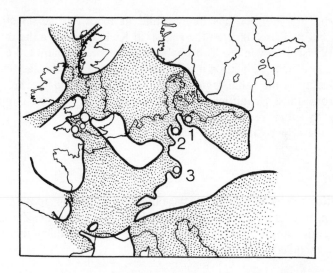

Fig. 16. Palaeogeography and insect localities of Europe in
the Lias. Based on Bubnoff. 1, Dobbertin (Mecklenburg);
2, Brunswick; 3, Bamberg. [An additional locality has been
added to Hennig's original map. Adrian Pont.]

In England the insect-bearing strata are restricted to the area of Somerset,
Gloucestershire, Warwickshire, and Worcestershire, according to Zeuner
(1939). The lower Lias fossils of Gloucestershire, Warwickshire, and Worcestershire are very close to the base of the Lias, according to Zeuner. Some are
in an excellent state of preservation (e.g. Zeuner, 1959a). More recently,
further lower Lias fossils have been found in Dorset (Lyme Regis: Gardiner,
1961; Charmouth: Zeuner, 1962). Zeuner (1962) has discussed in more detail
the conditions under which these deposits were laid down and has compared
them with conditions on the present coast of Bombay and the Gulf of
Cambay.

According to Zeuner (1939), upper Lias fossils from the same areas of
England are of the same age as the *Posidonia* shales of Brunswick and
Dobbertin (see above).

Asia. Martynova (in Rodendorf, 1957b) has listed several groups of Asian
localities.

The fossils from western Siberia (Sukhomesovo in the region of Chelyabinsk) and Kazakhstan (Kushmurun) are comparatively unimportant. On
the other hand, very important fossils are known from the Osh region of
Kirghiziya (Fergana: Kyzyl-Kiya, though Sulyukta is not important) and from
the Leninabad region of Tadzhikistan (Shurab). Martynov (1937) reported on
both these important localities in a large work that also dealt with general
zoogeographical problems (see below, at the conclusion of this section on the
Jurassic).

Only one grasshopper has been found at a lower Lias locality in the Kuznetsk basin (Korchakol, in the region of Kemerovo). On the other hand, the locality Ust'-Baley in the Yenisey basin (Irkutsk region on Lake Baikal) is important and has been known and investigated since the last century. A second locality in the same region, Iya, is of no importance. The age of the insects from these localities has been considered to be 'Dogger', but according to more recent Russian work (Rodendorf, 1957b, 1962) they belong to the upper Lias[89].

b. Dogger

In addition to Ust'-Baley which, as stated above, belongs to the upper Lias, Handlirsch only listed a few localities in England. However, Zeuner (1939) did not mention any English localities from the Dogger and so these too must have been incorrectly dated[90].

c. Malm

'One of the most extensive transgressions in the geological past took place during the Callovian, between the Lias and the Malm' (Bubnoff, 1949; p. 435). 'The geographical situation during the Malm was therefore quite different: a broad east–west sea covered the entire area between the Volga region and England. The south German area can now be regarded as a marginal part of Tethys. It was connected with the main part of Tethys via Poland, for a short time also via Saxony and Moravia, and in the west via Burgundy and the Straits of Poitou. The Bohemian–Ardennes island, the Central Massif, and Armorica (Brittany–Cornwall) lay between them as land-areas' (Bubnoff, 1949, p. 436). On the whole the climate was 'warmer and drier than at the height of the Jurassic. There was a pronounced boreal region in the north' (Schwarzbach, 1950).

Europe (Fig. 17). English localities (Durlston Bay, near Swanage, Dorset, and Dinton, Wiltshire: lower and middle Purbeck according to Zeuner, 1939) are not particularly productive. The Bavarian fossils from Solnhofen and district (e.g. Eichstätt) are much more famous. The lithographic limestone of Solnhofen belongs to the lower Portland and consists of 'shallow water deposits in lagoons lying between the mainland to the north-west and the edge of a reef which fringed the South German basin with Tethys in the south' (Bubnoff, 1949, p. 437). Zeuner (1939) has given a detailed account of the way it originated, basing himself mainly on Abel (1922). More recent accounts have been given by Kuhn (1963, 1966). In his comprehensive species list and bibliography of the plant and animal kingdoms from the Solnhofen slate, Kuhn (1961) has included 141 species of insect. It is an indication of how equivocal such statistical reviews of the number of fossil insects can be

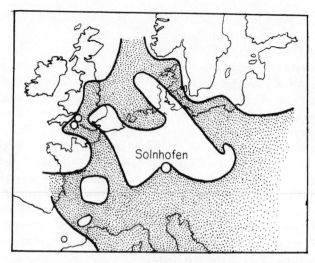

Fig. 17. Palaeogeography and insect localities of Europe in
the Malm. Based on Bubnoff. [An additional locality has
been added to Hennig's original map. Adrian Pont.]

that the 13 species of the genus *Pseudosirex* (Hymenoptera) listed by Kuhn
belong to only two different species according to Carpenter (1932a).

Three insects are known from Spain (Sierra de Montsech, Lérida province). These belong to the Kimmeridgian and are therefore rather older than
the fossils from England and Solnhofen (Condal, 1951).

Asia. The most important and productive localities are in Karatau
(Kazakhstan), between the Aral Sea and Lake Baikal, and the most
important individual locality is Galkino. In the earliest papers the age of these
fossils was given as 'lower Dogger', but they are now assigned to the Malm
(Rodendorf, 1957b, 1962). Rodendorf (1947) has given a general characterization of the dipterous fauna of this region. The same author (Rodendorf,
1957b) has also listed three individual localities from the Kuznetsk basin
('Upper Jurassic' or 'Dogger'), but nothing further seems to have been
published about them.

Fossil insects have been known from Transbaikalia (Buryat Mongolian
Republic, with the individual localities of Byrka, Tovega and Turga) since
Middendorff's expedition of 1843–44. One neuropteron is known, but the
most famous species are the giant mayflies of the genus *Ephemeropsis* (see
Demoulin, 1956b, 1954a). These have also been found in Mongolia (south of
Ulaanbaatar = Urga, Nalaihin Formation; Anda-Khuduk, Ondai Sair
Formation), Chinese Turkestan (T'u-lu Fan = Turfan, Sinkiang–Uighur
Autonomous Region: Ping, 1935) and Ch'eng-te-Shih (= Jehol). Some of
these localities were previously assigned to the Lower Cretaceous, but all of
them are now regarded as Upper Jurassic (see Chernova, 1961). Cockerell

(1925, 1927b) has also described some insects from Mongolia (Gobi Desert, Mt. Uskuk: Ondai Sair Formation).

Martynov (1937) compared the Lias fauna then known from Shurab and Kyzyl-Kiya with that from other faunal regions. He came to the conclusion that in the lower Lias the insect fauna of this area was still essentially an Angaraland fauna which differed from the preceding fauna primarily by the strong infusion of an Australian element. For example, he suggested that the Paratrichoptera (the stem-group of the Diptera!) arose in 'Indo-Australia' and penetrated into Turkestan as early as the Triassic. According to Martynov, it is also possible to discern faunal movements in the opposite direction but, nevertheless, the faunas of Angaraland, Europe, and Australia were still rather different during the Lias. However, many Angaraland elements penetrated into Europe during the upper Lias and even more so during the Upper Jurassic, and there were probably movements in the opposite direction too. In this way the faunas of West Asia and Europe came to resemble each other more and more.

It remains to be seen how far this view is still valid, especially as the Triassic faunas of Australia and Angaraland and the Jurassic faunas of Europe (the lower Lias of Lower Saxony) and of Asia are now much better known than they were 30 years ago. For example, the Paratrichoptera (*Pseudodiptera gallica*) have since been found in the Lower Triassic of the Vosges! Nonetheless, the idea of an increased faunal exchange between Angaraland and Europe during the Jurassic is still a logical one: for the channel in the area of the Urals that had previously separated both regions no longer existed during the Jurassic[91].

Revisionary notes

86. Bradbury and Kirkland (1967) have described the first Jurassic insects from North America. These belonged to the Hemiptera (Hydrocorisae) and were from the Todilto Formation (Upper Jurassic) of New Mexico (Ojo del Espiritu Santo Grant). [Willi Hennig.]

87. There are still virtually no Jurassic insects from Australia. Riek (in CSIRO, 1970, p. 171) wrote: 'A specimen from near Mudgee, New South Wales, was described (wrongly) as a cicada and referred to the Jurassic. A wing recorded by Etheridge from South Gippsland probably came from younger strata'. [Dieter Schlee.]

88. Carpenter (1970) and Tasch (1973) have described two Jurassic insects from lacustrine beds in the Antarctic (southern Victoria Land, Carapace Nunatak): the anisozygopteron *Caraphlebia antarctica* Carpenter and a beetle. Descriptions of further Antarctic insects are also in Carpenter (1970). [Rainer Willmann.]

89. Lin (1965) recorded two 'orthopterous' insects from the Lower Jurassic of Inner Mongolia. Imamura (1974) discussed a cockroach from the Nishinakaya Formation (Lias, Toyora group) of Ishimachi (Honshu, Japan; see also Fujiyama, 1974). [Rainer Willmann.]

90. Lin (1976) has recently described 20 genera and 22 species from Middle and Upper Jurassic sediments in the west of Liaoning province, China. The orders Plecoptera, Odonata, Blattoidea, Heteroptera, Psocoptera, Mecoptera, Diptera, Coleoptera, and Hymenoptera were represented. Lin (1978a) has also described some blattoids from the Middle Jurassic of China at Ning-hua (Fukien province), Wu-wei (Kansu province), and Yixian (Liaoning province), and from the Inner Mongolia Autonomous Region (Dongsheng). [Dieter Schlee.]

91. See note 87. [Dieter Schlee.]

3. Cretaceous
(Fig. VI)

One of the most regrettable gaps in our knowledge of insect phylogeny is the almost complete absence of fossils from the Cretaceous[92]. Individual insects have been found at various localities, but in no way does the extent of these discoveries reflect the importance of the events which must have taken place during the Cretaceous[93].

According to Schuchert and Dunbar (1945), the greatest submergence of the continents and the most extensive advances of the epeiric (epicontinental) oceans that have been recorded in the geological past took place during the Cretaceous. Deciduous trees suddenly became important in the early Cretaceous, although they must have originated earlier[94]. According to Schuchert and Dunbar, magnolias, figs, sassafras, and poplars appeared during the middle of the Lower Cretaceous. Forests of the Middle Cretaceous were probably fundamentally modern in character and consisted of beeches, birches, maples, oaks, walnuts, plane trees, tulip trees, breadfruit trees, ebonies, and *Sequoia*, with shrubs such as laurel, ivy, hazel, and holly. It is striking that Schuchert and Dunbar have given the same explanation for the comparatively sudden appearance of this modern flora that other authors (p. 59) have given for the sudden change in the insect fauna at the transition of the Carboniferous to the Permian: the resting stage of the Angiosperms is an unequivocal adaptation to adverse seasonal climatic conditions, either to aridity or to the cold of winter[95]. These forms probably originated in upland areas where the climate was cool and the periods of growth were restricted to particular seasons. If this was so, then according to Schuchert and Dunbar they would not have left any traces as fossils until they had moved down to the lowlands where sediments were being laid down.

Kossmat (1936) described the end of the Cretaceous, that is to say the transition from Mesozoic to Coenozoic, as 'a critical epoch of the first order' and said that it was characterized by various kinds of profound change. However, the retreat of the epicontinental seas was the only one that he actually mentioned.

It is therefore highly regrettable that so little is known about Cretaceous insects. In view of the sporadic occurrence[96] and, with few exceptions, the minor importance[96] of the fossils that have been found, it is not worth discussing the palaeogeographical and palaeoclimatic conditions of the Cretaceous in more detail.

A review of Cretaceous insects was given by Handlirsch (1939), but this was not quite complete and also assigned certain localities from Mongolia and China to the Cretaceous which are now considered to be Upper Jurassic.

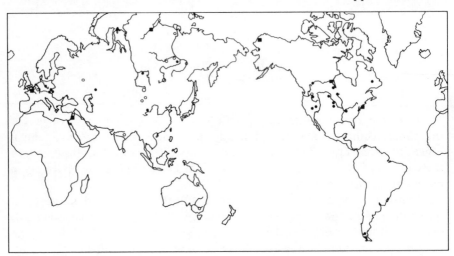

Fig. VI. Major insect localities in the Cretaceous. For further details, see Zherikhin (1978) and Schlee and Glöckner (1978). White circles: Lower Cretaceous compression fossils. Black dots: Upper Cretaceous compression fossils. Black squares: Cretaceous ambers [Original. Adrian Pont.]

a. Lower Cretaceous compression fossils

Long ago some beetles were described by Heer from the Lower Cretaceous of Greenland (Kome and Ivanguit), and these localities were mentioned by Handlirsch (in Schröder, 1920–21). Handlirsch also mentioned a cockroach from Montana (Great Falls coalfields, Cascade County) as the only fossil from the Lower Cretaceous of North America.

Only a few undescribed beetles have been listed from England (Isle of Wight; Swanage Bay; Bexhill; Govers Cliff and East Cliff, near Hastings).

Zeuner (1939) has listed the Vale of Wardour, Wiltshire, as a further locality ('Wealden', but actually Purbeck). Handlirsch mentioned one cicada from Bernissart (Belgium). His 'English' locality Lottinghem is actually in the Pas-de-Calais.

Zalesskiy (1953) described a dragonfly (*Aeschnidiella kabanovi*) from the Volga region (Ul'yanovsk district: Aptian). Some fossils from the Yenisey basin (Kempendyayi in the Yakutsk district) have apparently not been worked out yet, nor have those from Transbaikalia[97]. Some fossils have been described from Mongolia and from Ch'eng-te-Shih (= Jehol), but they are now considered to belong to the Upper Jurassic. I do not know whether the same applies to fossils from the Shantung Peninsula (Lai-yang Formation) (Grabau, 1923; Ping, 1928; see Handlirsch, 1939). In Australia, unimportant insect fossils have been known for about 85 years from northern Queensland (Flinders River: see Riek, 1954a)[98].

b. Upper Cretaceous compression fossils

North American localities in Colorado, Kansas, Nebraska, Manitoba (Millwood), and Montana (see Handlirsch, in Schröder, 1920–21), South Dakota (Fox Hills Formation: Northrop, 1928) and Utah (Brown, 1941) belong to the Upper Cretaceous. Handlirsch did not mention the amber from Tennessee (Eutaw Formation of Coffee Bluff, Hardin County), also assigned by Northrop to the Upper Cretaceous[99].

For the most part only beetles and problematic fossils have been found in the Upper Cretaceous of Czechoslovakia (localities Budyně nad Ohří, Vyšehořovice, Louny, Chuchle, Lidický dvur, Kounice, Bohdánkov, and Lipenec, according to Handlirsch, in Schröder, 1920–21, and Handlirsch, 1939), Saxony (Roschütz) and Lebanon (localities in Handlirsch, in Schröder, 1920–21, and Handlirsch, 1939).

Some of the fossils from Asia are still unworked, but they are equally unimportant. Rodendorf (1957b) has listed localities in Kazakhstan (continental deposits from the Cenomanian in the Terekty-say valley, a tributary of the upper Emba), in the Yenisey basin (Krasnoyarsk region: what was until recently the oldest known species of the Hymenoptera-Aculeata was found here), and in the region of Yakutsk, Khabarovsk, and the Maritime province: these have all produced some fossils, particularly beetle elytra, but none of them has yet been worked up[100].

Kuschel (1959) has described a beetle (Curculionidae; Maastrichtian) from Chile (Magallanes: Cerro Guida, Seno Ultima Esperanza).

c. Cretaceous amber fossils

Insect inclusions in fossil resin ('amber') are becoming increasingly important for the Cretaceous, and for the Tertiary which is not dealt with in further detail in this book. As we get closer to the present, we find that characters that cannot be seen on simple impressions of the body become more and more important for assessing the fossils. The characters of at least the exoskeleton can often be seen just as clearly in insects in resin as in modern specimens.

According to McAlpine and Martin (1966), the Canadian amber that has been found in the basin of the Saskatchewan River in Canada (at Cedar Lake, Manitoba, and other localities) belongs to the Upper Cretaceous (Campanian = upper Senonian). This has already produced some extremely interesting fossils and it is to be hoped that more intensive collecting in the near future will produce results of great importance for our knowledge of Upper Cretaceous insects[101].

The exact age of another Cretaceous resin, found on the Arctic coastal plain of Alaska (Hurd *et al.,* 1958), is not known.

A fossil resin has been found in Lower Cretaceous strata (Neocomian) in the Lebanon[102], and more than 60 insect inclusions have been found so far, including some extremely important ones (unpublished)[102]. Significant results can also be expected from intensive collecting at these localities[103].

Revisionary notes

92. As regards 'the almost complete absence of fossils from the Cretaceous', there have been considerable changes over the last decade. An account of Cretaceous fossils has been published by Zherikhin (1978). A number of well preserved amber fossils from the Lower, Middle, and Upper Cretaceous have been discovered, ranging from Alaska and Canada, to France, Lebanon, and Siberia. For further details, see notes 101–103 below. [Dieter Schlee.]

93. There have been many recent discoveries of various groups of insects in fossil resins (ambers) of different ages, and their analysis is making it possible to draw several important conclusions about this biologically and geologically important period of the geological past. [Dieter Schlee.]

94. The evolution and age of the angiosperms has been dealt with by Hughes (1961). [Willi Hennig.]

95. See Casey and Rawson (1973) for discussion of the boreal Lower Cretaceous. [Willi Hennig.]

96. This statement about the sporadic occurrence and minor importance of the Cretaceous insect fossils is no longer true. See notes 101–103 below. [Dieter Schlee.]

97. Several of these have now been described, e.g. by Sukacheva (1968), Ponomarenko (1976) and Skalski (1979a). [Adrian Pont.]

98. Well preserved compression fossils have been found in the Lower Cretaceous lacustrine sediments of Victoria (Koonwarra). The fauna consists mainly of aquatic immature stages of mayflies, midges, stoneflies, dragon-flies, and beetles, but a few specimens of terrestrial species have also been found. Riek (1971) considered that these fossils were closely comparable with the modern shallow freshwater communities of south-east Australia. This fauna has not yet been described in detail. [Rainer Willmann.]

The Middle Cretaceous of Bohemia is said to contain compression fossils of some 'Muscidae' (Diptera) and of some galls probably produced by 'Tenthre-dinidae' (Hymenoptera) (Jeannel in Grassé, 1949). [Willi Hennig.]

99. There is a new North American Cretaceous insect locality in Minnesota (New Ulm), but it is unlikely to yield any well preserved specimens. So far only caddis fly cases have been mentioned (Lewis, 1970). [Rainer Willmann.]

McAlpine (1970) has described some alleged puparia of the 'Calliphoridae' (Diptera) from the Upper Cretaceous of Alberta (Canada). An Upper Cretaceous (Cenomanian) deposit has been discovered near Knob Lake in northern Labrador (Canada): see Dorf (1967), also Carpenter (1967) and Ponomarenko (1969). [Willi Hennig.]

100. A new Lower Cretaceous insect locality is now known from China (Kansu province), and a single beetle has been described (Chen and T'an, 1973). [Rainer Willmann.]

101. A full discussion of Canadian amber, giving all the salient facts, has been published by McAlpine and Martin (1969a, 1969b). A number of results, overlooked by Hennig, had previously been published by Carpenter *et al.* (1937).

Papers on the following groups have been published, in addition to those listed by McAlpine and Martin (1969a, 1969b): Collembola (Delamare Deboutteville and Massoud, 1968); Coccina (Beardsley, 1969); Auchenor-rhyncha-'Jascopidae' (Hamilton, 1971b), but Evans (1972) stated that the sole specimen upon which this family was based belonged to the Cicadellidae and was not a new family at all; Hymenoptera-Aculeata (Evans, 1969); Hymenoptera-Chalcidoidea (Yoshimoto, 1975); Diptera-Cecidomyiidae (Gagné, 1977); Diptera-Bibionidae (Peterson, 1975); Diptera-Stratiomyidae (Teskey, 1971); Diptera-Ironomyiidae (McAlpine, 1973); Diptera-Sciadoceridae (McAlpine and Martin, 1966), but note Hennig's (1971a) comments that these actually belong to the stem-group of the Phoridae rather than to the 'Sciadoceridae'; a larval head-capsule of the Lepidoptera (MacKay, 1970).

The following groups have been recorded from Canadian amber: Collem-bola, Psocoptera, Thysanoptera, Heteroptera, Auchenorrhyncha, Aphidina, Coccina, Hymenoptera (19 families), Coleoptera (4 families), Trichoptera, Lepidoptera, and Diptera (20 families).

A summary of Canadian amber has been given by Schlee and Glöckner (1978). [Dieter Schlee.]

The first Mesozoic ants were described by Wilson *et al.* (1967), from the Magothy Formation of New Jersey. [Willi Hennig.]

102. The Lower Cretaceous (Neocomian) amber from the Lebanon is geologically the oldest known fossiliferous amber.

In 1967 and 1968, when Hennig's book was in press, I made some collections of Lebanese amber which contained the first insects to be found in this amber (Schlee and Dietrich, 1970; Schlee, 1972). A geological analysis was given by Dietrich (1975, 1976), and a recent summary has been published by Schlee and Glöckner (1978).

Reports on the following groups have been published: stem-group of the Hemiptera-Aleyrodina (Schlee, 1970); Thysanoptera (Zur Strassen, 1973), but see the comments by Seeger on p. 370 of this book; Diptera-Empididae (Hennig, 1970, 1971b); Diptera-Cyclorrhapha (Hennig, 1971b); Diptera-Psychodidae (Hennig, 1972b); some micro-structures of the Diptera-Chironomidae (Schlee, in Schlee and Glöckner, 1978); other parts are in preparation. Some birds' feathers have also been described by Schlee (1973; in Schlee and Glöckner, 1978); and certain Lepidoptera are being studied by Mickoleit (in preparation).

Lebanese amber from other collections has included some Lepidoptera-Micropterigidae, which have been analysed by Whalley (1977, 1978). [Dieter Schlee.]

103. Since 1968, some other fossiliferous ambers have been discovered:

1. From the Middle Cretaceous (Cenomanian) of France (Kühne *et al.*, 1973; Schlüter, 1974, 1975). All the available information has been summarized by Schlüter (1978).
2. From the Upper Cretaceous of Siberia (Zherikhin and Sukacheva, 1973; Rodendorf and Zherikhin, 1974; a German summary in Schlee and Glöckner, 1978). A great deal of fossiliferous amber has been found at these localities, which have been explored since 1970. The most productive locality has been Yantardakh, which is a local word meaning 'amber mountain', indicating that the amber has been known to the local inhabitants for a long time. Yantardakh is situated on the River Maymecha, and the discoveries have been so extensive that Zherikhin and Sukacheva (1973) have even compared their amber with Baltic amber.

   More than 60 groups of animals have already been found in Siberian amber, and papers on the following groups of insects have been published: Ephemeroptera (Chernova, 1971); Aphidina (Kononova, 1975, 1976); Psocoptera (Vishnyakova, 1975), but see the comments by Seeger on p. 371 of this book; Neuroptera (Meinander, 1975); Hymenoptera-Ichneumonidae (Townes, 1973); Hymenoptera-Apocrita (Rasnitsyn,

1975a); Hymenoptera-Aculeata (Evans, 1973); Diptera-Chironomidae (Kalugina, 1976); Diptera-Empididae (Kovalev, 1974); and Acarina (Bulanova-Zakhvatkina, 1974). See also Zherikhin (1978).

3. It has recently been suggested that Burmese amber is also Cretaceous in age (Štys, 1969), but this does not appear to be based on any geological analysis and must therefore be regarded as very questionable. A summary of Burmese amber has been given by Schlee and Glöckner (1978).

4. Zherikhin and Sukacheva (1973) also considered Chinese amber (Fu-shun amber) to be Cretaceous. Until recently, the only fossil to have been described was a blattoid (Ping, 1931). A summary of Fu-shun amber has been given by Schlee and Glöckner (1978).

Hong (1979) has recently described an ephemeropteron from Fu-shun amber and has clarified its stratigraphic origin: the fossiliferous amber is Eocene (Guchengzi Formation). Hong has also alluded to the 'insects, spiders and plants' preserved in this amber.

A list of Cretaceous amber fossils has been published by Schlüter (1978) and the subject reviewed by Zherikhin (1978).

The *Tertiary* is not included in this book. So far as the amber fossils are concerned, there are some new detailed studies as well as Hennig's own studies on the Diptera (see bibliography). (1) A palaeobiological study of Baltic amber, by Larsson (1978). (2) A comparative morphological and analytical review of world ambers and their inclusions, by Schlee and Glöckner (1978); this contains much new information on Dominican amber and records some 60 groups for the first time from this amber (based on collections in the State Museum of Natural History, Stuttgart); a new edition of this paper will be published in 1981, and will include an enlarged faunal list. (3) Another booklet with some 120 colour photomicrographs, dealing primarily with Dominican, Baltic and Lebanese ambers and the structure of the fossil inclusions (Schlee, 1980). (4) A continuing series of analyses of the fossils in Dominican and other ambers in *Stuttgarter Beiträge zur Naturkunde*, Series B (1978–1980). [Dieter Schlee.]

CHAPTER THREE

The Phylogenetic Development of the Insecta

A. The ground-plan, origin (sister-group) and age of the Insecta[104]

There have been many different theories about the origin and relationships of the insects. Most of these are now only of historical interest, and the only one that can be seriously considered is that the Insecta and the Myriapoda form a monophyletic group of higher rank, the Tracheata or Atelocera[105].

This theory has appeared in several forms: either the myriapods as a whole or one of their subgroups is considered to be the sister-group of the insects. The second alternative is the one that is now generally accepted. Previously it was thought that the Opisthogoneata (Chilopoda) were the sister-group of the insects (e.g. Verhoeff, 1926), but the Progoneata or even the Symphyla, a subgroup of the Progoneata, are now usually considered to be this sister-group. However, there is no evidence for this. Imms (1945) has listed a number of characters common to the Symphyla and Diplura, which are usually thought to be the most primitive group of insects. However, these characters are all symplesiomorphies and cannot be used as evidence that there was once a single stem-species from which only the Symphyla and the insects have descended. Discussions of this question have suffered a great deal from conceptual confusion, and it is possible to find contradictions in a single paper. For example, Snodgrass (1938) has published a phylogenetic tree in which the Symphyla and Hexapoda are shown to be sister-groups. However, in the text he stated that two lines have descended from the hypothetical *Proto*symphyla: one included the progoneate modern Symphyla, Pauropoda, and Diplopoda, and the other the opisthogoneate Hexapoda. This can only mean that he believed there to be a sister-group relationship between the Progoneata and the Insecta (= Hexapoda).

In fact, there are good reasons for believing that the Progoneata are a monophyletic group to which the Symphyla also belong. Evidence for this is provided by the position of the genital orifice which has moved forward secondarily, the absence of palpi on both maxillae, and the absence of the plates on the second maxilla (the labium) (see Dohle, 1965). On the other hand, not a single derived character is known that would suggest that there is a close relationship (sister-group relationship) between one group of the Myriapoda (Progoneata or Opisthogoneata) and the insects.

This means that we should investigate whether the Myriapoda as a whole may not be the sister-group of the Insecta. The first step is to see if there is any evidence that the Myriapoda are in fact a monophyletic group. There is a general tendency now to reject this idea. For example, Weber has retained the name 'Myriapoda' just as a 'collective name for the first two classes of the Antennata', and most modern authors have expressed similar opinions.

However, there are some characters that may indicate that the Myriapoda are a monophyletic group, and the most striking of these is the absence of the ocelli. Since the ocelli as well as the compound eyes are part of the ground-plan of the Mandibulata, or even of the entire Arthropoda, and since they appear to have been retained in the insects, they must have been lost in the Myriapoda. The same is true, in a rather more limited sense, of the compound eyes. Most of the Myriapoda have a few or several simple eye facets and these have probably arisen through reduction of the compound eyes[106].

The only Myriapoda to have complete compound eyes are the Notostigmophora (genus *Scutigera*, etc.). There is no doubt that this group belongs to the Chilopoda, and its species have a number of striking derived characters; it is probably because of this that their compound eyes have usually been thought to have arisen secondarily ('pseudo-compound eyes'). However, in addition to their derived characters, the Notostigmophora have some other characters in which they appear more primitive than all the other Myriapoda, or even Tracheata. Boettger (1958) has pointed out that the Notostigmophora are the only group of the Tracheata to have retained a ventral artery into which the blood flows from the heart through 1–3 lateral loops. As far as I know the Notostigmophora are the only Myriapoda, apart from the Symphyla, which have retained paired claws (see below). In view of this it is very likely that the compound eyes too have been retained in a relatively primitive form in the Notostigmophora (Fahlander, 1938; see Siewing, 1960). They must have been independently reduced (reduced to simple eyes) in the remaining Chilopoda and in all Progoneata, just as the paired claws have been lost (reduced and fused) in most Chilopoda and Progoneata. The Permian diplopod *Pleuroiulus levis* Fritsch provides some evidence for this as it has conspicuous compound eyes (Siewing, 1960). With all this in mind, I am inclined to believe that the ocelli too may have been independently reduced in various groups of the Myriapoda and that their absence does not indicate that the group is monophyletic. This is certainly a possibility as the ocelli have been lost independently in several groups of insects. However, just because a particular character can be shown to have arisen several times through convergence, it does not follow that it must have arisen independently on every occasion. Nevertheless, it is still important to find further synapomorphies that will provide evidence that the Myriapoda are a monophyletic group[107].

Snodgrass (1952) drew attention to two characters common to the Opisthogoneata and Progoneata which seemed to him to provide decisive evidence that these two groups are closely related: the musculature of the lacinia mobilis of the mandibles, and the supporting apparatus of the hypopharynx which consists of a pair of transverse fulturae anchored to the head capsule. Both of these features, or at least the first of them, still need to be studied further in case they should prove to be primitive characters in the ground-plan of the Tracheata, and therefore symplesiomorphies of the Myriapoda[108], which have been lost or developed further in the insects.

To summarize so far, the following points have been established.

There are very good reasons for regarding the Progoneata (including the Symphyla) and the Opisthogoneata (Chilopoda) as monophyletic groups. There is no real evidence that either of them is more closely related to the insects, which should undoubtedly be treated as another monophyletic group. The characters indicating that the Myriapoda are monophyletic are modest and need to be checked further. Nevertheless, the theory that there is a sister-group relationship between the two monophyletic groups of the Tracheata (Myriapoda and Insecta) is the only one for which any evidence has been produced so far, and consequently it is the only one that we should accept, until there is some reason for rejecting it. It agrees with the phylogenetic tree published by Ross (1965) in his Textbook of Entomology.

It is not possible to give any accurate information about the age of the insects, that is to say to pinpoint the date when the sister-group relationship between the insects and myriapods originated. Neither the stem-group of the Insecta nor that of the Tracheata has been found in the fossil record. However, there are reliable direct or indirect records of several groups of insects from the Middle Devonian (p. 106), and so the group as a whole must have originated before this.

The sister-group of the Tracheata (Myriapoda + Insecta) appears to be the Crustacea (see below), and the Tracheata + Crustacea form a single monophyletic group, the Mandibulata. The oldest Crustacea (*Protacaris*) are known from the Lower Cambrian. These species have a carapace and stalked eyes: these two characters have a functional association according to Siewing (1960), and from this it is clear that *Protacaris* really does belong to the Crustacea and not to the stem-group of the Mandibulata. There is certainly no reason for suspecting that carapace and stalked eyes have been lost in the Tracheata. As subgroups of the Crustacea have been recorded as early as the Upper Cambrian, we would probably have to go back into the Pre-Cambrian to find the actual point in the phylogenetic tree of the Mandibulata when the two sister-groups Crustacea and Tracheata separated. The division of the Tracheata into two sister-groups (Myriapoda and Insecta) must therefore have taken place between this epoch and the Lower Devonian at the latest[109].

There may have been a very long interval between the origin of the Tracheata and their separation into the Myriapoda and Insecta. During this period the stem-group of the Tracheata, for which Tillyard (1931a, 1932a) coined the name Protaptera, may not have been terrestrial. This question has been discussed many times with reference to the 'ancestors of the insects', but a lack of clarity about what is meant by this phrase is always disturbingly evident. Davey (1960) has investigated the role of the spermatophores in fertilization and has referred to Tiegs, who suggested that the insects descended from an aquatic ancestor that released its sperm into the water. Nielsen (1961) believed that the Arthropoda were originally terrestrial, and that the Crustacea and Trilobita became aquatic secondarily. If this view is accepted, the stem-group of the Tracheata must also have been terrestrial and the characters common to the Myriapoda and Insecta that might otherwise be interpreted as adaptations to terrestrial life must be regarded as primitive ground-plan characters of the Arthropoda. Gilyarov (1960) was right to point out that this theory is untenable. Both Ross (1965) and Sharov (1966b) were convinced that the Crustacea were originally aquatic and that the stem-group of the Mandibulata must have been too. If the question of whether the insects have descended from aquatic or terrestrial ancestors is to mean anything at all, it can only be expressed as follows: was the latest common ancestor of the Insecta + Myriapoda (the stem-species of the *Tracheata) still aquatic or had it already become terrestrial?

Most of the derived characters common to the Myriapoda and Insecta are adaptations to terrestrial life, and the next task is to see whether they could have arisen independently in each group.

Tiegs and Manton (1958) have emphasized that an intestinal gland ('liver') is present in the Crustacea and Chelicerata but absent in the Tracheata. If this gland is homologous in the Crustacea and Chelicerata, it must have been part of the ground-plan of the Euarthropoda that was retained in the Mandibulata and then reduced in the Tracheata. According to Sharov (1966b), Beklemishev considered that elongation of the intestine and the expanded function of the salivary glands led to reduction of the 'liver'. This could well be connected with the change to a terrestrial existence.

As far as I know, no reasons have been suggested for the loss of the 2nd antenna. It may have been lost for much the same reason as the exopodites on the body appendages were lost: these swimming organs became superfluous and were reduced when the terrestrial life was adopted. In this context it does not matter whether the 2nd antennae are interpreted as prostomial organs, as Ferris and Henry have done (see DuPorte, 1957, 1958b), or as the appendages of a true, originally post-oral body segment, which is the prevailing opinion. The fusion of the second maxillae into a single labium, which acts as a lower lip to close the mouth-opening posteriorly, may also be connected with the terrestrial mode of life. It may also be significant that there is a mouth-chamber in the terrestrial Chelicerata (Arachnida), although the 'lower lip' is formed differently in the individual subgroups. The fact that a mouth cavity did not develop immediately in the terrestrial Chelicerata, as part of their ground-plan, may be because they only imbibe liquid food, that is to say food liquefied by enzymes before it reaches the mouth.

The development of cursorial legs must also be regarded as a characteristic derived ground-plan feature of the Tracheata. It is connected with movement over a substrate, but not necessarily with a terrestrial mode of life. The Crustacea also have a walking leg (endopodite). The obvious question now is whether the swimming leg (exopodite) was a new acquisition in this group or whether it was lost in the Tracheata. In other words: were the common ancestors of the Crustacea and Tracheata terrestrial species and have the Crustacea subsequently become more or less pelagic, or have the Crustacea retained the mode of life of their common ancestors and was it the Tracheata which first became terrestrial? This question assumes that there really is a sister-group relationship between the Tracheata and Crustacea, as has usually been thought. However, some doubts have recently been raised about this, based amongst other things[110] on the morphology of the mandibles (in particular, see Tiegs and Manton, 1958; Manton, 1960; DuPorte, 1962). These are said to have originated independently in the Crustacea and Tracheata. Manton (1960) has stated that the Crustacea bite with a proximal endite whereas the Tracheata bite with the tip of an appendage, and, according to DuPorte, this difference explains the complete absence of a mandibular palp in the Tracheata. On the basis of this alleged difference in the structure and function of the mandibles in the crabs and the Tracheata, and on the basis of features allegedly shared by the Tracheata and Onychophora, these authors have suggested that there is a close relationship between the Tracheata and Onychophora[111]. I think that Siewing (1960) was right to reject this view, and he also set out the reasons for considering the Arthropoda and the Mandibulata to be monophyletic groups. Amongst other things, he pointed out that Snodgrass found extensive agreements in the musculature of the mandibles in the Crustacea and Myriapoda. He also emphasized correctly that the Crustacea and Tracheata do not only agree in having mandibles but that they have also modified the two following appendages (1st and 2nd maxillae) into mouth-parts. Moreover, the posterior border of the gnathocephalon in the Mandibulata (behind the 2nd maxillae) corresponds to the posterior border of the proterosoma in the Chelicerata. It is clear that the Mandibulata show a considerable amount of agreement in the subsequent development of the anterior section of the body, particularly when they are compared with their sister-group, the Chelicerata.

Manton (1960) has given a spirited defence of Tiegs' recognition of the three groups Monognatha (Onychophora), Dignatha (Pauropoda, Diplopoda; also Pentastomida according to Osche), and Trignatha (Chilopoda, Symphyla, Insecta). However, this is a typological distinction, as is Sharov's (1966b) division of the Tracheata ('Atelocera') into Monomalata and Dimalata. It is based on the unfounded premise (see above) that the Symphyla are more closely 'related' to the insects than are the other Myriapoda. Trignathy appears to be part of the ground-plan of the Mandibulata. The dignathy of the Pauropoda-Diplopoda originated from this (see Dohle, 1965). The monognathy of the Onychophora and the dignathy of the Pentastomida (Osche) have nothing to do with this and probably arose independently.

If the modification of the appendages of the gnathocephalon into mouth-parts (mandibles, 1st and 2nd maxillae) is considered to be an event that took

92

place in the stem-group of the Mandibulata [see Siewing's (1960) illustration of this], then the schizopodal nature of these three pairs of mouth-parts, and particularly of the mandibles(!), suggests that the ancestors of the Mandibulata were capable of swimming from time to time: the development of a swimming leg on appendages that were already part of the feeding mechanism could hardly be explained in any other way. Otherwise it would be necessary to suggest, for example, that the schizopodal character first originated on the true body appendages and was transferred secondarily to the mouth-parts. Such *ad hoc* hypotheses are of little value.

I think that Sharov (1966b) was absolutely right to suggest that in the Crustacea the exopodite on the third pair of legs (basipodite) is the same as the epipodites of the first two pairs of legs (precoxopodite and coxopodite), which have a respiratory function (Fig. 18). It did not become a 'swimming leg' until later, when the Crustacea developed from terrestrial into swimming species: the common ancestors of the Crustacea and Tracheata were terrestrial (aquatic) species, and the Tracheata continued this terrestrial life,

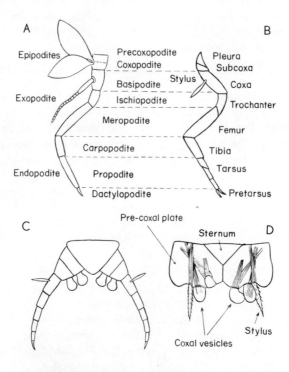

Fig. 18. Abdominal appendages of the Anaspidacea (Crustacea: A), 'Thysanura' (B), Protosymphyla (hypothetical: C), and *Machilis* sp. (abdomen: D). From Sharov (1966b). If Sharov's interpretation is correct, the coxa must be included in the 'pre-coxal plate' of *Machilis* (D).

whereas some or all of the Crustacea became pelagic. However, this does not explain why there is a marked difference between the epipodites of the first two leg segments and the exopodite (= epipodite of the basipodite) in all known species. Unlike the epipodites, the exopodite is always segmented, even on the mandibles (!), and this segmentation resembles that of the endopodite, though it is not identical. The fate of the two proximal epipodites and of the exopodite is also different in the Tracheata: the epipodites have developed into eversible coxal vesicles whilst the exopodite has become a coxal style (stylus). Surely this must indicate that they functioned slightly differently, at least in the latest direct ancestors of the Crustacea and Tracheata, and that the exopodite assisted with occasional swimming as well as with respiration. The possibility of a sensory function should also be considered. I think that we also need to explain why only the three proximal epipodites have been retained in the Mandibulata, when epipodites were originally present on almost all the leg segments according to Sharov, and why the most distal of these, the 'exopodite' of the basipodite, has acquired a different function from that of the two proximal ones, which were undoubtedly respiratory. Whatever the explanation of this, it is probable that the latest common ancestors of the Crustacea and Tracheata were species that swam only occasionally[112]. The earliest members of the stem-group of the Tracheata may have been too.

It is probable that the latest common ancestors of the *Tracheata had already become terrestrial. Evidence for this is provided by the modification of the two proximal epipodites into eversible coxal vesicles in the ground-plan of this group. According to Nielsen (1961), it has been proved experimentally that the coxal vesicles can absorb water when it is only present as a thin film and when 'drinking' is not possible (also Tiegs and Manton, 1958).

One of the reasons put forward as evidence that the Symphyla and Insecta are 'closely related' is that eversible coxal vesicles only occur in the Symphyla and in certain insects. However, the correct explanation of this is different. The modification of the proximal epipodites into coxal vesicles is a derived ground-plan character of the *Tracheata and is connected with their change to a terrestrial mode of life. Within the *Tracheata themselves, however, the presence of vesicles is a symplesiomorphic condition and cannot be used as evidence that the species retaining them are closely related. In fact, they have been retained only in relatively primitive species of each sister-group: the Symphyla (Myriapoda) and the primitive Entognatha and Ectognatha (Insecta). The more derived species of each group have lost the vesicles independently, and this appears to be the result of their increasingly impermeable cuticle (Nielsen, 1961).

The structure of the tracheal system and the distribution of the spiracles are more difficult to explain, and in these characters the Symphyla, which only have one pair of head spiracles, do not agree with the most primitive insects. It is therefore possible that the Myriapoda and Insecta each acquired a tracheal respiratory system independently of the other. Some support for this is found amongst the Chelicerata, where tracheae have clearly arisen independently in several groups (certain spiders, the harvestmen, mites, and their relatives). They also occur in the terrestrial Crustacea (Isopoda). It is a striking fact, however, that the most primitive species of the two main insect lineages (Entognatha: Diplura; Ectognatha: all groups) agree in having one pair of spiracles on most of the body segments, and in this respect they agree with most of the Myriapoda. The Collembola, which have no spiracles on the body segments, are very derived dwarf forms. Some of them have a pair of spiracles in the region of the neck: these are not the most primitive species, and so this character is considered to be a new acquisition. This may provide a clue as to how the Symphyla should be assessed, which are also dwarf forms. I can see no real reason for not accepting that a tracheal system with one pair of spiracles on most of the body segments is part of the ground-plan of the Tracheata. The tracheal system could have been reduced at first in certain dwarf forms, and then for some reason could have been replaced by a new formation in the region of the head (Symphyla) or of the neck (some Collembola)[113].

The structure of the pretarsus could also give a lead as to the mode of life of the primitive *Tracheata and particularly the stem-form of the insects. According to Snodgrass (1958b), the Tracheata (Myriapoda + Insecta) differ from all the other Arthropoda in having only one muscle in the pretarsus (dactylopodite): the dorsal muscle (levator) is absent. This seems to be a derived character but no plausible explanation has been given for it.

The presence of paired claws in addition to the third unpaired 'claw', which corresponds to the apical segment of the original dactylopodite, is characteristic of most Tracheata. It is interesting that paired claws are also characteristic of terrestrial Chelicerata. Their origin appears to be connected with movement over a solid substrate. Since paired claws are present in primitive Insecta and Myriapoda (Symphyla, Notostigmophora), they could be regarded as part of the ground-plan of the Tracheata and could be used as evidence that at least the latest common stem-forms of this group were terrestrial.

However, there is a weakness in this theory because paired claws are absent in most Myriapoda (except for the Symphyla and Notostigmophora) and also in some insects (Protura, Collembola, the Palaeozoic Monura). There is some evidence that paired claws have been secondarily lost in the Protura and Collembola, but none that this is the case in the Myriapoda. There is no doubt that the formation of a pretarsus, with or without paired claws, and of segmented tarsi is connected with the way in which the legs are actually placed on the ground. Unfortunately there has been no comparative study of the methods of locomotion in all these groups and of the way in which they are correlated with leg structure. Such a study might provide some evidence that the absence of paired claws in most Myriapoda is a derived character.

Sharov (1966b, Fig. 74) has given a heuristically stimulating interpretation of leg development in the Tracheata, but he failed to take any account of the connections between this development and the individual branches in the phylogenetic tree. His ideas are very reminiscent of the explanation for the shift in the position of the mammalian legs: so that the body could be raised from the ground, these moved from their characteristically 'reptilian' lateral position to a ventral position beneath the body. According to Sharov, the Tracheata achieved this by an S-shaped bend of the legs. As part of this process, the 3rd leg segment (basipodite = coxa in the insects) adopted a vertical position beneath the body and the two proximal segments (pre-trochantin, which eventually formed the pleura, and trochantin) became part of the body. The femur, which was originally the fifth leg segment, and the tibia, which was held vertically like the coxa, both became elongated. The tarsus was placed parallel to the substrate, became segmented, and acquired a pair of claws. The original function of the pretarsus was taken over by apical tibial spurs and the pretarsus itself was reduced. It has been suggested that the acquisition of paired claws is part of the ground-plan of the Tracheata, but I believe that it is one of the last steps in this sequence of development.

The Palaeozoic Monura make it very difficult to decide which stage in this sequence had been reached in the ground-plan of the *Tracheata and of the *Insecta. This is an example of how ideas that are well founded in functional–morphological terms still seem unrealistic if they fail to shed any light on a particular 'development' as an actual historical process, with all its potential parallel developments and occasional reversals, even though they take careful account of the genealogical relationships amongst the species that have reached different stages in the developmental sequence. One of the fundamental tasks for morphologists of the Tracheata, and particularly of the Insecta, must be to investigate this problem and, if possible, to confirm Sharov's ideas in actual historical terms.

Most accounts of the origin and relationships of the Insecta suffer from the defect that they treat agreements and differences between the individual arthropod groups in a purely statistical way. It will be possible to understand the actual historical development of these groups only by making a careful analysis of the heterobathmy of their characters. In the course of this, the mosaic-like distribution of primitive and derived character states will lead to an intelligible picture in which all the characters will take their logical place. The following synthesis is the result of a comparative study of the contra-dictory theories concerning the origin and relationships of the insects. It shows how all the contradictions discussed in the last few pages can be reconciled.

The stem-forms of the Mandibulata lived in water and had schizopodal legs. This means that the epipodite of the third leg segment had differentiated from the two proximal epipodites and had become an 'exopodite', and probably had a rather different function. There were three pairs of appendages on the gnathocephalon, to which the first three post-tritocerebral, ventrally situated, neuromeres are assigned; these appendages were already part of the feeding mechanism, and the second pair had been modified into mandibles. At first these mandibles retained their schizopodal character, but this was subsequently lost through parallel development.

These stem-forms also possessed spermatophores and at first the function of these was just to concentrate the spermatozoa. In this context, the theory that the ancestors of the insects were aquatic can be retained (Davey, 1960), although it refers to a very remote past.

When the Mandibulata separated into two sister-groups, which must have taken place in the Pre-Cambrian, the Crustacea remained aquatic whereas their sister-group moved exclusively along the ground and eventually became terrestrial, although at first it lived in very moist environments. Associated with this change, a mouth cavity developed. This was closed ventrally by a lower lip (labium), which was formed by fusion of the second maxillae. This gave rise to a more pronounced division between head and body: this had already been initiated in the ground-plan of the Mandibulata but had not developed very far.

For obvious reasons the swimming leg (?, exopodite) on all appendages was lost. The 2nd antenna was also lost, possibly for the same reasons, and the intestinal gland ('liver') was reduced.

The dorsal muscle (levator) in the pretarsus was lost. Claws may also have been acquired, which were already paired, and these must then have been lost secondarily in most of the Myriapoda. There must have been 14 body segments (excluding the telson), as this number is characteristic of the primitive Myriapoda (Symphyla, also Pauropoda) and insects.

Eversible coxal vesicles, derived from the two proximal epipodites, made it possible for small amounts of water to be absorbed from the ground (Nielsen, 1961).

The origins of the tracheal system and internal fertilization, and the transposition of the excretory system, were correlated with the change to a terrestrial mode of life. Except perhaps for the tracheal system, these and possibly other features undoubtedly originated in the stem-group of the Tracheata.

Outgrowths in the intestine (Malpighian tubes) replaced the primitive excretory organs (antennal and shell glands) which had been derived from nephridia. Only the labial gland, which was homologous with the shell gland of the Crustacea, was retained as a derivative of the original segmental organ. According to Beklemishev, the cells of the fat-body, a tissue that is characteristic of the Tracheata, originated as nephrocytes and apparently as storage kidneys (Sharov, 1966b).

Internal fertilization was first accomplished by means of spermatophores. According to Davey (1960), their function was to release the sperm into the female gonoducts and to prevent sperm loss by blocking the female genital orifice.

The division of the stem-group of the Tracheata into its two recent sister-groups, Myriapoda and Insecta, must have taken place in the Lower Devonian or even earlier, since several subgroups are known from the Middle

Devonian. At first these two sister-groups retained the fundamental ground-plan characters of the Tracheata. This means that the most primitive recent species of each group still have a considerable overall similarity: these groups are the Symphyla, and in other characters the Notostigmophora (Myriapoda); and the Diplura and 'Thysanura' (Insecta). However, the Myriapoda must have specialized in a cryptic mode of life very early: it is hardly possible to explain the loss of the median eyes (ocelli) and the extensive reduction of the compound eyes in any other way. The increased number of body segments and the associated elongation of the body itself are particularly striking features in the subsequent development of the Myriapoda. The start of this development can be seen in the otherwise primitive Symphyla. However, in the Opisthogoneata and the Progoneata it must have taken place independently, as a parallel development.

All the discussion will now be focused on the insects, which, as the sister-group of the Myriapoda, retained a much less specialized mode of life and also preserved various ground-plan characters which were lost in the Myriapoda. As a result, the right prerequisites for the origin of wings were found only in the insects, and this was the reason for their subsequent success.

A reduction in the number of posterior body appendages is a derived character of the Insecta. Appendages were retained as walking legs only on the first three post-cephalic segments, and as a result of this the body was divided still further. The original division into head and body was retained from the ground-plan of the Tracheata, and the Insecta were the only group in which the body was subdivided into two sections, the thorax (3 segments) and the abdomen (11 segments and telson). According to Sharov (1966b), the appendages of the abdominal segments ('styli' = telopodites = endopodites) were retained at first as supporting organs, because the legs could raise only the anterior part of the body off the ground and not the long diffuse abdomen: like tail-skids, they reduced friction as the abdomen was dragged over the ground.

The organs of locomotion and their muscles were contained in the thorax, and some additional vital functions in the abdomen: this was of great importance for the subsequent development of the insects.

According to Snodgrass (1958b) the insects also differ from the Myriapoda and Crustacea by having only six leg segments (including the coxae). Sharov (1966b) has published a very convincing illustration which shows that the two basal segments, the pre-coxopodite and coxopodite, form part of the body wall as the pretrochantin (i.e. 'pleura') and trochantin (see Fig. 18). The first free leg segment of the Insecta ('coxa') is homologous with the basipodite of the Crustacea. I agree with this interpretation of Sharov's, and not with that of Siewing (1960).

Exopodites ('coxal styles') are retained on all legs in the Symphyla and so appear to be part of the ground-plan of the Tracheata, but in the Insecta they have only been preserved in the Archaeognatha (Machilidae) where they are found on the last one or two pairs of legs (Sharov, 1966b). As far as I can ascertain, nothing is known about the function of these 'coxal styles' in the insects. Moreover, they always lack muscles.

The reduction of abdominal legs in the insects has sometimes been interpreted as an example of 'neoteny', for example by de Beer[114]. A great deal of harm has already been caused by this dubious concept. I think that Sharov (1966b) and other authors were right to reject 'neoteny' as an explanation for this phenomenon.

The cerci are another derived character of the insects, and it is generally accepted that they originated from the appendages of the 11th abdominal segment. Great importance has often been attached to the similarity between the 'cerci' (adenopodites) of the Symphyla and the cerci of many insects, which are also provided with spinning glands, but at the same time the presence of similar cerci in the Diplopoda has been overlooked. Daiber (1913) has referred to *Craspedosoma* in this context, and Verhoeff (1926) found similar cerci which he called 'pygopodites' in an entire section of the Diplopoda, the suborders Proterandria and Nematophora. Verhoeff thought that these pygopodites, like the adenopodites of the Symphyla, were 'appendages of the anal segment which had moved to a dorsal position or perhaps had always been there'. Ravoux (1948) has questioned whether the adenopodites are in fact appendages at all, and Snodgrass (1952) doubted whether they are homologous with the cerci of the insects. The close similarity between the adenopodites of the Symphyla and the non-segmented cerci of the Japygidae (Diplura) does not seem to be particularly important. There is no longer much doubt that the Entognatha and Ectognatha are sister-groups. The similarities between the long, multi-segmented filiform cerci present in the Ectognatha and the majority of the Diplura, which also appear to be the most primitive Entognatha on the basis of other characters, are too great to be regarded as the result of convergence, although they would have to be if the short unsegmented cerci of the Japygidae are considered to be primitive. These long, filiform, multi-segmented cerci are present in most of the Diplura (Campodeidae and Projapygidae), all the primarily wingless insects, and many primitive winged Ectognatha (Ephemeroptera, Plecoptera): they must be a derived ground-plan character in the ground-plan of the Insecta, and must have been retained even after the Insecta divided into Entognatha and Ectognatha. Their simple structure in the Japygidae (*Anajapyx*: Projapygidae) must therefore be a secondary character.

It is still not known whether a paracercus (terminal filament) is part of the ground-plan of the insects as a 'third caudal filament'. This is a matter of importance for the interpretation of certain fossils (Monura) and will be discussed in the section dealing with the ground-plan of the Ectognatha.

There is also some dispute about the external genitalia of the 8th and 9th abdominal segments in the insects. Matsuda (1958) did not accept that they have anything to do with the appendages. However, it is difficult to believe that some existing structures in both sexes were not used when these organs were being formed, and these can only have been primordia of the appendages. This was also Sharov's (1966b) view[115].

Revisionary notes

104. Chapter 3 is discussed and complemented by Kristensen (1975) and Boudreaux (1979). Dr N. P. Kristensen has informed me that he is preparing a review of insect phylogeny for the 1981 volume of *Annual Review of Entomology* (26).

This opening section, on the ground-plan and origin of the Insecta, should be read in conjunction with Boudreaux (1979) and Gupta (1979). [Adrian Pont.]

105. In the Diptera, the heart has 8 ventricles, and there are 8 pairs of ganglia and 8 pairs of spiracles. We still need to establish what the number of these organs was in the ground-plan of the Insecta and what the number was from which the reductions (?) in the Diptera originally arose.

Dahl (1969) has discussed the insects as arthropods and Marcus (1958) the respiratory organs of the Tracheata. Attention should also be paid to Manton's (1973, 1977) most recent views. [Willi Hennig.]

106. See Hoffmann (1964) and note 107. [Willi Hennig.]

107. Hoffmann (1964), like Eggers before him, has stressed that chordotonal organs appear to be completely absent in the Myriapoda and Chelicerata, whereas they are present in crabs and insects where 'even their anatomical details are in close agreement'.

It may be that chordotonal organs are a ground-plan character of the Antennata (Mandibulata). Their absence (reduction!) could then be regarded as a derived ground-plan character of the Myriapoda! Since the Notostigmophora and the fossil Progoneata *still* have compound eyes but no median eyes (ocelli), it would be possible to interpret the simple eyes of the majority of the Myriapoda simply as compound eyes that have broken up into single facets and to treat the absence of ocelli (median eyes) as a derived ground-plan character. See also Paulus (1972a, 1972b, 1974), who has discussed the structure and function of the ocelli in the Mandibulata and the 'Apterygota'. [Willi Hennig.]

108. For a discussion of the mandibles of the Myriapoda, see Lauterbach (1972a). [Willi Hennig.]

109. A discussion of the key events in the evolution of the Euarthropoda has been given by Lauterbach (1973). [Willi Hennig.]

110. Mariammal and Sundara Rajulu (1975) have concluded that there appears to be a serological relationship between the Insecta and Myriapoda, and between the Crustacea and Chelicerata. However, it could be suggested that the first two groups are not closely related to the other two, and that there are no grounds for recognizing the taxon 'Mandibulata' (= Insecta + Myriapoda + Crustacea). [Michael Achtelig.]

111. Lauterbach (1972a) has argued convincingly against this view. He has shown, for example, that the mandibular palp is absent in certain Crustacea and can also be suppressed during ontogeny in the Tracheata. [Michael Achtelig.]

112. The fore legs of Crustacea Malacostraca and Insecta have been discussed by Carpentier and Barlet (1959), and the respiratory organs of the Tracheata by Marcus (1958).

A further question is whether the Tracheata have dispensed with the pelagic larval stage or whether this larval stage (nauplius) is a derived character of the crabs. The latter appears to be the case, because the Chelicerata too have no pelagic larva and because the nauplius eye is a derived character of the crabs. [Willi Hennig.]

113. See Marcus (1958) for respiration in the Tracheata. [Willi Hennig.]

114. This is paedomorphosis rather than neoteny! (Insects could have arisen from polypod larvae.) In my view the application of the term paedomorphosis to reductive characters is very dubious. There are no ocelli in Myriapoda (or larvae), but the reverse is the case in insects. [Willi Hennig.]

115. See also Smith (1969). [Willi Hennig.]

B. The history of the Insecta[116] from their origin to the close of the Palaeozoic

The stem-group of the Insecta has not yet been found in the fossil record. The oldest insects are from the Middle and Upper Devonian, and they belong to various groups that are still in existence. We need to know what are the relationships amongst the oldest monophyletic groups of insects, so that we can make a correct assessment of these fossils, and this can be achieved only by making a careful analysis of the heterobathmy of their characters.

There is no longer any good reason for doubting that there is a sister-group relationship between the Ectognatha and Entognatha. The Entognatha and the primitive Ectognatha used to be grouped together as the 'Apterygota', but this name can now only be used as a term of convenience to cover all the primarily wingless insects and not as the name for a monophyletic group.

At the 10th International Congress of Entomology (1958), papers were read by Boettger, Handschin, and Tuxen on the relationships of the individual monophyletic groups of the 'Apterygota'. None of these was able to achieve complete theoretical clarity, but what they did show was that all objections to the relationships I am suggesting in the following sections are either theoretically unsound or lack adequate supporting evidence. Possibly the most interesting thing to emerge was that Boettger, the speaker who most emphatically supported a formal classification based on 'convenience', put forward the very ideas in his phylogenetic trees which I consider to be correct (except for the suggested relationship of the Insecta with the Symphyla and Myriapoda).

Additional note

116. Two short synopses have been published by Rodendorf (1969a, 1969b). [Dieter Schlee.] An account has also been given by Mackerras (in CSIRO, 1970). [Willi Hennig.] A book on the historical development of the Insecta has been published by Rodendorf and Rasnitsyn (1980). [Adrian Pont.]

1. Entognatha

The most important derived ground-plan characters of this group are the entognathy (retraction of the mouth-parts into the head capsule), and the reduction of the compound eyes[117] and Malpighian tubes. It is clear that the origin and development of the group was connected with specialized feeding methods and that these resulted from entognathy. Entognathy gave rise to further peculiarities as the body size became smaller, and most of these were in the form of reductions.

Tuxen (1959) has recently made a detailed study of entognathy. He has established that in the head of the embryo a fold (the plica oralis) develops on each side between labrum and labium, and eventually fuses with the labium. As a result, one or two pouches are formed on each side and these enclose the mandibles and maxillae. These in turn are withdrawn deeper into the head capsule, and in addition they move from their original vertical position into a horizontal position. In this way the originally hypognathous head becomes prognathous. However, in the Collembola there are various species with a (secondarily?) hypognathous head. Struts that form a V- or Y-shaped structure develop on the internal margins of the pouches formed by the plicae orales, and some of the maxillary muscles are attached to this. According to Tuxen, this structure is not homologous with the tentorium of the Ectognatha. It is a special formation of the Entognatha, an autapomorphic character. However, according to Snodgrass (1958a), structures homologous with the anterior tentorial arms of the Ectognatha are present in the

Myriapoda, and so they must be part of the ground-plan of the Tracheata! If they are absent in the Entognatha, then they must have been lost secondarily (see Matsuda, 1965b). However, Gouin (1968) disagreed with Tuxen's results and asserted that the Entognatha do possess a true tentorium.

There are also some peculiarities in the mouth-parts of the Entognatha, but it is still difficult to assess clearly their extent and importance. I still reject Sharov's (1966b) suggestion that the entognathy of the Entognatha has arisen polyphyletically, as there is absolutely no evidence for this[118].

There appear to be two sister-groups in the Entognatha, the Diplura and Ellipura, both of which originated before the Middle Devonian.

Additional notes

117. The phylogenetic significance of the ommatidia in the 'Apterygota' has been dealt with by Paulus (1974). Lauterbach (1972b) has discussed the origins of entognathy. [Willi Hennig.]

118. Janetschek (1970) has supported the polyphyletic origin of entognathy. [Willi Hennig.]

1.1. Diplura

The Diplura include about 400 recent species, according to Denis (in Grassé, 1949). They have retained unaltered a comparatively large number of characters from the ground-plan of the Entognatha: multi-segmented antennae; a relatively well preserved labium; the primitive number of 12 abdominal segments (including the telson); the full complement of styli on the abdominal segments; a heart with 9 ostia; division of the tarsus and tibia; paired pretarsal claws (as well as a single median claw); and the full number of thoracic and abdominal spiracles.

This is a large number of primitive characters and could well confirm the initial impression of considerable overall similarity with the Archaeognatha and Zygentoma, both of which belong to the Ectognatha. This similarity led earlier authors to consider that these three groups were closely related and to call them the 'Thysanura'.

There are several derived characters that indicate that the Diplura are monophyletic: the complete absence of compound eyes and ocelli; and the fusion of the abdominal 'coxopodites' (the basal three leg segments) with the sternal plates to form coxo-sternites. These also occur in other groups but have apparently arisen independently.

1.2. Ellipura

Several derived characters are present in the Ellipura Börner (= Panpro-tura Crampton). The antennae are reduced, and at the most consist of four segments (Collembola). Only the small triangular part of the labium bearing the palpi is preserved, because the lateral parts of the cranium have fused together below, resulting in labial degeneration (Tuxen, 1958). The tracheal system has been reduced: in particular, the abdominal spiracles are all absent. The cerci and the Malpighian tubes have been completely reduced. The tibia and tarsus are usually fused on at least the 2nd and 3rd pairs of legs. The pretarsus has been modified and has only one distinct claw. Reduction in the number of abdominal segments has begun.

Some of these characters, if not all of them, appear to be correlated with the reduced body size, and some of them need to be examined in more detail. For example, in the Collembola the number of abdominal segments has been reduced to six. This is probably a case of 'phylogenetic dilation' in the sense of Weber or of partial neoteny. Formation of the segments during embryonic growth is so slow when compared with the development of other characters that the full number is not acquired until the individual becomes sexually mature or approaches the end of its life. Some support for this idea is provided by the Protura, the sister-group of the Collembola, where the full number of twelve abdominal segments (including the telson) is acquired only during postembryonic growth ('anamorphosis'). In newly hatched individuals only nine abdominal segments are present. Lameere (1935–36) has suggested that abdominal segments 9–11 are not true segments, homologous with segments 9–11 in other insects, but that they have arisen through secondary division of the 8th segment. This interpretation appears to have been overlooked by subsequent workers but would make it possible to reconcile the position of the gonopore in the Protura and Diplura: in the Diplura, as in the Ectognatha, it is situated behind the 8th segment but in the Protura it is behind the 11th segment. According to Lameere's interpretation, the gonopore is in fact in the same position in all these groups because the 11th segment of the Protura, as well as the 9th and 10th segments, is simply a subdivision of the 8th segment. On the other hand, there is a striking agreement in the number of segments present in the Protura, at the conclusion of their postembryonic growth, and in the ground-plan of the Insecta or even of the Tracheata. Lameere's theory may be based on too rigid a concept of homology, since it only takes account of the 'criterion of position', which in this form is particularly unsuitable in serially homologous organs. In the embryo, the abdomen acquires its segmentation through division of the part in front of the telson, and it may be a matter of pure chance whether this takes place in front of the gonopore or behind it. Suppression of the segmentation in the region immediately in front of the telson is in fact the start of a process of phylogenetic development that culminates in those Collembola where the formation of segments ceases 'prematurely'. The formation of segments has been almost entirely suppressed in one group of the Collembola, the Symphypleona.

The Ellipura have only a single claw in the pretarsus. In addition there is a so-called 'empodial appendage'. Tuxen (1958) considered this to be a primitive character and an earlier stage in the formation of the pretarsus in the Myriapoda(?) and other insects. However, it seems much more likely that it is a rudiment of the 'unpaired claw', that is of the distal section of the dactylopodite, as it is in the Ectognatha(!). Denis (in Grassé, 1949) has reported that Ewing (1928) and earlier authors believed something like this. The single 'claw' of the Ellipura would therefore have to be the result of fusion of the paired claws. These claws are certainly part of the ground-plan of the Insecta, and probably of the Tracheata too, and in the Entognatha they have been retained by the Diplura. The structure of the pretarsus in the Ellipura would therefore be a derived character and would have to be seen as a further development from the more primitive stage that is still retained by the Diplura. I think that this is very probable, but it needs to be confirmed by a comparative functional–morphological study of the legs of the primitive Tracheata and Insecta.

There are two monophyletic subgroups of the *Ellipura, the Protura and the Collembola, and these have a sister-group relationship.

1.2.1. Protura[119]

The Protura include about 140 species, according to Tuxen's (1963) catalogue. In their ground-plan they are more primitive than the Collembola, particularly in the segmentation of the abdomen; the presence of free coxopodites on the anterior abdominal segments; the more extended nervous system; the presence of a tibio-tarsal articulation on the fore legs; and probably the less reduced tracheal system.

Their derived characters are the small head (microcephaly); the complete absence of compound eyes, ocelli and antennae (the 'pseudoculi' may be rudiments of the antennae, according to Tuxen (1958) but not according to others); the development of external genitalia in a form peculiar to this group (Tuxen, 1958); and some other less striking features (transfer of the function of the antennae to the fore legs, etc.). These characters provide strong evidence that the Protura are monophyletic, though this has never been questioned.

Additional note

119. New monographs have been published by Tuxen (1964) and Janetschek (1970). [Dieter Schlee.] The morphology of the head has been discussed by François (1969), and the legs and phylogeny by Rusek (1974). [Willi Hennig.]

1.2.2. Collembola[120]

The Collembola include about 2000 recent species, according to Sabrosky (1953). They are more primitive than the Protura in having ocelli, 'compound eyes' and antennae, as well as certain other characters which can be interpreted as primitive states of the derived characters of the Protura (see above). In fact, the antennae have no more than four segments, and the 'compound eyes' consist of no more than eight facets which are loosely fused to form an 'eye spot'.

In comparison with the Protura, the derived characters of the Collembola are the reduction of the abdomen to six segments; the absence of the tibio-tarsal articulation, even on the fore legs; the concentrated nervous system which consists of only three thoracic ganglia and in which, according to Denis (in Grassé, 1949), all the abdominal ganglia have fused with the metathoracic ganglion; the complete reduction of the tracheal system, which is occasionally replaced by a new formation; and the specialization of the rudimentary appendages on the 1st, 3rd, and 4th abdominal segments, which are the only abdominal appendages to be retained in both sexes. The 'coxites' on all three pairs of appendages have fused together medially to form 'syncoxae'. On the 1st abdominal segment they form the ventral tube, which has the coxal vesicles at its free end; on the 3rd segment, the retinaculum (tenaculum) whose two branches are homologous with the styli; and on the 4th segment, the manubrium of the spring-tail (furca) whose styli branch into dens and mucro[121].

Apparently Wille (1960) considered the form of the Malpighian tubes to be primitive: they are represented by small papillae, as they are in many Diplura. However, it is more probable that this is a derived (reduced) character. The same may be true of the gonads. Wille (1960) also reported that the Collembola differ from all other insects in having the germaria of the ovaries and of the testes in a lateral rather than an apical position. The eggs are holoblastic with a complete cleavage.

Additional notes

120. A new monograph has been published by Schaller (1970). [Dieter Schlee.] The fine structure of the ocelli of the Collembola, together with their significance for the phylogeny of the Insecta and the Mandibulata, has been dealt with by Paulus (1972a, 1972b). [Willi Hennig.]

121. For the musculature of the abdomen, see Pistor (1955) and Pistor and Schaller (1955). [Willi Hennig.]

Appendix: fossil Entognatha

This detailed discussion of the presumed relationships amongst the monophyletic groups of the Entognatha has been necessary because no adequate assessment of the fossils could be made without it.

Rhyniella praecursor Hirst & Maulik (1926), from the Middle Devonian of Scotland (Figs. 19 and 20), is the oldest known insect, alongside *Rhyniog-*

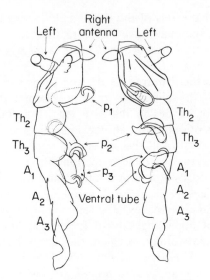

Fig. 19. *Rhyniella praecursor* Hirst
& Maulik (Collembola), from the
Middle Devonian of Scotland. From
Scourfield (1940).

natha (see p. 118). Moreover, it is almost the only fossil entognathan, since no others are known until the Upper Cretaceous (Canadian amber) and the Tertiary. The discovery of *Rhyniella* is important because its assignment to the Collembola proves that the Entognatha as a whole, and consequently their sister-group the Ectognatha must also have been in existence during the Middle Devonian. Moreover, the Protura and Diplura must also have existed as separate groups alongside the Collembola. These are the only possible conclusions that can be drawn from what has just been said about the relationships of the Collembola, Protura, Diplura and Ectognatha (Fig. 21), and no other theory about their relationships is anything like as plausible as this one. *Rhyniella* thus indicates a time when some of the earliest divisions in the phylogenetic tree of the insects must already have been completed.

Because of the great importance of this fossil, it is perfectly understandable that Carpenter (1947, quoted by Laurentiaux in Piveteau, 1953) has demanded more definite evidence for the assignment of *Rhyniella* to the Collembola. However, the work of Tillyard (1928b), and particularly that of Scourfield (1940), leaves little doubt that this assignment is correct[122]. Admittedly, it is true that the abdomen is incomplete and that some of the most important diagnostic and constitutive characters of the Collembola, such as the retinaculum and furca, cannot be definitely seen. However, the 4-segmented antennae exclude *Rhyniella* from any other group of primitive insects, except for the stem-group of the Ellipura which may well have had 4-segmented antennae. Even if we were to use very stringent criteria and

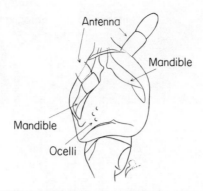

Fig. 20. Head of *Rhyniella praecur-sor* Hirst & Maulik. From Scour-field (1940).

regard the alleged traces of the ventral tube and of the retinaculum in *Rhyniella* as uncertain (Jeannel in Grassé, 1949), we would still have to admit that it could belong to the stem-group of the Ellipura. The mandibles described by Tillyard (1928b) can only be those of an entognathan.

It would not be necessary to make any great changes in our ideas about the latest possible time for the first divisions in the phylogenetic tree of the insects even if *Rhyniella* were to be assigned to the stem-group of the Ellipura, especially as the discovery of a winged insect in the Upper Devonian proves that the Ectognatha too must have separated into several subgroups by the Devonian. If Jeannel (in Grassé, 1949; based on Tillyard) was correct in his analysis of the development of the antennae in the Insecta, then the conclusions which can be drawn from the low number of antennal segments in *Rhyniella* would also be untenable. However, Jeannel's views are only true of insects with annulated antennae (Ectognatha). In the Entognatha, which have segmented antennae, the multi-segmented antennae of the Diplura must be regarded as primitive and the 4-segmented antennae of the Collembola as reduced.

Some authors have even assigned *Rhyniella* to one of the subgroups of the Collembola[124]. Delamare Deboutteville (discussant's comment on Hand-schin, 1958) considered that *Rhyniella* belonged to the Pseudachorutini, and he drew the logical conclusion that the other recent subgroups of the Collembola must already have 'separated' by the Devonian. Denis (in Grassé, 1949) believed that *Rhyniella* was a species of the Arthropleona where the prothorax also appears to be slightly reduced. Finally, Laurentiaux (in Piveteau, 1953) also assigned *Rhyniella* to the 'suborder' Arthropleona and thought that it agreed with the recent Entomobryomorpha in having the first two thoracic tergites fused. He assigned it to a group 'Protentomo-bryomorpha', together with the genus *Protentomobrya* described from Cana-dian amber (Upper Cretaceous). Martynova (in Rodendorf, 1962) has followed Laurentiaux. However, there is certainly no evidence for such an

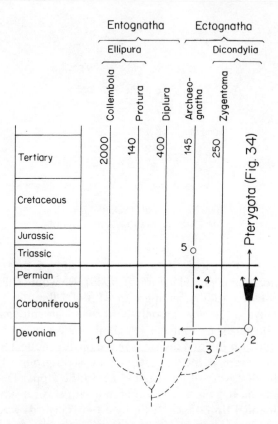

Fig. 21. The earliest events in the history of the insects. The arrows from the individual fossils intersect the monophyletic groups whose synchronous existence can be inferred from these fossils. 1, *Rhyniella*; 2, *Eopterum*[123]; 3, *Rhyniognatha*; 4, Monura; 5, *Triassomachilis*.

assignment, and it is misleading. The generally accepted division of the Collembola into Arthropleona and Symphypleona is a typological one and should therefore not be used for interpreting fossils. The Symphypleona are probably a monophyletic group. On the other hand, the diagnostic characters of the Arthropleona, when compared with those of the Symphypleona, are all plesiomorphic, and it must be assumed that they were present in the stem-group of all the recent Collembola. The assignment of fossils to the Arthropleona is therefore meaningless.

It will be necessary to make a careful division of the Collembola into monophyletic subgroups, and to define the sister-group relationships amongst these groups, before any proper assessment can be made of *Rhyniella* or of any other fossils that may be found. For several reasons it is clear that by the Middle Devonian the insects had already separated into several of the monophyletic groups that are still in existence. This could even be true of the Collembola, but so far there is no reliable evidence for this. Tillyard's view, put forward when the first imperfect fossils were found, was that *Rhyniella* belonged to the stem-group of the Collembola ('Protocollembola'), and this view has never been definitively refuted.

The genus *Rhyniognatha*, which Martynova also assigned to the Protentomobryomorpha, is almost certainly not a member of the Entognatha (p. 118).

Revisionary notes

122. A re-assessment of *Rhyniella* has been made by Massoud (1967) and Delamare Deboutteville and Massoud (1967), who have assigned it to the recent family Neanuridae. As already mentioned (note 48), Crowson (1970) has expressed some scepticism over the true age of this fossil and has considered it to be a more recent intrusion into Devonian rocks. [Willi Hennig.] *Rhyniella* is now considered to be Lower Devonian: see note 57. [Adrian Pont.]

123. Fossil no. 2 (*Eopterum*) and its arrow in the diagram should be deleted since it is now known that *Eopterum* is part of a crustacean and is not an insect at all. [Dieter Schlee, Rainer Willmann.]

124. See note 122. [Willi Hennig.]

2. Ectognatha

a. Ground-plan and general remarks on the classification

There are not many derived characters in the ground-plan of the Ectognatha: annulated antennae[125], a more fully developed tentorium, and a female ovipositor. It is difficult to say whether the paranota of the three thoracic segments and the paracercus are also part of the ground-plan. The first step towards the development of an embryonic membrane may be a further character (Sharov, 1966b).

Imms was the first to recognize that annulated antennae differ from segmented antennae in having two sharply differentiated sections: a basal segment (scape) which still contains muscles, and a distal section which does not. As the absence of muscles is characteristic of the apical segment of primitive segmented antennae, it may well be that the Ectognatha are derived from a stem-form in which the antennae were reduced to two or, more probably, three segments. In this case the multi-segmented antennal flagellum present in the most primitive Ectognatha would have developed by secondary annulation of the last of these segments, which naturally contained no muscles. The absence of muscles in the 2nd antennal segment (the 'pedicel') can be explained by the development of Johnston's organ, which has displaced all the muscles here (Snodgrass, 1958b). It would therefore be most appropriate to use the term 'flagellum' for the 3rd and subsequent antennal segments, and not to include the pedicel as part of the flagellum as has sometimes been done. This distinction may be important for the interpretation of the Palaeozoic fossils.

It is still not certain what form the tentorium took in the ground-plan of the Ectognatha. It now consists of two endoskeletal apodemes which originate ontogenetically from invaginations in the wall of the head capsule. According to Snodgrass (1958a), the two 'anterior arms of the tentorium' are present in the Myriapoda and did not originate in the Ectognatha. The homology of the 'anterior arms of the tentorium' in the Myriapoda and Ectognatha has been disputed by other authors (see Matsuda, 1965b), because they are absent in the Entognatha. As sui generis structures they would then be an apomorphic ground-plan character of the Ectognatha. The posterior arms of the tentorium, which have fused together and form a single transverse apodeme at the back of the head capsule, are characteristic of the Ectognatha and are apparently a derived ground-plan feature of the group. In the Dicondylia, which include most of the Ectognatha, the anterior and posterior arms of the tentorium are linked by a median longitudinal apodeme, the corpotentorium, and form a single X-shaped structure. The corpotentorium is absent in the Archaeognatha and, like Verhoeff, Snodgrass (1960) considered this to be the primitive state. On the other hand, Denis (in Grassé, 1949) believed that the corpotentorium had been lost secondarily in the Archaeognatha and that this was connected with the displacement of the compound eyes and the antennae. However, there are some Triassic and possibly Carboniferous fossils which may belong to the Archaeognatha and which have the compound eyes and antennae in their normal position: this problem could be solved if the endoskeleton of the head in these fossils could be examined, but it is unlikely that this will ever be possible.

The paranota are lateral expansions of the three thoracic terga, from which the wings of the Pterygota have developed. There is some doubt whether they are newly acquired, and therefore derived, characters of the Ectognatha. According to Sharov (1966b), flexible paratergal lobes are part of the ground-plan of the Arthropoda. Similar structures are found in Palaeozoic Pterygota, on the abdominal segments as well as on the three thoracic segments, and so it is possible that paratergal lobes are part of the ground-plan of the Insecta. They would then have been completely reduced in the Entognatha, but only reduced on the abdominal segments in the Archaeognatha and Zygentoma. However, it is not certain that the paranota really are homologous with the paratergal lobes of the arthropod ground-plan.

According to Snodgrass (1958b), the female ovipositor originated amongst the ancestors of the Machilidae, which at the same time were the ancestors of all the Ectognatha. It is therefore a derived character in the ground-plan of the Ectognatha. The generally accepted view is that the appendages of the 8th and 9th abdominal segments formed the basis for the three pairs of valves from which the ovipositor was formed. This view was also shared by Scudder (1961, 1964) and Sharov (1966b), but Matsuda (1958) has challenged it, possibly because he has overrated the embryological evidence. It is difficult to imagine that such an important organ, which arose so early in the history of the Insecta, could have originated as something completely new without making use of existing structures. However, our assessment of the ovipositor as a characteristic ground-plan feature of the Ectognatha is not influenced at all by the way in which it may actually have originated morphogenetically. According to Sharov (1966b), the 1st and 2nd pairs of valves are homologous with the epipodites of the 8th and 9th abdominal segments; the 3rd pair of valves, which is also part of the 9th abdominal segment, has been firmly absorbed into the structure of the ovipositor in the Pterygota, and this has been achieved in various ways throughout the group.

Compared with their sister-group, the Ectognatha have been an outstandingly successful group and this prompts the question as to the reasons for this success. It can hardly be explained by the very modest number of derived ground-plan characters. It is much more likely that the earliest species in the direct ancestral line of the *Ectognatha ensured the success of their descendants because they retained the primitive characters of the insect ground-plan and did not adopt a specialized mode of life. The direct ancestors of the *Entognatha did this, and it led to an irreversible specialization of the mouth-parts and the reduction of various organs.

It is clear from the subsequent development of the group that the paranota were the most significant acquisition of the Ectognatha—if we could be certain that they were actually a new acquisition and were not derived from the paratergal lobes which Sharov (1966b) considered to be part of the ground-plan of the Euarthropoda. In this case, the question would be why they were retained in some groups but lost in others.

It is also not clear whether the 3rd single 'caudal filament' (paracercus, terminal filament) in the Ectognatha is a new acquisition. According to Sharov (1966b), there is no doubt that a long flagellate 'telson' was present in the 'first Myriapoda' and also in the stem-group of the Insecta. Its function was locomotory, like the long tail of the reptiles. Sharov's view seems to be that the paracercus was reduced in the Entognatha, but was retained in the Ectognatha from the ground-plan of the Insecta.

In the Archaeognatha, which in many ways are the most primitive Ectognatha, the paracercus is a jumping organ[126]. It would be useful to establish whether this was its original function in the ground-plan of the Ectognatha. The paranota would then have been important for gliding back to earth after a leap into the air. The Zygentoma do not jump, although they too have a paracercus. Unlike the Archaeognatha, however, they are secretive creatures in which the ocelli and the compound eyes have been extensively reduced: it is therefore conceivable that their ability to jump was lost secondarily. The direct ancestors of the Pterygota, and consequently the common ancestors of the Zygentoma and Pterygota, can hardly have led such a secretive mode of life since there would have been no incentive to fly. Certain Pterygota, such as the Ephemeroptera(!), have retained the paracercus although they do not use it for jumping. This is obviously because there was no longer any need to jump once the ability to fly had been acquired. The idea that the ability to jump was one of the original features of the Ectognatha, and that the origin of both the paracercus and the Archaeognatha was correlated with this, need not be rejected because the Archaeognatha are the only group that has retained the ability to jump. This is all pure speculation, but it may have a certain heuristic value by stimulating a very specific investigation into this problem.

Two sister-groups can probably be recognized in the Ectognatha, the Archaeognatha and the Dicondylia.

The Archaeognatha† are more primitive than the Dicondylia because their mandibles have only the single original condyle which is part of the ground-plan of the Mandibulata. The femoro-tibial articulation also has only a single condyle (Sharov, 1966b). On the other hand, recent Archaeognatha (families Meinertellidae, Praemachilidae, Machilidae, according to Denis, in Grassé, 1949) have several derived characters which indicate that there was an ancestor common only to this group and that the group is monophyletic: the antennae are close together, and the compound eyes touch along the dorsal mid-line of the head; the spiracle of the 1st abdominal segment is absent; the maxillary palp has seven segments, which Denis (in Grassé, 1949) believed to be the highest number recorded in the Insecta[127]. It has often been stated that there is no anastomosis in the tracheal system of the Machilidae, but according to Hinton (1958) this is incorrect: there are in fact connections between the right and left sides of the body as well as between the individual segments.

The only derived ground-plan characters of the Dicondylia are the presence of a second condyle on the mandibles, a second condyle in the femoro-tibial articulation (Sharov, 1966b), and the structure of the female ovipositor.

According to Scudder (1961), the fundamental new character in the ovipositor of the Dicondylia is the gonangulum, and its role is to effect the movements of the two pairs of gonapophyses relative to each other. According to Scudder it has arisen from the second gonocoxa: the anterodorsal angle of the second gonocoxa 'separates' from the rest of the second gonocoxa, and connects with the first gonapophysis, the 9th tergum and the main part of the second gonocoxa. On the other hand, Sharov (1966b) interpreted the gonangula as vestiges of the 9th abdominal sternum: the second pair of valves moved forwards and its base divided the 9th sternum. It does not matter which view is correct, for the important role of the gonangulum in the

---

† Ross (1965) used the name Microcoryphia because Verhoeff's paper proposing this name was published on 22 April, 1904, a few weeks earlier than Börner's paper (3 May, 1904) proposing the name Archaeognatha.

structure and function of the ovipositor shows that it should be regarded as a derived character of the Dicondylia. According to Sharov (1966b), soft-shelled and malleable eggs are necessary for the ovipositor to function. The ovipositor is always reduced in insects that have hard-shelled eggs.

The development of a second condyle on the mandibles has led to a profound change in function and a drastic reorganization of the muscles (see below), and as long as there are no cogent reasons for thinking otherwise there can be no question of it having arisen convergently. It has sometimes been stated that the mandibles of ephemeropterous larvae are still monocondylic (see Ross, 1965). However, according to von Kéler [1956, under the entry 'Oberkiefer' (upper mandible)], this is not true: 'In ephemerid larvae, the primary position of the pleurostomal basal margin of the mandible has moved forward because the anterior condyle is placed deeper (in lateral view) than the posterior one, whereas in all other Pterygota the condyles on the biting mandibles are at the same level'. In other respects the primitive Dicondylia (Zygentoma, of which the family Lepismatidae is best known) are so similar to the Archaeognatha that both are still usually included as the 'Thysanura'. When this name is used, it is not clear whether the author regards the 'Thysanura' as a monophyletic group, which must imply that he believes that the second condyle of the mandibles, the second condyle in the femoro-tibial articulation, and the gonangulum in the ovipositor have all arisen convergently; or whether he is using the name 'Thysanura' in a purely typological sense, for a group of species with very great overall similarity, even though he may believe that the Zygentoma are more closely related to the Pterygota than are the Archaeognatha.

In view of the small numbers of recent species in the Archaeognatha and Zygentoma, this distinction may appear to be very trivial, but it is actually of some importance for the appraisal of the fossils.

Revisionary notes

125. Callahan (1975) has made an important study of insect antennae and their sensilla. [Willi Hennig.]

126. According to Weidner's revision of Weber (1974), the terminal filament is not a jumping organ. The Archaeognatha can jump just as well when it has been amputated. In the light of this, my remarks on the development of the ability to fly need to be revised.

Bitsch (1974) has discussed the function and ultrastructure of the eversible abdominal sacs of the Machilidae, and Weyda (1974) the coxal sacs. [Willi Hennig.]

127. The bee *Andrena grossella* Grünwaldt also has a 7-segmented maxillary palp. [Erberhard Königsmann.]

b. The oldest fossils and their systematic position

A single winged insect (*Eopterum devonicum*)[128] is known from the Upper Devonian and it can only be assigned to the Pterygota. This means that some primarily wingless Ectognatha must also have been in existence at the same time and should be regarded as the direct ancestors of the recent Archaeognatha and Zygentoma, and of only these two groups. There is now little doubt that the Dicondylia are a monophyletic group and that the Zygentoma are therefore more closely related to the Pterygota than are the Archaeognatha. Consequently, there must have been at least three separate lineages in the Upper Devonian, the Archaeognatha, Zygentoma, and Pterygota (Fig. 21). On the other hand, if the Thysanura are considered to be a monophyletic group, then the logical conclusion is that the second condyle on the mandibles, the second condyle in the femoro-tibial articulation, and the gonangulum must be convergent characters in the Zygentoma and Pterygota: in this case the discovery of a pterygotan in the Upper Devonian would only prove that a more extensive lineage, the Thysanura, was in existence. It would be impossible to say whether the ancestors of the recent Archaeognatha and Zygentoma had already separated.

Fossils belonging to the primarily wingless Ectognatha are not known from the Devonian, but two species of the genus *Dasyleptus* have been described from the Upper Carboniferous (Commentry) and the Lower Permian (Kaltan in the Soviet Union: Fig. 22). Sharov (1957d) based the 'order Monura' on these species, which he considered to be close to the 'Thysanura'. It is clear from his discussions and diagrams (Sharov, 1959b and subsequent papers) that he is a follower of typological systematics. He has never discussed whether the Zygentoma or the Archaeognatha are most closely related to the Pterygota, as he has not considered it to be a relevant problem. Consequently, he has failed to make use of the Monura to solve certain intractable problems. His conclusions are vague because he has only compared the 'Monura' with the 'Thysanura', which are probably a paraphyletic group. Unfortunately, it is unlikely that it will ever be possible to determine from the fossils themselves whether the mandibles of the Monura had one or two condyles. Sharov himself has neither posed nor answered this question in any of his papers. Consequently, it is necessary to examine other characters.

Sharov (1957d) has listed the following characters in which the Monura are more primitive than the recent 'Thysanura' (= Archaeognatha + Zygentoma): greater development of the rudiments of the maxillary and labial segments; more markedly homonomous legs, which have a closer resemblance to those of immature 'Thysanura'; narrower abdominal tergites, which are not curved round on to the ventral surface and do not overlap with the edges of the 'sternites'; coxal styles on nine abdominal segments; greater development of the 14th body segment, which is not fused with the terminal filament (paracercus); narrow 1st body segment.

These primitive characters could support the assignment of the Monura to

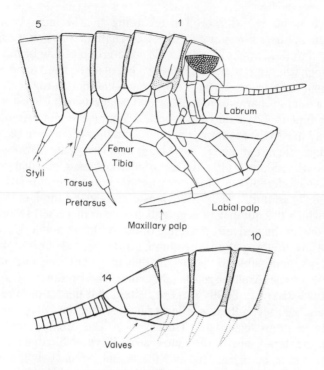

Fig. 22. *Dasyleptus brongniarti* Sharov ('Monura', probably Archaeognatha), from the Lower Permian of the Kuznetsk basin. From Sharov (1966b).

the stem-group of the Ectognatha. This assignment need not be rejected simply because they also have some strikingly derived characters (autapomorphies) which completely rule out the possibility that they could have given rise to certain recent insects: reduction of the cerci and the structure of the tarsus, which is unsegmented and has one simple claw. Sharov considered this to be a primitive character, but paired claws and a segmented tarsus (with three segments) are probably part of the ground-plan of the Insecta, and perhaps even of the Tracheata (see above). They are certainly in the ground-plan of the Ectognatha. Consequently, the Monura could only be a specialized side-branch of the stem-group of the Ectognatha, which survived until the Permian and then became extinct. However, the primitive characters of the Monura do not necessarily prove that they *must* have belonged to the stem-group of the Ectognatha. They are so trivial that it is easy to imagine them developing into more derived states convergently. Wygodzinsky (1961) has shown that several relatively derived characters, such as the reduction of the coxal styles, formation of coxosternites, reduction of ocelli, development of body hairs into scales, must have arisen independently on several occasions in the Archaeognatha and Zygentoma.

As there is no real reason for regarding the Monura as a surviving side-branch of the stem-group of the Ectognatha, we should examine the possibility that they might belong to the stem-group of one of the subgroups of the Ectognatha. In practice only the Archaeognatha need to be considered. The fact that the paracercus has been retained, although the cerci have been reduced (a unique character in the Insecta), suggests that it must have had an important function. It could have been a jumping organ. Amongst recent species only the Archaeognatha can jump, and they often have the paracercus much longer than the cerci. The long maxillary palps also suggest the Archaeognatha. Sharov (1957d) regarded the longer maxillary palps as a primitive character of the Monura, since they are more greatly developed than in the 'Thysanura'. He has obviously overlooked the fact that elongate palps, in which the number of segments has been increased secondarily, are probably one of the derived characters of the Archaeognatha. Unfortunately, he did not say anything about the number of palpal segments in the Monura. This makes it impossible to decide whether they could have belonged to the stem-group of the Archaeognatha. Nor does the presence of a paracercus provide any assistance. Firstly it is not certain that the Monura really used it for jumping, and secondly it is at least possible that the Dicondylia have lost the ability to jump secondarily (see above). This ability could have been present in the stem-group of the Ectognatha. It is possible that a more precise comparative morphological study of the recent Archaeognatha and Zygentoma could shed some light on this. For the present, no definite decision can be made about the position of the Monura, but it seems slightly more likely that they belonged to a specialized side-branch of the stem-group of the Archaeognatha, as Ross (1965) has shown in his phylogenetic tree. In this case Sharov was wrong to consider the tarsal characters to be primitive; moreover, the presence of only one condyle in the femoro-tibial articulation (according to Sharov) does not support the assignment of the Monura to the Dicondylia.

There is no doubt that the Monura must belong to the stem-group of one of the recent monophyletic groups. However, the 'dialectical method' recommended by Sharov will not help at all in deciding which one this is: this can only be done by carefully analysing the characters of the Monura, using the methods of phylogenetic systematics. If the fossils are so poorly preserved that no such analysis can be made, then it will be impossible to decide what is their true systematic position. By insisting that the problem can be solved by means of the 'dialectical method', Sharov is using a cheap slogan to conceal his own inability to make any progress, and this attitude will actually impede the fresh studies that are needed to solve this problem.

At present there is no better prospect for making a more adequate assessment of the Triassomachilidae (*Triassomachilis uralensis* Sharov) from the Upper Triassic of the Bashkirskaya Republic in the southern Urals, in which the compound eyes are separate and situated on the sides of the head as in the Monura. According to Wygodzinsky (1961), the Triassomachilidae resemble the recent Archaeognatha in having the coxopodites ('precoxal plates') of the abdominal segments in a sublateral position, whilst the sternites are reduced in size. This suggests that the Triassomachilidae really do belong to the stem-group of the Archaeognatha[129].

Revisionary notes

128. *Eopterum* belongs to the Crustacea: see note 57. [Dieter Schlee, Rainer Willmann.]

129. Paclt (1972) has recently discussed the relationship of *Triassomachilis* to the Archaeognatha. [Willi Hennig.]

c. Classification of the Ectognatha

2.1. **Archaeognatha (Microcoryphia)**[130]

There are two sister-groups in the Ectognatha, and the Archaeognatha have all the characters of a relic group, a 'side-branch' of the main insect lineage. Their constitutive characters have been listed above on p. 112. Apart from the Monura, which I have discussed in the preceding section, fossils are only known from the Mesozoic and they still seem to belong to the stem-group (see p. 116). I have no idea when the latest common stem-species of the recent species may have lived, and so I do not intend to discuss the group in any more detail here. There are about 145 recent species, according to Denis (in Grassé, 1949).

Revisionary note

130. Birket-Smith (1974) has dealt with the abdominal morphology of the Archaeognatha and Zygentoma.

The abdominal musculature of the Ephemeroptera is almost identical with that of *Petrobius* (Archaeognatha), whereas *Lepisma* is very different. This could be an autapomorphy. However, the same could be true of the monocondyly of the Archaeognatha!

Tuxen (1972) has given a phylogenetic tree in which the 'Thysanura' (Archaeognatha + Zygentoma) are shown as the sister-group of the Pterygota.

See also note 129. [Willi Hennig.]

2.2. **Dicondylia**
Ground-plan, classification, and oldest fossils

As I have already mentioned, only three derived characters have been found in the Dicondylia: the development of a 'gonangulum' in the female ovipositor (discussion on p. 112), a second condyle in the femoro-tibial articulation (according to Sharov, 1966b), and a second condyle on the mandibles (dicondyly). According to Snodgrass (1938) and Weber (1933), the second condyle gave the mandible a long axis of attachment with a definitely limited transverse movement of abduction and adduction, and this involved changes in the muscles (Fig. 23A). The dorsal muscles took over the main

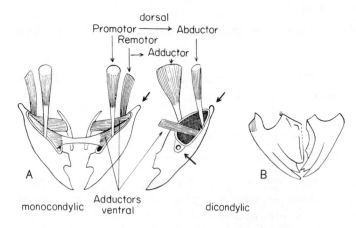

Fig. 23. A: The muscles of monocondylic and dicondylic mandibles. The thicker arrows point to the condyles. From Snodgrass (1935). B: *Rhyniognatha hirsti* Tillyard, from the Middle Devonian of Scotland. From Scourfield (1940).

function: the promotor became the abductor and the remotor became a much stronger adductor. The ventral adductors that were part of the ground-plan of the Insecta were lost: the posterior pair, which was attached to the median tendon, appears to be absent in the ground-plan of the Dicondylia, whereas the anterior pair is still distinct in some primitive species (*Lepisma*, ephemeropterous larvae, and 'locustids', according to Weber). In all other characters the Dicondylia are more primitive in their ground-plan than the Archaeognatha.

Two monophyletic sister-groups can easily be recognized in the Dicondylia: the Zygentoma and the Pterygota. The oldest fossil is probably *Rhyniognatha* from the Middle Devonian of Scotland (Fig. 23B), and this was contemporary with *Rhyniella*. It is difficult to make a realistic assessment of this species because the only organs preserved are the mandibles, which appear to be dicondylic. The next fossil is *Eopterum devonicum*, from the Upper Devonian. This belongs to the Pterygota and cannot possibly be assigned to the stem-group of the Dicondylia, and so it is clear that the two sister-groups Zygentoma and Dicondylia must already have separated by the Upper Devonian (Fig. 21)[131].

This observation sheds new light on the two controversial mandibles described as *Rhyniognatha hirsti* Tillyard, which were found in the Middle

Devonian of Scotland (Rhynie) with *Rhyniella*. Their shape suggests that they should be dicondylic, as Laurentiaux (in Piveteau, 1953) also pointed out, although this was not mentioned in the description. Moreover, they do not resemble the mandibles of the Zygentoma. The discovery of the Upper Devonian *Eopterum* provides definite(?) evidence that the Dicondylia and their two monophyletic subgroups must have arisen at an even earlier period[131]. This agrees with the conclusions that I drew from the Middle Devonian Entognatha (*Rhyniella*: pp. 106 and 36).

Revisionary note

131. These remarks are no longer valid because *Eopterum* has been shown to belong to the Crustacea (see note 57). [Dieter Schlee, Rainer Willmann.]

2.2.1. Zygentoma†[130]

The Zygentoma include about 250 recent species and should be treated rather like the Archaeognatha (p. 117). However, it is very difficult to find evidence that the Zygentoma are a monophyletic group because nearly all the characters in their ground-plan are primitive when compared with the Pterygota. The only derived character is the reduction of the compound eyes to a few primitive facets. The ocelli are usually absent, but they have been retained in *Tricholepidion gertschi* which Wygodzinsky (1961) has recently described from California. The reduction of the eyes indicates that the entire group specialized early in a more cryptic mode of life, certainly more cryptic than that of the common ancestors of the Zygentoma and Pterygota. No fossil Zygentoma are known until the Tertiary.

2.2.2. Pterygota

a. Ground-plan characters: the origin and structure of the wing

Most insects belong to the Pterygota. The one definite derived ground-plan character of this group is the presence of two pairs of wings, which have developed from the paranota of the mesothorax and metathorax. Primitive Palaeozoic Pterygota also have well developed paranota on the prothorax and even on the abdominal segments, but these must have been reduced independently on several occasions[132]. A great deal has been written about the origin of the wings. It was the

---

† Ross (1965) has restricted the name 'Thysanura' to this group. However, this name has been used in the past, and is still generally used, as a collective term for the Archaeognatha (Microcoryphia) + Zygentoma. It is therefore ambiguous, and should no longer be used as the name for a monophyletic group.

subject of an interesting discussion at a meeting of the Royal Entomological Society of London, in which many English authors took part (see Wigglesworth et al., 1963)[133]. In the course of this, several highly fanciful ideas were put forward: the original function of the paranota was to prevent the insects from sinking into water (Rees); to protect the legs (Wallwork); to act as epigamic structures by providing species-recognition signals (Alexander and Brown, quoted by Leston); to provide stabilizers (hydrofoils) when the ancestors of the insects still lived in water, and to do the same as they raced from pool to pool (Broughton). On the basis of his experience with grasshoppers, Bekker (1958, quoted by Scudder, 1964) suggested that the paranota originated in jumping insects as functional reinforcements of the outer thoracic wall. In other words, they arose in response to the increased development of the leg muscles.

All these theories, if they are to be taken seriously, deal with the origin of the paranota themselves and should really be discussed with reference to the ecological conditions that could have given rise to paranota in the stem-group of the Ectognatha: this assumes that the paranota were not derived from the paratergal plates that were already present in the ground-plan of the Euarthropoda! However, the direct ancestors of the Dicondylia and the Pterygota inherited paranota from their remote ancestors, that is to say from the stem-group of the Ectognatha or even earlier species, and so the only problem that needs to be discussed here is the nature of the conditions that led the stem-group of the Pterygota to modify their long-established immobile paranota into flapping wings.

Only one hypothesis can be seriously considered, that the direct ancestors of the Pterygota must have been using their paranota for gliding, irrespective of how early they actually acquired them. Wigglesworth (1963; in Wigglesworth et al., 1963) regarded the ecological significance of such passive gliding surfaces as facilitating dispersal over great distances. He pointed out that according to Southwood this must have been of importance principally to the inhabitants of 'temporary habitats'. The ability to migrate from localities that the adverse Devonian climate, with its enormous deserts, had made uninhabitable must have been of far-reaching importance: it was then possible for convection currents and rain-bearing winds to transport insects to more favourable habitats. Southwood and Johnson agreed with Wigglesworth to a certain extent, whilst Johnson emphasized that longer aerial journeys must have been more feasible with flapping 'wings' rather than with immobile gliding surfaces. Hinton (supported by Haskell) pointed out how important it must have been for the early insects to escape rapidly from predators. He thought that during the Devonian these must have been mainly spiders. He associated the origin of paranotal flexibility with the advantages that a controlled attitude must have had during a fall, because it enabled the insect to escape from a predator or to take flight after an accidental fall from a plant or some other raised object.

Even if we could decide whether it was the need to escape from predators or the advantages of dispersal by air currents during unfavourable climatic conditions that initiated or encouraged the origin of paranota and their subsequent active mobility, we would still have to determine how insects got into the air in the first place, for it is most improbable that air currents could have lifted them off the ground.

The most widely accepted view is that of Snodgrass (1958b), who considered that insects first developed the habit of climbing and then launched themselves as gliders from elevated positions such as plants; Hinton's view is only a variation of this. It is often linked with the origin of tall plants growing in the Devonian swamps (e.g. Zeuner, 1940a). However, Snodgrass's drawing shows that his gliding insects have cerci but no paracercus, although this character was certainly retained in the Dicondylia and then the Pterygota from the ground-plan of the Ectognatha. Handlirsch (1937) may have been right when he suggested that even the abdominal tergites originally had lateral expansions ('paranota'), because these were still present in many Palaeodictyoptera. However, this is not certain. It is therefore worth noting that Hartley (in Wigglesworth et al., 1963), and earlier Forbes (1943), suggested that insects may have become airborne by jumping. In this connection it should be borne in mind that the Archaeognatha are also able to jump: in many ways they are the most primitive Ectognatha, and they were the first to acquire paracercus and paranota. The Collembola too, which are so often found in the aerial plankton, are jumping species. Admittedly, they do not have any paranota, but they are very small when compared with their closest relatives. It seems likely that the direct ancestors of the Pterygota must have been about as large as the recent 'Thysanura', that is to say the Archaeognatha and Zygentoma, as Leston (in Wigglesworth et al., 1963) was right to emphasize. This has been confirmed by the aerodynamic experiments of Flower (1964).

The modification of the paranota on the mesothorax and metathorax into movable surfaces was the earliest event in the phylogenetic development of the Pterygota. The features that were most important in determining the outstanding success of the Pterygota during their subsequent history were the acquisition of a true articulation at the base of the paranota, the development of special flight muscles which were modified from muscles that originally had other functions, and the origin of the wing veins.

Usually it is only the wing venation that is preserved in the fossils, and even this is seldom complete. There is only one fossil in which the basal articulation is so distinct that its details can be studied (Ostrava: P. 150)[134]. The muscles are, of course, never preserved[135]. For this reason, it is of the greatest importance to study this whole complex of characters in recent insects because the fossils can often only be interpreted when the differences between the individual monophyletic groups have been worked out[136]. As the wing venation plays a special role in this, Forbes (1943) has tried to correlate its development with that of the basal articulation and the wing muscles. I think that Forbes' paper provides a good basis for further discussion of this extremely important problem, irrespective of whether his individual conclusions are correct.

According to Forbes, the basal articulation of the wing must fulfil two functional requirements: it must first provide a pivot for the vertical movements of the wing which we assume in steering; and then a longitudinal hinge to serve the wing stroke as soon as true flight is developed. These requirements were fulfilled by the shortening of the long, chitinous wing base and by the dechitinization of the line where the wing attaches to the tergum (notum) or pleura. As the pleura developed from the first pair of legs, there must already have been a membraneous area between pleura and wing, but between tergum and wing it must have been newly formed.

For the wing to be effectively mobile, it had to have a firm but articulating attachment to tergum and pleura. I think it is particularly interesting that Forbes has correlated the origin ('retention' might be more accurate) of firm attachments across the membraneous areas with the origin of the wing folds and the position of the most important longitudinal veins in relation to these folds: there has been a great deal of discussion by morphologists, palaeoentomologists, and phylogeneticists of whether it is important or even possible to differentiate between convex and concave veins (see Fig. 24).

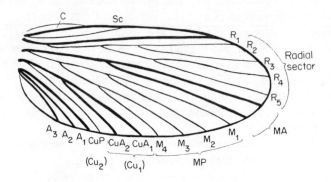

Fig. 24. Diagram of the wing veins of the Pterygota. From Séguy (1959). The 'convex' veins are shown with thick lines, the 'concave' veins with thin lines[137].

According to Forbes, the attachment to the tergum is taken by the radius (R_1) and the basal part of this has been expanded to form the radial plate which has remained attached to the tergum. As a result, the radius (R_1) has risen and become a convex vein which is situated mainly on the dorsal surface of the wing, on a convex fold. On the other hand, the attachment to the pleura is taken by the media, and so most of this is situated on the lower surface of the wing, in a concave furrow ('concave veins'). The basal plate of the media is attached to the 'pleural wing process' (Forbes' 'fulcrum') which is formed from the pleural ridge, and this process serves to resist the downstroke of the wing.

It was functionally necessary for the two fixed points in the basal articulation of the wing (radius–tergum and media–pleura) to be as close together as possible. According to Forbes, this was also important for the subsequent fate of the stems of other longitudinal veins, but this varies from group to group in the Pterygota and will be discussed later.

Forbes' views conflict with those of Snodgrass and Weber. These authors thought that it is the radial vein, that is to say its basal plate ('radial plate'; Weber's 2nd pteralium) that rests on the pleural wing process, and not the medial vein (or its basal plate). To uphold Forbes' theories, we would have to assume that the '2nd pteralium' represents the fused basal plates of both radius and media. One of the most urgent tasks for future phylogenetic research in entomology is a careful comparative functional–morphological investigation of the basal articulation of the wing in all orders, and it is surprising that no more than the first steps for this have been carried out so far. Such a study would have to take into account the most recent biophysical investigations of insect wings (Nachtigall, 1968).

Two small pleural sclerites, the basalar and subalar sclerites, are also part of the basal articulation of the wing, and they are independent of the episternum and epimeron in all recent Pterygota. According to Forbes, the two direct flight muscles are attached to these sclerites. The basalar muscle is the principal direct flight muscle and also achieves the down-warping of the wing on landing; the (pleural) basalar muscle makes its connection with the subcosta and causes it to become a concave vein (like the media!). On the other hand, the vein in front of this, the costa, makes connection with the tergum through the tegula, by means of a small muscle which pulls the wing forward, and is therefore a convex vein according to Forbes. Behind the pleural fulcrum (and in Forbes' terminology this would mean behind the media), Forbes believed that the development of the veins and the sclerites of the basal articulation has followed completely different directions in the various orders. The cubital vein appears to have been a convex vein originally, but its base and the bases of the remaining longitudinal veins have been expanded into a single plate. So far as is known, this plate is fused with the medial plate in all recent Pterygota and the two are separate only in the palaeodictyopterous genus *Ostrava* (p. 150).

The 'anal process', or '3rd pteralium' of Weber, is a particularly important element in the basal articulation of the wing. It is assigned to the anal or axillary veins and serves as a point of attachment for a pleural muscle. Forbes did not discuss whether the anal or axillary veins were originally convex or concave. Unlike other authors who believed that there was a strict alternation of convex and concave longitudinal veins, he regarded all the veins that are not connected with dorsal or pleural elements of the exoskeleton as 'neutral'. Nonetheless, he did emphasize that the very strict alternation of convex and concave veins serves to strengthen the wing.

The longitudinal veins themselves are thickenings of the wing surface to accommodate the essential blood-space, tracheae, and nerves. Forbes was right to suggest that the veins are not places where the cuticle was thickened but places where it has not thinned as much as it has between the tracheal paths.

I shall discuss the number and distribution of the longitudinal veins in the Pterygota in more detail later, under each individual group. It seems that in the ground-plan of the Pterygota there was no fixed number of tracheal stems close to the anterior and posterior wing margins, but in all groups most of the stems can be homologized with some certainty.

Long ago Berlese suggested that the wing and its veins could be divided into four areas that lay one behind the other: anteala, preala, interala, and postala. He thought that these corresponded to the division of each thoracic segment into four metameric parts. This theory is no longer of any importance, but I have alluded to it here because Haupt (1949), apparently independently of Berlese, has recently attempted to give a 'new interpretation' of the wing venation along similar lines. He has distinguished three groups of longitudinal veins, and these are supposed to correspond to the 'pretergite, tergite, and posttergite' of a thoracic segment. However, Haupt's views, like those of Berlese, are based on incorrect morphological premises and are therefore completely untenable.

There is a dense, irregular network of cross-veins, the archedictyon, between the principal longitudinal veins in primitive Pterygota and certainly in the ground-plan of this group. However, it is not absolutely certain whether a system of relatively simple cross-veins or a network of numerous small cells is part of the ground-plan of the Pterygota. In derived Pterygota the archedictyon is reduced and only single regular cross-veins are preserved. However, apparent cross-veins arise through fusion of the branches of longitudinal veins.

The structure of the very complicated basal articulation of the wing, the venation and the flight muscles shows considerable agreement throughout the Pterygota, even in details, and this provides evidence that the Pterygota are a monophyletic group, as Wigglesworth and others were right to emphasize. However, some authors have challenged this (Lemche, 1940; Smart, in Wigglesworth *et al.*, 1963), but none of them has put forward any convincing reasons for doing so.

Revisionary notes

132. Lombardo (1973) has shown that there are two coxal proprioreceptor organs in all the orders of the Pterygota. [Willi Hennig.]

133. At a later symposium on insect flight, Wigglesworth (in Rainey, 1976) attempted to show that there is considerable physiological, ecological, and other evidence for deriving the wings from coxal styli. A recent study of the origin and evolution of insect wings has been published by Kukalová-Peck (1978); see also Hamilton (1971a). [Michael Achtelig.]

134. The wing base of *Eugereon* (Palaeodictyoptera) has recently been studied in detail by Müller (1977b). [Dieter Schlee.]

135. It has been possible to observe some muscles in specimens preserved in Lower Cretaceous amber from the Lebanon (Schlee, 1970), and it is always possible that these and other minute features will be visible in any older ambers that may be discovered. [Dieter Schlee.]

136. Kukalová-Peck (1974) has published a study on the basal wing articulation of the Palaeodictyoptera, Megasecoptera, and Diaphanopterodea (all Palaeoptera). Kukalová-Peck and Peck (1976) have also published an informative SEM photomicrograph of the base of the sub-imaginal fore-wing of *Moravia convergens* Kukalová (Palaeodictyoptera, Calvertiellidae). [Michael Achtelig.]

137. The archetypal insect wing has been discussed by Edmunds and Traver (1954) and by Hamilton (1972a). [Willi Hennig.]

b. Classification of the Pterygota

No one knows for sure what the earliest events in the phylogenetic development of the Pterygota were. After the origin of the Pterygota themselves, I think that the first event that can be recognized with some certainty is the origin of the sister-group relationship between the Palaeoptera (Ephemeroptera + Odonata) and the Neoptera (all other recent Pterygota). However, this classification has given rise to many contradictory and confused ideas. If the aim of this book is to be achieved, it is essential that I discuss the various possible groupings in detail.

The Ephemeroptera, Odonata, and Neoptera (and their fossil stem-groups) are well founded or adequately founded monophyletic groups of the Pterygota. There is scarcely any disagreement about this. It is the relationship between these three groups that is in dispute. If it is accepted that each one is monophyletic, then there can only be three possible relationships between them:

1. The Ephemeroptera + Odonata are the sister-group of the Neoptera. If we accept this, we are following Martynov in distinguishing the 'Palaeoptera' (= Ephemeroptera + Odonata) and the 'Neoptera'.
2. The Ephemeroptera are the sister-group of the Odonata + Neoptera. If we accept this, we are following Börner in distinguishing the 'Archipterygota', with the single order Ephemeroptera, and the 'Metapterygota' (= Odonata + Neoptera). The name Archipterygota has had an unfortunate history. It was first used by Börner (1909) for the Ephemeroptera which he separated from all other Pterygota, his 'Metapterygota'. Later, Martynov (1924a, 1924b, 1925a) and Crampton (1924, 1931) published

independently and almost simultaneously their division of the Pterygota into two groups, Ephemeroptera + Odonata and all other orders: Martynov gave these the names Palaeoptera and Neoptera, and Crampton the names Archipterygota and Neopterygota. Still later, Weber (1949) used the name 'Archipterygota' for the hypothetical stem-group of the Pterygota.

3. The Odonata are the sister-group of the Ephemeroptera + Neoptera. If we accept this, we are following Shvanvich in distinguishing the 'Orthomyaria' (Odonata) from the 'Chiastomyaria' (= Ephemeroptera + Neoptera), or Lemche who called these groups 'Plagioptera' (Odonata) and 'Opisthoptera' (= Ephemeroptera + Neoptera).

Unfortunately, the arguments about which of these three possibilities is historically correct have frequently been complicated by the fact that some authors have depicted the second one in their phylogenetic trees but have continued to group the Ephemeroptera and Odonata together as the Palaeoptera, for typological reasons. So the name 'Palaeoptera' has been used to cover different ideas about the actual course of phylogeny. This is unfortunate, because it is essential to avoid misunderstandings by maintaining the greatest possible clarity in the terminology and nomenclature of groups.

As already mentioned, I consider it most likely that the Palaeoptera and Neoptera are sister-groups (possibility 1 above). As other opinions are often put forward, both sides of the argument must be discussed in detail.

No one disputes that the Ephemeroptera and Odonata are alone amongst the recent Pterygota in being unable to flex their wings back over the abdomen: according to Weber they have 'non-flexing wings'. Most authors regard this as a primitive character. This is correct, for it is clear that the earliest wings to have developed from paranota must have been simple plates that could only make limited vertical movements. Flexing the wings backwards, on the other hand, requires a more complicated articulation and a special arrangement of the muscles, and these cannot have been present originally.

However, even if 'non-flexing wings' were an original feature of the first winged insects, they are not necessarily primitive in the groups where they are now known to occur: recent and fossil Odonata and Ephemeroptera, Palaeodictyoptera, and one section of the Megasecoptera. Such wings could have developed secondarily from neoptery (the ability to flex the wings back over the abdomen). Rodendorf, and Schmidt (1963) in his popular account based on Rodendorf, both appear to have believed this because the oldest known winged insect (*Eopterum devonicum*) has wings that are set rather obliquely back (see Fig. 26)[138].

Shvanvich (according to Wille, 1960) even considered neoptery to be primitive, whereas Martynov and others did not. However, it seems unlikely that it is primitive: firstly, there is no morphological evidence to support this view; secondly, it is striking that in the various groups with non-flexing wings there are modifications for moving the wings from their rigid lateral resting position, because the search for a place of concealment and other activities would have been hampered by such a rigid posture. In the Odonata this movement is achieved by tilting the thoracic segments, whilst in the Ephemeroptera the wings are held vertically upwards above the thorax with their dorsal surfaces touching. So the view that the non-flexing wings of the Ephemeroptera and Odonata really are a primitive character can be upheld. Sharov (1966b) has recently introduced a fresh point of view into the discussion. He has suggested that in the stem-group of the Pterygota the wings were set obliquely backwards, like the wing pads of many nymphs. However, these partly flexed wings had nothing to do with 'neoptery'.

The wings of the Ephemeroptera and Odonata do not flex back over the abdomen because certain articular sclerites in the wing base (axillaries) are firmly attached to the longitudinal veins. These sclerites appear to be no more than the expanded bases of their respective longitudinal veins. It is only in the Neoptera that they have become free from the veins and as a result have made it possible for the wing to be flexed backwards over the abdomen. The firm attachment of the axillaries to the longitudinal veins, from which the axillaries are apparently derived, also indicates that the non-flexing wings of the Ephemeroptera and Odonata are primitive.

According to Forbes (1943), a further primitive character of these two groups is that the convex radial vein (R_1) is wholly free and unbranched. Unlike other authors, Forbes believed that the radius (R_1) and the two branches of the so-called radial sector (R_{2+3} and R_{4+5}) were originally the stems of tracheae and longitudinal veins that entered the wing base independently; in most groups these branches did not become fused with the radius (R_1) to form the radial complex until later, and they are still separate in the Odonata and Ephemeroptera. However, Forbes' interpretation is disproved by the Palaeozoic genus *Triplosoba*. The generally accepted opinion, which is certainly correct, is that *Triplosoba* belongs to the Ephemeroptera (see p. 136), but its radial sector originates from the radius as in most Pterygota.

If these characters really are primitive features of the Pterygota which have only been retained in the Ephemeroptera and Odonata, then they are 'symplesiomorphies' and cannot be used as evidence for the relationship that these two groups have with each other and with the Neoptera. It is necessary to see if two of these three groups have any derived characters which could be interpreted as 'synapomorphies' and which would indicate that the two groups had an ancestor that was common only to them.

The following are the derived characters which the Ephemeroptera share with the Neoptera and which could indicate that there is a close relationship between these groups (possibility 3 on p. 126): the structure of the basal articulation of the wing, the wing muscles, and the ontogenetic origin of the wings.

Shvanvich (1943 and subsequently) and Snodgrass (1958b), who quoted several authors but did not mention Shvanvich, distinguished two different types of wing movement. The indirect method is used by the Ephemeroptera and Neoptera, but not by the Odonata: the upstroke of the wing results from the depression of the notum by notosternal or other vertical muscles; and the downstroke is produced by lengthwise compression and arching of the notum by contraction of the longitudinal dorsal muscles. So that the notal plates should really arch and should not simply be pulled together during contraction of the muscles, the intersegmental membranes between the mesonotum and metanotum, and between the metanotum and 1st abdominal tergite, have become sclerotized. They form the 'postnotal plates' of the mesothorax and metathorax. The pleural wing process, which supports the 'radial plate' (2nd axillary) of the wing, is simple in these species, but the basal articulation itself has a rather complicated structure. It consists of individual sclerites which are derived from the notum, the pleura, and the wing veins.

The direct method of wing movement is most efficiently developed in the Odonata. The dorsal longitudinal muscles are relatively small or, in derived species, completely absent. The upper part of the pleural ridge, which forms the wing process (fulcrum), curves inwards and divides the dorso-ventral muscles into two groups: those attached laterad of the fulcrum have become wing depressors and those mesad of it have become wing elevators. There are only two large sclerites in the basal articulation of the wing (Fig. 25D) and Tannert (1958, 1961) has called these the costal plate (the humeral plate of other authors) and the radio-anal plate (the axillary plate). The pleural wing process is forked: one arm supports the costal plate, and the other the radio-anal plate.

Tannert (1961) has tried to show that these two forms of basal articulation are two sharply opposed types: 'pteralia or axillary type' in the Ephemeroptera and Neoptera, and 'basal-plate type' in the Odonata. Snodgrass (1958b), too, was inclined to regard the direct method of wing movement of the Odonata and the indirect method of the Ephemeroptera + Neoptera as alternative and mutually exclusive realizations of two functional possibilities.

I think that this view is based on an over-hasty typological assessment of morphological features, of the sort that has so often been detrimental to the progress of phylogenetic research.

So far as Tannert's two 'types' of basal articulation are concerned, I shall show later that the 'basal-plate type' of the Odonata must be interpreted as a specialized development in this group (an 'autapomorphy') and that if Tannert's 'axillary type' also includes the Ephemeroptera it is not even a uniform functional type. In fact, these two types of basal articulation, as defined by Tannert, cannot be used as evidence of a sister-group relationship between the Odonata and the Ephemeroptera + Neoptera. The same is true of the methods of direct and indirect wing movement. The direct wing movement of the Odonata shows such significant features of unilateral specialization and involves such obviously derived characters, such as the structure and position of the pleural wing-process, that it must be an

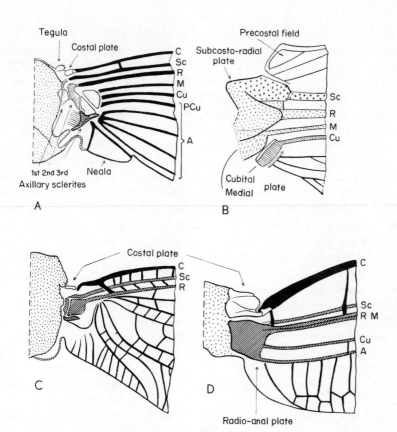

Fig. 25. Wing-base of the Neoptera (A), Palaeodictyoptera[139] (B: *Ostrava nigra* Kukalová), Ephemeroptera (C), and Odonata (D). From Snodgrass (1935) (A, C), Séguy (1959) (D), and Kukalová (1958d) (B).

'autapomorphy' of this group, just as the basal-plate type of basal articulation is too. Snodgrass even had to assign an intermediate position to the Blattopteroidea (cockroaches, praying mantises, termites), since they have not fully developed either the direct or the indirect method of wing movement. This shows that these two methods did not immediately give rise to two separate groups when flexible wings were first acquired. In fact, more recent investigations have shown that the situation is much more complicated than Snodgrass thought. Direct wing muscles even play a fundamental role in the Coleoptera (Pringle, 1957; Nachtigall, 1968).

Lemche (1940) believed that he had found a fundamental difference in the ontogenetic development of the wing pads between the Odonata, which he grouped together with the Palaeodictyoptera as the Plagioptera, and the remaining Pterygota which he called Opisthoptera (= Ephemeroptera + Neoptera). In the Plagioptera the wing pads are said to develop as lateral outgrowths from the tergum whereas in the Opisthoptera they are said to be directed backwards. However, even in the Odonata (Plagioptera) they eventually come to be directed backwards, but in this case their fore-margins touch along the mid-line. Lemche regarded this as evidence that the backwardly directed position is secondary, and he even suggested that the Plagioptera and Opisthoptera originated diphyletically from the wingless insects. However, his conclusions are invalidated by the fact that the wing pads of the 'Saltatoria' (Opisthoptera) are in exactly the same position as are those of the Odonata. Lemche himself attempted to work out some kind of distinction and to point out some fundamental characters common to the 'Saltatoria' and other Opisthoptera, but to the unprejudiced observer his work only shows how easy it is for convergent developments to take place. To conclude, there is no soundly based evidence to support the view that there is a close relationship between the Ephemeroptera and Neoptera, or that there is a sister-group relationship between this group (Chiastomyaria or Opisthoptera) and the Odonata (Orthomyaria or Plagioptera).

Until recently it was thought that there was a sister-group relationship between the Ephemeroptera and the Odonata + Neoptera (possibility 2 on p. 125) because the Ephemeroptera had two characters in which they were more primitive than the Odonata and the Neoptera: they appeared to be the only group in which the adult was preceded by a sub-imaginal winged instar and in which the wing tracheation still preserved its primitive form.

The absence of the sub-imago can no longer be regarded as a synapomorphy of the Odonata and Neoptera, since Sharov (1957a) has shown that there was a sub-imago in *Atactophlebia termitoides* Martynov, a 'protoperlarian' that appears to belong to the Neoptera.

This just leaves the wing tracheation. According to Weber (1949), the Ephemeroptera only have a single tracheal trunk: this comes from the leg trachea and corresponds to the trachea of the paranotal lobes in the Archaeognatha and Zygentoma, which are homologous with the wings. In the other Pterygota there is a second posterior trunk that appears to be formed by anastomosis of the leg trachea with the following spiracular trunk. This united tracheal arch can be interrupted medially, with the result that there are two separate alar tracheal trunks. However, Forbes (1943) has pointed out that a posterior alar trunk is actually present in the Ephemeroptera, though it is so small that it is almost always overlooked. This indicates that in this character too the Ephemeroptera are not more primitive than the other Pterygota and that these apparently primitive features are in fact reductions.

It is still stated from time to time that the larvae of the Ephemeroptera differ from all other Pterygota by retaining monocondylic mandibles, but this has already been shown to be incorrect (p. 113).

There are some other characters in which the Ephemeroptera appear to be more primitive than all the other Pterygota. However, the Odonata have several very striking adaptations (autapomorphies) and these make it extraordinarily difficult to determine whether they also have any other derived characters which they share (synapomorphically) with the Neoptera. The only way to show that there is a sister-group relationship between the Ephemeroptera and the Odonata + Neoptera is to find such synapomorphic characters. However, no reliable characters have yet been found (possibility 2 on p. 125).

The Ephemeroptera and Odonata have the following derived characters in common, and this suggests that they have a sister-group relationship with the Neoptera (possibility 1 on p. 125):

1. The short bristle-like flagellum of the adult antenna, which induced Lameere (1935–36) to propose the name Subulicornia for the Ephemeroptera + Odonata.
2. The intercalary veins in the adult wing, which arise between the true longitudinal veins as a result of modifications in the archedictyon.
3. The fusion of galea and lacinia into a single lobe in the larval maxillae (Crampton, 1938).
4. The aquatic larvae.

These characters are relatively trivial, and nothing is known about their functional significance. This is hardly surprising. If non-flexing wings are regarded as a primitive character of the Pterygota, there must originally have been a very large number of insects with non-flexing wings in the stem-group of the Pterygota, although most of them became extinct towards the end of the Palaeozoic. The dominant group now consists of descendants which had acquired the characters of the Neoptera. The other group, which includes the Ephemeroptera and Odonata, retained their non-flexing wings, and this was probably because their larvae became aquatic whereas the larvae of the Neoptera remained primarily terrestrial. The ancestors of the modern Palaeoptera were thus able to avoid competition from their sister-group, which had gained its selective advantage mainly or exclusively by modifying the basal articulation of the adult wing. This sister-group relationship must have originated very early in the history of the Pterygota because some fossils are known from before the Upper Carboniferous which prove that these two sister-groups existed at that time and which can even be assigned to several subgroups of the Neoptera that are still in existence. At first the two subgroups of the Palaeoptera were characterized only by their retention of primitive characters, but once their larvae had adapted to an aquatic life they must soon have followed separate paths of development because they now have very little in common.

Unfortunately, it is impossible to decide whether the development of the antennal flagellum into a bristle has taken place convergently in the Odonata and Ephemeroptera, because the structure of the antennae in Palaeozoic fossils that can be reliably assigned to the stem-groups of these orders is still inadequately known (Carpentier and Carpentier, 1949).

Not all authors are as convinced as I am that the aquatic mode of life is a derived character in the larvae of the Palaeoptera, but there is no doubt that in itself it is a derived character in the insects. Rudimentary spiracles and tracheae leading to them have been found in the larvae of Ephemeroptera and Odonata (Calvert, 1929, quoted by Emden, 1957), which indicates that this mode of life is secondary but otherwise does not mean very much. However, the Ephemeroptera and Odonata are hemimetabolous groups and during their ontogenetic development several adult characters are prefigured in the larvae. It would be conceivable that groups containing such species developed a purely terrestrial mode of life from ancestors whose young larvae lived in the water; we would only need to imagine that individuals took to the land during progressively earlier instars, and that the functioning tracheal system of the adult was also being developed by progressively earlier instars. This only goes to show that in such cases the arguments should not be over-simplified. In fact, there is no reason to suppose that anything like this ever took place in the Pterygota. I find it impossible to believe that the larvae of the Pterygota and the Neoptera were primarily aquatic, without developing elaborate secondary hypotheses, and this should never be done without compelling reasons. Until there is some evidence to the contrary, the aquatic mode of life of the larvae of the Ephemeroptera and Odonata should be regarded as a derived character, in the sense that it is a synapomorphy of the Ephemeroptera and Odonata.

After carefully examining the three possible ways in which the relationships between the Ephemeroptera, Odonata, and Neoptera can be assessed (p. 125), I think that the most soundly based hypothesis is that the Ephemeroptera and Odonata form a single monophyletic group (the Palaeoptera) and that the Palaeoptera are the sister-group of the Neoptera. This relationship has also been shown in the phylogenetic trees of Martynov (1938b) and Jeannel (in Grassé, 1949). The origin of the sister-group relationship between Palaeoptera and Neoptera is thus the earliest recognizable event in the phylogenetic history of the Pterygota.

Revisionary notes

138. This sentence is no longer valid because *Eopterum* belongs to the Crustacea (see note 57). [Dieter Schlee, Rainer Willmann.]

139. Details of the wing base of *Eugereon* (Palaeodictyoptera) have been illustrated by Müller (1977b, Fig. 3). [Dieter Schlee.] The pteralia of the Palaeodictyoptera, Megasecoptera, and Diaphanopterodea have been described and illustrated by Kukalová-Peck (1974). [Willi Hennig.]

c. The oldest fossils and their systematic position

The oldest pterygote fossil is *Eopterum devonicum* Rodendorf (Fig. 26), from the Upper Devonian of Ukhta in the Komi Republic (Soviet Union). It was described only a few years ago (Rodendorf, 1961b). The fact that the wings are set rather obliquely backwards has given rise to a great deal of confusion in assessing its systematic position: it was first thought that it should be assigned to the Neoptera or even to one of the subgroups here (Blattopteroidea), and 'neoptery' was considered to be more primitive than non-flexing wings. Once again it was Sharov (1966b) who solved the enigma (see p. 127). In his view, there was no reason for not assigning *Eopterum* to the 'Archaeoptera', the stem-group of the Pterygota. He has suggested that the Upper Carboniferous Paoliidae also belonged to the Archaeoptera, but this is a comparatively unimportant point because they are no older than several subgroups of the *Pterygota and should therefore be regarded as a surviving 'side-branch' of the stem-group[140].

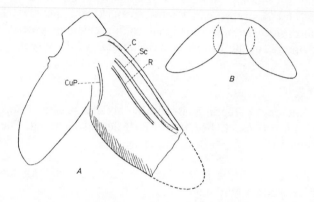

Fig. 26. *Eopterum devonicum* Rodendorf, from the Upper Devonian of Ukhta. A: Interpretation of the venation. B: Suggested position of the non-flexing wings at rest. Both from Sharov (1966b)[141].

The Paoliidae were previously assigned to the Palaeodictyoptera. There is no doubt that the Palaeodictyoptera are the best known of all the fossil insects. Up until 1937, Handlirsch thought of them as the stem-group of the entire Insecta, not just of the Pterygota. I should point out that Handlirsch's concept of a 'stem-group' is rather different from my own. As he based his work entirely on typological principles, he retained several groups in the Palaeodictyoptera which he believed, sometimes incorrectly, to be more closely related to *different* recent orders such as the 'Orthoptera', Hemiptera, Odonata, and Ephemeroptera. According to Handlirsch (1937), Lameere was totally opposed to these views: he stated that the Palaeodictyoptera consisted of 'a mixture of Subulicornia (Ephemeroptera and Odonatoptera), Rhynchota (Hemiptera), and some Orthoptera', and for this reason he wanted to break up the group completely.

On the whole, the general trend of palaeoentomological opinion has followed Lameere, even though many of his individual views about the relationships of some subgroups of the 'Palaeodictyoptera' have not stood the test of time any more than those of Handlirsch. The follower of phylogenetic systematics would now take all the groups that are thought to be more closely related to particular recent groups than they are to each other or to the rest of the 'Palaeodictyoptera', and would assign them to the appropriate recent groups: this applies particularly to the 'Orthoptera', Odonata, Ephemeroptera, and some of the Plecoptera. It is entertaining to see what subtle arguments are put forward to achieve this, even by authors who support typological systematics in their classification of recent insect groups and in their theoretical work.

In dealing with the history of the Pterygota, I think it would be best for practical and methodological reasons to see how far the monophyletic groups Palaeoptera and Neoptera can be traced back into the past. I shall then discuss what position should be assigned to the remaining species, such as the Palaeodictyoptera, in the historical development of the insects.

Revisionary notes

140. This opening paragraph is totally invalid because *Eopterum* is a crustacean, not an insect. See note 57. [Michael Achtelig, Dieter Schlee, Rainer Willmann.]

141. This figure should be deleted: see preceding note. [Michael Achtelig, Dieter Schlee, Rainer Willmann.]

2.2.2.1. Palaeoptera† (Fig. 34)

There is very probably a sister-group relationship between the Ephemeroptera and Odonata, which are the only groups of recent insects that have retained non-flexing wings, and in the preceding section I have discussed in detail the evidence for this assertion. This relationship must have arisen before the Upper Carboniferous because the oldest fossils from this period include both Ephemeroptera and Odonata. However, if the Ephemeroptera and Odonata really are sister-groups, then their recent species have diverged considerably from each other and from their common ancestors, both in their mode of life and in their morphological characters.

2.2.2.1..1. Ephemeroptera (mayflies)[142]

All recent Ephemeroptera pass the greater part of their life in the larval stage. The adults have reduced mouth-parts (they are agnathous): they do not feed, and live for only a very short time. The fore-wings are the main organs

---

†The oldest name for this group is Subulicornes Latreille, 1807 (or Subulicornia Burmeister, 1839). This is the name that should be used if the Law of Priority is followed, as it was with the Archaeognatha–Microcoryphia. Lameere (1935–36) called the group Palaeoptilota.

of flight: as a result, the mesothorax is very strongly developed, whereas prothorax and metathorax are degenerate. The hind-wings are reduced or, in derived species, are completely absent. The fore-wings are triangular in shape[143]. Palmén's organ is a characteristic derived feature in both larval and adult stages: this is probably a static sense organ, and is located close to the transverse anastomosis of the cephalic tracheal system, in front of the cerebral ganglion.

The paracercus and the sub-imaginal winged instar are primitive characters of the Ephemeroptera. The order appears to be the only group to have retained the paracercus from the ground-plan of the Pterygota. In recent Pterygota, the ecdysis that marks the passage from the fully winged sub-imago to the imago only takes such a primitive form in the Ephemeroptera. According to one theory, the sub-imago of the Ephemeroptera corresponds to the pupal stage of the Holometabola[144]. A further primitive character, according to Wille (1960), is the three-lobed hypopharynx.

It is not known whether certain other characters, such as the paired gonopores in both sexes, are primitive or derived. The male has paired penis lobes. The absence of the female ovipositor is certainly a derived character, because an ovipositor is present in the Zygentoma, in the Palaeozoic Palaeodictyoptera and Megasecoptera, in a reduced form in the Odonata, and also in the Neoptera: consequently, it is undoubtedly part of the ground-plan of the Pterygota as well as of the Palaeoptera. For this reason I think that the paired female oviducts and their position between the 7th and 8th abdominal sternites is a derived character[145]: development of the single gonoduct in female Ephemeroptera has probably been secondarily suppressed, and this may also be true in the male. I must also draw attention to Carpenter's (1939) observation that *Permohymen schucherti* Tillyard from the Lower Permian of Kansas, a species of the Megasecoptera (subgroup Protohymenoptera), also has paired penis lobes. At present I am not sure what to make of this agreement with the Ephemeroptera. Even the male

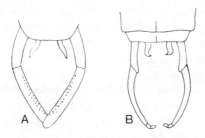

Fig. 27. External male genitalia of *Permohymen schucherti* Tillyard (A: Megasecoptera), from the Lower Permian of Kansas, and *Hexagenia* sp. (B: Ephemeroptera; recent). From Carpenter (1939).

claspers of *Permohymen* agree with those of certain Ephemeroptera (*Hexagenia* sp., Fig. 27).

It is still not clear how the wing base and the basal articulation should be interpreted (Fig. 25C). The number of anal (or axillary) veins that reaches the wing base is relatively large, and this may be a primitive character. According to Forbes (1943), these veins are gathered on a basal transverse chitinization, which is (still?) rather amorphous but which corresponds to the 3rd axillary of the Neoptera. The other sclerites of the basal articulation have not yet been interpreted. According to Grandi (1947a, 1947b, quoted by Snodgrass, 1958b), there are several small sclerites in the axillary region ('pseudopteralia'), but Matsuda (1956) stated that in primitive species (*Siphlonurus*) these correspond to the usual three axillaries of the Neoptera. The 3rd axillary has already been identified; and the 1st axillary of the Neoptera, which is represented in the Ephemeroptera by a sclerite at the base of the wing, is an offshoot of the notum. The vital question, therefore, is whether the third of the axillaries distinguished by Matsuda corresponds only to the 2nd axillary of the Neoptera, or whether it incorporates this and the so-called median plate. This is an important problem which needs to be more carefully studied in the Ephemeroptera, and I shall deal with it again in the discussion of the Neoptera.

There is still a great deal of uncertainty over the interpretation of the wing venation in the Ephemeroptera and Odonata. According to Tillyard and Fraser (1938–40), the base of MA^+ is attached to the radial sector. However, Forbes (1943) believed that the vein which Tillyard and Fraser called MA^+ was originally a free convex vein which is attached to the media and forms its anterior branch in certain fossil Pterygota (see below), but in the Ephemeroptera, Odonata, and Neoptera it is attached to the radial sector and should be called R_{4+5}. This character is a parallel development in these three groups, as is shown by the Upper Carboniferous *Triplosoba* (Fig. 28) where this vein is (still?) fused with MP^- and is entirely free from the radial sector. For this reason, Forbes (1943) rejected the idea that there is any relationship between *Triplosoba* and the Ephemeroptera, and suggested that this genus, which has always been regarded as a member of the stem-group of the Ephemeroptera, does in fact belong to the Palaeodictyoptera.

The one-segmented tarsus and the absence of paired claws in ephemeropterous larvae are said to be derived characters. The position of the mandibular articulation may be relatively primitive (see p. 113).

Very few of these characters are actually visible in the fossils, and it is impossible to say when they originated and in what sequence. There are some Palaeozoic species which almost certainly belong to the stem-group of the Ephemeroptera: the oldest species is *Triplosoba pulchella* Brongniart (Fig. 28), from the middle Upper Carboniferous (Stephanian) of Commentry (France), and there are also about 20 species from the Lower Permian of Kansas and the Lower and Upper Permian of other regions[146]. The Permian species have been grouped together as the Permoplectoptera, whilst *Triploso-*

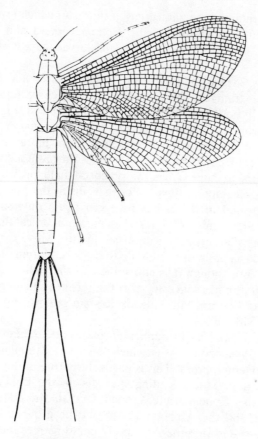

Fig. 28. *Triplosoba pulchella* Brongniart (Protephemeroptera; reconstruction), from the Upper Carboniferous of Commentry. From Demoulin (1956a).

ba pulchella is the only species of the Protephemeroptera. In certain characters of their wing venation, the Palaeozoic species are more primitive than most of the Mesozoic and recent species, and so it is not easy to show that they really do belong to the stem-group of the Ephemeroptera. Their characters could equally well show that they belong to the direct stem-group of the Palaeoptera (Odonata + Ephemeroptera), and I would not hesitate to assign them here if they had been found in the Lower Carboniferous or Devonian. However, there were some Upper Carboniferous species contemporary with *Triplosoba* and the Permoplectoptera, which had some derived characters that are found only in the more recent Odonata and which certainly belong to the stem-group of this order (Protodonata: see p. 142) *Triplosoba* and the Permoplectoptera should therefore be assigned to the stem-group of the Ephemeroptera alone. Unlike all other known Palaeozoic fossils, they have retained the paracercus, a character which I would expect to find in all the direct ancestors of the Ephemeroptera and of the recent Palaeoptera, etc. Some permoplectopterous larvae have also been found (Demoulin, 1966).

Triplosoba (Fig. 28) has the hind-wing slightly reduced and there is also an indication of the triangular wing shape that is characteristic of recent Ephemeroptera. In both characters the Permoplectoptera (Fig. 29) are more primitive. Their wings are evenly rounded and almost equal in size. However, in the structure of the costa the Permoplectoptera more closely resemble Mesozoic and recent Ephemeroptera. In front of the basal section of the costa *Triplosoba* appears to have a narrow precostal field provided with cross-veins. The Permoplectoptera have no cross-veins in the precostal field, and the costa is connected to the radius by a cross-vein only shortly before it enters the fore-margin: this forms a Y-shaped 'costal brace' that lies horizontally along the fore-margin of the wing-base. In most Mesozoic species and all recent ones, the costa has moved right up to the fore-margin of the wing, even at its base (Fig. 30), though it is still joined to the radius by the 'humeral cross-vein'. Demoulin has suggested that the recent Ephemeroptera must be derived from two different late Palaeozoic stem-groups, but this is still not certain (see pp. 348–350).

In addition to *Triplosoba* (Protephemeroptera) and the Permoplectoptera, there are other Palaeozoic fossils which have been thought to be closely related to the Ephemeroptera. This is particularly true of the 'Archodonata': Demoulin (1954b, 1958b), a leading specialist in the Ephemeroptera, has suggested that the Ephemeroptera, including all the Palaeozoic species discussed above, and the Archodonata should be grouped together as the 'Archiptera'. He stated that the Archiptera could be recognized by having a paracercus and intercalary veins. However, he also stated that the only archodonatan that is at all well preserved is *Permithemidia caudata* Rodendorf, and this has no paracercus! Other species, such as *Rectineura*, are only known from wing fragments. I am therefore unable to understand what basis there is for saying that the Archodonata are really most closely related to the Ephemeroptera. The oldest species, known from a single wing fragment, is *Rectineura lineata* Bolton from the middle Upper Carboniferous, and it is no older than *Triplosoba*: consequently, the problem of whether the Archodonata really do belong to the stem-group of the Ephemeroptera, perhaps as a side-branch, hardly matters at all. The remaining species, of which there are no more than half a dozen, have been described from the Permian of Kansas and the Soviet Union, and were more or less contemporary with the Permoplectoptera. In addition, the Archodonata appear to lack the precostal field, and in this respect they are less primitive than the oldest known Ephemeroptera.

The same is probably true of the systematic position of the Syntomopteridae, with three species from the middle Upper Carboniferous of North America. These appear to have intercalary veins and the base of MA$^+$ fused

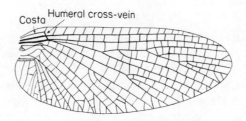

Fig. 29. Hind-wing of *Misthodotes obtusus* Sellards (Permoplectoptera), from the Lower Permian of Kansas. From Rodendorf (1962).

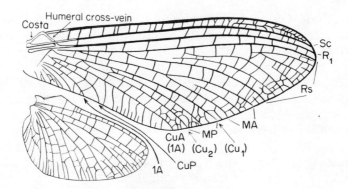

Fig. 30. Wings of *Siphlonurus aestivalis* Eaton (Ephemeroptera; recent). From Despax (in Grassé, 1949). Nomenclature of the wing-veins from Séguy (1959), with Despax's names in parentheses.

with R, but there is no precostal field. It is not known whether the paracercus was present or absent. Laurentiaux (in Piveteau, 1953) believed that this group too was related to the Ephemeroptera.

Laurentiaux and others have said that some or all of these groups are 'still' Palaeodictyoptera but are close to the origin of the Ephemeroptera. Such statements are meaningless. However, my concern in this book is not to assess the relationships of the fossils *per se*. These controversial species are no older than others which undoubtedly belong to the stem-groups of the Ephemeroptera and of the Odonata, and they need not be discussed in more detail.

In a later part of this chapter (pp. 348–350) I shall discuss whether any subgroups of the *Ephemeroptera were already in existence by the late Palaeozoic (Permian).

Additional notes

142. A new monograph on the Ephemeroptera has been published by Illies (1968). [Dieter Schlee.] The classification of recent and fossil Ephemeroptera has been discussed by Chernova (1970). Edmunds (1965) discussed the problems raised by the different character gaps amongst taxa in the adult and larval stages. [Willi Hennig.]

143. The wings of the Ephemeroptera have been dealt with in an important paper by Brodskiy (1970). [Willi Hennig.]

144. For a possible explanation of the sub-imago, see Schaefer (1975). [Willi Hennig.]

145. Abul-Nasr (1954) considered the position of the oviducts in the Ephemeroptera and Dermaptera to be primitive. See note 180, under the Dermaptera. [Willi Hennig.]

146. A species has been described by Kinzelbach (1970) from the lower Rothliegende of the Saar–Nahe–Palatinate region (near Bad Kreuznach), and tentatively assigned to the family Misthodotidae of the Permoplectoptera. Guthörl (1965) has described *Protereisma rossenrayensis*, from the Zechstein of the Lower Rhine. Demoulin (1954c) gave a review of the fossil history of the Ephemeroptera, and more recently (Demoulin, 1970a) has discussed the Permian 'ephemeropterous' larvae. [Willi Hennig.]

2.2.2.1..2. **Odonata (dragonflies)**[147]

Recent Odonata are characterized by a large number of derived characters which leave no doubt that this is a monophyletic group. Most or all of these characters are connected with the predaceous habits of the larvae and adults. In the mouth-parts this is true of the large dentate mandibles, and particularly of the labium, which consists of a median lobe (the fused glossa and paraglossa) and a pair of lateral lobes (labial palpi). According to Chopard (in Grassé, 1949), the glossa and paraglossa are still separated in the labium of *Hemiphlebia*, and so fused glossae and paraglossae are not yet part of the ground-plan of the Odonata. The prehensile mask of the larva has the same structure, but submentum and mentum (Weber's mentum and postmentum) are very long and powerful. The median lobe is fused with the mentum. The submentum is attached to the mentum by a powerful hinge, and when at rest it is directed backwards. The labial palpi act as grapplers for impaling prey. The maxillae are poorly developed, with their palpi reduced to one segment[148].

The fusion of galea and lacinia in the maxillae is part of the ground-plan of the Palaeoptera, as it has taken place in the Ephemeroptera as well as in the

Odonata. The same is probably true of the short bristle-like antennal flagellum. The great development of the compound eyes is, of course, associated with the dragonflies' predatory behaviour and their skilled flight. The same is true of the structure of the thorax: the prothorax is greatly reduced, whilst the mesothorax and metathorax are fused to form a synthorax which is strongly sclerotized and has an oblique position. As a result, the attachments of the strongly bristled legs have shifted forwards to a position suitable for catching and holding prey ('feeding basket'). When at rest, the wings can be folded over the abdomen, except in the Anisoptera, which are more derived and have lost this ability. In fact, the wings are only being raised with their dorsal surfaces touching, as in the Ephemeroptera, and are not being flexed backwards as in the Neoptera.

I have already mentioned that the wings are moved only by direct muscles and that the dorsal longitudinal muscles have been reduced. I need only recapitulate that the upper part of the pleural ridge with the wing process (fulcrum) curves inwards and divides the dorsoventral muscles into two opposing groups: those attached laterad of the wing process become depressors and those mesad of it are elevators. Furthermore, the wing process is forked. There are only two large sclerites in the basal articulation of the wing, the costal or humeral plate and the radio-medio-cubital plate (Fig. 25D). Each plate is supported by an arm of the pleural wing process. These are highly derived features and I shall discuss them again in the section on the Neoptera, as that is the most appropriate place to deal with features of the wing base.

The wings themselves are long and narrow[149], and this constriction led to a fusion and reduction of the longitudinal veins which is evident even in the oldest species, from the lower Upper Carboniferous. These have the precostal field preserved only as an expansion on the proximal fore-margin of the wing. According to Forbes (1943), who regarded the free radius (R_1) of the Palaeoptera as a primitive character, the radial sector (R_{2+3} and R_{4+5}), media, and cubitus are fused at base into a single stem in the Odonata. This character has also been found in the oldest fossils from the Upper Carboniferous, or at least in those species where the veins can be clearly traced as far as the wing base, and results from constriction of the wing and particularly of the wing base. According to Forbes, all the anal (axillary) veins originate from a single stem, and their crowding together at the wing base is another result of this constriction.

All recent Odonata have some other characteristic features, such as the nodus, the discoidal cell, and the attachment of the anal veins to the cubitus. The acquisition of these can be followed step by step during the Palaeozoic. There are very considerable differences of opinion in the interpretation of details of the venation, but these hardly affect our assessment of the relationships themselves.

In addition to the narrow wings, the constriction and elongation of the abdomen are also correlated with the mode of flight. The gonopods (or parameres, according to Snodgrass' controversial views) and the male penis are extensively reduced. Instead there is a set of secondary copulatory organs on the 2nd and 3rd abdominal sternites, a peculiar and characteristic derived feature of the Odonata[150]. In both sexes of all species, there are two lobe-like appendages on the 10th tergite. Cerci and paracercus are degenerate and are no longer present as long segmented appendages. Odonata are also more primitive than Ephemeroptera in having a short female ovipositor.

According to Lemche (1940), the growth of the wing pads is characteristic of the Odonata: at first they grow out postero-laterally, but later they are directed over the abdomen so that their morphological fore-margins touch.

Finally, according to Henriksen (1932; quoted by Wille, 1960), the Odonata differ from all other recent insects in the way they moult. The usual manner in insects is for the skin to split along the middle of the thorax. In Odonata, however, the splitting of the thorax is confined only to the prothorax and the anterior part of the mesothorax. The ecdysial line then branches towards the bases of the wings and runs backwards and outwards on each side.

There are many other characters in the thoracic endoskeleton and internal anatomy that cannot be discussed here, and careful study would certainly show that they are associated with the striking external adaptations of the Odonata to their mode of life. Unfortunately, most of them are not preserved in the fossils.

The oldest fossil that definitely belongs to the stem-group of the Odonata is *Erasipteron larischi* Pruvost, from the Namurian of the Upper Silesian coal basin (Fig. 31). It is known only from a single incomplete wing (a fore-wing?),

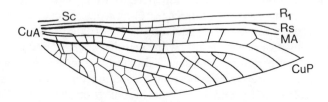

Fig. 31. Wing (probably a fore-wing) of *Erasipteron larischi* Pruvost, from the lower Upper Carboniferous (Namurian) of Czechoslovakia. From Carpenter, based on Laurentiaux (in Piveteau, 1953).

which can be recognized as belonging to the stem-group of the Odonata by the absence of MP and the virtual absence of CuA, which is retained only as a vestige. *Erasipteron* and several other Palaeozoic fossils have been assigned to the Protodonata (= Meganisoptera) and are characterized by having a wing base which is still (i.e. primarily) comparatively broad and in which the various changes characteristic of later Odonata, and particularly of recent species, have not yet taken place. The precostal field is only developed as a callus, but it is still rather long. Carpenter (1954b) stated that this group already had large eyes, large mandibles, obliquely directed thoracic segments, and spinose legs. None of these characters is visible in the oldest fossil which, as stated above, is a single wing fragment, and so it is not known when these characters originated or in what sequence. According to Laurentiaux (in Piveteau, 1953), *Meganeurula gracilipes* Handlirsch is the only species for which any precise anatomical details are known. This species is from the Stephanian of Commentry, and Carpentier and Lejeune-Carpentier (1949) have pointed out that the name is a junior synonym of *Meganeura* [*Meganeurula*] *selysii* Brongniart. According to these authors, the bodies of this and other species (Carpentier and Carpentier, 1949) were not actually as slender as they appeared to be in earlier reconstructions.

It would also be important to obtain some information about the abdominal appendages. It is clear that long, segmented cerci and a similar paracercus were no longer present. Laurentiaux stated that paired cerci and a paracercus were present, as in the larvae. However, I think that the paired lateral appendages might actually be lobes of the 10th tergite, as they are in recent species. It may be that the apparently single median lobe is not really a single structure but represents one of the cerci superimposed over the other, but it is not possible to tell whether this is so from Carpentier's drawing. However, it is a suggestion that should be borne in mind.

Apart from *Erasipteron larischi* from the Namurian of Upper Silesia, which is the oldest species, a few Protodonata are known from the middle and upper Upper Carboniferous of Commentry, England, and North America (Rhode Island), but there are less than a dozen species, including several dubious fragments. The celebrated locality Commentry has yielded *Meganeura monyi* Brongniart, which is the largest European species and is actually the largest known European insect. According to Carpenter (1947), it had a wing-span of 66 cm. Permian fossils are slightly more numerous, and about 20 species are known from the Lower to Upper Permian of Kansas and Oklahoma, the Urals region and Arkhangelsk, and Germany (*Ephemerites rueckerti* Geinitz). *Meganeuropsis americana* Carpenter and *M. permiana* Carpenter, from the Lower Permian of North America (Kansas and Oklahoma), are even larger than the European species and had a wing-span of about 75 cm, according to Carpenter (1947). For this reason, Carpenter called the Protodonata the dinosaurs of the insect world, although not all species were as large as this. Earlier (Carpenter, 1939) he had included the families Protagriidae and Calvertiellidae in a group 'Euprotodonata' and had assigned them to the Protodonata because they still retained MP and CuA. However, according to the Russian Textbook of Palaeontology (Rodendorf, 1962), both families belong to the Palaeodictyoptera. This problem need not concern us here because some of the undoubted Protodonata (= Meganisoptera) are older than either of these two families.

It is particularly interesting that some alleged Protodonata have also been found in the Mesozoic: *Triadotypus guillaumei* Grauvogel & Laurentiaux (Lower Triassic of the Vosges and Savoy), *Thuringopteryx gimmi* Kuhn (middle Bunter Sandstone of Saalfeld), *Reisia gelasii* Reis (Muschelkalk of Lower Franconia), *Piroutetia liasina* Meunier (Rhaetic of Fort Mouchard), and *Liadotypus relictus* Martynov (lower Lias of Turkestan). These species are all represented by wing fragments. Particularly strict criteria should be used in making identifications and assignments here because the survival of insect groups from the Palaeozoic into the Mesozoic is such a critical problem. In fact, Tillyard and Fraser (1938–40) regarded the assignment of *Reisia* and *Piroutetia* to the Protodonata as uncertain. In the Russian Textbook of Palaeontology (Rodendorf, 1962), *Piroutetia* was assigned to the Anisozygoptera, a subgroup of the *Odonata (see p. 354). Nevertheless, in view of the evidence provided by *Triadotypus* and the other species, Carpenter (1953) may well be right in regarding the Protodonata as the only one of the 'extinct orders' to survive long into the Mesozoic. Some support for this comes from the existence of some other groups such as the Protomyrmeleontidae (p. 356), which are even more closely related to the latest common ancestors of the *Odonata than are groups like the 'Protodonata' which survived from the Palaeozoic into the Mesozoic and then appear to have become extinct.

The Protanisoptera are a small Permian group that includes about ten species from the Lower Permian of Kansas, Siberia, and the Urals region, and from the Upper Permian of Australia, Arkhangelsk, and the Urals region. Like the Protodonata, they have retained a broad non-petiolate wing base and a very well developed precostal field. A cross-vein ('brace-vein') extends

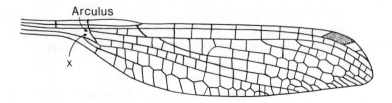

Fig. 32. Wing (probably a hind-wing) of *Permagrion falklandicum* Tillyard, from the Upper Permian of the Falkland Islands. At the point marked × recent Odonata (or *Odonata) have a cross-vein that closes the discoidal cell proximally; see Fig. 33B. From Tillyard (1928a).

from the distal end of the precostal field across the subcosta to the radius, rather as in many Ephemeroptera. This character is not found elsewhere in the Odonata.

In addition to these species, there are some Permian species in which the wing base is strongly constricted or petiolate (Fig. 32). One result of this constriction is that the basal section of the anal vein is fused with the cubitus, and these two longitudinal veins no longer issue independently from the wing base. The precostal field is degenerate and is present only as a thickening of the costa at the extreme wing base. The humeral cross-vein ('brace-vein') is absent: it may have been lost when the precostal field was reduced. The nodus and pterostigma are further derived features that are characteristic of all recent Odonata, and which are found in Permian species. The nodus is a cross-vein that connects the apical section of the subcosta with the radius, and it often continues as a cross-vein ('subnodus') between the radius and the radial sector. The fore-margin of the wing is often constricted at the nodus, but not invariably so. This is a comparatively simple venational feature, and it is conceivable that nodus-like structures may have arisen independently on several occasions. The same may be true of the pterostigma, a sclerotized plate at or in front of the tip of the radius, between the radius and the costa. Martynov thought that the pterostigma was generally a primitive character in insects. This could have an important bearing on the position of the Palaeozoic Odonata in relation to the direct ancestral line of recent Odonata.

There are about 16 Permian species that have petiolate wings with the basal sections of CuP and A_1 fused, and they have been called the 'Protozygoptera'. They are from the Lower Permian of Kansas, Oklahoma, and the Urals, and from the Upper Permian of the Urals region, Arkhangelsk, and the Falkland Islands. Individual species, such as *Permagrion falklandicum* Tillyard from the Upper Permian of the Falkland Islands (Fig. 32), have even been assigned to the recent suborder Zygoptera. However, this gives a distorted view of the actual relationships. It is still customary to divide the recent Odonata into two or three suborders, Zygoptera and Anisoptera, or Zygoptera, Anisozygoptera, and Anisoptera. If the Anisozygoptera (one recent species) and Anisoptera are grouped together, the diagnostic characters that are generally given appear to indicate that the Zygoptera and the Anisozygoptera + Anisoptera are monophyletic sister-groups which have inherited the ground-plan characters of the Odonata and have developed them further in an alternating sequence[151].

Fraser (1954) has shown convincingly that odonatan wings with a broad base have developed from narrow petiolate wings resembling those of the recent 'Zygoptera'. This means that fossil wings which agree formally with those of the recent Zygoptera could just as well be placed in the stem-group of all the recent Odonata.

None of the described Palaeozoic Odonata, including *Permagrion* and the Protozygoptera, has acquired a discoidal cell, and in this respect their venation is more primitive than that of all the recent and most Mesozoic Odonata. The distal border of the cell is actually present in these and other Palaeozoic species: it is formed by the 'arculus', which consists of the basal part of MA^+ and a cross-vein between this and CuP. However, the proximal border of the discoidal cell, a cross-vein between the arculus and CuP, is always absent. It is therefore highly unlikely that the recent Odonata are derived from *different* Palaeozoic forms, and we should continue to assign *Permagrion* and the Protozygoptera to the stem-group of the Odonata.

Another problem is whether any descendants of the various Palaeozoic 'Protozygoptera' survived into the Mesozoic or even later. It has been suggested that the 'Archizygoptera', which include five species from the

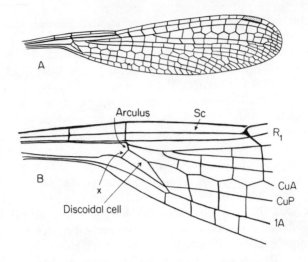

Fig. 33. Fore-wing of *Lestes virens* Charpentier (Odonata; recent). × marks the cross-vein that closes the discoidal cell proximally. From Séguy (1959).

Upper Triassic and from the Lias, had different Permian ancestors from recent and most Mesozoic Odonata, except possibly for the 'Protodonata' (see above). There is no definite evidence for this. The wing has a pterostigma, which is a characteristic feature of most Mesozoic and all recent Odonata. However, there are such striking reductions in the basal part of the wing and around the anal veins that the absence of a closed discoidal cell should probably be regarded as a further reduction. The most recent Palaeozoic species had still not developed a closed discoidal cell, nor has it been proved that it was closed in the oldest Mesozoic species. So it is possible that the Archizygoptera descended from the same early Mesozoic ancestors as all the recent and most (or all?) of the other Mesozoic Odonata.

Revisionary notes

147. A new monograph on the Odonata has been published by Quentin and Beier (1968). [Dieter Schlee.]

148. Jurzitza (1969) has dealt with the structure and function of the pterostigma in the Odonata. [Willi Hennig.]

149. Quentin (1969) has discussed the characters in their venation that the Odonata share with other insect orders. The wing base of the Odonata was described by Grandi (1947a) and compared with that of the Ephemeroptera (Grandi, 1947b). [Willi Hennig.]

150. A study of the comparative functional anatomy of the secondary copulatory organs has been published by Pfau (1971). [Dieter Schlee.]

151. According to Pfau (1971), *Epiophlebia* (Anisozygoptera) and the Anisoptera are sister-groups, but it is still not clear whether the 'Zygoptera' are monophyletic. [Dieter Schlee.]

Appendix to the Palaeoptera: the Palaeodictyoptera and their systematic position[152]

The Palaeodictyoptera and Megasecoptera are generally assigned to the Palaeoptera[153]. Forbes (1943) is the only author to have challenged this, and I shall have to come back to his views later.

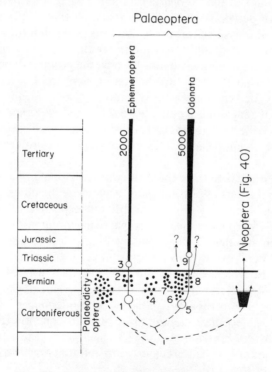

Fig. 34. Phylogenetic tree of the Palaeoptera. 1, *Triplosoba pulchella* Brongniart; 2, Permoplectoptera; 3, so-called *Ephemeroptera from the Lower Triassic (see p. 349); 4, Archodonata (systematic position uncertain, possibly Ephemeroptera); 5, *Erasipteron larischi* Pruvost; 6, Protodonata (Meganisoptera); 7, Protanisoptera; 8, Protozygoptera; 9, Upper Triassic species from the stem-group of the Anisozygoptera + Anisoptera (see p. 354).

Few other groups show the potentialities and the limitations of palaeoentomology as clearly as the Palaeodictyoptera. To the end of his life Handlirsch thought of them as *the* original insects—the stem-group of all the insects and not just of the Pterygota. He believed that winged insects developed directly from the trilobites and that all other insect groups evolved from the Palaeodictyoptera.

It is wrong to regard the Palaeodictyoptera as the stem-group of *all* the insects. It is inconceivable that they could be the ancestors of the Entognatha, Archaeognatha, and Zygentoma, groups which are generally regarded as primarily wingless, and correctly so.

There are both practical and formal aspects to the problem of whether the Palaeodictyoptera should be regarded as the 'stem-group' of at least the Pterygota. It is *a priori* probable that the earliest and most primitive members of all the subgroups of the Pterygota resembled each other more closely than do their modern descendants. So if we assign all these early species to one group solely on the basis of this great similarity and call it the Palaeodictyoptera, then of course the Palaeodictyoptera are the 'stem-group' of most or all the winged insects, depending on how broadly the group is defined. However, this is precisely what I call an 'invalid stem-group' (p. 33), and such groups are found only in typological classifications, which means that they are artefacts and not genealogical units. If we so wished, we could recognize such stem-groups as collective groups of fossils, but this would depend very much on what we are trying to achieve in our phylogenetic studies and classifications.

The modern trend is to trace the history of individual monophyletic groups as far back into the past as possible, and to associate the oldest direct relatives of each group in some way or other with their respective groups. No matter how consistently we pursue these aims, we reach an impasse when the fossils are so fragmentary that it is absolutely impossible to make any definite assessment of their relationships. For example, Carpenter (1954b) assigned the Paoliidae, to which some of the oldest known winged insects belong (*Ampeliptera*, *Stygne*), to the Palaeodictyoptera. Kukalová (1958a, 1958b) treated them as 'Protorthoptera'. Sharov (in Rodendorf, 1962) considered them to be 'Paraplecoptera' (Neoptera), but subsequently (Sharov, 1966b) included them with *Eopterum*[154] in the 'Archaeoptera', the stem-group of the Pterygota. Haupt (1944) had previously insisted that *Stygne* had indisputable fulgoroid features[155]!

Apart from some dubious species and a large number of fragments that cannot be interpreted, the 'Palaeodictyoptera' now include only some Upper Carboniferous fossils concentrated in the 'families' Dictyoneuridae, Lithomantidae, and Spilapteridae. I shall now discuss whether these 'true' Palaeodictyoptera really do form a monophyletic group and, if so, how their relationships should be assessed.

In complete impressions, the wing position shows that some of these species had non-flexing wings (palaeoptery)[156]. With *Ostrava nigra*, which I shall discuss in more detail below, it is even possible to show this by study of the basal articulation of the wing itself. Other species are thought to have had non-flexing wings because their venation agrees so well with that of species that definitely did have non-flexing wings. By definition, however, non-flexing wings cannot be folded back over the abdomen as in the Neoptera, and this is a primitive character. It is therefore wrong to conclude that all the species with this character belong to a single monophyletic group, or that they are more closely related to recent 'palaeopterous' groups (Ephemeroptera and Odonata) than to the Neoptera.

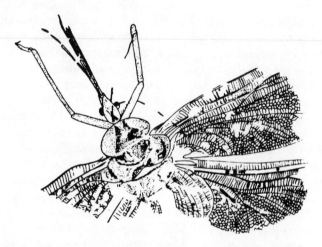

Fig. 35. *Eugereon boeckingi* Dohrn (Palaeodictyoptera), from the Lower Permian of Thuringia. From Laurentiaux (in Piveteau, 1953, based on Dohrn)[157].

First we need to establish whether the Palaeodictyoptera were a monophyletic group. The presence of sucking mouth-parts (a true proboscis) could be evidence for this. Handlirsch (1937) always believed that *Eugereon boeckingi* Dohrn (Fig. 35), the famous 'dragonfly with a bug's proboscis' from the Rothliegende of Birkenfeld on the Nahe[158], was 'the only undisputed example of sucking mouth-parts amongst Palaeozoic insects'. He thought that it marked 'the point of origin of the hemipterous orders'. This is incorrect[159]. Carpenter (1954b) has even distinguished an order 'Protohemiptera' from the Palaeodictyoptera: he stated that it possessed a proboscis, and he assigned several families to it in addition to the 'Eugereonidae' with the sole genus *Eugereon*. Apart from *Eugereon*, a proboscis has in fact only been found in *Lithomantis carbonarius* Westwood and *Mecynostoma dohrni* Brongniart, according to Laurentiaux (in Piveteau, 1953). Other species appear to have been assigned to the 'Protohemiptera' on the basis of similarities in their wing venation. This must also be true of *Lycocercus goldenbergi* Brongniart: Laurentiaux attributed a short proboscis to this species, but Demoulin (1960) asserted that no proboscis is actually visible. It is therefore highly significant that *Stenodictya lobata* Brongniart also has a long proboscis (Laurentiaux, in Piveteau, 1953). Aubert has described this as the most famous and most characteristic of the Palaeodictyoptera, and its venation shows that it belongs to the Dictyoneuridae, which was the largest and, according to Laurentiaux, the most representative family of the Palaeodictyoptera. There is therefore some justification for Demoulin's (1958b) suggestion that the mouth-parts of the 'Palaeodictyoptera' were adapted for sucking rather than chewing, and for Rodendorf's brief statement in the Russian Textbook of Palaeontology (1962) that the entire 'order Palaeodictyoptera' had sucking mouth-parts.

Once it has been established with some certainty that the Palaeodictyoptera were a monophyletic group, then an investigation into their history and the reasons for their extinction becomes a more meaningful exercise. The oldest described species are *Patteiskya bouckaerti* Laurentiaux (1958), from the lower Upper Carboniferous (Namurian, Sprockhöveler Beds) of the Ruhr coal measures, and *Ostrava nigra* Kukalová (1958d), from the lower Upper Carboniferous (Namurian C) of Upper Silesia. About 150 species, excluding a large number of dubious fragments, have been described from the middle and upper Upper Carboniferous. Europe has produced most of these species: not more than a dozen are known from North America, and only two from the Kuznetsk basin in Siberia. Apart from *Stenodictya lobata*, one of the famous giant species has been described from the Upper Carboniferous (Stephanian) of Europe, from the famous inland basin of Commentry: *Homoeophlebia gigantea* Agmis, with a wing-span of 41 cm. Another striking group is the Spilapteridae. These have brightly coloured wings, and Kukalová (1958c, 1961) has reported that some particularly beautiful examples have been found in the Upper Carboniferous (Westphalian) of Nýřany (western Bohemia). The Palaeodictyoptera became extinct at the end of the Palaeozoic, and only a few Permian species are known: *Dunbaria fasciipennis* Tillyard, from the Lower Permian of Kansas (see below), *Oberia longa* Kukalová, from the Lower Permian of the Boskovice trough, two species of *Eugereon* (according to Haupt, 1951) from the lower Rothliegende and lower Zechstein of Germany, four species from the Lower Permian of Kargaly and the Kama basin in the Urals region, and three species (*Kamia angustovenosa* Martynov, *Thnetodes craticius* Martynov, and *Spongoneura incerta* Martynov) from the Upper Permian (Kazanian Stage) of Tikhiye Gory, which is also in the Kama basin[160]. However, Handlirsch (1937) had some doubts about the assignment of these Upper Permian species to the Palaeodictyoptera, and he may well have been right.

The discovery of the wing base, with its articulating sclerites, of *Ostrava nigra* is of particular importance for assessing the phylogenetic position of the Palaeodictyoptera[161]. According to Kukalová's (1958d) illustration, all the major groups of longitudinal veins issue from large plates with which they do not appear to articulate (Fig. 25B). Kukalová distinguished a subcosto-radial plate, medial plate, cubital plate, and anal plates. This is a primitive condition that could really have been predicted and from which the various conditions present in modern Pterygota can be derived. A surprising feature is that no costal plate was apparently found, although it is present in all recent Pterygota and has been called the humeral plate by Snodgrass and Weber; it has been particularly enlarged in the Odonata, as a buttress for the anterior arm of the pleural wing process. A second puzzling feature is Kukalová's assertion that *Ostrava* had a combined subcosto-radial plate[162]. The basal articulation is not known in any fossil apart from *Ostrava*, but in recent insects the subcosta does not appear to be connected to any of the axillary plates and the plates themselves appear to have developed from the bases of the longitudinal veins. According to Weber, there is a process connected to the subcosta, the 1st axillary of Snodgrass or 1st pteralium of Weber, which is said to be attached to the notum.

This problem is not particularly important, and in other respects the basal articulation of the wing of *Ostrava* appears to be more primitive than that of all other known insects. Kukalová, with peculiar reticence, only stressed that the basal articulation of the Odonata could easily be derived from that of *Ostrava*. However, I think that the basal articulation of all recent Pterygota must have developed from a forerunner like that present in *Ostrava*. In this connection Kukalová's description of the (subcosto-)radial, medial, and cubital plates is particularly important. In the Odonata, these three plates, and possibly the anal plates too, though this is not important, appear to have fused into a single radio-anal plate, to use Tannert's term. The same may be true in the Ephemeroptera, but here the 'anal plates' are free. The *Palaeoptera seem to be more derived than *Ostrava* in having the (subcosto-?)radial and cubital plates fused into a single articulating process[162].

In the Neoptera this process has been divided into the 2nd axillary and the so-called median plate, and in this respect their basal articulation is even more derived. I shall discuss this in more detail in the section on the Neoptera. It is possible that both the Palaeoptera (Ephemeroptera + Odonata) and the Neoptera have descended from a common ancestor in which the radial, medial, and cubital plates, or the expanded bases of all the veins of the radial, medial, and cubital complexes, were fused into a single plate. They are still separate in *Ostrava*. If we believe that *Ostrava* really does belong to a monophyletic group 'Palaeodictyoptera', then the structure of the basal articulation as described by Kukalová would be provisional evidence that the Palaeodictyoptera really do belong to the stem-group of the Pterygota. It would also indicate that the Palaeodictyoptera had diverged from the main pterygote stem before the lower Upper Carboniferous and before the origin of the latest common stem-form of all recent Pterygota (Palaeoptera + Neoptera).

At the time that the Palaeodictyoptera were living, the Odonata, Ephemeroptera, and several subgroups of the Neoptera were also in existence. The Palaeodictyoptera can therefore only have been descendants from the stem-group of the Pterygota, and their proboscis indicates that they were probably descendants from a side-branch. They would still have to be regarded as a side-branch even if they should prove not to belong to the stem-group of the Pterygota but to that of the Palaeoptera, where they have usually been assigned (even by Sharov, 1966b), or to that of the Neoptera (Forbes mentioned that the Holometabola might have originated from the Palaeodictyoptera!). No matter what their systematic position is eventually considered to be, they cannot give us any insights as to when the division in the insect lineage took place that could not have been deduced without them. I do not know of any evidence to support Ross (1965), whose phylogenetic tree (his Fig. 346) showed the Palaeodictyoptera as a side-branch of the Ephemeroptera.

Including the Megasecoptera in the Palaeodictyoptera in the broad sense does not substantially alter this picture. Handlirsch (1937) described this group as a branch of the Palaeodictyoptera that led to a dead-end. Carpenter (1954b) distinguished them from the Palaeodictyoptera by the presence of distinct cross-veins that were not irregular or reticulate. It is also interesting that he expressly mentioned the absence of a paracercus as a diagnostic character. No paracercus has ever been noted in Palaeodictyoptera that still have the abdomen preserved. The paracercus is definitely part of the ground-plan of the Pterygota and has been retained in the Ephemeroptera. Its absence in the Palaeodictyoptera and Megasecoptera could be a derived character common to these groups (synapomorphy) which indicates that they are closely related. However, atrophy of the paracercus can easily take place convergently. It is absent in all Pterygota except for the Ephemeroptera, and so its absence in the Megasecoptera would be compatible with almost all the various suggestions that have been made about the relationships of the group. *Dunbaria fasciipennis* Tillyard (1924), from the Lower Permian of Kansas, is said to be the only species with a paracercus[160]. This is extremely surprising as it is thought to be one of the latest survivors of the Palaeodictyoptera. Handlirsch (1937) has written: 'The long cerci are ephemerid-like. They are situated behind the 10th segment, and between them a process can be seen which is presumably a terminal filament'. However, is it possible to interpret this process differently? Laurentiaux (in Piveteau, 1953) thought that it was a short ovipositor, and I am inclined to agree with him.

The Megasecoptera also seem to have had a proboscis, though only a short one, and this feature has an important bearing on their relationships. Kukalová (1961) reported that she had found two complete specimens in the Permian of Czechoslovakia 'which established for the first time that the Megasecoptera had a broad rostrum; a curved sabre-like ovipositor which had been ruffled and on which forwardly directed hairs could be detected; and possibly a ceruminous gland'. However, Laurentiaux (in Piveteau, 1953) had already pointed out that a short ovipositor and a short 'rostrum' had both been noted in certain Megasecoptera, and he suggested that they were characteristic of the entire group. It has also been stated that mandibles were present in *Psilothorax*, but this has not been confirmed (Carpenter, 1951).

The Megasecoptera may therefore have had sucking mouth-parts like the Palaeodictyoptera.

The oldest species are from the middle Upper Carboniferous, and about 55 have been described. None are known from the lower Upper Carboniferous, but this is not important when we recall that it is only recently that the first two species of Palaeodictyoptera have been described from this horizon. As with the Palaeodictyoptera, most of the Carboniferous Megasecoptera have been described from Europe, but curiously none has been found in the Saar–North France–Limburg–Belgium coal basin where so many Palaeodictyoptera have been found. Three species are known from the Kuznetsk basin in Siberia, and five from North America. Like the Palaeodictyoptera, the Megasecoptera became extinct at the end of the Palaeozoic. However, a relatively large number of species is known from the Lower Permian: about 30 species from North America (Kansas and Oklahoma) and about a dozen from the Pechora basin and the Urals region (Kama basin) of the Soviet Union. Seven species have been described from the Upper Permian (Kazanian Stage) of Iva-Gora and the Kama basin. Unlike the Palaeodictyoptera, there are no striking giant forms amongst the Megasecoptera. The largest species belong to the genus *Mischoptera* and Laurentiaux (in Piveteau, 1953) stated that they had a wing-span of 168 mm.

The Megasecoptera are particularly interesting because they included one group of species that could flex their wings backwards. In this respect they resembled the Neoptera, at least externally. Carpenter (1945) has discussed how two different phylogenetically uniform groups ('phylogenetic lines') can be distinguished in the 'Megasecoptera': 'Protohymenoptera' with non-flexing wings (families Aspidothoracidae, Corydaloididae, Mischopteridae, Protohymenidae), and 'Eumegasecoptera' (families Diaphanopteridae, Prochoropteridae, Elmoidae, Martynoviidae, Asthenohymenidae) which could flex their wings back over the abdomen like the Neoptera. At first (Carpenter, 1945), he was unable to decide whether the Eumegasecoptera or even the entire Megasecoptera belonged to the Neoptera, but later (Carpenter, 1954b) he suggested that the Eumegasecoptera had a different mechanism from the Neoptera for flexing their wings backwards. This was also Zalesskiy's view (1958). If this were correct, then I would follow Handlirsch, who suggested that the Megasecoptera were a monophyletic group; that their proboscis indicates that they were closely related to the Palaeodictyoptera; that both groups belonged to the stem-group of the Pterygota; and that a later branch independently acquired the ability to flex the wings back over the abdomen but became extinct during the Permian.

Forbes' (1943) views were different from those of any other author. He considered that the most important feature in the Odonata and Ephemeroptera (though not in *Triplosoba*, p. 136) was the development of the radius (R_1) as an independent vein, whereas in all other insects the convex R_1 is attached to the first concave longitudinal vein (R_{2+3}) that lies behind it. As R_1 was not free in the Palaeodictyoptera and Megasecoptera, Forbes thought that these two groups were more closely related to the Neoptera than to the Palaeoptera (Ephemeroptera + Odonata). He even assigned the Megasecoptera to the Holometabola, but he said nothing about whether the Palaeodictyoptera were more closely related to the Holometabola than to other groups of the Neoptera. For him the vital feature was the fate of the convex vein which lay between the concave media and the concave branch (R_{2+3}) of the radial sector. He believed that this was originally an independent convex longitudinal vein: in many groups it is attached to the radial sector, in which case he called it R_{2+3}, but in other groups it is attached to the media, and he then called it MA. According to Forbes, the way in which this convex longitudinal vein developed is of fundamental importance for interpreting the relationships between the various groups of winged insects. I shall deal with this view again in the general discussion of the Neoptera, but for the present I shall just point out that Forbes was forced by his own interpretation of the venation to suggest that neoptery has developed polyphyletically amongst the recent Neoptera: the ability to flex the wings back over the abdomen must have arisen independently and convergently in the Holometabola, via the Megasecoptera and possibly the Palaeodictyoptera, and in the Polyneoptera + Paraneoptera, which Forbes called the 'Orthopteroidea'.

The uniform structure of the basal articulation of the wing in all recent Neoptera is a strong argument against this view. Moreover, Forbes was compelled to reject any idea that the Protephemeroptera (*Triplosoba*) are closely related to the Ephemeroptera, which is supported by a great deal of evidence (p. 136), and actually had to assign *Triplosoba* to the Palaeodictyoptera: these are further arguments against the great importance which he attached to the independence of the radius in the 'Palaeoptera' and which made it impossible for him to treat the Palaeodictyoptera and Megasecoptera as members of the stem-group of all the Pterygota. It will be necessary to find a different explanation for the development of the pterygote wing venation from Forbes'. This does not affect the problem of whether the Eumegasecoptera belong to the Neoptera or even to the Holometabola. If they do, then the characters that they share with the Protohymenoptera (of Carpenter, 1945) would be symplesiomorphies or the results of convergence. This problem will not be solved until there is some information about the structure of the basal articulation of the wing in the Eumegasecoptera.

Many suggestions have been made as to why the 'Palaeodictyoptera' and 'Megasecoptera' became extinct. However, such speculation is pointless if we do not know the extent to which these groups are monophyletic, if at all, because a historical development in any real sense can only take place in monophyletic groups. According to Laurentiaux (in Piveteau, 1953), Martynov suggested that the aquatic mode of life of their larvae was responsible for the extinction of the Palaeodictyoptera and Megasecoptera, but it has not been proved that either group had aquatic larvae. Handlirsch (up to 1937), too, thought that their larvae were aquatic, but Carpenter (1948a) has shown that there is no evidence for this. It is very likely that typological analogies have been responsible for this: both groups have non-flexing wings (and the ability of some Megasecoptera to flex their wings back over the abdomen was only discovered much later); as a result they were assigned, and are usually still assigned, to the 'Palaeoptera' alongside the Ephemeroptera and Odonata which also have non-flexing wings; and as all recent 'Palaeoptera' have aquatic larvae, it was thought that this must have been true for the Palaeodictyoptera and Megasecoptera—an unwarranted assumption. Sharov (1966b) has recently discovered some nymphs in the Upper Carboniferous of Siberia, which undoubtedly belong to the Palaeodictyoptera. He has reported that they were not aquatic[163].

Weber believed the exact opposite of this. He thought that of all the groups with non-flexing wings, the Ephemeroptera were most able to survive because they spent the greatest part of their lives as larvae (in water!). He believed that the Palaeodictyoptera, in which he apparently included the Megasecoptera, 'possibly' died out because of their non-flexing wings which left them less able to compete with the Neoptera in the struggle for existence. This could not have been true of the Eumegasecoptera. Laurentiaux (in Piveteau, 1953) suggested that the proboscis which he attributed to all 'Palaeodictyoptera' was a specific adaptation to particular plants: when these died out, the Palaeodictyoptera followed them into extinction. However, it is not known how the Palaeodictyoptera used their proboscis. Carpenter (1953, 1954a) gave no opinion as to whether they fed on plant juices or on the blood of amphibians and reptiles.

We really need to place this problem in a much broader context. Very many groups of animals, and not just insects, died out during the Permian. Although a fair number of Neoptera are known from the Palaeozoic, relatively few survived into the Mesozoic. The Permian was clearly a period of drastic environmental change, and it is therefore much more meaningful to enquire what were the special features that enabled particular groups to *survive*, and not to worry too much about the nature of the disadvantages which condemned those like the Palaeodictyoptera to *extinction*. The survivors radiated with such vigour that their success only accelerated the final decline and extinction of these other groups.

Revisionary notes

152. Some new observations and fresh discussion of the Palaeodictyoptera have been published by Müller (1977b). [Dieter Schlee.]

153. Riek (1973b) described 'a species of Palaeodictyoptera with some features of the Megasecoptera' (Riek, in CSIRO, 1974, p. 28) from Upper Carboniferous shales in Tasmania. [Dieter Schlee.]

154. *Eopterum* is not an insect: see note 57. [Dieter Schlee.]

155. For further discussion of *Stygne*, see Schwarzbach (1939). Pruvost (1927) stated that *Ampeliptera* would belong to the Protorthoptera sensu Pruvost or to the Palaeodictyoptera sensu Handlirsch. [Willi Hennig.]

156. Müller (1977b) has emphasized that the absence of axillary sclerites and the large, broad wing base are further indications that these were non-flexing wings. [Dieter Schlee.]

157. Müller (1977b) has given a detailed analysis and new illustrations of this fossil. [Dieter Schlee.]

158. A single specimen of *Eugereon* is still all that is known, a mould and cast from a nodule of clay ironstone. See the preceding note. [Dieter Schlee.]

159. Müller (1977b) has also emphasized that the similarities between *Eugereon* and the Hemiptera are the results of convergence. [Dieter Schlee.]

160. Kukalová-Peck (1972a) has published a detailed redescription of *Dunbaria*, from which it appears that there was no paracercus, and (1975) has described some new Megasecoptera from the Lower Permian of Moravia. [Willi Hennig.]

161. The basal articulation of the wing of *Eugereon* has been dealt with by Müller (1977b), and that of the Palaeodictyoptera of Commentry by Kukalová (1969a, 1969b, 1970). See also Kukalová-Peck (1974). [Dieter Schlee.]

162. There is a subcosto-radial plate in *Eugereon* too (Müller, 1977b). [Dieter Schlee, Rainer Willmann.]
These results could indicate that this is a special character and might perhaps be a synapomorphy of the Palaeodictyoptera, or of a subgroup of the Palaeodictyoptera. This problem will only be solved by further comparative studies of well preserved fossils. [Dieter Schlee.]

163. Carpenter and Richardson (1972) have described what they believed to be an undoubted palaeodictyopteran nymph. They could find 'no indication that the nymph was modified for an aquatic existence' (Carpenter and Richardson, 1972, p. 271). [Willi Hennig.]

2.2.2.2. Neoptera

Carpenter (1953, 1954a) estimated that 97% of recent insect species belong to the Neoptera. Handlirsch, even in his last paper (1937), never accepted that the Neoptera are a monophyletic group, but his position has two fundamental weaknesses.
The first is that he attempted to derive the Neoptera from *different*

'Palaeodictyoptera' and at the same time gave the Palaeodictyoptera a completely artificial definition. It sometimes happens that several fossil species are included together in a single group because of their great similarity, but it also escapes attention that this similarity is based only on a number of shared primitive characters ('symplesiomorphies'). This is the case with Handlirsch's 'Palaeodictyoptera', which includes the stem-species of various subgroups of a single recent monophyletic group. These stem-species still resemble each other very closely. This kind of fossil group is what I have called an 'invalid stem-group'. However, if we then go on to say that a particular group of recent organisms is polyphyletic because it has arisen from *different* 'subgroups' of an invalid stem-group, this is certainly a circular argument and is logically incorrect. It is wrong to believe that a recent group is not monophyletic because its subgroups have arisen from *different* elements of an invalid stem-group (see Fig. 8 on p. 34).

Handlirsch's second error is an overhasty typological assessment of 'neoptery' because, if 'neoptery' is understood to mean no more than the ability to flex the wings backwards, it could have developed independently on several occasions, that is to say polyphyletically. However, even·if this could be proved, it does not follow that the *group* in which this particular *character* has arisen polyphyletically is in itself polyphyletic. It is possible that other characters could prove that the group is monophyletic.

The main problem with the Neoptera is that the 'Eumegasecoptera', a group of the Palaeozoic Megasecoptera, appear to have the character of 'neoptery'. However, they have other characters that appear to indicate that they are not closely related to the other 'neopterous' insects (the true Neoptera). There is no evidence that the 'neoptery' of the Palaeozoic Eumegasecoptera is really the same as that of the true Neoptera and it could well be that they used a different mechanism for flexing their wings backwards. This would simply imply that the relationships of the fossil Eumegasecoptera remain completely unclear.

The structure of the basal articulation of the wing which enables the wing to be flexed backwards agrees so closely in all the recent Neoptera, and probably in the fossil groups that have been assigned here too (except for the Eumegasecoptera), that it is difficult to imagine that it could have arisen polyphyletically.

In order to understand the structure of the basal articulation of the wing in the Neoptera, mention should be made of the basal processes once again. It should first be recalled that some of these appear to have developed originally from the enlarged bases of the longitudinal veins. The enlarged bases of individual groups of veins are still rather conspicuously separated from each other in *Ostrava nigra* from the lower Upper Carboniferous (Fig. 25B on p. 129), which is the oldest winged insect where the basal articulation is known. On the other hand, in recent Palaeoptera the enlarged bases of the radial, medial and cubital complexes appear to have fused into a single plate, whilst in at least the Ephemeroptera the enlarged bases of the anal veins are still more or less separate. In the Palaeoptera the basal plates are still firmly attached to the longitudinal veins from which they originated.

The basal articulation of the Neoptera (Fig. 25A) differs fundamentally from that of the Palaeoptera in three respects:

1. The presence of a single large and powerful process (3rd axillary, Weber's 3rd pteralium), which appears to have developed from the enlarged bases of the anal veins.
2. The presence of two processes instead of the single radio-medio-cubital plate. One of these processes (2nd axillary, Weber's 2nd pteralium) is attached to the radius, whilst the other (Weber's ancillary plate, Snodgrass' anterior median plate) is attached to the media and the cubitus.
3. These processes are not firmly attached to the longitudinal veins.

These characters can only be explained if their interactions and functional significance are understood. They also need to be considered as further developments of conditions that have been partly retained in the Palaeoptera.

A single large '3rd axillary' has apparently developed from the enlarged bases of the anal veins (character 1) because it is the point of attachment for the muscle which flexes the wing back over the abdomen, into its resting position. This muscle is also present in the Palaeoptera (Snodgrass' Fig. 103D, according to Forbes, 1943). In the Palaeoptera it is attached to the almost united axillary sclerite and pulls it downwards, whereas in the Neoptera it continues to contract and pulls the wing downwards and backwards in a way that has been described in detail by Snodgrass. This can only be done if the firm attachment between the longitudinal veins and their basal plates is loosened (character 3). Forbes stated that in the Neoptera the longitudinal veins flex directly behind the point where they issue from the basal plates and that they are often entirely membraneous behind the radius. The division of the single radio-medio-cubital plate (character 2) is probably associated with the need to make the wing base more flexible along its transverse axis.

In the wing base of the Neoptera, the 2nd axillary and the ancillary plate are separate: the 2nd axillary is connected to the radius, and the ancillary plate to the medio-cubital complex. It is tempting to treat this as a relatively primitive state (when compared with the Palaeoptera) and as a direct development from a forerunner such as the lower Upper Carboniferous *Ostrava*. The fact that the 2nd axillary is connected to the 3rd axillary behind, as well as to the radius in front, is an argument against this. As a result its position is not *in front of* but proximally *alongside* the ancillary process. Even the fact that the 2nd axillary rests on the pleural wing process may indicate that this axillary and the ancillary plate have arisen through division of what was originally a single plate: Forbes (1943) believed that the pleural wing process originally supported the wing in the area of the media and not in the area of the radius. It is also possible that part of the 3rd axillary, which Snodgrass called the posterior median plate, developed from the radio-medio-cubital plate which was originally a single plate. Its attachment to the sclerites that form most of the 3rd axillary would then have taken place independently. Finally, I should point out that the basal articulation of the wing has been modified still further in various groups of the Neoptera and that it never agrees exactly with the simplified scheme just described.

The name Neoptera did not originally refer to features in the basal articulation of the wing or to the ability to flex the wing backwards. Martynov (1924b, 1925a) recognized this group of the Pterygota because all the species have a characteristic section of the wing which he called the 'neala' and which is supposed to be absent from the Palaeoptera.

Unfortunately, there has been some confusion over the term neala, which has been applied to different areas of the wing, and the nomenclature of the associated longitudinal veins. If we are not aware of this confusion, we shall make mistakes when comparing the views of different authors.

Forbes (1943) pointed out that there is a furrow that is characteristic of the Neoptera, and he particularly stressed that it is a furrow and not a fold, as has often been incorrectly stated. He called it the 'anal furrow' and stated that it separates the anal lobe from the rest of the wing. Snodgrass called the anal lobe the 'vannus', and its longitudinal veins the vannal veins (V_1, V_2, etc.). However, these particular veins have usually been called anal veins (1A, 2A, etc., or Weber's An_1, An_2, etc.). Directly in front of the anal furrow there is a concave vein that Snodgrass called the 'postcubitus' (PCu) and this should not be confused with the posterior cubitus (CuP), the concave posterior branch of the cubital vein. The postcubitus is said to correspond to a tracheal stem that originally entered the wing base independently and, like the media and the cubitus, it is sometimes connected to the 'ancillary plate' (Snodgrass' anterior median plate). Ragge (1955a) rejected the terms 'postcubitus' and 'vannal veins' and, following Comstock, Needham, and others, preferred to call the postcubitus the '1st anal vein'. He and authors following him have called the various 'vannal veins' the 2nd, 3rd, etc., 'anal veins'.

A second furrow, that Martynov (1925a) called the 'jugal furrow' or 'anojugal fold', separates another smaller area from the anal lobe. This area is called the jugum (= jugal lobe) or neala (Martynov, 1925a). It has two longitudinal veins which are known as jugal veins, or vena arcuata and vena cardinalis. Sometimes, however, the entire anal lobe has been called the 'neala', and in some papers such as the Russian Textbook of Palaeontology the term jugum has even been used in the cockroaches for this lobe. This is not the original meaning of these terms and should be avoided. Forbes (1943) could find no obvious function for the anal furrow in the adult wing, and suggested that its development might have been connected with the packing of the unexpanded wing in the pupal sheath. It is a striking fact that the anal and jugal furrows form the anterior and posterior borders respectively of an area in the wing where the longitudinal veins, Snodgrass' 'vannal veins', usually (and probably originally) issue from a single stem, which is connected to the 3rd axillary. Furthermore, at its anterior end the anal furrow meets the edge of this wing process, which is such an important feature of the Neoptera. Unfortunately, no functional connections can be deduced from the kind of description that is usually published. Nevertheless, it does seem rather likely that the development of the anal and jugal furrows is connected with the ability of the Neoptera to flex their wings back over the abdomen.

Rodendorf and others have emphasized that the ability to flex the wings back over the abdomen was a definite advantage for the primitive Neoptera, because it enabled them to conceal themselves in narrow places. This in turn provided a stimulus for diversification in wing function. In several groups the fore-wings developed into protective covers for the hind-wings and even for the abdomen, whilst the function of flight was taken over almost entirely by the hind-wings. In other groups the fore-wings became the main or sole organs of flight.

In a more recent work, Snodgrass (1957) has stated that a further characteristic feature of most Neoptera is that the claspers of the male genitalia, which developed from the styli, have been reduced. They may have been replaced by analogous structures that have developed from the phallic primordia, though this feature is not yet present in the ground-plan. Moreover, Snodgrass believed that in such species the primary phallic lobes have divided into two secondary lobes: the median pair ('mesomeres') have united to form the aedeagus, whilst the lateral lobes ('parameres') have developed into claspers which in their turn may have become two-segmented secondarily. *Grylloblatta* (Notoptera) is the only group in which claspers homologous with the styli are said to have been retained. Should this be true, then the secondary reduction of the styli and their replacement by the parameres must have taken place several times in the Neoptera. However, Sharov (1966b) believed that Snodgrass' interpretation was wrong, and he considered correctly that the male claspers of all Dicondylia developed from appendages of the 9th abdominal segment[164].

The oldest fossils that can be reliably assigned to the Neoptera are known from the lower Upper Carboniferous. They belong to various subgroups of the *Neoptera, which must have arisen at an even earlier period.

Unfortunately, there is still considerable doubt about the monophyletic subgroups of the *Neoptera. Martynov distinguished three groups, Polyneoptera, Paraneoptera, and Oligoneoptera (= Holometabola), and these divisions are still often used. There is little doubt that the Holometabola and even the Paraneoptera (or at least their basic groups) are monophyletic. The Polyneoptera are much more difficult to assess. They are not shown as a monophyletic group in Martynov's (1938b) phylogenetic tree, nor by Jeannel (in Grassé, 1949) or the Russian Textbook of Palaeontology (Rodendorf, 1962), both of which follow Martynov closely. Moreover, it is still not certain whether the Paraneoptera are more closely related to the Holometabola or to the Polyneoptera, should the latter prove to be a monophyletic group. In the following discussion I am accepting the first of these possibilities, with certain reservations. I suggest that the first event that can be recognized in the phylogenetic development of the Neoptera is the origin of the sister-group relationship between the Paurometabola and the Eumetabola (= Paraneoptera + Holometabola). The Plecoptera and the Paurometabola have usually been called the Polyneoptera, but at present the relationships of the Plecoptera are unclear.

Additional note

164. The various theories about the gonopods have been summarized by Tuxen (1969) and Scudder (1971). [Willi Hennig.]

2.2.2.2..1. Plecoptera (stoneflies)[165]

Ground-plan and classification

If the Embioptera are left on one side for the time being, there are two different views about the relationships of the Plecoptera, but I have not been able to decide which of them is correct. The first view is that the Plecoptera are closely related to the orthopteroid insects, particularly to the grasshoppers and their relatives (Martynov, 1938b; Jeannel, in Grassé, 1949; Rodendorf, 1962), and the other that they are closely related to the Paraneoptera + Holometabola (Ross, 1955, 1965). There is no definite evidence for either of these views[166].

I do not know of any derived characters common to the Plecoptera and the orthopteroid orders (Paurometabola), except possibly for the enlarged anal fan of the hind-wings, which are the main organs of flight. However, this is a character that could very easily have arisen convergently, and in any case it is uncertain whether it is in the ground-plan of the Paurometabola. The enlargement of this fan reflects the increased number of anal veins. Unfortunately, it is not known how many anal veins there are in the ground-plan of the Neoptera (or Pterygota). It has been suggested that originally there were only three, as in the fore-wing of the Plecoptera, and that the presence of more indicates that a secondary increase has taken place. This increase could also have taken place independently in different groups: the fore-wing of the Ephemeroptera, for example, also has more than three anal veins.

Sharov (1966b) believed that the enlargement of the anal fan is a derived character of the 'Polyneoptera', to which he assigned the Plecoptera, but he weakened his argument with a whole series of equivocal and contradictory statements. For example, at one point (his p. 119) he tried to derive the Paraneoptera from the Protoblattodea (in other words, the 'Polyneoptera'), but elsewhere he stressed the fact that the 'Polyneoptera' and Holometabola agree closely in the (derived!) mode of their embryonic development. On the other hand, the Paraneoptera and the Palaeoptera were said to have a different but equally derived mode of embryonic development. At the same time he mentioned that the Plecoptera, which he regarded as a group of the 'Polyneoptera', were more primitive than *all* the other Pterygota in their embryonic development. Little can be made of statements such as these which fail to make any sharp distinction between typological and genealogical principles of 'derivation'.

Ross (1955) has mentioned the mesotrochantin as evidence that the Plecoptera are more closely related to the Paraneoptera and Holometabola. In these three groups, the mesotrochantin is supposed to have become a slender band-like sclerite, which at first was unattached in the leg membrane. In the 'orthopteroids', on the other hand, it has remained essentially a triangular sclerite as in the Palaeoptera. These conclusions have not been confirmed by Matsuda (1960). It is more tempting to regard the fusion of the trochantin with the katepisternum as a derived character common to these three groups. The trochantin is part of the subcoxal articulation, but we still know much too little about it to draw any far-reaching conclusions from the differences and agreements amongst these groups[167].

Adams (1958) agreed that there is a close relationship between the Plecoptera and Holometabola. He stated that the acquisition of a 3rd coxal articulation, between the furcasternum and coxa, is probably the most characteristic feature in the thorax of the Holometabola. He described how a process issues from each side of the furcasternum in 'some' Plecoptera (*Taeniopteryx*) and approaches very close to the coxae. He referred to Hanson, who stated that these furcasternal arms do not articulate with the coxae but provide them with firm support when the legs are moved. Adams considered that this feature of the Plecoptera was a direct forerunner of the sternocoxal articulation of the Holometabola. However, not much can be made of his statements because they are too vague. It is not at all clear whether the furcasternal arms are part of the ground-plan of the Plecoptera or are only 'autapomorphies' that have developed in some of the more derived species of this group[168].

The Plecoptera also have some characters that are undoubtedly so primitive that it would be reasonable to suggest that there is a sister-group relationship between the Plecoptera and the rest of the Neoptera. According to Sharov (1957a, 1957c), the Plecoptera are the only Pterygota that have retained the primitive mode of embryogenesis of the 'Apterygota': the germ-band is curved ventrally and immersed in the yolk at the base of a sac-like invagination (amniotic depression). According to Snodgrass (1935), the original division between anapleurite and coxopleurite has only been retained in the prothorax of the Plecoptera. In all other groups these pleurites have fused into a single plate[169].

If we wanted to reconstruct a hypothetical ancestor from which all the Neoptera could be derived, it would not be very different from the most primitive Plecoptera. This makes it very difficult to find the stem-group of the Plecoptera and to determine the age of the group: Palaeozoic fossils which are only slightly more primitive than recent Plecoptera could belong to the stem-group of all the Neoptera, or of the Plecoptera, or even of other groups of the Neoptera.

The recent Plecoptera have a number of characteristic derived ground-plan features. The larvae are aquatic and have tracheal gills, at least some of which appear to have developed from rudiments of the abdominal appendages. According to Snodgrass (1961), the larval musculature described by Wittig (1955) also shows some striking specializations. Larvae and imagines have only three tarsal segments. The female has no ovipositor: Quadri (1940) found that primordia of the anterior valves occur only in young nymphs and do not develop further. Finally, the adults lack the 1st abdominal sternite which has been retained in many Paurometabola[170].

As in other groups, the wing venation is particularly important because it is usually the only feature that is preserved in the fossils. It is generally accepted (Illies, 1965) that the Plecoptera (Fig. 36) are characterized by having MP^- in the fore-wing completely fused with CuA^+, so that MA^+ is free. In the hind-wing the basal section of MA^+ is fused with the basal section of R and RS; as a result the only free branch of the media (MA^+) appears to issue from the radial sector, and the root of MP^-, which is also called the 'arculus' in both wings, still issues from the common stem of R and M.

Séguy (1959) has put forward a different interpretation. In his view there is no difference between the fore-wing and the hind-wing. He believed that both wings have MA^+ attached to the radial sector and that both wings have branches of MP^- that terminate freely in the wing margin[171].

It is really the fate of the media that has given rise to these differences of opinion. They can be resolved, and their full importance understood, only if they are considered in relation to the general trends that the development of the wing venation has followed in the Pterygota, or at least in the Neoptera.

In the Pterygota, the media and its branches are at the middle of the wing, between the radial and cubital complexes. It is obvious that any reduction in wing size, and particularly any concentration of the articulating processes of the wing base and the associated constriction of the wing base, must have caused sections of the media to fuse with branches of the radius in front, or of the cubitus behind. It is generally believed that the medial complex originally had the same structure as the two adjacent groups of veins: it consisted of an anterior convex element (MA^+) and a posterior concave one (MP^-).

According to Forbes (1943), all Neoptera have the anterior convex branch (MA^+) attached to the radius or radial sector, and only the posterior concave branch (MP^-) is free. On the other hand, Sharov (1966b) believed that all Pterygota have MP^- fused with CuA^+ and that only its basal section is retained as an apparent cross-vein between M and Cu. The branches of the media that reach the wing margin should then be called MA.

Few Paurometabola have a venation that is well enough developed to be interpreted. I do not think that there is any convincing evidence that they have MA attached to the radial sector, and Ragge's (1955a) suggestion that MA is free is much more likely. On the other hand, in the Plecoptera (Séguy,

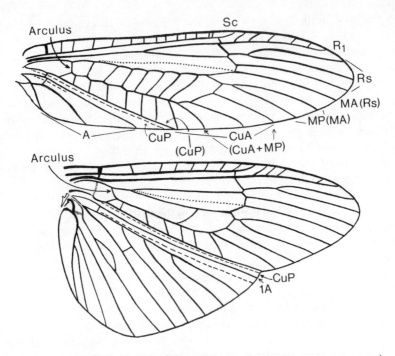

Fig. 36. Wings of *Dinocras cephalotes* Curtis (Plecoptera; recent).
Séguy's (1959) interpretation of the venation has been followed. In the
fore-wings, the customary nomenclature, as used by Sharov, has been
added in parentheses. I would like to suggest that in both wings MA
has fused with the posterior branch of the radial sector (RS).

1959, see above), Paraneoptera, and Holometabola, MA really does seem to
be attached to the posterior branch of the radial sector (R_5) and does not
reach the wing margin.

Carpenter (1966) recently dealt with this question in some detail. He
emphasized that the current interpretation of the wing venation, which is
usually attributed to Comstock and Needham, is actually based on Redtenbacher who was the first to recognize the important difference between
convex (+) and concave (−) veins. Moreover, Carpenter pointed out that the
tracheal paths are not as important as has usually been thought for homolo-
gizing the individual veins. He quoted the work of Holdsworth (1940, 1941)
and others who have shown that it is only late on in ontogenesis that the
tracheae grow into the blood lacunae which are along the sites of the
presumptive adult veins. When extending through the wing, the tracheae
follow the lines of least resistance.

Carpenter established that it is not possible to make any definite distinction
between MA and MP in the 'orthopteroids' because the media never actually
has a convex anterior branch. Carpenter's 'orthopteroids' include the
Paurometabola of the present work and the Palaeozoic fossils which Sharov

and others considered to be the closest relatives of the Plecoptera. Other methods have been used to identify an anterior branch of the media as MA. In the 'Protorthoptera', which are particularly important when considering the origin of the Plecoptera (p. 168), many species have this branch free, whilst in others it fuses with the radial sector at one point or for a short distance. Fusion for a short distance is a characteristic feature of the Oedischiidae which are generally and correctly assigned to the stem-group of the Orthopteroidea (in my sense: see pp. 211–212).

Carpenter has suggested that the connection between MP and CuA has arisen in various ways and has taken place on several occasions independently. I am not convinced by all the details of his interpretation. I would prefer to follow Sharov in interpreting the 'cross-vein' connecting MP and 'CuA$_2$' as the base of a posterior branch of MP that has fused with CuA, although Carpenter called it 'CuA$_1$' because it is convex. On the other hand, I think that Sharov was wrong to regard the remaining branches of the media to reach the wing margin as elements of MA. I believe that Carpenter's interpretation of this feature is probably correct.

It has been suggested that the posterior branch of MP, not the entire vein itself, was fused with CuA in the ground-plan (in the stem-group) of all Pterygota, or at least of all Neoptera (see Fig. 54). The 'arculus' of the Plecoptera is the base of this branch. I do not think that this suggestion can be refuted, and consequently there is no support for the claim that certain Carboniferous fossils should be assigned to the stem-group of the Plecoptera on the basis of this character (see below).

There is no precostal field in the Plecoptera. The tip of the subcosta is forked and, according to Séguy (1959), one branch (Sc$_1$) reaches the costa whilst the other (Sc$_2$) terminates in the radius. This also appears to be a feature of the ground-plan of the Paraneoptera and Holometabola.

If we want to trace the Plecoptera as far back into the past as possible, the first step must be to find the oldest fossils in which all, or as many as possible, of the derived characters of recent Plecoptera are present. The smaller the number of derived characters common to fossils and recent species, the more doubtful is the assignment of fossils to the stem-group of the Plecoptera. However, no assignment can be refuted or proved, even in fossils where all the visible characters are more primitive than in the recent Plecoptera.

For this reason, it is highly significant that several species which undoubtedly belong to the Plecoptera have been found in the Lower Permian of the Urals region (Martynov, 1940) and of the Kuznetsk basin (Sharov, in Rodendorf et al., 1961). Most of them are represented only by fore-wings, but *Perlopsis filicornis* Martynov (Fig. 37) and *Palaeotaeniopteryx* are known from fore- and hind-wings and these agree very closely with the wings of recent Plecoptera. In *Perlopsis filicornis* even the tarsi are preserved, and these have three segments as in recent Plecoptera[172].

In the light of these discoveries, it scarcely matters whether any other Permian fossils actually belong to the stem-group of the Plecoptera. Carpen-

166

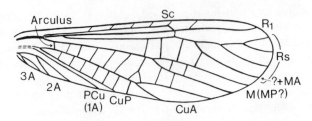

Fig. 37. Fore-wing of *Perlopsis filicornis* Martynov
(Plecoptera), from the Lower Permian of the Urals.
From Rodendorf (1962).

ter (up to 1954) believed that the only group closely related to the Plecoptera
was the Protoperlaria, to which he assigned the single 'family Lemmatophor-
idae' (Fig. 38). This family includes 10 or more species from the Lower
Permian of Kansas and Oklahoma, 1 species from the Lower Permian of the
Urals, and possibly 1 very problematic species from the Permian of Thuringia
(*Germanoprisca zimmermanni* Zeuner)[173].

According to Carpenter, the larvae of the Lemmatophoridae were aquatic,
and this is evidence that the group should be assigned to the stem-group of the

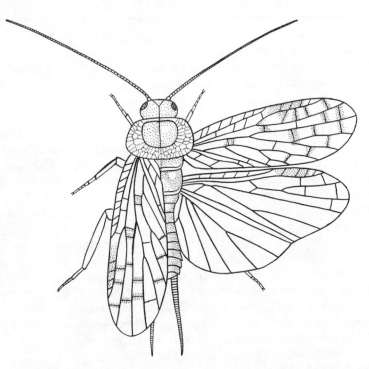

Fig. 38. *Lemmatophora typa* Sellards (Protoperlaria), from the
Lower Permian of Kansas. From Rodendorf (1962).

Plecoptera unless it is regarded as a convergent development. On the other hand, several primitive characters have also been retained: lateral lobes were often present on the pronotum and, according to Carpenter (1935), *Artinska* had five-segmented tarsi. More recently, Carpenter (1966) assigned the Lemmatophoridae to the 'Protorthoptera'. However, if they are supposed to have been most closely related to the Plecoptera, they can only have been surviving 'side-branches' from older elements in the stem-group of the Plecoptera.

It seems most likely that the Plecoptera arose before the lower Upper Carboniferous. Fossils from this epoch are known for all the groups which might prove to be the sister-group of the Plecoptera. It is logical to believe that the stem-group of the Plecoptera must also have been in existence at the same time, although it may not actually be possible to prove it[175].

The Carboniferous 'Paraplecoptera'

It is particularly important to know whether any species from the stem-group of the Plecoptera can be identified amongst Carboniferous fossils from the northern hemisphere, which is where virtually all the Carboniferous fossils have been found. A further question, touched on above, is whether the Plecoptera are more closely related to the Paurometabola, or to the Eumetabola, or to one of the subgroups of the Eumetabola (Paraneoptera + Holometabola). It is remarkable that no fossils that definitely belong to the Paraneoptera and Holometabola are known from the Carboniferous, although a large number of groups have been found in the Lower Permian. Various explanations have been put forward for this, one of which is that the Paraneoptera and Holometabola could have originated in regions where no fossils have yet been found: for example, Jeannel (in Grassé, 1949) suggested Gondwanaland. These explanations would be redundant or would have to be modified if the Plecoptera really are most closely related to the Eumetabola and if the Carboniferous fossils from the northern hemisphere belong to the stem-group of the Plecoptera, or even of the Plecoptera + Eumetabola (or of the Plecoptera + one subgroup of the Eumetabola). On the other hand, the discovery of Carboniferous fossils from the stem-group of the Plecoptera would not have so many far-reaching implications if the Plecoptera are more closely related to the Paurometabola ('Polyneoptera'). This is because various groups of the Paurometabola, such as the Orthopteroidea, are known to have existed in the Upper Carboniferous, and this would prove that the Plecoptera were also in existence at that time, for the Plecoptera could only be the sister-group of the entire Paurometabola. It is still not clear whether the Plecoptera are more closely related to the Paurometabola or to the Eumetabola, and in view of the far-reaching consequences of any decision we should be extremely cautious in appraising the Carboniferous fossils.

Sharov (1961b, in Rodendorf, 1962) suggested that the Carboniferous Cacurgidae and Narkemidae (Fig. 39) were closely related to the Plecoptera.

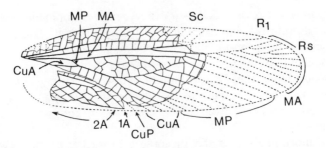

Fig. 39. *Narkema angustifrons* Sharov (Narkemidae), from the Upper Carboniferous of the Kuznetsk basin. From Illies (1965, based on Sharov).

Both 'families' are characterized by having 'MP' fused for a short distance with CuA. This could be regarded as a forerunner of the complete fusion of these two veins which is generally thought to be a characteristic feature of the Plecoptera (see above). However, Sharov himself (1966b) cast doubt upon the correctness of his interpretation by suggesting that all Pterygota have 'MP' and CuA partly fused like this! If this is correct, then no close relationship between the Cacurgidae + Narkemidae and the Plecoptera can be inferred from the venation.

As discussed above, Séguy interpreted the plecopterous wing as having MA fused with the last branch of the radial sector, MP largely free, and only one of the more posterior branches of the media fused with CuA. Even if we were to follow Séguy's interpretation, it would hardly be possible to derive the plecopterous venation from that of the Cacurgidae and Narkemidae.

Sharov (1961b, in Rodendorf, 1962) included these two Carboniferous 'families' and a number of others in his 'Paraplecoptera', which he regarded as the stem-group of the Plecoptera and Embioptera (see the section of his phylogenetic tree reproduced by Illies, 1965). However, it seems very unlikely that the Plecoptera and Embioptera form a single monophyletic group and shared a stem-group that was common only to them.

Previous authors (Laurentiaux, in Piveteau, 1953; Carpenter, 1954b, 1966) assigned all of Sharov's 'Paraplecoptera' to the 'Protorthoptera', together with his Protorthoptera and his Protoblattodea. Sharov's division of the 'Protorthoptera' shows him taking sides in a very old controversy. Handlirsch broke with the principles that he had defended in his treatment of the 'Palaeodictyoptera' by assigning certain species which he believed to be closely related to the cockroaches, grasshoppers, or stoneflies into separate groups (Protoblattoidea, Protorthoptera, Reculoidea-Hapalopteroidea). Later authors found it impossible to assign individual fossils to a particular group. As a result, there was a gradual tendency to group all these species together as the 'Protorthoptera'. It is obvious that this is an 'invalid stem-group' of several insect orders. Sharov (1961a), like Martynov (1938b), attempted once again to distinguish the Protoblattodea, Protorthoptera, and

Paraplecoptera as the stem-groups of the cockroach-like, grasshopper-like, and stonefly-like insects. He was right in suggesting that the stem-group of these three groups had probably separated by the Upper Carboniferous. He was also correct in recognizing that the characters of his Protorthoptera and Protoblattodea really do indicate that there is a close relationship between these two groups and the modern Orthoptera and Blattodea, respectively. The fact remains, however, that he has only defined the Paraplecoptera with primitive characters. There are always insuperable difficulties in the way of assigning fossil wing fragments to one or other of these groups, when similarities in the venation are the only features that are used. Apart from this, species with the characters of Sharov's Paraplecoptera really could belong to the stem-groups of different recent groups and none of the characters in his diagnosis would rule out this possibility. Sharov (in Rodendorf, 1962) also stated that the Paraplecoptera were prognathous, but this was certainly an unwarranted assumption because no unambiguously prognathous species have ever been found. Carpenter (1954b) stated that the 'Protorthoptera' are 'hypognathous'! In this group he included the Lemmatophoridae ('Protoperlaria'), where there is nothing to suggest that the species were not hypognathous, and also all the species in Sharov's Paraplecoptera. If prognathy really was a character of the Paraplecoptera, it would actually indicate that the group should *not* be assigned to the stem-group of the Plecoptera, because the most primitive recent Plecoptera have rather round heads with mouthparts directed downwards. Flattening of the head and more or less conspicuous prognathy, but with no true gula, are only found in groups of the Plecoptera that have other derived characters.

The absence of the precostal field in the 'Paraplecoptera' could be of some importance, because a precostal field, or at least a weakly developed one, is probably part of the ground-plan of the Pterygota and of the Neoptera. However, the precostal field is absent in very many Neoptera, apart from the Orthopteroidea, where it has been enlarged, probably secondarily. On its own this reduction can give no clear indication of the relationships of the Paraplecoptera.

At least 250 species of Paraplecoptera have been described, which is many more than the Palaeodictyoptera, and it would be interesting for several reasons to prove that at least the majority of them do actually belong to the stem-group of the Plecoptera. It would mean, for example, that during the late Palaeozoic the stem-group of the Plecoptera would have been just as important as any other group, even though the Plecoptera themselves occupy only a modest place in the modern fauna. It would be clear that this stem-group had developed far greater ecological diversity than have the recent Plecoptera. For example, the Cnemolestidae have the fore legs modified into raptorial limbs. In some species (Ideliidae), a long ovipositor must also have been present. However, the wing base of this group, illustrated by Sharov as the 'Ideliidea', shows clearly that they cannot have belonged to the stem-group of the Plecoptera.

Sharov's Paraplecoptera are very probably an 'invalid stem-group', and my verdict on their composition and status is not so very different from that on the Palaeodictyoptera–Megasecoptera. It is not clear whether they include any species that actually belong to the stem-group of the Plecoptera.

Because the limits of the 'Paraplecoptera' have been so labile, there is little point in discussing whether they became extinct at the end of the Palaeozoic or whether some of their descendants survived into the Mesozoic or later. There cannot have been many that did this, but they must have included the *Plecoptera, whose ancestors are certainly present amongst the Permian 'Paraplecoptera' if not amongst the Carboniferous ones. Sharov has mentioned a few 'Paraplecoptera' from the early Mesozoic, but these are very controversial.

In addition to the 'Paraplecoptera', the Russian Textbook of Palaeontology (Rodendorf, 1962) included the Miomoptera and Caloneurodea as fossil 'orders' in the 'Plecopteroidea'.

Schmidt (1962) reported that Martynov originally included the Protoperlaria and a few other fossil families in his Miomoptera. Subsequently, Martynov (1938b) restricted the name to the two families Palaeomantidae and Delopteridae. The Russian Textbook of Palaeontology added the family Archaemiopteridae, which includes one species from the Upper Carboniferous (Stephanian) of the Saar region and one species from the Upper Permian of the Kuznetsk basin.

It does not really matter how the Miomoptera are interpreted. They are of no greater relevance to the problems I am examining in this book than the Paraplecoptera.

The same is true of the Caloneurodea, from the Stephanian of Commentry. The hind-wings have a weakly developed anal fan and are very similar to the fore-wings. The relationships of the group have been shown by Martynov (1938b), Sharov's phylogenetic tree (reproduced by Illies, 1965), and a phylogenetic tree in the Russian Textbook of Palaeontology (Rodendorf, 1962, Fig. 27, which contradicts the text!). These have all shown the Caloneurodea as the side-branch of a stem-group from which they and the Glosselytrodea (see p. 212), and the 'Saltatoria' + Phasmatodea but not the Plecoptera, are supposed to have arisen. Sharov (1966a) made a detailed study of the Lower Permian fossils and concluded that the Caloneurodea and Glosselytrodea belong to the 'Neuropteroidea' (Holometabola)!

The interpretation of the Permian 'Paraplecoptera', Miomoptera, and Caloneurodea is really of secondary importance. It is much more important to decide whether the 'true' Plecoptera, which are definitely known from the Permian (*Perlopsis* and other genera, p. 165, still belong to the stem-group of the recent species; or whether there is any evidence that several subgroups of the recent Plecoptera had already separated in the Permian. Illies (1965) published a phylogenetic tree of the Plecoptera, and if this and the position that he assigned to the fossils are well founded, then at least five of the modern subgroups must have been in existence during the late Palaeozoic.

The origin of these groups would then have to be placed far back in the Palaeozoic.

In order to confirm this, the classification of the recent Plecoptera must be examined[174].

For many years two groups have been recognized: Holognatha (or Filipalpia) and Systellognatha (or Setipalpia). It could have been predicted that the Systellognatha (Setipalpia) were a monophyletic group and could be retained, whereas the Holognatha (Filipalpia) were based solely upon primitive characters and sooner or later would have to be divided.

Illies (1965) recently removed the Eustheniidae and Diamphipnoidae from the Filipalpia and assigned them to his Archiperlaria. He regarded the Archiperlaria as the sister-group of all the other Plecoptera, which consisted of the remaining Filipalpia + Setipalpia (see his phylogenetic tree). Unfortunately, however, this division is not soundly based because Illies listed only primitive characters for the Archiperlaria. Nor did he mention any derived characters common to the members of their alleged sister-group, the remainder of the old Filipalpia + Setipalpia. The matter is made even more complicated by the fact that Illies regarded the evenly rounded hind-margin of the hind-wing of the Eustheniidae as a primitive character. The incision at the boundary between the anal lobe and the main part of the wing would then have had to have arisen convergently in the Diamphipnoidae, which Illies also assigned to the Archiperlaria, and in the rest of the Plecoptera. I do not think that this is likely. An incision between the anal fan and the main part of the wing is found in many different groups of the primitive Neoptera, such as the recent and Permian grasshopper- and cockroach-like insects, and in the 'Paraplecoptera'. It could therefore have been present in the direct common ancestors of the Neoptera. The smooth margin of the hind-wing could well be a derived character of the Eustheniidae.

A further puzzling feature is why the Austroperloidea belong to the Filipalpia and not to the Archiperlaria. Illies regarded them as the most primitive group of his Filipalpia. However, he did not explain why the Filipalpia should be a monophyletic group once the Archiperlaria had been removed. It is striking that both the Austroperloidea and the Archiperlaria are groups with an amphinotic distribution. Illies suggested that they both dispersed independently across Antarctica. This then raises a problem that he appears to have overlooked: why did the Setipalpia become extinct in the southern hemisphere, where they must certainly have been present if his phylogenetic tree is correct? The same is also true to a certain extent of the Taeniopterygidae. The situation would be much clearer, and perhaps much simpler too in the context of Illies' ideas on distribution, if the Archiperlaria + Austroperloidea were distinguished from the rest of the Plecoptera as a southern monophyletic group.

Until these difficulties have been resolved, we should be extremely careful when assigning Palaeozoic fossils to groups of the *Plecoptera, especially as most of the available fossils are only fore-wings[175]. The consequences of any error could be far-reaching. Illies has been very careless in some respects. For example, in his text he assigned *Stenoperlidium* to the 'Eustheniidae' and *Palaeotaeniopteryx* to the 'Taeniopterygidae', as did Sharov. His phylogenetic tree, however, showed these fossils in the *stem-groups* of the Archiperlaria and Taeniopterygoidea, respectively.

The most pressing need is for a careful analysis of the heterobathmy of as many adult and larval characters as possible. This would give a good idea of the relationships amongst the various recent groups of the Plecoptera, and these results could lead to a rational understanding of the development of the venation, especially in the fore-wing[174]. It is well known that parallel developments in the venation are very common and can easily lead to false conclusions, yet at the same time these developments can often be useful for phylogenetic research. This is because they affect different elements of the venation, which follow the same general direction when changing but which do not follow the same sequence in the various groups. In the light of this I find it difficult to accept that formal agreements between the wings of Lower Permian fossils and those of recent groups indicate any close relationship or prove that the recent groups of the Plecoptera separated so early on. Nevertheless, I am most impressed with the fact that *Stenoperlidium*, from the Upper Permian of Australia, is said to agree very closely with the recent genus *Stenoperla*. Even if this resemblance is based on symplesiomorphies and is partly the result of convergence, it would be perverse not to believe that the modern Australian species, not just the genus *Stenoperla*, have really descended from ancestors that lived there during the Permian and which had characters that have scarcely changed since then[172]. There is a parallel case in the Mecoptera.

It cannot be ruled out that several subgroups of the *Plecoptera that still exist in the modern fauna had already separated by the Lower Permian. At present there is no fully satisfactory evidence for this, and it is highly desirable that some should be found[175].

Revisionary notes

165. When Hennig's account was first published, an independent study of the phylogenetic classification and relationships of the Plecoptera had just been completed and was published shortly afterwards (Zwick, 1969, 1973). A further comprehensive synopsis has just been published (Zwick, 1980a). [Peter Zwick.]

166. Unfortunately, the relationships of the Plecoptera with other orders are still not clear, but some of the suggested affinities with individual subgroups of the Neoptera can now be safely discarded.

Relationship with the Embioptera has been suggested, but at best this is unproved. It has been confirmed that there are similarities between the two orders (Bitsch and Ramond, 1970), but these are all symplesiomorphies.

Critical analysis has indicated that the Embioptera are most closely related to the Phasmatodea rather than to the Plecoptera (Rähle, 1970).

In my view the most probable hypothesis is that there is a sister-group relationship between the Plecoptera and the rest of the Neoptera. Unfortunately, many of the derived characters shared by the other Neoptera are of a kind that could easily have arisen convergently, and so it has not been possible to show convincingly that the probable sister-group of the Plecoptera is monophyletic. [Peter Zwick.]

167. Plecoptera have a slender trochantin which lies freely in the membrane around the coxa on all segments. However, this is apparently a primitive state (Matsuda, 1970) and cannot be used for working out the relationships of the stoneflies. [Peter Zwick.]

168. Some stoneflies have so-called furcasternal arms that look like, and may perhaps function like, a sternocoxal articulation, but this character does not indicate any close relationship to the Holometabola. Hennig was right in his conjecture that this would not be part of the ground-plan of the Plecoptera, for it has now been shown that it has arisen secondarily and only in one derived subgroup, the superfamily Nemouroidea, where it occurs in varying degrees of development. [Peter Zwick.]

169. Some characters of the Plecoptera are so primitive that the order could be the sister-group of the rest of the Neoptera. Hennig mentioned the pleural structure and embryogenesis as examples of this. Further primitive characters have been found (Baccetti *et al.* 1970; Moulins, 1968; Zwick, 1973): the metameric structure of stonefly testes; the presence of two separate penial openings; the presence of an internal transverse muscle in the stipes, which has not been found in any other Pterygota but which is known in the Machilidae and Zygentoma; and the retention of accessory filaments in the tail of the spermatozoon as simple hollow microtubuli, a condition that is more primitive than that found in any other insects. [Peter Zwick.]

170. An independent, sclerotized 1st abdominal sternite is found in the Pteronarcyidae, Austroperlidae, and some of the larger Gripopterygidae. However, even in these species, the longitudinal muscles from the thorax pass over the sternite and are attached as usual to the antecosta of sternite 2. [Peter Zwick.]

171. No new data have come to hand to indicate which of the controversial interpretations of the plecopteran venation is correct. [Peter Zwick.]

172. So far as the fossils are concerned, the situation is most unfortunate. The Permian and Mesozoic fossils have been assigned to a few extinct families of uncertain affinities, and to the recent families Eustheniidae, Gripopterygidae, Pteronarcyidae, Perlidae, and Taeniopterygidae. However, *none* of these assignments can survive a rigorous examination, for none is based on the presence of the constitutive characters of the families in question. Characters that are supposed to be diagnostic are usually plesiomorphic character states, and are only diagnostic if the comparisons are limited *a priori* to certain groups. For example, a Cretaceous fossil from Labrador (Canada) has been described as a pteronarcyid because it has reticulate wings. However, the families Eustheniidae, Diamphipnoidae, Austroperlidae, and Gripopterygidae also include species with a similar venation, but they are now restricted to the southern hemisphere and were not mentioned (Rice, 1969).

Stenoperlidium is said to be closely related to *Stenoperla* (Tillyard, 1935b), and this would mean that the subfamily Stenoperlinae of the Eustheniidae has existed in Australia since the Permian. Riek (1956) has expressed some doubts as to the family assignment of this genus, but subsequently pointed out that the fore-wings are very similar, except for some differences in the cross-veins (Riek, in CSIRO, 1970). However, the wings of *Eusthenia*, large Austroperlidae, Gripopterygidae, and even several Arctoperlaria do not differ significantly from this. This assignment of *Stenoperlidium* was influenced by the discovery of a fragmentary larva from the same site that was thought to be conspecific with the adult. This larva has finger-shaped abdominal gills represented only by short stubs, and one of these gill remnants is said to have the moniliform annulations that are diagnostic for the Eustheniidae. The Stenoperlinae have these gills on the first five segments and the Eustheniinae on the first six. The fossil larva lacks segments 1–3, but gill remnants are visible on segments 4 and 5. However, finger-shaped gills that are generally similar to these are probably part of the ground-plan of the Plecoptera and in the recent fauna have only been retained by a few groups, the Eustheniidae, and the North American perlodid *Oroperla*.

Some authors have not given any family assignment for the fossils that they have described, but names such as *Mesoleuctra, Mesonemoura, Sinonemoura*, and *Sinoperla* give an indication of what they had in mind. *Sinoperla* has subsequently been assigned to the Perlidae and the other three to the Taeniopterygidae, apparently on the basis of symplesiomorphies. However, *Sinoperla* is said to have a small exposed triangular clypeus, and in this case it would not belong either to the Perlidae or to the Systellognatha. None of the three 'taeniopterygids' has three long equal tarsal segments, which is a constitutive character of the family Taeniopterygidae. I think that these fossils look like, and may even belong to, the superfamily Nemouroidea, but on the published evidence it would be unwarranted to say more.

The situation is different with fossils from Baltic amber, which are available in large numbers and which are often well preserved. This Oligocene fauna appears to have been similar to the recent Holarctic fauna at the family level. The most interesting specimen was a species of *Megaleuctra*, a genus which is now restricted to western North America (Ricker, 1935; Illies, 1967). Several generic names, such as *Leuctra*, *Nemoura*, *Perla*, and *Perlodes*, have been used in their former broad sense for the fossils. Using this fine material, it should be possible to form some interesting conclusions about stonefly evolution at the levels of genus and possibly subfamily, and these should be of particular value for the notonemourid–nemourid lineage. The methods and criteria now used to distinguish between recent genera and species will have to be applied to the study of fossils in order to reach any conclusions, but this has not yet been done. [Peter Zwick.]

173. A redescription of *Germanoprisca* has been published by Müller (1977b). [Rainer Willmann.]

174. My new phylogenetic classification of the Plecoptera is set out in the table overleaf. It appears to have been generally accepted by students of the order, and has recently received support from studies on stonefly drumming behaviour (Rupprecht, 1976). It differs radically from earlier classifications, both in the basic division of the order into suborders and in the assessment of the relationships of the individual families. This is, incidentally, exactly what Hennig felt to be required. There are only two family-group taxa (Perlodidae and Notonemouridae) for which no, or no fully convincing, synapomorphies are known. Constitutive characters for taxa of all ranks from subfamily to suborder have been found in various features of the external and internal morphology.

The wing venation varies greatly within the order and within individual families. No constitutive characters have been found in the venation for any taxon above the level of a genus, and it is only possible to work out diagnostic venational characters for the modern fauna in limited geographical areas, such as individual continents. In fact, unless a new species has the constitutive characters of one particular subgroup, such as an individual family, it can only be assigned to a suborder after study of a number of subtle details. The only characters that will lead to a reliable subordinal assignment at all times are to be found in the prothoracic muscles. [Peter Zwick.]

175. Several recent families of the Arctoperlaria are known to have existed during the Eocene, and from their positions in the phylogenetic tree it can be safely inferred that all the families of the Arctoperlaria must have been distinct by then. However, for reasons given below, I cannot accept any claim that individual modern genera were already in existence during the Mesozoic or even during the Permian. As no useful data can be derived from the fossils, we can only work out the timing of the evolutionary process by analysing information on the course of evolution and on the geographical areas where this is supposed to have taken place, together with palaeogeographical reconstructions. The former can be derived from the phylogenetic classifica-

Phylogenetic classification of the Plecoptera, and the geographical distribution of the families. The number of species is given in parentheses.

Zoogeographical regions: A, Australian region (excluding New Guinea); B, Neotropical region; C, Afrotropical region; D, western Palaearctic region (Europe, North Africa, Asia Minor); E, eastern Palaearctic and Oriental regions; F, Nearctic region.

Key to notes: 1, only in the south; 2, mainly in the south, with only one species north of 10°S, and absent from Central America; 3, absent from the Oriental region, or only present in areas adjacent to the Palaearctic region; 4, only the subfamily Acroneuriinae; 5, only *Neoperla*, absent in Madagascar; 6, only Korea and Japan.

		Geographical region					
		A	B	C	D	E	F
I.	*ANTARCTOPERLARIA:*						
I.A.	EUSTHENIOIDEA:						
I.A.a.	Eustheniidae (16)	×	×[1]				
I.A.b.	Diamphipnoidae (5)		×[1]				
I.B.	GRIPOPTERYGOIDEA						
I.B.a.	Austroperlidae (13)	×	×[1]				
I.B.b.	Gripopterygidae (120)	×	×[2]				
II.	*ARCTOPERLARIA:*						
II.A.	SYSTELLOGNATHA:						
II.A.a.	Pteronarcyidae (14)					×[3]	×
II.A.b.1.	Peltoperlidae (35)					×	×
II.A.b.2.	Perloidea:						
II.A.b.2.-1.	Perlodidae (218)				×	×[3]	×
II.A.b.2.-2.1.	Perlidae (353)		×[4]	×[5]	×	×	×
II.A.b.2.-2.2.	Chloroperlidae (111)				×	×[3]	×
II.B.	EUHOLOGNATHA:						
II.B.a.	Scopuridae (2)					×[6]	
II.B.b.	Nemouroidea:						
II.B.b.-1.	Taeniopterygidae (70)				×	×[3]	×
II.B.b.-2.1.1.	Notonemouridae (59)	×	×[1]	×			
II.B.b.-2.1.2.	Nemouridae (337)				×	×	×
II.B.b.-2.2.1.	Capniidae (203)				×	×[3]	×
II.B.b.-2.2.2.	Leuctridae (169)				×	×	×

tion and the latter from the present distribution, and I have shown both in the table opposite.

The Plecoptera can be divided into an exclusively southern suborder, the Antarctoperlaria, and a northern suborder, the Arctoperlaria. Some of the Arctoperlaria are now found in the southern hemisphere, but their relationships prove that they are secondary immigrants from the north. In my view, current interpretations of the oldest fossils are based on extremely flimsy evidence (see note 172), and I believe that the break-up of Pangaea into Laurasia and Gondwanaland, which probably took place in the Jurassic, led to the separation and independent evolution of the two main lineages of the Plecoptera. Stonefly subgroups appear to be much less ancient than has generally been assumed.

There are no real problems with the Antarctoperlaria. With the clarity of a textbook example they provide a perfect hierarchical sequence, with multiple sister-group relationships between taxa in South America, New Zealand, and Australia. The group must have begun to divide up on the southern continent (Gondwanaland) before this eventually broke up as a result of continental drift. They are absent from Africa and India, but this can be explained by extinction, which is neither an unusual nor an improbable hypothesis. It is unfortunate that there is no satisfactory classification of the Gripopterygidae, the only reasonably diverse family. Two different subfamily divisions have been made (Illies, 1963; McLellan, 1977), but both were partly based on characters that have undoubtedly arisen convergently (Zwick, 1973, 1980b). There is no simple solution to the problems raised by the structural uniformity of the widely disjunct gripopterygid fauna.

Most families of the Arctoperlaria are also very straightforward, and they do not provide any clues as to the timing of stonefly evolution. They are Holarctic in distribution, except for a few relict groups (Scopuridae, Pteronarcyidae, Peltoperlidae). There are considerable resemblances between the North American and East Asian faunas, with many genera and even a few species in common, and the European fauna is relatively distinct, having only a few genera in common with Asia and even fewer with North America: these features are probably the result of extinctions in Europe and faunal exchanges across the Bering Strait, both taking place during the Pleistocene. They may also reflect the Pleistocene separation of Europe and Asia by the Turgai Strait. The families Perlidae and Notonemouridae need to be examined separately.

The majority of genera and species in both subfamilies of the Perlidae are Holarctic–Oriental in distribution. One genus of the Perlinae, *Neoperla*, is found in Africa south of the Sahara, but must have arrived after the isolation of Madagascar. There are many species in Asia and some also occur in North America. *Neoperla* may have entered Africa during the Tertiary, when the Afrotropical region became accessible to the large mammals of Asia, or even later. However, the presence of the subfamily Acroneuriinae in South America is less easily explained. The Andean genera, all of which belong to the Acroneuriini, are fairly similar to their Nearctic and Asian relatives and might have entered South America after the comparatively recent re-connection of the Americas. On the other hand, the tribe Anacroneuriini, which is endemic in Brazil and Central America, is too diverse to have evolved in such a limited time. Stark and Gaufin (1976) suggested that towards the end of the Cretaceous Asian or European perlid stock might have entered South America via Africa, which was still connected to South America.

By far the most difficult problems are posed by the Notonemouridae. This family is restricted to the most southern areas of Africa (including Madagascar), South America, Australia, and New Zealand. It now looks as though future studies will confirm that notonemourid distribution is not the result of a single event: the family is probably a paraphyletic assemblage of lineages surviving from an archaic pre-nemourid stock. In anticipation that such confirmation will be forthcoming, I have already treated two ostensibly monophyletic groups as separate units. Additional data on the internal anatomy conflict with some of my latest views (Zwick, 1980a). The difficulty is that there are two types of ovipositor, and at first these appeared to indicate that each group is monophyletic. However, these types represent two alternative technical solutions to the problem of forming an ovipositor, and both types have developed convergently in different families. For this reason, convergence amongst the Notonemouridae is always a possibility, particularly as not all the species have an ovipositor.

Other features of the internal and external morphology need to be found that will support or refute the present classification, but these are not available for all genera. As a result, there is still no sound basis for the kind of zoogeographical and evolutionary considerations that might shed some light on the history of the Plecoptera, and it is not possible to present more than an incomplete picture of stonefly evolution. [Peter Zwick.]

2.2.2.2..2. Paurometabola

The groups that I am including in the Paurometabola are markedly terrestrial, with fore-wings modified into almost rigid tegmina and the main flight function taken over by the hind-wings. The wings have often been completely reduced, but some groups like the migratory locusts have developed considerable powers of flight secondarily. According to Ross (1955), reduction of the 1st abdominal sternite is characteristic of the whole group[176]. Snodgrass (1937) found the presence of accessory glands in the male genitalia to be a further characteristic feature. Usually insects have only one pair of these glands, but the Paurometabola as defined here have an irregular cluster.

Basically the 'Paurometabola' correspond to the 'Polyneoptera' of Martynov and other authors who have followed his classification of the Neoptera. I have decided not to use the name Polyneoptera here because Martynov also included the Plecoptera in his Polyneoptera, and there is at most a sister-group relationship between the Plecoptera and the Paurometabola. The only derived ground-plan characters of the Paurometabola are the cluster of accessory glands in the male genitalia and the terrestrial mode of life, which is reflected morphologically in the structure of the wings. These characters are not present in the Plecoptera and indicate that the Paurometabola are monophyletic. On the other hand, the aquatic larvae of the Plecoptera are a derived character of this group, for the immature stages of the Paurometabola appear to be primarily terrestrial. The name Polyneoptera could be retained for the Plecoptera + Paurometabola if these prove to be a monophyletic group. So far, however, there is no evidence that the Plecoptera are more closely related to the Paurometabola than to the Paraneoptera and Holometabola.

The cockroach-like and grasshopper-like insects are two obvious monophyletic groups in the Paurometabola. Three further groups contain fewer species and are undoubtedly monophyletic, but their relationships are

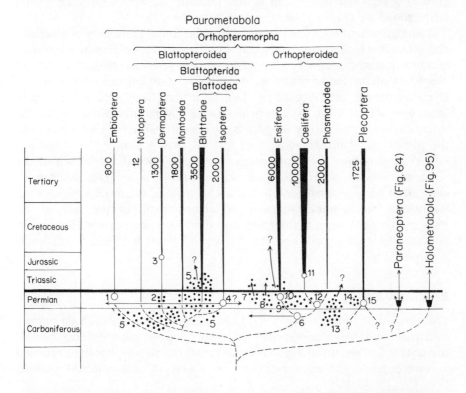

Fig. 40. Phylogenetic tree of the Paurometabola. 1, *Sheimia sojanensis* Martynova; 2, Protelytroptera; 3, *Mesoforficula*, etc.; 4, *Puknoblattina*; 5, Palaeozoic 'Proto-blattoidea' and Blattodea; 6, *Oedischia*; 7, Glosselytrodea; 8, Sthenaropodidae; 9, Oedischiidae and Elcanidae; 10, *Tettavus*; 11, *Triassolocusta*; 12, *Tcholmanvissia*; 13, Sharov's 'Paraplecoptera'; 14, Protoperlaria; 15, *Perlopsis* and other undoubted Plecoptera from the Lower Permian[166].

very uncertain: Embioptera, Dermaptera, and Notoptera, the Embioptera being the most problematic. It has been suggested that the Embioptera are related to the Isoptera, the Zoraptera, or the Plecoptera, but there is no sound evidence for any of these views. The most widely accepted view is that the Embioptera are most closely related to the Plecoptera, and the two groups have often been called the Plecopteroidea. This is also unfounded. The characters common to the Embioptera and Plecoptera are all relatively primitive (symplesiomorphies). There is not a single derived character common to these two groups which would indicate that the 'Plecopteroidea' is a monophyletic group. Furthermore, it is logically incorrect to claim that the Embioptera and Plecoptera can both be derived from the Palaeozoic Paraplecoptera (of Sharov) or Protorthoptera (of Carpenter) and must therefore be closely related to each other. It is possible, and even probable, that the 'Protorthoptera' or 'Paraplecoptera' include species belonging to the stem-groups of the Embioptera and Plecoptera, but they are both invalid stem-groups. It does not follow that groups formally 'derived' from such invalid stem-groups are closely related to each other (p. 24).

Apart from their terrestrial mode of life, the Embioptera have two derived characters that indicate their relationships: the reduced 1st abdominal sternite and the clustered accessory glands of the male genitalia, both of which are characters of the Paurometabola. A reduced 1st abdominal sternite is also characteristic of the Plecoptera, but as it has been retained in many Paurometabola it must have been independently reduced on several occasions. Clustered male accessory glands are known only in the Embioptera and the rest of the Paurometabola, and for this reason I think that Ross was correct in regarding the Embioptera as a highly specialized branch of the 'Orthoptera-stem'. The Embioptera have followed a separate direction in specialized wing development, and show no trace of the modification of the fore-wings into tegmina which is so characteristic of the rest of the Paurometabola. I suggest therefore that there is a sister-group relationship between the Embioptera and the rest of the Paurometabola ('Orthopteromorpha').

Revisionary note

176. According to Kristensen (1975), the mention of the reduced 1st sternite as a paurometabolan apomorphy must be a lapsus, since Hennig himself correctly stated shortly afterwards (above) that this sternite has been retained in some orders of the Paurometabola. See also p. 163 and note 170. [Adrian Pont.]

2.2.2.2..2.1. Embioptera[177]

There are only about 150 recent Embioptera[178] and they are characterized by a large number of derived features. The head is prognathous and a gula is present. Ocelli are absent and the compound eyes poorly developed. The legs are short, with three-segmented tarsi. The metatarsi of the fore-legs possess silk glands with which the animals construct silken tunnels in which to live. The hind femora are thickened. Females are always wingless, while the wings of the male are reduced and narrow, secondarily homonomous, and without an anal fan. The venation is very reduced. The cerci are reduced to two segments, and in the male combine with parts of the tergites to function as genitalia. Females have no ovipositor.

This large number of derived characters makes it absolutely certain that the Embioptera are monophyletic, but makes it extraordinarily difficult to work out their history. Various Palaeozoic fossils have been assigned to the Embioptera, but none from the Mesozoic. The most striking fossil is *Protembia permiana* Tillyard, from the Permian of Kansas. This was assigned unhesitatingly to the Embioptera by Davis (1940a, 1940b), a leading authority on this group, but Carpenter (1950) did not accept this.

Sheimia sojanensis Martynova (Figs. 41 and 42), from the Upper Permian of Arkhangelsk, has been described as an embiopteron. In certain characters it is more primitive than all the recent species of this group. Sc is complete, whereas in all recent species it is short and does not reach the wing margin (Fig. 43); the wings are longer; the cerci are multi-segmented; and a short ovipositor appears to be present. It is also interesting that it is a winged female, for the females of all recent Embioptera are wingless. On the basis of these characters *Sheimia* could well belong to the stem-group of the Embioptera. However, in certain characters of the venation *Sheimia* (Fig. 42) is

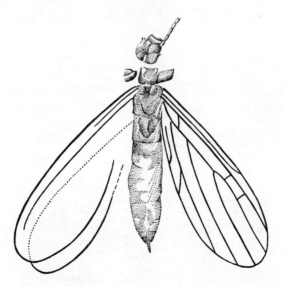

Fig. 41. *Sheimia sojanensis* Martynova ('Embioptera'), from the Upper Permian of Arkhangelsk. From Rodendorf (1962).

182

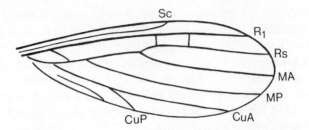

Fig. 42. Wing of *Sheimia sojanensis* Martynova. The interpretation of the venation follows Martynova (in Rodendorf, 1962), but is still unclear. I think that 'MA' represents the fusion of MA and the last branch of the radial sector and that 'CuP' should be interpreted as CuA_2.

undoubtedly more strongly derived than some of the recent species (Fig. 43). Unlike the ground-plan of the *Embioptera, the radial sector has two branches, not three; the media ('MP') is unforked instead of forked; and 'MP' is fused with the radial sector for a short distance rather than being free. If *Sheimia* really is an embiopteron, then it must belong to a 'side-branch' of the stem-group. However, these three derived characters in the venation do actually occur in certain recent species. It might be possible to show that these are more closely related to *Sheimia*, but it would then follow that all the primitive characters of *Sheimia* listed above must have developed indepen-

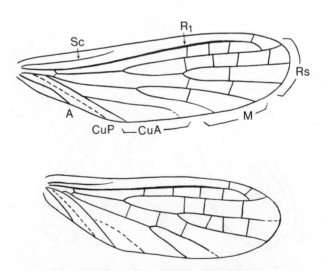

Fig. 43. Wings of *Donaconethis abyssinica* Enderlein (Embioptera; recent). From Enderlein (1912).

dently and convergently into the derived states now found in the *Embioptera. This is most unlikely and in any case is not supported by any evidence.

Fundamentally, the problem of whether *Sheimia* or any other Permian fossils really belong to the Embioptera is unimportant. If the Embioptera really are the sister-group of the rest of the Paurometabola, then they must have been in existence as a discrete group by the Upper Carboniferous, though in a much more primitive form, because several subgroups of the Orthopteromorpha have been reliably recorded from this epoch. Zalesskiy (1938) has described some fossil fore-wings from the Lower Permian (*Tillyardembia*) and the Upper Carboniferous (Spanioderidae). It would be possible to 'derive' *Sheimia* and the recent Embioptera formally from these fossils which really could belong to the stem-group of the Embioptera, although there is no evidence for this. I do not anticipate that fossils will ever shed any light on the relationships of the Embioptera.

There is no reason for believing that the recent Embioptera could have descended from *different* Palaeozoic species.

Revisionary notes

177. A new monograph on the Embioptera has been published by Kaltenbach (1968). [Dieter Schlee.]

For a review of embiopteran biosystematics, see Ross (1970). Rähle (1970) has suggested that the Embioptera and the Phasmatodea are sister-groups. [Willi Hennig.]

178. There are about 800 recent Embioptera (Ross, 1970). [Willi Hennig.]

2.2.2.2..2.2. Orthopteromorpha

The characteristic feature of this group is the modification of the fore-wings into tegmina, which has not taken place in the Embioptera.

Unfortunately, the relationships between the monophyletic subgroups of the Orthopteromorpha are still not very clear.

The Blattopteroidea (praying mantises, cockroaches, and termites) are a relatively well founded monophyletic group, to which the Notoptera and Dermaptera are evidently closely related. These three groups can be called the Blattopteriformia. The 'Saltatoria' and Phasmatodea are probably a second monophyletic group, the Orthopteroidea.

As a working hypothesis, but not as an established fact, I would suggest that there is a sister-group relationship between the cockroach-like species (Blattopteriformia) and the grasshopper-like species (Orthopteroidea). Sharov (1966b) suggested that the development of the ovipositor has followed different directions in the Blattopteriformia and the Orthopteroidea. According to him, the Blattopteriformia have the 3rd pair of valves formed mainly from the precoxal plates, and the styli have been reduced. The Orthopteroidea, on the other hand, have the 3rd valves formed mainly from the styli, and the precoxal plates have been reduced to 'valvifers'. This is an interesting idea, but it loses much of its force because the groups whose relationships are most unclear, such as the Dermaptera and Phasmatodea, have reduced ovipositors.

The common stem-group of the Orthopteromorpha is not definitely known from the fossil record. It is possible that surviving 'side-branches' of this stem-group are to be found amongst Sharov's 'Paraplecoptera', but this can hardly be proved. The actual ancestors of the Orthopteromorpha must have lived before the Upper Carboniferous because the fossils from this epoch include some undoubted representatives of subgroups of the Orthoptero-morpha.

2.2.2.2..2.2..1. **Blattopteriformia**

The overlapping prosternum, mesosternum, and metasternum may poss-ibly be a derived character of this group. The enlarged coxae could be another. In this case, the structure and position of the coxae in the Dermaptera would be the result of a secondary change. I shall discuss the arguments in favour of this in the section on the Dermaptera (p. 186). On the other hand, it is always possible that enlarged coxae should be regarded as a primitive character of the Paurometabola. This is something that urgently needs to be clarified.

The Blattopteroidea are one monophyletic group of the Blattopteriformia. The relationships of the Notoptera and Dermaptera to the Blattopteroidea are not clear.

2.2.2.2..2.2..1.1. **Notoptera (Grylloblattodea)**

This is a small group which includes only 12 species from Palaearctic Asia and North America. They have several derived (autapomorphic) characters: the wings and the ocelli are absent, and the compound eyes are only weakly developed.

In addition, the Notoptera have a relatively large number of primitive characters. Some of these occur individually in other groups, but they are only found together in the Notoptera. Several attempts have been made to work out the relationships of the Notoptera from the distribution of charac-ters, but none has succeeded. This is because most characters have only been sampled in a random way and because no clear distinction has been made between primitive and derived characters.

Giles (1963) attempted to do this and analysed 283 individual characters. His conclusion was that the Dermaptera and Notoptera had common ancestors, by which he presumably meant 'ancestors common only to them'. I do not regard this as a final conclusion because his method was far too statistical. His list appears to contain only two derived characters in which the

Dermaptera agree with the Notoptera, out of 283 characters analysed! These are the absence of the mesothoracic sternellum and the almost horizontal metathoracic episternum and epimeron. However, these are very simple features and could easily have arisen convergently. This view is supported by the presence of a peculiar derived character in which the Dermaptera agree with the Blattopteroidea but which is (still!) absent in the Notoptera: the 7th abdominal sternite of the female has been modified into a long subgenital plate over which the genital atrium is situated. I think that this is a forerunner of the genital atrium that is so characteristic of the Blattopteroidea. The Notoptera have a long ovipositor. According to Sharov (1966b), who has followed Makhotin, the 3rd valves in the ovipositor of the Notoptera are supposed to have developed from the precoxal plates and the styli have been lost. In these respects it differs from the ovipositor of the Orthopteroidea but agrees with the reduced ovipositor of the Blattopteroidea. This would be important evidence that the Notoptera are more closely related to the Blattopteroidea than to the Orthopteroidea, but it does not alter the fact that the structure of the female abdomen is more primitive than that of the Dermaptera and Blattopteroidea.

If we follow Giles in believing that the Dermaptera and Notoptera had ancestors common only to them, then the characters shared by the Dermaptera and Blattopteroidea must have arisen convergently. This is very improbable. Until there is some evidence to the contrary, I would suggest that the two characters that Giles found to occur only in the Dermaptera and Notoptera have arisen convergently. Giles also listed other characters common to these two groups which occur haphazardly in other groups of the Paurometabola: these too have probably arisen convergently.

Sharov (1966b) suggested that the Notoptera (Grylloblattodea) are more closely related to the Blattopteroidea than are the Dermaptera, but he gave no evidence for this. This contradicts my own views and those of Giles.

No Mesozoic or Palaeozoic fossils are known that could be assigned even doubtfully to the Notoptera.

It is unlikely that palaeontology will ever be able to provide any information about the historical development of this group. As the recent Notoptera are apterous, we have no idea what the wings of their ancestors may have looked like. Indeed, we would hardly be able to recognize these ancestors for what they are even if they were represented amongst the Palaeozoic fossils, as they may actually be. It would be easy to suggest that some of the Protoblattoidea, or 'Protorthoptera' or 'Paraplecoptera', belong to the stem-group of the Notoptera. However, all such assignments are pure speculation and have no empirical value. The only evidence available for determining the age of the Notoptera is the existence of Lower Permian fossils belonging to all the groups that could be considered as the sister-group of the Notoptera. This means that the oldest members of the Notoptera must have been in existence at least during the Carboniferous. I shall discuss this problem again in the sections on the Dermaptera and the Blattopteroidea.

2.2.2.2..2.2..1.2. **Dermaptera (earwigs)**[179]

Recent Dermaptera are characterized by a number of derived features which are partly correlated with their cryptic and nocturnal habits. Popham (1962) thought that they were also correlated with the change to predaceous feeding habits. However, the connection between characters and habits is not always obvious, nor is it certain that the predaceous habit ('carnivory') is part of the ground-plan of the group.

The head is prognathous. One result of this, according to Popham (1962), is that the brain has been rotated upwards and backwards, so that the upper surface of the cerebral ganglia is directed downwards and the posterior surface forwards. The eyes and the antennae have moved forward to take up positions just behind the mouth. As a result they are joined to the protocerebrum and deuterocerebrum by particularly long nerves. The maxillae are elongated and the mandibular teeth are directed forwards for capturing and holding prey. These features are less distinctly developed in the relatively primitive species.

The ocelli are absent. Sometimes, as in the Russian Textbook of Palaeontology, they are said to have been retained in primitive species, but Giles (1963) stated that this is an error. The labium has no glossae according to Giles.

The fore-wings have been reduced to short 'elytra'. They have no more than a rudimentary venation, and the distal margin is truncated or even concave. The anal fan of the hind-wing has been strongly enlarged at the expense of the remigium, and the wing can be folded up in a characteristic way. The tarsi have been reduced to three segments. The coxae are particularly problematic, and I shall discuss them again when dealing with the age of the Dermaptera.

In the abdomen the tergites overlie the lateral margins of the sternites. The 1st sternite is completely absent. The cerci are 1-segmented and have been modified into claspers. In some larvae of the genera *Diplatys* and *Karschiella*, the cerci are multi-segmented, according to Chopard (in Grassé, 1949), and during metamorphosis it is the basal segment that becomes modified into the adult clasper[180].

The two species of the Oriental genus *Arixenia* which are closely associated with bats are correctly placed in the Dermaptera, and so is *Hemimerus talpoides* Walker. This species lives in the fur of African *Cricetomys* rats, and recently Popham (1961) has tried once again to give it ordinal rank. I agree with Giles (1963) in regarding this as unjustified[181].

Several Mesozoic fossils have been described which definitely belong to the Dermaptera. Unfortunately, the classification of the group has not been adequately worked out. As a result it is not possible to tell whether the fossils still belong to the stem-group of the Dermaptera or whether they are more closely related to one of their subgroups. Two suborders have usually been recognized, the Eudermaptera and Protodermaptera, but it is doubtful whether these really are sister-groups. In particular, the Protodermaptera seem to include all the most primitive species. The new classification suggested by Popham (1965) is no better founded. Martynova (in Rodendorf, 1962) has assigned *Semenoviola obliquotruncata* Martynov, from the Malm of Karatau (Fig. 44), to the Labiidae, but this does not mean a great deal. The hind-margins of the elytra have been produced into a point and may be more primitive than in all the recent species. Nothing is known about the structure of the coxae. The species was originally described as a beetle.

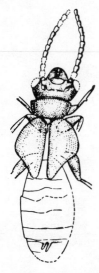

Fig. 44. *Semenoviola obliquotruncata* Martynov (Dermaptera), from the Upper Jurassic of Kazakhstan. From Rodendorf (1962).

It is very doubtful whether there is any greater justification for the assignment of *Mesoforficula sinkianensis* Ping to the Forficulidae. This species was described from the Upper Jurassic of T'u-lu Fan (Turfan) in Chinese Turkestan, though Martynova (in Rodendorf, 1962) inadvertently listed it as 'Cretaceous'. *Protodiplatys fortis* Martynov (Fig. 45) is particularly interesting. It has rather long fore-wings, segmented cerci, and five-segmented tarsi. Furthermore, the tergites do not appear to overlie the edges of the sternites. *Protodiplatys* is from the Malm of Karatau and is therefore no older than *Semenoviola* and *Mesoforficula* which agree closely with recent Dermaptera. It can only be a survivor (a 'side-branch') from the stem-group of the Dermaptera. It is also interesting because in some respects it bridges the morphological gap between recent Dermaptera and the Palaeozoic Protelytroptera.

The Protelytroptera (Fig. 46; Carpenter and Kukalová, 1964) include about 20 species from the Lower Permian of Kansas, 1 species from the Lower Permian of the Kama basin (Chekarda), and also what Haupt (1952) considered to be the fragment of a hind-wing (*Archelytron priscus†*) from the

---

† *Archelytron* Haupt, 1952, is not identical with *Archelytron* Carpenter, 1933.

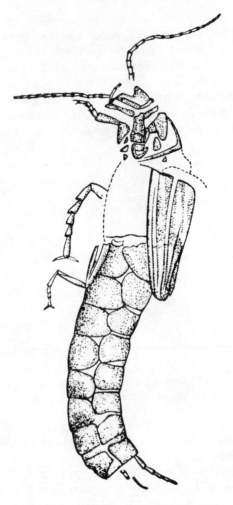

Fig. 45. *Protodiplatys fortis* Martynov, from the Upper Jurassic of Kazakhstan. From Rodendorf (1962).

Lower Permian of the Thüringer Wald. In these species the fore-wing is still long and reaches beyond the tip of the abdomen. Nevertheless, it is already rigid and has the venation reduced. The venation and the method of folding the hind-wing can be seen as taking a different direction from that of the 'cockroaches' and foreshadowing the situation in recent Dermaptera. The pronotum does not cover the head. No one doubts that the Protelytroptera belong to the stem-group of the Dermaptera, though as a 'side-branch'. However, several Upper Carboniferous species are known which belong to the stem-group of the Blattopteroidea (Mantodea + Blattodea). The Dermaptera must be just as old as the Blattopteroidea, their presumed sister-

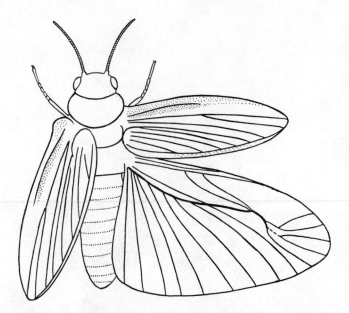

Fig. 46. *Protelytron permianum* Tillyard (Protelytroptera; reconstruction), from the Lower Permian of Kansas. From Carpenter and Kukalová (1964).

group, and consequently we would have to go back as far as the middle or lower Upper Carboniferous to find the common ancestors of the Dermaptera and Blattopteroidea. In the light of this, the interpretation of the Protelytroptera is not as important as appears at first sight. Moreover, they provide no information about two characters that are important for assessing the relationship of the Dermaptera to the Blattopteroidea.

The first of these is the anal furrow of the fore-wing. All Blattopteroidea are characterized by a smoothly curved anal furrow along which vein CuP (Cu$_2$) runs and which forms a well defined anterior edge to the anal lobe. Unfortunately, it is not known whether the ancestors of the Dermaptera also had such a curved anal furrow.

The second important character is the coxa. Long conical coxae with a partial or complete suture are characteristic of the Blattopteroidea (cockroaches, praying mantises, termites), and contrast with the rounded and apparently primitively simple coxae of the Phasmatodea and 'Saltatoria'. According to Chopard (in Grassé, 1949), the coxae of the Notoptera are similar to those of the cockroaches, but they do not have a suture: this could be a primitive state, but it could also be secondary and correlated with the atrophy of the wings. The suggestion that relatively long conical coxae are a derived character in the ground-plan of all the cockroach-like insects is well founded. It is supported by the fact that cockroach-like species with elongate coxae, and even some that have retained a long ovipositor[182] (Fig. 47), are known as early as the Palaeozoic, at least from the Lower Permian. It is

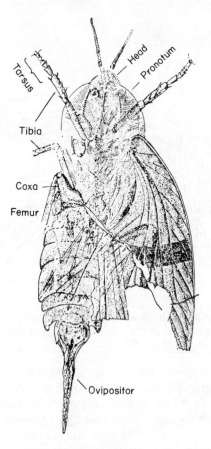

Fig. 47. *Kunguroblattina microdictya* Bekker-Migdisova & Vishnyakova (Blattopteroidea), from the Lower Permian of the Urals. From Roden-dorf (1962). Note the ovipositor and elongate coxae[182].

uncertain whether elongate coxae can be used as a diagnostic character for the Blattodea and Protoblattoidea, as was done in the Russian Textbook of Palaeontology, because most fossils are known only from fore-wings. How-ever, there is some evidence that all these species may have had comparative-ly long conical coxae. The Dermaptera do not agree with this, but this lack of conformity may only be apparent. Giles (1963) stated that all three pairs of coxae are broader than long and he regarded this as a character in which they differ from the Blattopteroidea but agree with the Plecoptera, Embioptera, Phasmida, 'Saltatoria', and, so far as the mid and hind coxae are concerned, with the Notoptera. This would make it difficult to derive the Dermaptera from Palaeozoic cockroach-like forms and raises the question of whether the short coxae of the Dermaptera could not be a derived character. Secondary reduction could be correlated with compression of the body and reduction of the legs. Certain changes in the thorax are also correlated with this, as is shown by the horizontal position of the pleural suture. Giles' illustrations give the impression that the coxae have been strongly twisted so that the anterior part of the upper surface takes a more forward and downward direction, and the lower surface a more backward and upward direction. This shows clearly that it is often not enough to compare the superficial similarity between certain features, such as the proportions of coxal length and width. It is also necessary to enquire how this similarity came about. This could be clarified by a more precise morphological investigation.

The present state of knowledge gives no support to the belief that the Dermaptera arose as a separate group before the Upper Carboniferous. However, no final conclusion can be reached because it is still not definitely known what is the sister-group of the Dermaptera.

Revisionary notes

179. A new monograph on the Dermaptera has been published by Günther and Herter (1974). [Dieter Schlee.] For a zoogeographical check list, see Steinmann (1973). [Willi Hennig.]

180. Abul-Nasr (1954) suggested that in the ground-plan of the Insecta (?Pterygota) the female abdomen has the common oviduct invagination behind the 7th sternite, the spermathecal invagination behind the 8th sternite, and the accessory glands invagination behind the 9th sternite. He considered the situation in the Ephemeroptera (Fig. VIIA) and Dermaptera (Fig. VIIB) to be primitive, but at the same time pointed out that these characters are the same in the 'Thysanura' and the Pterygota (Fig. VIIC). [Willi Hennig.]

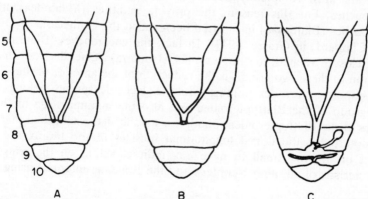

Fig. VII. Stages in the evolution of the efferent genital system in the Ephemeroptera (A); Dermaptera (B); Thysanura, Orthoptera, Isoptera, Hemiptera, and Hymenoptera (C). From Abul-Nasr (1954).

181. Popham (1973) has re-stated his belief that *Hemimerus* should be excluded from the Dermaptera. Giles (1974) gave further reasons for rejecting this, and recognized three suborders in the Dermaptera: Hemimerina, Arixeniina, and Forficulina. The suborders Hemimerina and Arixeniina have been revised by Nakata and Maa (1974), who also considered them to be Dermaptera. [Willi Hennig.]

182. Vishnyakova (1971) has discussed the abdominal appendages of fossil 'cockroaches'. [Willi Hennig.]

2.2.2.2..2.2..1.3.   **Blattopteroidea† (praying mantises, cockroaches, termites)**

A characteristic feature of this group of cockroach-like insects is the deposition of egg-cases contained in an 'ootheca' that is formed in the female bursa (vaginal pouch). The bursa is closed ventrally by the 7th abdominal sternite (the 'subgenital plate') and its roof is formed by the reduced 8th and 9th sternites. It is a further development of the genital atrium of the Dermaptera. As in the Dermaptera, the female gonopore is behind the 7th sternite and the ovipositor is reduced.

Sharov (1966b) stated that in the insects the primitive position of the female gonopore is in the 7th abdominal segment. This contradicts Weber's (1949) statement that its 'typical' position in the Pterygota is behind the 8th sternite. The ovipositor is formed from derivatives of the appendages on the 8th and 9th abdominal segments and is part of the ground-plan of the Pterygota. This is one reason why I find Weber's arguments more convincing than Sharov's. Another reason is that the forward shift of the female gonopore behind the 7th sternite in the Dermaptera and Blattopteroidea may well be a derived character connected with reduction of the ovipositor. The egg-cases and the bursa are most completely developed only in the praying mantises and cockroaches. For this reason, the praying mantises (Mantodea) and the cockroaches (Blattariae) have often been called the Oothecaria or Dictyoptera (Chopard, in Grassé, 1949). In fact, the cockroaches (Blattariae) are more closely related to the termites (Isoptera). One derived character common to these two groups is the reduction of the anterior ocellus. This is very important for determining which Palaeozoic species belong to the stem-group of the Blattopteroidea and also the minimum age of the sub-groups of the Blattopteroidea. Snodgrass (1958b) has established that in the Blattopteroidea the dorsal longitudinal muscles of the thorax are either absent or are too small to have any indirect action on the wings as is characteristic of all other Neoptera and the Ephemeroptera. Arising on the

---

† I am using the name Blattopteroidea in the sense of Grassé (1949). In the Russian *Textbook of Palaeontology* (Rodendorf, 1962), the 'super-order Blattopteroidea' also includes the Dermaptera. Wille (1960) has called this group the 'Dictyoptera', but Grassé used this name for the Mantodea + Blattariae!

notum of each alate segment of the Blattopteroidea are numerous muscles which are mostly leg muscles. According to Snodgrass, these muscles should elevate the wings by depressing the notum, in the manner of the notosternal muscles of insects with a typical indirect flight mechanism. Lateral muscles attached to the basalar and subalar sclerites probably lower the wing.

Snodgrass believed that the Blattopteroidea had ancestors that lacked any well developed indirect flight mechanism and he doubted whether these ancestors ever flew any better than the modern cockroaches and praying mantises. It is not clear how these views can be reconciled with what is known in general about the relationships of the different groups of the Pterygota, or with the assignment of the Blattopteroidea to the Paurometabola. Moreover, Snodgrass thought that the flight mechanism of the Blattopteroidea was not yet fully understood. This matter needs urgent clarification, along with the structure of the coxae: these are probably derived characters, as are the equivalent characters in the Odonata though the situation is not exactly the same.

In estimating the minimum age of the Blattopteroidea, the most important fact could be the discovery of 'oothecae' in the middle Upper Carboniferous (upper Westphalian) of northern France, Illinois, etc. (Laurentiaux, in Piveteau, 1953). If these identifications are absolutely correct, then we can be certain that at least the stem-group of the Blattopteroidea (Mantodea + Blattodea) must have been in existence at this time. This would mean that the stem-groups of the Embioptera, Notoptera, and Dermaptera must also have been in existence as separate groups at the same time, because, unlike the Blattopteroidea, these groups cannot be derived from ancestors that produced oothecae.

This is another example of the extraordinarily far-reaching consequences that may result from the correct or incorrect interpretation of a single fossil or, as in this case, of a group of similar fossils. In such cases, only the very highest criteria of proof should be accepted. For this reason, it is highly significant that Brown (1957) has stated that none of the Palaeozoic 'oothecae', which Sellards was the first to describe, can be recognized with absolute certainty as being the egg-cases of cockroach-like insects. Brown has reported that at first Pruvost (1919) regarded the only one of several such fossils that he described as the tooth of a fish (*Ctenoptychius* sp.). Only later did he consider it to be the ootheca of a cockroach. Brown believed that Pruvost's earlier opinion was correct—and apparently applied to all the Palaeozoic egg-cases!

Laurentiaux (1960a) made no reference to Brown's objections, but continued to take it for granted that egg-cases of cockroach-like insects were reliably known from the Upper Carboniferous. I shall deal later with the conclusions that he has drawn from this assumption. As these conclusions are open to dispute, however, I must first discuss how far other characters can be used to trace the subgroups of the Blattopteroidea, that is to say the groups which are characterized by the production of egg-cases, back into the

Palaeozoic. Only then can any assessment be made of the oldest cockroach-like fossils.

2.2.2.2..2.2..1.3.1. Mantodea (praying mantises)[183]

The praying mantises are characterized by several apomorphic features which are correlated with their predatory habits: the fore-legs have been modified into raptorial limbs; the prothorax is elongate; the head is extremely mobile; the mandibles have been enlarged; and, according to Popham (1961), the eyes are broadly separated to give better stereoscopic vision.

Sharov (1966b) suggested that the stem-group of the 'Polyneoptera' was predaceous, and that the Mantodea preserved this mode of life whereas the other groups acquired omnivorous or herbivorous habits. I am not convinced by this. The adaptations of the Mantodea to their predaceous life are so striking that they can only be regarded as derived characters.

In some characters the Mantodea are more primitive than the Blattodea (cockroaches and termites): the anterior ocellus has been retained and the head is apparently primarily orthognathous. Some characters in the wing venation are also primitive (Fig. 48): the subcosta is long, and most of the anal veins terminate freely in the hind-margin of the fore-wing. Compared with the Blattariae, these two characters are primitive and, according to Holmgren (1909), they are also present in the ground-plan of the termites (Isoptera). However, this is open to doubt.

Unfortunately, the ground-plan of the wing in the Mantodea has still not been clearly worked out. It is probable that the fore-wing is very narrow and has the distal half strikingly elongate (see Ragge, 1955a, Fig. 105). This may have given rise to the secondary elongation of the subcosta, its approximation to the radius, and the atrophy of the veinlets that run from the radius and subcosta into the costa in other Blattopteroidea.

No Palaeozoic or Mesozoic fossils have been found with the derived characters of the Mantodea: raptorial legs, head, prothorax. Tillyard (1922b) described *Triassomantis pygmaeus*, a wing fragment from the Upper Triassic of Australia which was supposed to belong to the Mantodea. Handlirsch (1939) was right to voice considerable scepticism about this, and he suggested that it was really a hind-wing with the anal fan missing. However, the Mantodea would be very difficult to recognize from a hind-wing. Sharov (1965) assigned the 'Triassomanteidae' to the Saltatoria ('Orthoptera') in his phylogenetic tree. Laurentiaux (in Piveteau, 1953) reported that several closely related Mantodea together with some oothecae had been found in the Lower Triassic of the Vosges, but apparently they have never been described. Finally, I have been unable to see what features in the fore-wing of *Archaeomantis fleischmanni*, from the Lower Permian of the Thüringer Wald, led Haupt (1952) to assign it to the Mantodea.

In the light of these dubious fossils, the only evidence that can be used to determine the age of the Mantodea is that all the Mesozoic cockroach-like

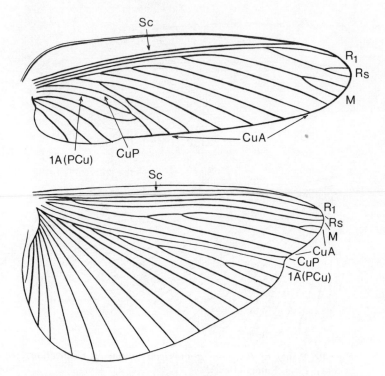

Fig. 48. Wings of *Sphodromantis viridis* Forskål (Mantodea; recent).
From Ragge (1955a).

species have the characters of the Blattodea: the subcosta is short and several anal veins terminate in the anal furrow. Fossils with the same derived characters are also known from the Upper Carboniferous. As a result, we shall probably have to go back at least as far as the Upper Carboniferous to find the common ancestors of the Blattodea and Mantodea, if we consider them to be sister-groups. The probable age of the Mantodea can therefore only be inferred by indirect methods. There is absolutely no evidence that subgroups of the *Mantodea had already arisen by the Palaeozoic.

Additional note

183. A new monograph on the Mantodea has been published by Beier (1968a). [Dieter Schlee.]

2.2.2.2..2.2..1.3.2. Blattodea (cockroaches and termites)[184]

The absence of the anterior ocellus is the only undoubted derived character common to the cockroaches and the termites. Two further characters may be derived: the short subcosta (?) and the arrangement of the anal veins. In all Blattodea, as well as the Mantodea, the 1st anal vein (= postcubitus,

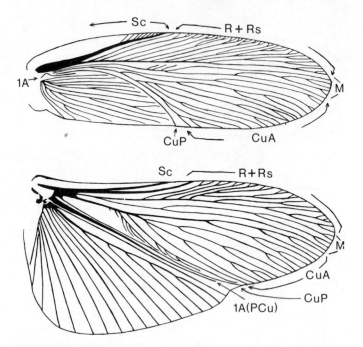

Fig. 49. Wings of *Periplaneta americana* Linnaeus (Blattariae; recent). From Smart (1951).

1A = PCu) of the fore-wing is short and never reaches the hind-margin (Fig. 49). In the fore-wing of the Mantodea, almost all the other anal veins reach the hind-margin, except in a few species that are probably derived, but it seems that they never do in recent Blattodea. In this group the majority of the anterior anal veins reach the anal furrow. This character is no longer distinct in the termites, and this is probably due to the very striking changes that have taken place in the wing: these are very characteristic of the termites and have also affected the anal lobe. In many ways *Mastotermes* is the most primitive living termite, and the anal lobe of the fore-wing could very well have developed from a forerunner that has been fully realized in the cockroaches.

Holmgren (1909) stated that a long subcosta is part of the ground-plan of the termites and that the anal veins of the fore-wing reach the hind-margin. In his day, however, *Mastotermes* was only imperfectly known, and in any case Holmgren had apparently not seen it for himself. As these two characters are so important, a fresh study should be undertaken to establish whether the short subcosta and the termination of the anal veins in the anal furrow are really part of the ground-plan of the termites as well as the Blattariae. Our ideas about the relationships between living Blattariae (cockroaches) and Isoptera (termites) determine how we estimate the age of the Blattodea and their subgroups and how we assess the numerous Palaeozoic fossils.

Weidner (1966) has discussed this in a recent paper. He reported that

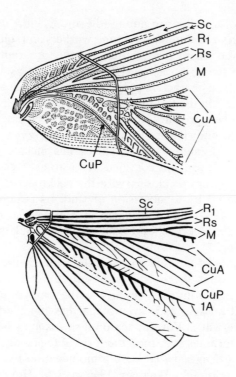

Fig. 50. Wing-bases of *Mastotermes dar-winiensis* Froggatt (Isoptera; recent). From Grassé (in Grassé, 1949).

Holmgren also dealt with this problem as early as 1909 and based his results on the analysis of about 60 characters. Holmgren found that 'each group has seven derived characters, which are present in a primitive state in the other group', and he concluded 'that the termites could not have descended from the cockroaches but that the two are sister-groups' (Weidner, 1966, p. 325). Holmgren's conclusions could be well founded if his assessment of these seven derived characters were correct, but there are a few difficulties that cannot be ignored.

On the whole, the characters of the termites (Isoptera) are much more strongly derived than those of the Blattariae. Their wings have become largely homonomous secondarily. The anal fan of the hind-wing has only been retained in *Mastotermes*, and here it is very reduced. Both pairs of wings have a basal suture along which the fracture and shedding of the wings subsequently takes place. According to Grassé (in Grassé, 1949) the thorax is highly desclerotized compared with the Blattariae: the sclerites are separated by large membraneous areas. The female bursa is reduced, and the female does not produce an ootheca like the Blattodea and Mantodea. In *Mastotermes*, however, the eggs are (still) laid in clusters of 16–24, cemented together with a gelatinous secretion. According to Grassé (in Grassé, 1949), the spermatozoa of termites are non-motile and lack the usual tail. Colonies and castes are a very characteristic feature; workers and soldiers are apterous.

All these derived characters of the termites are represented by more primitive states in the cockroaches. Holmgren's (1909) statement that there are also seven derived characters in the Blattariae, which are represented by more primitive states in the termites, is nothing like as well founded. Some of the characters are very trivial. Others have had to be assessed differently in the light of present knowledge of the genus *Mastotermes*, which was inadequately known in Holmgren's day. This genus includes a single species, *darwiniensis* Froggatt, which is confined to Australia. According to Weidner (1966, p. 325), it 'agrees so closely with the cockroaches in characters that have been discovered by a succession of students (Crampton, 1923; Rau, 1941; McKittrick, 1964) that the imago of *Mastotermes* could be assigned unhesitatingly to the cockroaches if no account were taken of the conformation of its fore-wing'[185].

'The conformation of the fore-wings' is the main character that Holmgren used as evidence that there is a sister-group relationship between the cockroaches and the termites. 'As early as the Carboniferous the blattoids had tegmina, and no species is known that has modified these into really membraneous fore-wings. For this reason it seems most improbable that the fore-wings of the termites ever took the form of tegmina. There is nothing in the termite wing to suggest that such a modification has ever taken place' (Holmgren, 1911; quoted by Weidner, 1966, p. 325). Holmgren was forced to conclude that the two sister-groups, cockroaches and termites, separated far back in the Palaeozoic. Weidner followed Rodendorf in suggesting that 'this may have taken place at the transition of the Devonian and Carboniferous'.

I think that Holmgren's suggestion that the 'membraneous' fore-wings of the termites could not have developed from 'tegmina' is highly questionable. On the contrary, tegmina appear to be part of the ground-plan of all Orthopteromorpha. Moreover, it is not improbable that in the stem-group of the Neoptera the fore-wings were originally tougher than the hind-wings and provided them with some protection.

At first sight, opisthognathy appears to be a derived character of the cockroaches (Blattariae). Popham (1961) found that two further characters are functionally linked with opisthognathy: a forward shift in the origins of the maxillary adductor muscles and a posterior shift in the circumoesophageal commissures. The tentorial body is anteriorly enlarged to provide points of origin for the maxillary adductor muscles, and there is a small foramen through which the nerve commissures pass. Compared with this, orthognathy (Popham's hypognathy) in the termites is a primitive character. However, Lameere has shown that the elongate labium of primitive Isoptera indicates that orthognathy in the termites developed from an earlier opisthognathous state (in the sense of Popham). This suggests that the short pronotum of the Isoptera, which does not cover the entire head as it does in the Blattariae, is not a primitive character, as is generally thought, but is in fact derived. It is also difficult to assess other 'more derived' characters in the cockroaches[186].

This discussion shows that it is no longer possible to accept unquestioningly that there is a sister-group relationship between the Blattariae and Isoptera. It is possible that one of the subgroups of the Blattariae is more closely related to the Isoptera than are the others.

Ahmad (1950) suggested that this was the case with the North American, subsocial, wingless, wood-feeding cockroach *Cryptocercus punctulatus* Scudder. This species is usually placed in a separate family, the Cryptocercidae[187], and for some time it has been prominent in discussions about the origin and relationships of the termites.

According to Weidner (1966, p. 325), this is because the nutritional physiology is so similar in *Cryptocercus* (Blattariae) and *Mastotermes* (Isoptera), and because 'its method of feeding has forced *Cryptocercus* into a primitive form of social behaviour'. After describing the nutrition and symbiotic fauna of *Cryptocercus*, Weidner (1966, p. 327) wrote: '*Mastotermes* digests wood in the same way, and has a fauna of flagellates that is very similar to that of *Cryptocercus*, particularly the genera *Oxymonas* and *Trichonympha*. So far as is known, *Cryptocercus* is the only cockroach genus in which flagellates are present. Moreover, it is linked to *Mastotermes* by the presence of intercellular [misprint for intracellular] symbiotic microorganisms which are common to all the cockroaches but which are only found in *Mastotermes* amongst the termites'. Weidner (1966) also reported that McKittrick considered *Cryptocercus* to be the living cockroach which has retained the greatest number of primitive characters. 'It occupies the same position among the cockroaches as *Mastotermes* does among the termites'. Weidner's subsequent discussion is not very clear but he appears to suggest that there is a sister-group relationship between the Blattariae and the termites and that the common ancestors of these two groups had a mode of life similar to that still followed by *Cryptocercus*. He also appears to suggest that there is a sister-group relationship between *Cryptocercus* and all the other recent Blattariae, and between *Mastotermes* and all the other recent Isoptera.

The development of the symbiotic fauna would reflect this, as was suggested by Grassé and Noirot (1959; quoted by McKittrick, 1964). A double symbiosis occurred in the common ancestors of the cockroaches and termites: intracellular microorganisms and xylophagous intestinal flagellates. Both groups of symbionts were retained by the cockroach *Cryptocercus* and the termite *Mastotermes*. The rest of the cockroaches only retained the intracellular symbionts whilst the rest of the termites only retained the intestinal flagellates.

I have serious misgivings about this simplistic idea. Neither Princis (1960) nor McKittrick (1964) considered *Cryptocercus* to be the sister-group of the other cockroaches (Fig. 121). According to McKittrick, the phylogenetic tree of the Blattariae has separated into two divergent phylogenetic lines, in other words into two 'sister-groups'. She called these the Blattoidea and

Blaberoidea. The two most primitive families, Cryptocercidae and Polyphagidae, bear certain resemblances to the termites and to each other, but both have characteristics linking them unmistakably to the derived species within their respective superfamilies: the Cryptocercidae to the Blattoidea and the Polyphagidae to the Blaberoidea. McKittrick's phylogenetic tree agrees with the conclusions given in her text.

Princis also recognized two main branches in the phylogenetic tree of the Blattariae. However, his distribution of families amongst these two sister-groups is different from that of McKittrick. For example, he stated that the Cryptocercidae are most closely related, or should be assigned, to the Blaberoidea and not to the Blattoidea.

Neither author considered the possibility that there may be a sister-group relationship between the Cryptocercidae and the rest of the cockroaches, corresponding to the sister-group relationship between the Mastotermitidae and the rest of the termites!

If either Princis' or McKittrick's ideas about the relationships amongst the recent Blattariae are correct, then there can only be three possible relationships between these groups:

1. The Blattariae and Isoptera are sister-groups. The mode of life and the double set of symbionts in *Cryptocercus* represent a primitive condition, present in the common ancestors of these two groups. The Blattariae must have lost this mode of life and nutritional physiology, several times and independently, for they have now adopted a less specialized mode of feeding (omnivory) and nutritional physiology which have made them independent of the wood-digesting symbionts.
2. The Blattariae as currently defined and Isoptera are not sister-groups; *Cryptocercus* is more closely related to the termites than to the rest of the 'Blattariae'. In this case it would be correct to consider that the mode of life and nutritional physiology of *Cryptocercus* is the same as that of the common ancestors of *Cryptocercus* and the termites. The Cryptocercidae would then have to be regarded as a relatively primitive side-branch of the 'remaining' Isoptera (including *Mastotermes*), as a sister-group that is still subsocial and has not developed a system of castes.
3. The Blattariae and Isoptera are sister-groups. Features in the mode of life and nutritional physiology that are shared by *Cryptocercus* and the Isoptera are the result of convergence.

It will only be possible to decide which of these three possibilities is correct if fresh studies are undertaken that are specifically directed at finding a solution.

It is *a priori* very unlikely that the first of these possibilities is correct, since it would mean that all the recent cockroaches must have developed from the specialized mode of life found in *Cryptocercus*: '*Cryptocercus* is entirely dependent on the presence of flagellates for its nutrition. It starves without them' (Weidner, 1966, p. 326). For this reason, the transfer of flagellates is very delicately linked to the seasonal rhythm. On the other hand, the third possibility, suggesting convergence, seems much more likely. There is now an enormous geographical area separating *Cryptocercus* (North America) and *Mastotermes* (Australia). In the Tertiary, however, species that are assigned to the Mastotermitidae were widely distributed in the northern hemisphere: Weidner listed 15 fossil species. It is likely that they lived with the ancestors of *Cryptocercus*, and in this way transfer of symbionts could easily have taken place.

In themselves these *ad hoc* hypotheses have no value. However, they can be useful, as here, when they are used simply to get round the difficulties that stand in the way of a suggestion which, for other reasons, appears to be entirely plausible.

Cryptocercus is apterous, and unfortunately this makes it extremely difficult to draw any further conclusions. According to McKittrick, there is one other group of cockroaches that agrees with *Cryptocercus* (and also with the termites) in many primitive characters. This is the Polyphagidae, or Polyphagoidea of Princis (1960), and this group has a hind-wing that in many ways is more primitive than all the other recent winged cockroaches. They are the only species that do not fold the anal fan (vannus) beneath the remigium of the hind-wing when at rest, and apparently this was also a feature of the Palaeozoic cockroaches. Laurentiaux (1960b) dealt with this character. His first step was to refute the old view, put forward by Tillyard, that the furrow separating the anterior part of the hind-wing ('remigium') from the anal lobe is actually in a different position in Palaeozoic and recent cockroaches (between different longitudinal veins). Laurentiaux followed Smart's (1951) new and painstaking study of the venation of recent cockroaches. He concluded that in all known species the furrow lies between the postcubitus (PCu; sometimes called '1st anal vein', see p. 159) and the 1st vannal vein ('2nd anal vein'). The three longitudinal veins CuP, PCu, and 1V (1st vannal vein, '2nd anal vein') define a particular area that is affected by the way in which the anal lobe (vannus) is folded up.

Laurentiaux stated that there is one fundamental difference between Palaeozoic and recent cockroaches. In all the Palaeozoic species where the hind-wing is known, the postcubitus has numerous branches. In recent species, however, this vein is almost always simple and is often short. A short apical fork has been retained only in the Polyphagidae. Certain Permian genera occupy a somewhat intermediate position because the branches of the postcubitus do not reach the wing margin as they do in all the Carboniferous species: some or all of them only reach the furrow where the hind-wing folds.

According to Laurentiaux, the remigium is strikingly narrow, and this is the result of (or has given rise to) a reduction of the postcubitus and its branches. The remigium forms a protective cover for the folded anal lobe, as the fore-wing does for the entire hind-wing. There would be no room beneath a narrow remigium for an anal lobe that was simply folded under. In recent species, therefore, the lobe is folded in a complicated, fan-like manner.

The Polyphagidae (Princis' Polyphagoidea) are an exception to this. They have the anal lobe folded simply beneath the remigium, as in the Palaeozoic 'cockroaches'. According to Laurentiaux, they have compensated for the reduction of the postcubitus by means of a secondary increase in the number of branches of CuA (Cu_1).

It is clear, therefore, that not all the characters of the Polyphagoidea are as primitive as in the Palaeozoic species. They have the postcubitus reduced, and this is undoubtedly a derived character in which they agree more closely with the rest of the recent cockroaches. Laurentiaux suggested that the enlargement of the wing in the region of CuA was a secondary feature, but it is still unclear how this should be assessed. I would hazard the guess that there were two possible ways in which the *Blattariae could have compensated for the reduction of the postcubitus: the Polyphagoidea enlarged the region around CuA and retained the original simple method of folding the anal lobe beneath the remigium; the rest of the *Blattariae acquired a more complex method of folding the anal lobe. This explanation would gain in probability if it could be shown that there is a sister-group relationship between the Polyphagoidea and the rest of the *Blattariae. However, neither Princis (1960) nor McKittrick (1964) showed this to be so. A fresh study needs to be made of this problem.

All Blattariae have the postcubitus reduced and straight (except for a short apical fork in the Polyphagoidea), and this is particularly important for assessing the relationship between the cockroaches and the termites. If the current interpretation of the venation is correct, the Isoptera have a postcubitus ('1A') with a large number of branches. This is certainly a primitive character. On the other hand, the Blattariae have a large anal lobe, which is a primitive character, whereas in the termites this is greatly reduced (*Mastotermes*) or completely absent. These two characters actually do enable us to deduce from the wing structure that there is probably a sister-group relationship between Blattariae and Isoptera. There is some doubt about *Cryptocercus*, which is completely apterous.

This heterobathmy of wing characters makes it possible to determine the period when the sister-group relationship originated.

Additional notes

184. New monographs have been published on the Blattariae by Beier (1974) and on the Isoptera by Weidner (1970a). [Dieter Schlee.]

Roth (1969) has discussed the tergal glands of male Blattariae. [Willi Hennig.]

185. For further discussion of the primitive characters of the termites, see Emerson (1961). Emerson (1965) has also given a review of the Mastotermitidae. Roonwal (1975) has published a phylogenetic tree of the termites. [Willi Hennig.]

186. For a discussion of phyletic relationships within the Blattidae, see McKittrick and Mackerras (1965). [Willi Hennig.]

187. Two further species are known from China. [Willi Hennig.]

Appendix to the Blattopteriformia: the so-called Palaeozoic 'cockroaches'

Large numbers of Palaeozoic fossils have been described and assigned to the 'cockroaches'. Like other authors, Carpenter (1934b) considered the abundance of the 'cockroaches' to be the most striking feature of the Carboniferous insect fauna. At that time more than 800 species had been described, and those amounted to about 60% of the described Carboniferous insects. This percentage is hardly likely to have changed a great deal since then. I should stresss that most of these 800 'species' have only been described from single fore-wings or even from wing fragments. The question of how many of them could actually be varieties of a single species has never been adequately studied. Nevertheless, even if a fresh investigation were to reduce the number of species, there would still be no disputing the abundance of the 'cockroaches' during the Carboniferous. Did the 'cockroaches' reach a climax during the Carboniferous, did they die back during the Permian, and have they reached another climax since the Mesozoic?

Before answering these questions, we should recall that the word 'cockroach' comes from a pre-scientific era of insect classification. It is a typological concept, based on a number of striking but superficial physical features. The name and concept of the cockroach was originally applied to animals which characteristically have no ovipositor but in its place have a vaginal pouch in which egg-cases are formed. To what extent are we correct in calling a fossil insect a 'cockroach' if it had an ovipositor and undoubtedly laid eggs singly?

According to Laurentiaux (1951), it has been known since Brongniart first described them in 1889 that some of the Palaeozoic 'cockroaches' possessed a long ovipositor. Zalesskiy (1939), too, has described such a species as *Uraloblatta insignis*. As discussed above, Laurentiaux (1960a) did not question the identification of certain Upper Carboniferous fossils as oothecae, although this has been disputed by others. He believed that there have been two distinct groups since the lower Westphalian: to one of these, the Eoblattodea containing the family Archimylacrididae[188], he attributed an ovipositor; he believed that the other group, the Neoblattodea with the families Mylacrididae, Poroblattinidae, and Mesoblattinidae, produced the so-called oothecae that have been found as early as the Westphalian. He did not believe that the specialized method of oviposition of the Neoblattodea would have given them any advantage over the Eoblattodea in the subtropical forests of the Upper Carboniferous of the northern hemisphere. In the adverse climatic conditions of the Permian, however, they would have had a definite advantage. This is why the Neoblattodea survived the Palaeozoic and developed into the modern cockroaches. However, these ideas are based on two very doubtful premises: the interpretation of the alleged Upper Carboniferous oothecae, and the lack of any fossil abdomina of the Palaeozoic Neoblattodea. Only when these doubts have been cleared up will it be possible to show that this group had no ovipositor and produced oothecae.

It needs to be stressed from the start that the 'cockroaches' with a well developed ovipositor cannot be assigned to the stem-group of the Blattariae, or even to that of the Blattopteroidea, at least not without the support of additional complicated arguments—which are pure speculation and therefore inadmissible evidence. In fact, all the later 'Paurometabola' could be derived from these 'cockroaches' and for this reason they are of no use for determining the age of any of the individual subgroups.

It is not certain whether the Palaeozoic fossils with no ovipositor preserved did actually have one. In any case, this character alone would be of little assistance and we need to look for others. The most important ones are the structure of the pronotum and the wing venation.

Several fossils have been found since the Upper Carboniferous in which the fore-wing has a short subcosta and several of the anterior anal veins only reach the anal furrow rather than the wing margin (Fig. 51). Handlirsch

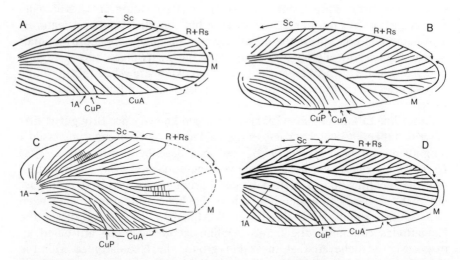

Fig. 51. Wings of various Palaeozoic and Mesozoic Blattopteroidea. A, *Sysciophlebia euglyptica* Germar (Spiloblattinidae), from the Upper Carboniferous (Ottweiler Beds, Saxony), with a long subcosta and free anal veins; B, *Autoblattina* sp. (same age and locality), with a short subcosta but free anal veins; C, *Dictyomylacris poiraulti* Brongniart (Dictyomylacrididae), from the Upper Carboniferous (Stephanian of Commentry), with a short subcosta and anal veins reaching CuP; D, *Triassoblatta typica* Tillyard (Mesoblattinidae), from the Upper Triassic of Australia (Ipswich). From Handlirsch (1906–1908) (A–C) and Tillyard (1937e) (D).

(1920–21) assigned them to the Neorthoblattinidae, Dictyomylacrididae, Neomylacrididae, Poroblattinidae and Diechoblattinidae. However, Martynova (in Rodendorf, 1962) stated that in the Poroblattinidae the anal veins do reach the hind-margin of the wing. All the species with these two derived characters can be assigned to the stem-group of the Blattodea, although it is possible that one or both of them originated independently on several occasions. The oldest genera appear to be *Acmaeoblatta* and *Dichronoblatta* from the upper Upper Carboniferous of Ohio, and *Nearoblatta* from the upper Upper Carboniferous (Ottweiler Beds) of Saxony. Handlirsch placed all three genera in the Mesoblattinidae. The upper Upper Carboniferous is therefore the earliest moment when we can show that the sister-groups Mantodea and Blattodea were probably in existence. No Upper Carboniferous fossils can be assigned to the Blattariae or Isoptera, however.

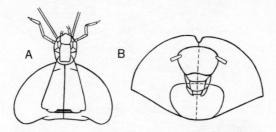

Fig. 52. A: Prothorax and head of an Upper Carboniferous archimylacridid (ventral view). B: Prothorax and head of *Oniscosoma grandicollis* Saussure (Panchloridae; recent, Australia). From Tillyard (1937e).

According to Tillyard (1937e), fossils in which the pronotum has begun to expand forwards to overlap the head (Fig. 52A) are not known before the Lower Permian. In these species, the development of this character has not progressed very far. Nevertheless, they could be assigned to the stem-group of the Blattariae if we could be sure that the termite pronotum, which does not overlap the head, is a primitive character. However, even this is not certain (see p. 198).

In the light of this, the only character that can be used to indicate the age of these two sister-groups is the development of the postcubitus in the hind-wing. If Laurentiaux (1960b) is correct (p. 201), all the Palaeozoic 'cockroaches' had the postcubitus of the hind-wing with numerous branches (as in Fig. 53, where it is labelled as '1A'), whereas in all the recent Blattariae (Fig. 49) the postcubitus has no more than a short apical fork. On the other hand, the postcubitus in the hind-wing of the Isoptera does not appear to be reduced (Fig. 50; PCu = 1A). If this is correct, it would be possible still to derive the Blattariae and the Isoptera from the most recent Palaeozoic fossils. Unfortunately, only a few fossils actually have the hind-wings preserved.

Many Palaeozoic and Mesozoic species have been described which are supposed to belong to the Isoptera, but all of these assignments are extremely doubtful. *Uralotermes permianus* Yu. Zalesskiy is a very poor fore-wing fragment but has been treated as a fossil termite by Weesner (1960). In the Russian Textbook of Palaeontology it was assigned to the Paraplecoptera *incertae sedis*, to which I would only add that the Paraplecoptera themselves are a very problematic group (p. 168).

The most important fossil is probably *Puknoblattina* (= *Pycnoblattina*) (Fig. 53), which was described from the Lower Permian of Kansas. The

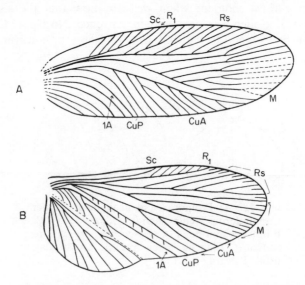

Fig. 53. A: Fore-wing of *Puknoblattina compacta* Sellards, from the Lower Permian of Kanas. B: Hind-wing of *Puknoblattina* sp. From Tillyard (1937e).

termite wing could be formally derived from the hind-wing of this genus, but its most important character is the extensive reduction of the anal lobe. This is no more strongly developed than in *Mastotermes* but the venation is rather more primitive. McKittrick (1964) believed that the termites had descended from a Lower Permian genus that was not very different from *Puknoblattina*.

However, *Puknoblattina* has at least one character that makes it doubtful whether it really belongs to the stem-group of the Isoptera: according to Tillyard (1937e; see also Fig. 53), all the anal veins of the fore-wing terminate freely in the wing margin. Holmgren (1909) stated that in the Isoptera this character is primitive, relative to the Blattariae. However, the published drawings of *Mastotermes* (cf. Fig. 50) seem to me to show that the anal veins only reach the anal furrow, as in the Blattariae. If we derive the termites from

Puknoblattina, this character must have arisen independently in the termites and the Blattariae. This is not impossible, but we must bear in mind that even the Mantodea have at least two anal veins that no longer reach the wing margin. According to Tillyard (1937e), the pronotum of *Puknoblattina* did not cover the head as it did in some of the contemporary 'cockroaches', and this may have been a primary condition. In this character *Puknoblattina* agrees with the Isoptera. However, it is not impossible that the head has become freed independently from the pronotum in recent termites. Tillyard (1936a) seemed to believe this, and I have also given the reasons for thinking it possible (p. 198). In these circumstances it is doubtful if the reduced anal lobe of the hind-wing (Fig. 53) is sufficient to establish that *Puknoblattina* belongs to the stem-group of the Isoptera.

In this book I am following the sound principle of placing the origin of any particular group *only* as far back in the past as is supported by well founded evidence. As a result, I can only conclude that at least for the present there is no justification for believing that the sister-groups Blattariae and Isoptera arose long before the upper Upper Carboniferous or even at the transition of the Devonian and Carboniferous (Weidner, 1966, following Rodendorf). It is not even certain that the Isoptera can be traced as far back as the Lower Permian.

If I were asked to explain how the generally accepted view of the Carboniferous as the 'Age of the Cockroaches' should really be understood, I would summarize the preceding discussion as follows.

During the Carboniferous there were undoubtedly many species that resembled the recent cockroaches in certain characters. However, if we restrict this term to a single genealogical unit, the Blattariae, then most of these Carboniferous species were not true 'cockroaches'. Some of them undoubtedly belonged to the stem-group of all the Paurometabola and were probably 'side-branches' of various archaic 'layers' of this stem-group. It is not known to what extent the stem-groups of the Notoptera, Dermaptera, Mantodea, and Blattodea were also in existence at this time as distinct groups, and it will probably never be possible to determine this. It is probable that they were distinct by the upper Upper Carboniferous (see under Mantodea, p. 194). It is also possible that the stem-groups of the Blattariae and Isoptera were also distinct at the same time, but this is not certain, even for the Permian.

During the Carboniferous, there were probably species with adaptations for particular biotopes that are unknown amongst the recent Paurometabola. Little is known about this, but Kukalová (1961) has described a species with long tarsi which she placed in the Archimylacrididae, a group which is supposed to have retained a long ovipositor. The tarsi 'probably enabled the insects to move over the soft surface of the Carboniferous swamps'.

At the beginning of the Permian many of the Carboniferous groups were probably eliminated. This was described by Laurentiaux (1960a) and is also known to have taken place in other groups of insects. It was mainly species of the stem-group of the Paurometabola that were most closely affected by this. It is not known what effect these changes during the Permian had on species of the stem-groups of the Notoptera, Dermaptera, Mantodea, and Blattodea, although some of these are known to have been in existence during the Upper Carboniferous. This is because there are so few fossils that can be assigned with any confidence to these stem-groups. Some of these groups, such as the Protelytroptera from the stem-group of the Dermaptera, must have undergone a new radiation during the Permian.

It is also difficult to decide how many Paurometabola were able to survive into the Mesozoic and how many of them belonged to the stem-groups of various subgroups that are still extant in the recent fauna. Bekker-Migdisova (in Rodendorf, 1962) divided the Mesozoic 'Blattidae' (*sic*) amongst the Diechoblattininae and Mesoblattininae, two distinct groups ('subfamilies') which were also in existence during the Palaeozoic. Furthermore, the 'family' Poroblattinidae is supposed to have lived from the Carboniferous until the Upper Triassic. The short subcosta shows that both the Diechoblattininae and the Mesoblattininae (Fig. 51D) belonged at least to the stem-group of the Blattodea. In other respects, it is doubtful if they are genealogical units. Almost 100 Mesozoic species of Mesoblattinidae have been described and, according to Bekker-Migdisova, the anal veins of the fore-wing reach the hind-margin. This means that they cannot belong to the Blattariae or the Isoptera, but could actually be survivors from the stem-group of the Mantodea which subsequently became extinct during the Mesozoic. In many species, however, it is not clear whether the anal veins do in fact reach the hind-margin of the wing as they are said to do by Bekker-Migdisova. Bekker-Migdisova included the genus *Kulmbachiellon* in the Mesoblattininae, but Zeuner (1939) preferred to regard it as a cricket. If it is possible to have such radically different assessments of a single species, then we should be extremely sceptical over the interpretations of others.

In the diechoblattinine *Kokandoblattina analis* Martynov, described from the lower Lias of Kyzyl-Kiya, the anal veins only reach the anal furrow of the fore-wing, and not the hind-margin. Bekker-Migdisova stated that this is 'often' the case in other Diechoblattininae. These species could belong to the stem-group of the Blattariae or the Isoptera. Unfortunately, no hind-wings are known for any of them.

To summarize, it seems certain that only a few of the species that were able to survive from the Palaeozoic into the Mesozoic are now represented by any descendants in the modern fauna. It is unlikely that the recent species of any of the so-called 'orders' of the Paurometabola have descended from *several* different Palaeozoic or early Mesozoic species. This is just as true of the Embioptera, Notoptera, and Dermaptera as it is of the Mantodea, Blattariae, and Isoptera. All these orders were represented by distinct stem-species during the Palaeozoic, and some even during the Upper Carboniferous. However, the latest common stem-species of the recent species of each order must have lived at some time during the Mesozoic.

Additional note

188. Laurentiaux (1967) has revised the genus *Archimylacris* and has described a new genus from the Westphalian. [Willi Hennig.]

2.2.2.2..2.2..2. Orthopteroidea

Sharov's (1967) studies have totally changed current thinking about the phylogeny and phylogenetic classification of the Orthopteroidea. Grasshoppers, locusts, crickets, etc. ('Saltatoria' of the old classification) together with the stick-insects and leaf-insects (Phasmatodea) belong to this group. Sharov stated that his results were still rather provisional. He referred to studies he had made of some newly discovered fossils, but said that their detailed study and description was a task that he had still to undertake.[†] However, his provisional conclusions are so convincing that I am regarding them as well founded and am using them as the basis for the discussion which follows.

The Ensifera, Caelifera, and Phasmatodea have long been recognized as well founded monophyletic subgroups of the Orthopteroidea. The main difficulty has been to work out the relationships of the Phasmatodea. Zeuner (1939) stated that the morphology and development of the Phasmatodea indicated that they were very closely related to the Orthopteroidea rather than to the cockroach-like insects. This reflected the generally accepted view, but it was too vague to be regarded as well founded. Ragge (1955b) suggested that there were no grounds for regarding the Phasmatodea as closely related to any of the 'orthopteroid orders' (amongst which he included the Blattopteroidea!). Ross (1955) believed that the Phasmatodea were more closely related to the Mantodea-Blattodea, because each phasmatodean egg is embedded in a secretion comparable to the ootheca of the Mantodea-Blattodea. Subsequently, Ross (1965) even assigned the Phasmatodea to an order Dictyoptera, together with the Mantodea and Blattodea.

On the other hand, Sharov (1967) believed that the Phasmatodea are most closely related to the Caelifera and that these two groups shared ancestors common only to them. If this suggestion is correct, certain characters that have been regarded as synapomorphies of the 'Saltatoria' (= Ensifera + Caelifera) will have to be re-assessed: the modification of the hind-legs into jumping legs would have to be a symplesiomorphy, and the reduction in the number of tarsal segments to four or less[189] would have to be convergence. It is not clear how some of the other characters would have to be interpreted. In these circumstances, it is difficult to list the derived ground-plan characters of the Orthopteroidea. One such character is probably the fusion of the labial stipes into a single plate[190].

A more striking character, and one that is almost always visible in fossils, is the enlarged precostal field that has arisen as a result of the costa moving

---

[†] This study has recently been published (Sharov, 1968).

away from the wing margin. It is known that the longitudinal wing veins form around the tracheae, and that the tracheae grow into the wing pads from the large alar trachea at the base of the wing. Consequently, a precostal field, even a small one, must be regarded as a primitive character in the Pterygota irrespective of whether we follow Forbes (1943), who thought that the costa was originally an independent longitudinal vein, or Sharov (1966a), who thought that it was derived from a branch of the subcosta. In fact, a precostal field is also present in the Palaeoptera, and in the stem-groups of the Ephemeroptera and Odonata it is possible to trace its gradual reduction, that is to say its complete approximation to the wing margin even at its extreme base. The Orthopteroidea are the only Neoptera in which the costa has become detached from the wing margin, giving rise to a precostal field. This must be regarded in the first place as a primitive character, but the displacement of the costa away from the wing margin, the *enlargement* of the precostal field, is certainly not primitive. It is tempting to correlate this development with the different resting positions of the fore-wings. In the Ensifera and Caelifera these have been modified into tegmina: they do not lie flat over the abdomen but are held in a more lateral position alongside the body which is laterally compressed rather than dorso-ventrally flattened. This must have allowed a rather differently shaped wing shape to develop, which in turn was furthered by the retention and enlargement of the precostal field. This interpretation is supported by the Palaeozoic Sthenaropodidae (Fig. 54),

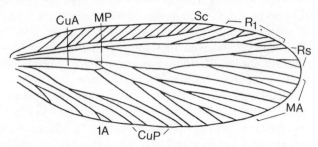

Fig. 54. Fore-wing of *Sthenaropoda bruesi* Meunier, from the Upper Carboniferous of Commentry. From Ragge (1955a). See also Fig. 55 for the interpretation of the venation.

which are generally and correctly regarded as the most primitive species in the stem-group of the Orthopteroidea. The abdomen in this family is not laterally compressed and the precostal field is very small, lacking both cross-veins and archedictyon (Sharov in Rodendorf, 1962).

On the other hand, Zeuner (1939) reported that the oldest known Gryllodea, which belong to a relatively subordinate group of the Orthopteroidea, are supposed to have held their wings flat over the abdomen. It is not easy to reconcile this with what is known about the historical development of the Palaeozoic Orthopteroidea, and Zeuner's conclusions need to be confirmed.

According to Ragge (1955b) the costa is entirely absent in the Phasmatodea, and the basal section of the fore-margin of the wing is not even enlarged. It is possible that this state developed from one in which the costa had moved away from the fore-margin, as in the Ensifera and Caelifera. Nevertheless, the Blattariae, which, like all the Paurometabola, never have a precostal field, sometimes have a reduced costa. However, Sharov (1967) has reported that fossils have been discovered recently in the Lower Triassic of Central Asia whose venation bridges the morphological gap between the Permian Tcholmanvissiidae and the recent Phasmatodea. This must mean that they still have a costa and a precostal field.

The oldest fossils that can be definitely assigned to the Orthopteroidea are the Oedischiidae and Sthenaropodidae. In the fore-wing they have one branch of the media fused for some distance with the anterior branch of CuA^+ (Cu_{1a}) (Figs. 54 and 55). This is undoubtedly a derived character and

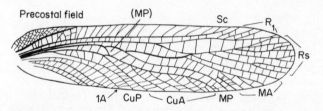

Fig. 55. Fore-wing of *Paroedischia recta* Carpenter (Oedischiidae), from the Lower Permian of Kansas. The interpretation of the venation is essentially that of Carpenter (1966). The section labelled MP may be identical with the 'arculus' of the Plecoptera (Fig. 36) or with the base of a branch of the media which is fused with CuA in all Pterygota or even Neoptera. This problem is very important for interpreting many of the fossils and for assessing the relationships amongst the orders of the Neoptera. See also p. 165.

must therefore be part of the ground-plan of the Orthopteroidea. It may even be part of the ground-plan of the Neoptera (p. 165).

In the Orthopteroidea the cerci are unsegmented, and not multi-segmented as in the Blattopteroidea. Two segments are present in the Tridactylinae, but this may not be a primitive state. The reduction of the cerci may therefore be another derived ground-plan character of the Orthopteroidea.

According to Sharov (1966b), there are certain derived characters in the structure of the ovipositor: unlike the Blattopteriformia, the Orthopteroidea have the third pair of valves formed largely from the styli, and the precoxal plates have been reduced to 'valvifers' (see p. 183–184).

The oldest fossils that can be definitely assigned to the Orthopteroidea are from the Upper Carboniferous. It is known that some of them had the base of the hind femur thickened: this is true of the Oedischiidae, for example *Oedischia williamsonii* Brongniart from the Stephanian of Commentry (Fig. 56). This was not the case in the Sthenaropodidae, a 'family' to which Sharov

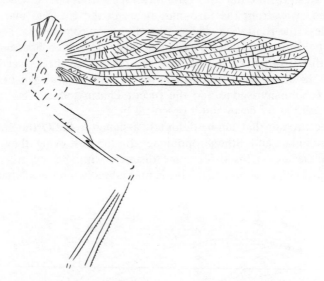

Fig. 56. *Oedischia williamsonii* Brongniart (Oedischiidae), from the Upper Carboniferous of Commentry. From Handlirsch (1906–1908).

(1961a) restricted the old name 'Protorthoptera' which had already been used in so many different senses. The Sthenaropodidae are also more primitive than the Oedischiidae in certain other characters, such as the small costal field. They are not of any real importance for determining the age of the Orthopteroidea because their oldest fossils are no older than the Oedischiidae. They can only be a surviving side-branch from an 'older layer' of the stem-group of the Orthopteroidea, and they only prove that the Orthopteroidea must be older than any of the oldest available fossils.

Riek (1953b) has described *Mesacridites elongata* from the Middle Triassic of Australia (New South Wales, Brookvale), as a Mesozoic species of the Sthenaropodidae. If this interpretation is correct, then this family must have survived into the early Mesozoic before it became extinct.

The Caloneurodea and Glosselytrodea are two problematic groups. Martynov (1938b) showed them as early side-branches of the 'Orthoptera-Saltatoria' in his phylogenetic tree, and this was followed in the Russian Textbook of Palaeontology (Rodendorf, 1962, Fig. 27). However, this figure contradicted Rodendorf's own text, in which the Glosselytrodea were assigned to the 'Orthopteroidea' and the Caloneurodea to the 'Plecopteroidea'.

Sharov (1966a, 1967) has recently attempted to show that these two groups belong to the 'Neuropteroidea', in the Holometabola. As they have both been found as early as the Carboniferous, this would be of great importance for our understanding of the phylogeny of the Holometabola (p. 281). However, neither group is of any importance for working out the phylogenetic development of the Orthopteroidea, even if earlier authors were correct in assigning them here.

The Caloneurodea are only known from the Palaeozoic. On the other hand, several fossils have been described from the Upper Triassic of Australia and Central Asia (Issyk-Kul') and assigned to the Glosselytrodea (*Polycytella*, and *Mesojurina sogjutensis* Martynova). Like the Sthenaropodidae, they are one of the few groups that are definitely known to have survived from the Palaeozoic (Upper Carboniferous) into the early Mesozoic, when they finally became extinct.

It has been accepted until very recently that there are two sister-groups within the Orthopteroidea: the 'Saltatoria' (Ensifera + Caelifera) and the Phasmatodea. However, Sharov (1965, 1967) thought that it was possible to show that there is a close relationship between the Caelifera and Phasmatodea. He suggested that there is a sister-group relationship between the Ensifera and the Caelifera + Phasmatodea (which would then have to be placed in a single group with a name like the 'Caeliferoidea'). The reasons he gave to support this suggestion cannot be rejected, but so far it seems that he has only mentioned the important fossils and has not yet described them.† I think that the best course for the present is to keep the Ensifera, Caelifera, and Phasmatodea as three distinct groups. This will leave all options open should an opportunity arise in the future to establish the true sister-group relationships.

Revisionary notes

189. *Tettigonia* has five-segmented tarsi (Henning, 1974). [Willi Hennig.]

190. The fusion of the labial stipes into a single plate cannot be regarded as a derived ground-plan character of the Orthopteroidea. [Eberhard Königsmann.]

2.2.2.2..2.2..2.1.   Ensifera[191]

Unlike the Caelifera and Phasmatodea, it is difficult to establish that the Ensifera are a monophyletic group. In almost all their characters they are more primitive than the other two groups. According to Sharov (1967), they have retained the predaceous mode of life that is characteristic of the ground-plan of the Orthopteroidea; only a few subgroups have become secondarily phytophagous.

---

† This study has just appeared (Sharov, 1968), but it has not been possible to give a detailed account of it here.

What is almost the only undoubted apomorphic character is in the hind-wing. According to Ragge (1955a), the anal fan includes at least the anterior and posterior cubital areas (Fig. 57). In the Caelifera (Fig. 61), on

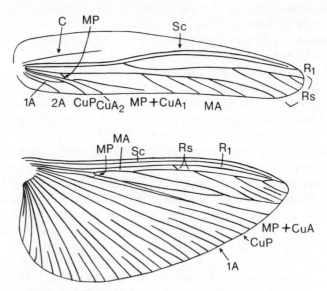

Fig. 57. Wings of the female of *Tettigonia viridissima* Linnaeus (Ensifera; recent). From Ragge (1955a).

the other hand, and apparently in the Phasmatodea too, the anal fan is confined to the anal lobe, as in the ground-plan of the Orthopteroidea. In his key, Ragge only mentioned this character for the Tettigonioidea, but this was simply the result of the arrangement of the key. Elsewhere in his paper Ragge stated that the Grylloidea also fold the hind-wing in the same way.

It is unfortunate that very few fossil hind-wings are known, and as a result it is impossible to establish whether this derived ensiferan character was present in any fossils.

In the fore-wing, the presence of a stridulatory apparatus could be a derived character of the Ensifera. According to Sharov (1967), there are several Permian fossils with a primitive apparatus for sound production, formed by the weakly sinuous anal veins. Unfortunately, he did not mention the names of any of these fossils, but if he is correct they must belong to the stem-group of the Ensifera. This would mean that the Lower Permian is the earliest date we have for the origin of the sister-group relationship between the Ensifera and Caelifera (? + Phasmatodea). Sharov has defined the family Oedischiidae to include the Permian fossils with a stridulatory apparatus, the stem-group of the Ensifera, together with Carboniferous species that apparently belong to the stem-group of the entire Orthopteroidea. His Oedischiidae are therefore an 'invalid stem-group'.

There is some further indirect evidence that the Ensifera arose in the

Lower Permian or earlier. This is provided by the contemporary Tcholman-vissiidae, which Sharov assigned to the stem-group of the Caelifera + Phasmatodea (see p. 220).

It is much more difficult to determine whether various subgroups of the *Ensifera also arose in the Permian. The answer depends on our ideas about the relationships between the subgroups of the *Ensifera that can definitely be recognized as monophyletic. However, all these ideas are highly controversial. If the recent fauna consisted only of the Tettigonioidea and

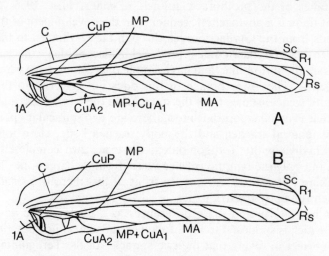

Fig. 58. Fore-wings of the male of *Tettigonia viridissima* Linnaeus (A, right wing; B, left wing). From Ragge (1955a). Sharov (1967) has suggested a different interpretation of the venation (see p. 216).

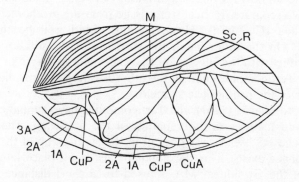

Fig. 59. Fore-wing of the male of *Eneoptera* sp. (Gryllidae; recent). From Ragge (1955a).

Grylloidea, we would have to accept that they are sister-groups. But in addition to these two undoubted monophyletic groups, there are the Schizodactyloidea, Gryllacridoidea, and Prophalangopsoidea, and there are strong differences of opinion about the relationships of these groups. In order to avoid too many complications in what follows, I am basing my discussion on the working hypothesis that there is a sister-group relationship between the Tettigonioidea and the rest of the Ensifera, which I am calling the 'Gryllodea'. This agrees with Sharov's (1967) views.

Zeuner, on the other hand, suggested that the Tettigonioidea are more closely related to the Grylloidea than are the Gryllacridoidea. He based this on his studies of the prothoracic spiracle (Zeuner, 1934, 1936, 1939). He found that there is a phylogenetic sequence in the development of this feature which leads from the Gryllacridoidea through the Grylloidea to the Tettigonioidea. In the Gryllacridoidea (and Schizodactyloidea) the prothoracic spiracle is small, button-like, and has a simple operculum, from which issues a single trachea. The lower branch of this does not give rise to the femoral trachea until some distance from the spiracle. In the Grylloidea, and basically in the recent Prophalangopsoidea too, there are two spiracular opercula from which the femoral trachea and the body trachea issue, close together but distinct. Finally, in the Tettigonioidea, there are two completely separate spiracles, but only the femoral trachea issues from the larger one. This shows some striking differential features. In front of this there is a small spiracle, from which the body trachea issues.

Zeuner has interpreted these differences as a phylogenetic development sequence which is supposed to have followed the course just described. He is certainly correct in saying that the two spiracles of the Tettigonioidea could only have developed from a state resembling that retained by the recent Prophalangopsoidea and Grylloidea. However, it is doubtful if the Gryllacridoidea and Schizodactyloidea really represent an even more primitive state, as the single spiracle of these groups could well be the result of a reduction. There is some support for this suggestion in Chopard's statement (in Grassé, 1949) that two separate tracheal stems issue from the prothoracic spiracle in the Acridoidea (Caelifera). However, this could well be the result of convergence. On the other hand, the structure of the prothoracic spiracle does not conflict with the suggestion that there is a sister-group relationship between the Tettigonioidea and all the other Ensifera. Sharov did not discuss the structure of the prothoracic spiracle.

According to Sharov (1967), the Tettigonioidea have a number of characteristic derived features in the venation of the fore-wing. The connection between M and RS, which was present in the stem-group of the Orthopteroidea, no longer exists. This development also took place later in certain subgroups of the Gryllodea. MA is fused for a short distance with MP + CuA_1 and the branches of MA have been completely reduced. The point at which Cu divides into CuA and CuP has moved towards the wing-tip, level with the point where MP fuses with CuA. As a result, MP + CuA_1 looks like a continuation of the media, and because of its position the base of CuA_1 looks like MP. Zeuner (1939) and Ragge (1955a) have in fact interpreted it as

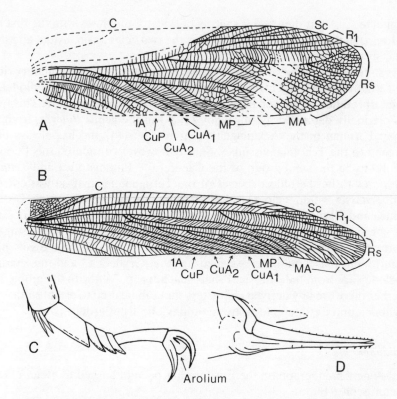

Fig. 60. A: Fore-wing of *Tettavus fenestratus* Martynov (family Tettavidae), from the Upper Permian of Tikhiye Gory. B: Fore-wing of *Pinegia longipes* Martynov (family Tcholmanvissiidae), from the Upper Permian of Chekarda. C: Hind tarsus of *Jubilaeus beybienkoi* Sharov (family Tcholmanvissiidae), from the Lower Permian of Chekarda. D: Ovipositor of the same species. All from Sharov (1968).

MP, but Sharov's study of the venation of the Upper Permian *Tettavus fenestratus* Martynov has shown that it is really the basal section of CuA_1.

Sharov's new interpretation of the stridulatory rib was also connected with his new interpretation of the basal part of the cubital area. He stated (1967) that in the Lower Permian 'Oedischiidae' several 'anal veins' were involved in sound production (see above), whereas there is only a single stridulatory rib in all recent Ensifera. Zeuner (1939) interpreted this stridulatory rib as the 1st anal vein (= postcubitus, PCu), and Ragge (1955a) as Cu_2 (= CuP). It is significant that neither author has disputed the homology of the stridulatory rib in the Tettigonioidea and Gryllodea (Figs. 58 and 59).

Sharov (1967) stated that in the Tettigonioidea the 1st anal vein (PCu) is the stridulatory rib whereas in the Gryllodea it is CuP. However, he did not give any real justification for this view. Anyone not specializing in the Orthopteroidea would readily conclude that the stridulatory rib is homologous in these two groups because its course is identical in both of them. First it runs parallel to CuA but then bends sharply backwards and fuses with the first two anal veins in the 'node'; shortly after this the two veins again divide, and their distal sections remain independent.

As stated above, Sharov probably based his view that the stridulatory rib of the Tettigonioidea is not identical (homologous) with that of the Gryllodea on his interpretation of the venation of *Tettavus fenestratus* Martynov. This fossil was originally described by Martynov (1928) in the genus *Pinegia*, from the Upper Permian of the Arkhangelsk region (Fig. 60), and has always been assigned to the Tcholmanvissiidae. Sharov however considered this Permian 'family' to be the stem-group of the Caelifera + Phasmatodea and *Tettavus fenestratus* to be the oldest species of the Tettigonioidea. If he was correct, then *Tettavus* would help to establish that the two sister-groups Tettigonioidea and Gryllodea had separated by the Upper Permian. This is possible and not at all improbable, but I do not think that it can be proved using the information currently available about *Tettavus*. No Palaeozoic fossils have been found that could be assigned to the Gryllodea, and the earliest Gryllodea are from the Triassic. I shall deal later (p. 360) with the problem of whether there are any derived characters that can be used to establish that the Gryllodea and Tettigonioidea are monophyletic sister-groups.

Additional note

191. A new monograph on the Ensifera has been published by Beier (1972). [Dieter Schlee.]

2.2.2.2..2.2..2.2.   **Caelifera**[192]

There are a number of derived characters that show that the recent Caelifera are a well founded monophyletic group.

The antennae are short. There are no more than 30 segments.

The number of tarsal segments has been reduced to three. The presence of two pairs of 'pulvilli' on the lower surface of the basal segment indicates that this consists of three fused segments. The presence of four tarsal segments in the ground-plan of the *Ensifera could indicate that reductions in the number of tarsal segments first took place in the common ancestors of the Ensifera and Caelifera, and that the reduction has been taken further in the Caelifera, as it has been in the crickets (Ensifera). However, late Palaeozoic fossils appear to show that the tarsi of the stem-groups of the Ensifera and Caelifera had five segments (see below), and that consequently the reduction in both groups has taken place independently.

In the fore-wing (Fig. 61) the media has only two branches: Ragge (1955a) interpreted these as MA and MP, whereas Sharov (1967) regarded them both as branches of MA. Sharov's suggestion that MP, or a posterior branch of the media, is fused with CuA_1 in the Caelifera is certainly correct. He also

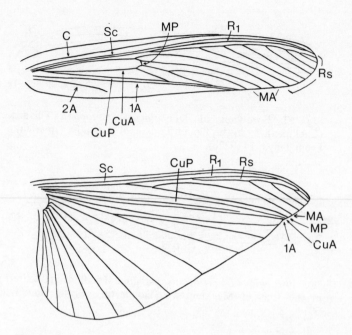

Fig. 61. Fore- and hind-wings of the Acridoidea. From Ragge (1955a).

suggested that the free basal section of 'MP' noted by Ragge as absent has actually developed into a 'cross-vein'.

It is possible that a number of other undoubtedly derived characters are part of the ground-plan of a group of higher rank, the 'Caeliferoidea', to which the Phasmatodea and the Caelifera belong. Such characters are the reduction of the proventriculus, the absence of styli on the 9th abdominal sternite of the male, and the short female ovipositor which is armed with spines and is used for digging. The absence of these spines in the Phasmatodea would be a secondary reduction and the result of different reproductive habits (see p. 222).

Stridulatory organs and the auditory organs on the 1st abdominal segment do not seem to be part of the ground-plan of the Caelifera or of a group of higher rank, the Caeliferoidea[193]. Possibly the most interesting derived character is the strongly developed arolium between the claws. This is also present in the Phasmatodea and according to Sharov shows that these two groups are closely related.

The Locustopsidae (Figs. 62 and 63) are a group of fossils that definitely belong to the Caelifera. In certain characters they are even more primitive than recent Caelifera. For example, the media (Sharov's MA) had three free branches, and in some species the antennae retained 40–50 segments (Sharov, 1967). On the other hand, 'most of them' had three segmented tarsi like the recent Caelifera (Sharov, 1967).

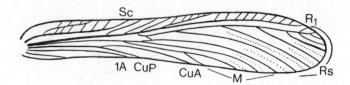

Fig. 62. Fore-wing of *Triassolocusta leptoptera* Tillyard (Caelifera), from the Upper Triassic of Australia (Ipswich). From Tillyard (1922b).

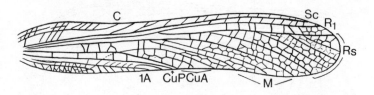

Fig. 63. Fore-wing of *Locustopsis magnifica* Handlirsch (Caelifera), from the Lias of Mecklenburg (Dobbertin). From Handlirsch (1939).

The Locustopsidae may be an 'invalid stem-group': species that belong to the stem-group of all the recent Caelifera have been grouped with species that belong to the stem-groups of individual recent families. Sharov's account (1967) seems to indicate that this is what has happened. However, this problem only affects the history of the Mesozoic Caelifera (see p. 368).

The oldest members of the Locustopsidae, and of the Caelifera, have been described from the Upper Triassic of Australia (*Triassolocusta*). The Locusta-vidae were more or less contemporary: Sharov (1967) proposed describing this family on some future occasion.† However, the Locustavidae are of no fundamental importance for determining the age of the Caelifera. The wings are known, but have not yet been described. According to Sharov they are important because they have some of the derived characters of the Locustop-sidae and their descendants as well as some more primitive characters. In these respects they bridge the morphological gap between the Locustopsidae and the Palaeozoic Tcholmanvissiidae. The precostal field is narrow and long, and is filled with a comb-like series of costal branches. The subcosta almost reaches the wing-tip. The media divides into two main branches which are long and fork at their tips as in the Palaeozoic species. According to Sharov (1967), the three-branched media of the Locustopsidae developed from this forerunner: the forks of the 1st main branch (MA_1) were suppressed, and the point of bifurcation of the 2nd main branch moved proximally.

The Tcholmanvissiidae are known from the Upper and Lower Permian, and according to Sharov they belong to the stem-group of the Caelifera +

---

† Subsequently described by Sharov (1968).

Phasmatodea. They had five-segmented tarsi, as Sharov (1967) has been able to prove with a photograph of one of the Lower Permian species.

The vital feature that indicates that the Tcholmanvissiidae really do belong to the stem-group of the Caelifera + Phasmatodea is the strongly developed arolium. Sharov (1967) suggested that the arolium is a derived character, connected with the ability of individual groups to climb up plants. It is not present in the Ensifera. Species in this group that can also climb plants have different adaptations for accomplishing this.

The Caelifera and Phasmatodea should therefore form a monophyletic group, the 'Caeliferoidea'. Early on in their history, and before they separated into their recent subgroups, they must have adapted to living on plants and to phytophagy, whereas the Ensifera at first remained predaceous. Sharov reported that the ovipositor of the Tcholmanvissiidae was heavily spinose and adapted for digging. There are five distinct tarsal segments: this is a primitive character and in itself is of no significance. However, it has been used to show that the Phasmatodea as well as the Caelifera may be derived from the Tcholmanvissiidae since the Phasmatodea are the only group of the Orthopteroidea to have retained five-segmented tarsi.

There is some further evidence that at least one other group existed alongside the Ensifera during the Lower Permian from which either the Caelifera or the Caelifera + Phasmatodea are derived. This evidence is provided by the existence of species that are said by Sharov to have possessed a stridulatory apparatus on the fore-wing, even though it was a very primitive one. The *Ensifera can be derived from such species, but the Caelifera and Phasmatodea cannot.

Even if it can be confirmed that the Caelifera and Phasmatodea belong to a single monophyletic group of higher rank, there is no evidence that their ancestors had developed into separate groups by the late Palaeozoic.

Additional notes

192. A new monograph on the Caelifera has been published by Beier (1972). [Dieter Schlee.]

193. An analysis of tympanal organs in the 'Acridomorpha' has been published by Mason (1969). [Willi Hennig.]

2.2.2.2..2.2..2.3. Phasmatodea (stick-insects, leaf-insects)[194]

Most Phasmatodea are arboreal, but a few have developed secondarily into savannah species. Their legs are not modified for jumping. These characters and the five-segmented tarsi could indicate that the Phasmatodea are more primitive than all the other Orthopteroidea (the 'Saltatoria'). However, this would mean that the common ancestors of the Phasmatodea and 'Saltatoria' must have been in existence before the Upper Carboniferous.

Sharov suggested that the Lower and Upper Permian Tcholmanvissiidae were the common stem-group of the Caelifera + Phasmatodea. If he was correct then the loss of the ability to jump and the reduction of the hind legs to simple walking legs would have to be regarded as a derived character of the Phasmatodea. There is a parallel to this in the ensiferan family Phasmodidae which, as their name indicates, are very similar to the stick-insects in appearance.

There are several other derived characters which may be synapomorphies of the Caelifera + Phasmatodea (see p. 219).

Moreover, the Phasmatodea have a large number of derived characters (autapomorphies), most of which should be regarded as adaptations to an arboreal life. Unfortunately, these have not yet been fully worked out. Any study of these characters should take into account the possible or probable relationship of the Phasmatodea and the Caelifera, because the interpretation of many characters depends on whether such a relationship really does exist.

The fore-wings of the Phasmatodea are short or absent. The venation is reduced and the costa is absent. Sharov (1967) has associated the parallel arrangement of the veins with the development of the protective adaptations that are so characteristic of the group. When vestiges of the fore-wings are present, they can only protect the wing base and basal section of the hind-wing. Their usual role has been taken over by the remigium of the hind-wing which is strongly sclerotized and coloured. The hind-wings are often reduced too. According to Günther (*in litt.*) the wing pads of the nymphs are not in the position that Lemche (1940) found to be so characteristic of the Ensifera and Caelifera. This could be a secondary character, a result of the reduction of the fore-wings. Nonetheless, any discussion of the relationships of the Phasmatodea should take account of this difference.

The head of the Phasmatodea is weakly prognathous, and this is probably a derived character. No work has been done on the differences in the structure of the prothorax and the prothoracic spiracle. The first abdominal segment is fused to the metathorax. The absence of styli on the 9th abdominal segment and the short female ovipositor may have originated amongst the common ancestors of the Caelifera and Phasmatodea(?). When the eggs were no longer laid in the soil, the ovipositor was not used for digging and the spines, a morphological expression of this function, were lost. Females release the eggs singly and they fall to the ground. According to Sharov (1967), the egg has a hard, thick, sculptured exochorion and closely resembles a seed.

No Mesozoic or Palaeozoic fossils with the distinct derived characters of the Phasmatodea have been found. Martynova (in Rodendorf, 1962) has grouped together the genera *Aeroplana*, *Aerophasma*, and *Chresmoda* in the Chres-mododea. Further genera in this group are *Raphidium* and *Chresmodella*, which contain five species (Bode, 1953). These species have no characters to suggest that they really do belong to the stem-group of the Phasmatodea. Ragge (1955b) was correct in stating that any relationship between this group and the Phasmatodea should not be looked upon as anything more than a possibility. Handlirsch (1906–08) reported that these species were very common in deposits that contained numerous marine fossils. He concluded from the leg structure that they lived on the water surface. This is hardly consistent with their assignment to such a strikingly arboreal and phytophagous group as the Phasmatodea. Handlirsch also stated that some of the recent Phasmatodea live like pond-skaters, but according to Uvarov (in Zeuner, 1939) this is not true. It is always possible that Handlirsch was wrong in his interpretation of the mode of life of the fossil genus *Chresmoda*. As far as I know, no recent author has challenged his views. For example, Carpenter (1932a) accepted without further comment that Handlirsch had conclusively proved that *Chresmoda* belonged to the Phasmatodea[195].

Sharov (1967) reported that a large number of very diverse Phasmatodea had recently been discovered in the Lower Triassic beds of Central Asia. He found that they bridged the morphological gap between recent Phasmatodea and the Permian Tcholmanvissiidae. It has not been possible to discuss Sharov's (1968) descriptions of these fossils.

Additional notes

194. A new monograph on the Phasmatodea has been published by Beier (1968b). [Dieter Schlee.]

For further discussion of the systematic position of the Phasmatodea, see Kristensen (1975). Kristensen also suggested that *Tinema* might be the primitive sister-group of the rest of the Phasmatodea. [Adrian Pont.]

195. Esaki (1949) said nothing about the habits or systematic position of *Chresmoda*. [Willi Hennig.]

General remarks on the Paraneoptera and Holometabola (Eumetabola?)

It is very difficult to decide whether the Paraneoptera and Holometabola form a monophyletic group (Fig. 64). Each is undoubtedly monophyletic, but it is not certain whether they have both descended from ancestors that were common only to them. Opinions about this are divided.

Forbes (1943) was undecided about the position of the Psocoptera, but he grouped the majority of the Paraneoptera ('Hemiptera') together with the 'Polyneoptera' under the name 'Orthopteroidea'. He considered it possible that the Holometabola had descended from 'the' Orthopteroidea and so it is not clear whether he regarded his 'Orthopteroidea' as monophyletic in my sense or not. Several of the published phylogenetic trees are also unclear, and Martynov's (1938b) tree is one of these. Martynov has been followed by Jeannel (in Grassé, 1949), Weber (1949), Wille (1960), Rodendorf (1962), etc. These trees show the Paraneoptera, Holometabola, and Polyneoptera issuing independently and at different times from a single 'stem-group', and sometimes they even show individual subgroups of the Paraneoptera and

Holometabola arising independently. Such trees are of no use to me: they do not attempt to show monophyletic groups and their genealogical connections, but make use of units which their authors consider to be uniform structural types.

Both Aubert (1950) and Ross (1955, 1965) suggested that the Paraneoptera and Holometabola had ancestors that were common only to them. The only character that Ross could find to support this view was the absence of ocelli in the immature stages, and Stannard (1956) agreed with this. This character shows how the morphological division between the larval and adult stages could be initiated, a division which then proceeded in different directions in the two groups.

This conflicts with Sharov's view (1957a, 1957c) that the hemimetabolous orders (Paurometabola, Plecoptera, and Paraneoptera) are characterized by increasing imaginalization of the immature stages whilst in the Holometabola larval and imaginal stages have followed more and more separate lines of development.

On the other hand, Sharov placed equally strong emphasis on agreements in the mode of embryonic development between the 'Polyneoptera' (Paurometabola + Plecoptera) and Holometabola on the one hand and the Paraneoptera + Palaeoptera on the other. I have already discussed the difficulties inherent in this assertion (pp. 161–162).

It is necessary to examine whether the imaginalization of the immature stages, which Sharov regarded as such a characteristic feature of the hemimetabolous insects, really is so distinct in the ground-plan of the Paurometabola and Paraneoptera that it can be regarded as a synapomorphy of these two groups. Sharov placed the Plecoptera with the Paurometabola in his Polyneoptera, but in this group the larval and imaginal stages have begun to follow separate modes of life, as in the Palaeoptera and Holometabola. There is no question of any 'imaginalization of the immature stages' having taken place here.

Moreover, it seems most unlikely that the resemblance between immature stages and imagines in the relatively primitive Paurometabola (Embioptera, Dermaptera) and Paraneoptera (Zoraptera) is due to imaginal characters occurring very early during post-embryonic development. It is more likely that the adults of these groups are still very close in general physical form to thysanuroid immature stages. This character is therefore primitive rather than derived in the ground-plan of the Paurometabola and Paraneoptera.

Hungerford (in Wille, 1960) has reported that the nymphs of certain Gelastocoridae (Hemiptera) have ocelli. However, this family occupies a relatively subordinate position in the phylogenetic tree of the Paraneoptera, and the fact that ocelli are present is in no way incompatible with the suggestion that the immature stages really do lack ocelli in the ground-plan of the Paraneoptera and Holometabola. The premature occurrence of this 'adult character' in gelastocorid nymphs could be a secondary acquisition.

In addition to the absence of ocelli in the immature stages, a further synapomorphic character of the Paraneoptera and Holometabola may be the reduced number of Malpighian tubes.

The situation as regards the wing venation is not at all clear. This is particularly true of the fate of the individual branches of the media. There is a great deal of confusion about this in the literature, and for the present this is best avoided by refraining from any attempt to distinguish between the two main branches, MA and MP. A much better course would simply be to establish how far the fusion of an anterior branch (MA or an anterior branch of MA) with the radial sector and the fusion of a posterior branch (MP or a posterior branch of MP) with the cubitus are part of the ground-plan of individual groups. The suggestion that all the Paraneoptera and Holometabola have a posterior branch of the media fused with CuA appears to be well founded. The result of this fusion is a 'basal cell' close to the wing base that is independent of any cross-veins, and this may be important for the functioning of the basal wing articulation. Sharov (1966b) stated that this feature is characteristic of all the Pterygota. However, I do not think that this has been proved, and it also contradicts Sharov's own earlier interpretation of the venation of the 'Paraplecoptera' and other groups (in Rodendorf, 1962). Nor do I think that his illustrations of the Paraplecoptera show that a posterior branch of the media has fused with CuA. These veins are also fused in the Orthopteroidea and Plecoptera, but in these groups it may be the result of convergence.

On the other hand, the Paraneoptera and Holometabola appear to have an anterior branch of the media fused with the radial sector. This may also be true of the Plecoptera, but it is almost certainly not part of the ground-plan of the Paurometabola.

In all the Paraneoptera the venation has been affected by extensive secondary changes, and this makes it extremely difficult to decide whether the Paraneoptera and Holometabola have any common derived ground-plan characters (synapomorphies) in their venation.

If the Paraneoptera and Holometabola really do form a monophyletic group (the Eumetabola), then their sister-group relationship must have arisen before the Lower Permian. This is because a considerable number of subgroups of the *Paraneoptera and *Holometabola are known from this epoch.

The Carboniferous fossils are very problematic. Different authors have assigned various Upper Carboniferous fossils to the Paraneoptera or to the Paurometabola, but there has been considerable dispute over these interpretations. It has been suggested that both groups lived during the Carboniferous in regions where no fossils have yet been found, and such a hypothesis is impossible to refute.

2.2.2.2..3. **Paraneoptera** (Fig. 64)

The only undoubted derived character in the ground-plan of the Paraneoptera is the concentration in the abdominal nerve cord. Only the most primitive species have retained two (Zoraptera) or one ganglionic mass. Even in these species, however, the abdominal ganglia are very close to the third thoracic ganglion. The reduced number of tarsal segments may also be a derived character. No recent subgroup of the Paraneoptera has more than three tarsal segments, but the Permian Permopsocida are said to have four or even five. It is usually accepted that the Permopsocida belong to a subgroup (Psocodea) rather than to the stem-group of all the Paraneoptera, but if this is so, then the number of tarsal segments in the Paraneoptera must have been independently reduced on several occasions. However, the relationships of

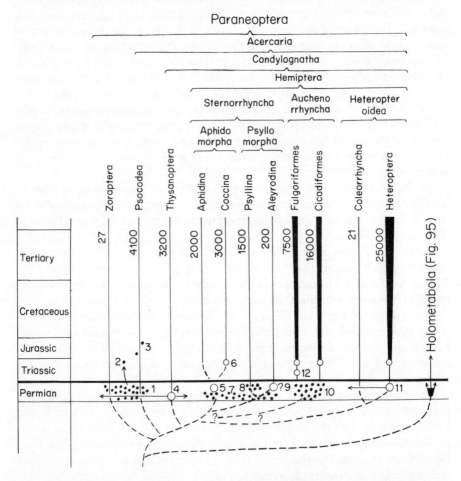

Fig. 64. Phylogenetic tree of the Paraneoptera[196]. 1, Permopsocida; 2, *Procicadellopsis*; 3, Archipsyllidae; 4, *Permothrips longipennis* Martynov; 5, *Permaphidopsis*; 6, *Mesococcus asiaticus* Bekker-Migdisova; 7, Archescytinidae; 8, Cicadopsyllidea; 9, *Permaleurodes rotundatus* Bekker-Migdisova; 10, Permian Auchenorrhyncha; 11, *Paraknightia*: 12, *Boreocixius*.

the Permopsocida are still not very clear (see below), and the study of this and other problems is made more difficult by the Zoraptera. This group probably has a sister-group relationship with the rest of the Paraneoptera, but some characters are so reduced that it is extremely difficult to make a comparison with other groups.

Additional note

196. *Aleuronympha bibulla* Riek (Permaleurodidae), from the Upper Permian of Natal (Riek, 1974c), should be added to this phylogenetic tree. [Willi Hennig.]

2.2.2.2..3.1. Zoraptera[197]

There have been several different views about the relationships of the Zoraptera. In particular, it has been suggested that they are related to the Embioptera or to the termites. I am convinced that some of the characters common to these groups are primitive ones which they alone have retained (symplesiomorphies) but which do not indicate any close relationship. Others are the results of convergence, because of the numerous reductions, and they too are not indicators of relationship.

In some characters the Zoraptera are more primitive than the rest of the Paraneoptera. They are the only group to have the maxillae developed normally. Wille (1960) described these as 'only slightly elongated'. Furthermore, they are the only group in which two discrete abdominal ganglionic masses are present, the cerci have been retained though they are one-segmented and strongly reduced, the 1st abdominal sternite is fully developed, and only six Malpighian tubes are present (Königsmann, 1960).

In other respects the Zoraptera are strongly specialized, and this is particularly true of the venation (Fig. 65). A characteristic feature of the Paraneoptera, present when the venation is not too reduced, is the so-called 'areola postica' of the fore-wing, formed by an apical fork of CuA (CuA$_1$ and CuA$_2$). This is still distinct in the fore-wing of the Zoraptera and, together with the reduction in the abdominal ganglionic cord, is particularly convincing evidence for the assignment of the Zoraptera to the Paraneoptera. The venation has been differently reduced in the Embioptera, and the Isoptera generally have a completely different venation.

Weidner (1969) has very recently published a paper, which I am unable to discuss here in any detail. Basing his arguments mainly on Delamare Deboutteville's studies of the skeletal morphology of the thorax, he resurrected the idea that the Zoraptera are the sister-group of the Isoptera (termites) and are not related to the Paraneoptera at all. I do not think that there is any basis for this, and he has made no attempt to reconcile it with the structure of the female reproductive organs.

No fossils have been found that could be assigned to the stem-group of the

228

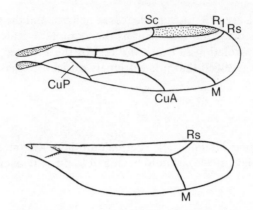

Fig. 65. Fore- and hind-wings of *Zorotypus snyderi* Caudell (Zoraptera). Not to the same scale. From Denis (in Grassé, 1949).

Zoraptera. However, it is always possible that some of the Permian Permopsocida belong to this stem-group (see p. 230).

Additional note

197. A new monograph on the Zoraptera has been published by Weidner (1970b). [Dieter Schlee.] See also Kristensen (1975). [Adrian Pont.]

2.2.2.2..3.2. Acercaria

Börner (1904) used the name Acercaria for the rest of the Paraneoptera. The cerci and the 1st abdominal sternite are absent. The abdominal ganglionic cord is reduced to a single ganglionic mass. Only four Malpighian tubes are present (Königsmann, 1960). Most striking of all, the lacinia of the maxillae has developed into a stylet and its base has been withdrawn deep into the head capsule. This is probably the first step towards the development of biting and sucking mouth-parts which take the form of a true proboscis in most Paraneoptera.

In the primitive groups ('Psocoptera') the lacinia is bristle-like or chisel-like in shape. The mode of life of these groups gives a clue to the conditions that probably gave the original impetus to the development of the proboscis of the Hemiptera. All the species in these primitive groups are small and cryptic, like the Zoraptera. However, the Zoraptera have primitive mouth-parts and are predaceous mainly on mites (Eidmann), whereas the Psocoptera feed on algae, moulds, fungal spores, lichens, and organic debris. The function of the bristle- or chisel-like laciniae is to support the mandibles whilst they scrape and fragment particles of food (Weber, 1933). The powerful dilator muscles of the pharynx and the large, strongly arched postclypeus to which they are attached show that these species suck up their food once it has been pulverized[198]. Amongst the Paraneoptera, the development of a piercing proboscis thus began amongst species where the parts that eventually became the stylets were first modified into chisel-like instruments. These supported the mandibles, which carried out the cutting and masticating.

There are no fossils which can be assigned with any confidence to the stem-group of the Acercaria (see the Permopsocida below). However, what is known about the age of individual subgroups of the Hemiptera makes it very probable that the Acercaria arose no later than the Upper Carboniferous.

There are two subgroups of the Acercaria in the recent fauna, the Psocodea and Condylognatha. They too must have arisen in the Upper Carboniferous, but since then they have followed very different lines of development.

Revisionary note

198. This sentence ('The powerful muscles . . . pulverized') should be omitted, because the function of these muscles is uncertain and the food is not sucked up. [Wolfgang Seeger.]

2.2.2.2..3.2..1. Psocodea[199]

In their form and mode of life, the Psocodea at first retained the basic features of the common ancestors of the Acercaria. I know of no derived character which can be used to show that the group is definitely monophyletic[200]. This is because the Psocodea also include the Phthiraptera, which are wingless and highly adapted to a parasitic life. The Phthiraptera have been treated as a separate order, or even as two orders (Mallophaga and Anoplura), because of the large number of their adaptations. There has in fact been little enthusiasm for working out the ground-plan characters which they share with the so-called 'Psocoptera' and which distinguish the Psocodea from other groups. However, all the characters of the Phthiraptera can easily be recognized as derived expressions of those that are present in more primitive form in the so-called Psocoptera. It is therefore likely that the ground-plan characters of the 'Psocoptera' are really the ground-plan characters of a group of higher rank that includes the 'Psocoptera' and the Phthiraptera[201].

This is particularly important for characters in the venation (Figs. 66 and 67), as these play a vital role in the interpretation of the fossils[202]. Even in the most primitive recent Psocoptera, the basal sections of M and Cu_1 are fused and the media has only three free branches. In this respect they are more derived than the most primitive Hemiptera, but on the other hand they are more primitive than the Hemiptera because they have the radial sector forked. This is the typical distribution of primitive and derived features which is characteristic of sister-groups. A further, probably derived, character of the Psocodea is the coupling of the fore- and hind-wings when at rest and when in

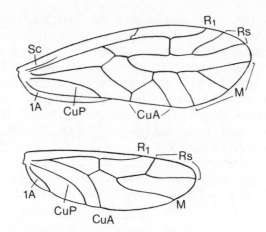

Fig. 66. Fore- and hind-wings of the female of *Stenopsocus stigmaticus* Imhoff & Labram (Psocoptera; recent). From Badonnel (in Grassé, 1951).

flight. This is accomplished by the 'stigmapophysis', a blunt chitinous projection at the base of the pterostigma of the fore-wing, which grips the fore-margin of the hind-wing when at rest. In flight the fore-margin of the hind-wing is linked to the fore-wing by a hook at the tip of CuP[203].

Unfortunately, these characters are not visible in any fossils[204]. The oldest Psocodea are the Permopsocida (Fig. 68). The two pairs of wings are homonomous, according to Bekker-Migdisova and Vishnyakova (in Rodendorf, 1962). In the fore-wing, Sc is long and the pterostigma is not sclerotized. The media has four branches. There are two free anal veins. The clypeus is flat. The tarsi have five segments, and cerci are present. Species with these characters are not only more primitive than all the recent Psocoptera and Phthiraptera, but they are also more primitive than the Zoraptera. It is therefore possible that the Permopsocida are an invalid stem-group. Species have been assigned here because of their general similarity, but this is the result of symplesiomorphies. Some of these species may actually belong to the

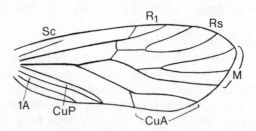

Fig. 67. Fore-wing of *Amphientomum* sp. (Psocoptera; recent). From Badonnel (in Grassé, 1951).

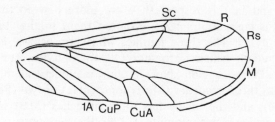

Fig. 68. Fore-wing of *Permopsocus* sp., from the Lower Permian of Kansas. From Carpenter (1954b).

stem-group of the entire Paraneoptera, whilst others may belong to the stem-groups of various subgroups of the *Paraneoptera. In a purely formal sense, the recent Condylognatha (Thysanoptera + Hemiptera) could also be derived from species in which all the primitive characters listed by Bekker-Migdisova and Vishnyakova are present. However, it is improbable that this group has descended from the Permian Permopsocida because other Permian fossils are known which can be assigned with much greater certainty to the stem-group of the Thysanoptera + Hemiptera.

In view of this, the 'Permopsocida' are almost of no use for determining the age of any recent insect groups. In fact, they highlight the difficulties of palaeoentomology with striking clarity. It is not particularly important that different authors have given different definitions of the group. Carpenter (1954b) gave it the broadest scope and he also included the Permembiidae, Palaeomantidae, and Delopteridae (Fig. 69). These three 'families' have M and CuA partly fused, as in recent Psocoptera. They also have a two-

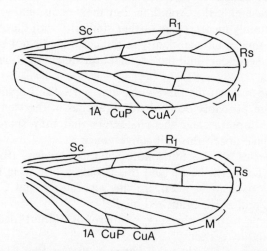

Fig. 69. Fore- and hind-wings of *Delopterum minutum* Sellards, from the Lower Permian of Kansas. From Carpenter (1933).

branched M, and in this respect they are more derived than the recent Psocoptera. Russian authors have assigned them to the 'Plecopteroidea' (order Miomoptera)! The three families Martynopsocidae, Permopsocidae and Dichentomidae, which Carpenter (1954b) and Russian authors have assigned to the 'Permopsocida', have M and CuA independent (Fig. 68). The radial sector has three branches (Martynopsocidae) or two (Permopsocidae, Dichentomidae), and the media has three branches (Martynopsocidae) or four (Permopsocidae, Dichentomidae). Like the Permopsocida, these three families probably form an invalid stem-group too: in a purely formal sense, the Phthiraptera and Zoraptera as well as all the recent Psocoptera could be derived from them. Carpenter (1933) also reported that the Dichentomidae had homonomous wings, four-segmented tarsi, and cerci.

Carpenter (1954b) also assigned the Zygopsocidae, Lophioneuridae, and Zoropsocidae to the Permopsocida, but Bekker-Migdisova and Vishnyakova (in Rodendorf, 1962) included them with certain recent families in a suborder 'Parapsocida' within the Psocoptera. There is no doubt that this produces a completely distorted picture of the actual relationships. Even if no account is taken of the Phthiraptera for the time being, Tillyard's (1926f, 1926h) division of the 'Psocoptera' into Parapsocida and Eupsocida probably does not show the true relationships. The 'Parapsocida' appear to be a paraphyletic group. There is no evidence that the Permian families Zygopsocidae, Lophioneuridae, and Zoropsocidae, and even the Permian Surijokopsocidae, which Russian authors have also assigned to the Parapsocida, are more closely related to certain recent families of the Psocoptera (or to the Phthiraptera) than to others. In these Permian families the media has only two branches. In the Lophioneuridae, the bases of M and CuA are fused with the base of R, so that M and CuA appear to issue from the radius (Tillyard, 1921a). In this character they resemble the Aphidina, but their occasional association with this group is certainly incorrect. A further similarity with the Aphidina is the fact that Sc terminates independently in the costa and is not attached to R_1. This is probably secondary and the original attachment to R_1 must have been lost.

In *Zoropsocus* at least, the hind-wings are short. The Lophioneuridae and Zoropsocidae are probably derived side-branches of the stem-group of the Psocoptera–Phthiraptera, and the characters that they share with certain recent species have arisen through convergence.

Only one species of the Surijokopsocidae is known, from the Upper Permian of the Kuznetsk basin, and this is only a fore-wing fragment. The only characters that can be seen are the separate M and CuA, as in the Permopsocidae, and a five-branched media!

It is not possible to trace the recent subgroups of the Psocodea back into the Permian with any certainty.

Another problem is whether the Permian species that survived into the Mesozoic included any species that were not the direct ancestors of the recent Psocoptera–Phthiraptera. The Jurassic Archipsyllidae are very important in this respect. As in the various 'Permopsocida', M and CuA are not fused and the media has four branches. In both characters the Archipsyllidae appear to be more primitive than any other Mesozoic or recent Psocoptera. On the other hand, they appear to have the anal veins (and CuP?) reduced. This would not rule out the possibility that they are closely related to the recent Psocoptera–Phthiraptera. Carpenter (1954b) assigned the Archipsyllidae to the 'Permopsocida', but it is still possible that they are more closely related to the recent Psocodea than are any of the Palaeozoic species.

Revisionary notes

199. A monograph on the Psocoptera has been published by Weidner (1972), and a bibliography and catalogue by Smithers (1965, 1967). Recent keys to Psocoptera are by Günther (1974) and New (1974). [Wolfgang Seeger.]

Timmermann (1957) has discussed Mallophaga and the classification of the Charadriiformes. Ludwig (1968) has given a synopsis of the number, occurrence, and distribution of the Anoplura. Symmons (1952) has discussed the comparative anatomy of the mallophagan head. See von Kéler (1957) and also the monograph by von Kéler (1969). [Willi Hennig.]

200. This sentence ('I know . . . monophyletic') now requires modification because recent work has demonstrated that the Psocodea do indeed have some derived characters that show that they are a monophyletic group. [Wolfgang Seeger.]

201. Seeger (1975) found that certain specialized structures in the antennal flagella of the Psocoptera and the Phthiraptera are a derived character common to all Psocodea. He described and analysed a structural mechanism for rupturing the antennae, something which is not known to occur in any other insects (Fig. VIII). Typically this mechanism consists of a collar-like fold of the epicuticle and exocuticle (but not of the endocuticle) at the base of the flagellar segments. In the suborders Trogiomorpha and Troctomorpha of the Psocoptera, it functions effectively as a protective device. It exists in a fully functional or partly reduced form only in the Psocoptera and Phthiraptera, but not even in other groups of the Paraneoptera, and it provides convincing evidence that the Psocodea are very probably a monophyletic group.

An additional character of the Psocodea, the polytrophic ovarioles, has been discussed by Kristensen (1975). [Wolfgang Seeger.]

202. This sentence ('This is particularly . . . fossils') should be rephrased as follows: 'It is very important to decide whether the characters in the venation that have been considered to be characteristic of the Psocoptera could also be constitutive characters of the Psocodea. They are very important for inter-

234

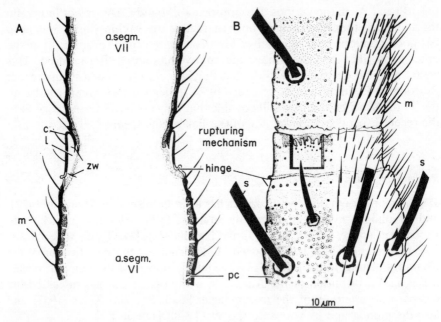

Fig. VIII. Mechanism for rupturing the antennae in *Cerobasis guestfalica* Kolbe (Psocoptera, Trogiomorpha): longitudinal section (A); surface view (B). a.segm. = antennal segment; c = collar; 1 = loosening of exocuticle and endocuticle; m = microtrichiae; pc = pore canal; s = seta; zw = zones of weakness. From Seeger (1975).

preting the venation (Figs. 66 and 67), which in turn plays a vital role in the interpretation of the fossils'. [Wolfgang Seeger.]

203. As the wings are absent in the Phthiraptera, it is impossible to compare wing characters within the Psocodea and consequently there remains some uncertainty as to exactly when the wing-coupling mechanism actually originated. It is possible that this mechanism is not a genuine (constitutive) character of the Psocodea (Psocoptera + Phthiraptera) but originated later in the stem-group of the Psocoptera (see also p. 137).

Certain apomorphic characters in the egg stage indicate that the Psocoptera are also a monophyletic group (Seeger, 1979): (1) the chorion is extraordinarily thin, and this is offset either by a thick serosal cuticle (suborder Trogiomorpha) or by additional suprachorionic layers such as coverings of silk and/or encrustations from anal secretions and other debris (suborders Troctomorpha and Psocomorpha); (2) there are no micropyles and aeropyles; (3) the embryo adopts an unusual (dorsal) position within the egg, although no rotation of the egg or the embryo can be observed; (4) the appendages of the embryo are folded in a characteristic manner, which is very different from the ground-plan of the Psocodea and from all the other Paraneoptera. [Wolfgang Seeger.]

204. This sentence ('Unfortunately . . . fossils') should be rephrased as follows: 'Because of their minute size, these characters are not visible in Palaeozoic compression fossils (summarized by Smithers, 1972), but they can be seen in amber fossils of the Mesozoic and Coenozoic (Schlee and Glöckner, 1978).' [Wolfgang Seeger.]

2.2.2.2..3.2..2. Condylognatha

The Condylognatha (Börner) are probably the sister-group of the Psocodea and consist of two subgroups, the Thysanoptera and Hemiptera. These are so dissimilar in general appearance that few authors have treated them as members of a single monophyletic group of higher rank. However, Ross (1955) and Pesson (in Grassé, 1951) have suggested that the Thysanoptera and Hemiptera had ancestors that were common only to them. In the phylogenetic tree given in the Russian Textbook of Palaeontology (Rodendorf, 1962), the Thysanoptera are shown as the sister-group of all the other Paraneoptera, whereas the Zoraptera, which I believe to be this sister-group, are shown to be most closely related to the Psocoptera. There is no evidence for this view.

The Thysanoptera agree with the Hemiptera in that the mandibles as well as the maxillary laciniae have been modified into stylets. Unfortunately it is not known whether the mouth-parts of the Condylognatha can be derived from those of the Psocodea. If they can, it would mean that one of the groups of the 'Psocodea' is more closely related to the Condylognatha than are the others. The lacinia has completely detached from the stipes of the maxillae, which is probably the case in all Acercaria, and Wille (1960) has listed the following additional derived features as characteristic of the 'Psocoptera': the suspensory sclerites and the sitophore (the basal sclerite of the hypopharynx) have been modified into a cup-shaped sclerite, which is connected to the large oval superlinguae by a branched filament; a process arising from the clypeal wall fits into the cup-shaped structure, thus forming a 'mortar-and-pestle' apparatus. If this peculiar structure is part of the ground-plan of the Psocodea and if the Phthiraptera are most closely related to one of the subgroups of the 'Psocoptera' (see p. 372), it must have been secondarily reduced in the Phthiraptera[205]. It is doubtful if this is also true of the Condylognatha. If it is primarily absent here, it must be a derived character of the Psocodea. Until there is a definite solution to this problem, we can have no clear understanding of the steps that have led to the origin of the mouth-parts of the Condylognatha.

In other respects, the distribution of derived and primitive characters amongst the Thysanoptera and Hemiptera is such that they must be regarded as sister-groups. Both groups are known from Lower Permian fossils. Unfortunately, the wings of the Thysanoptera are so strongly specialized that it is not possible to decide which characters in the venation must have been present in the common ancestors of the Thysanoptera + Hemiptera. It is therefore impossible to decide if the older fossils belong to the stem-group of the Condylognatha or to one of the two subgroups (Thysanoptera or Hemiptera).

Revisionary note

205. This sentence ('If this peculiar structure . . . Phthiraptera') should be rephrased as follows: 'If this peculiar structure is part of the ground-plan of the Psocodea, it must have been secondarily reduced in certain subgroups of the Phthiraptera (Anoplura, Rhynchophthirina)'. [Wolfgang Seeger.]

2.2.2.2..3.2..2.1. Thysanoptera (thrips)[206]

The Thysanoptera have a large number of derived characters. The mouthparts are strikingly asymmetrical, and the right mandible has been completely reduced. The wings are very narrow and strap-shaped, with greatly simplified venation and long marginal setae ('ptiloptery'). The tarsi are two-segmented[207]. The claws are rudimentary, but are normal in certain larvae. The arolium is a protrusible and retractile vesicle. The abdominal spiracles have been reduced to two pairs, on the 1st and 8th abdominal segments. Their metamorphosis, which has been called 'remetaboly', has certain elements of holometaboly which have undoubtedly arisen through convergence: there are two wingless larval stages, which are followed by a prepupal stage and then 1–2 inactive pupal stages. According to White (1957), the males may be haploid.

There is no doubt that some of these characters are linked with the small size of the thrips, most of which are 2–5 mm and only a few as much as 14 mm (von Kéler, 1955). This small size is in turn an adaptation to a highly specialized and cryptic mode of life. So far, however, these connections are not entirely clear. Rodendorf (in Rodendorf, 1962) believed that the Thysanoptera were originally adapted for sucking up minute plant cells (he must have been thinking of fungal spores or pollen grains), but Weber and Eidmann considered that they were originally predaceous.

The oldest fossil is *Permothrips longipennis* Martynov (Fig. 70), from the Lower Permian of the Urals. It has been assigned to the Terebrantia, like the Lower Jurassic *Liassothrips crassicornis* Martynov. However, the division of the recent Thysanoptera into the two suborders Terebrantia and Tubulifera does not show their true relationships, because the Terebrantia are probably a paraphyletic group: this means that the diagnostic characters of the 'Terebrantia' were also present in the common ancestors of *all* recent Thysanoptera. In fact, neither *Permothrips* nor *Liassothrips* shows any relationship with any particular monophyletic subgroup of the recent Thysanoptera[208].

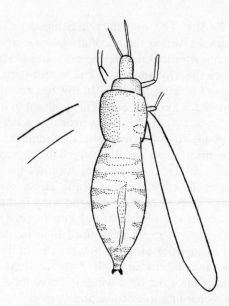

Fig. 70. *Permothrips longipennis*
Martynov (Thysanoptera), from the
Lower Permian of Chekarda. From
Rodendorf (1962).

Additional notes

206. A new monograph on the Thysanoptera has been published by Priesner (1968), and a catalogue by Jacot-Guillarmot (1970, 1971). See also notes 198, 369, and 370. [Dieter Schlee, Wolfgang Seeger.]

A new phylogenetic classification of the families has recently been proposed by Mound *et al.* (1980). [Adrian Pont.]

207. For the functional morphology of the pretarsus, see Heming (1971). [Willi Hennig.]

208. According to Sharov (1972), 'the Thysanoptera arose from the Psocoptera . . . probably no earlier than the beginning of the Triassic'. In the same paper, Sharov described the genus *Karataothrips*, from the Upper Jurassic of Karatau, and assigned it to the 'Terebrantia'. [Willi Hennig.]

2.2.2.2..3.2..2.2. Hemiptera

The Hemiptera are probably the sister-group of the Thysanoptera. They have a relatively large number of very characteristic derived features, and this shows that they are a monophyletic group.

In its ground-plan the rostrum is a 3- to 4-segmented organ formed from the 2nd maxilla (a 'labial proboscis'), but its morphological structure is still in dispute (Matsuda, 1965b). The maxillary and labial palpi are completely absent. The mandibles, as well as the maxillary laciniae, have been modified into stylets: this is a character that the Hemiptera inherited from the ancestors which they shared with the Thysanoptera. The mouth-parts are symmetrical, and the right mandible has been retained. In this respect, the Hemiptera are more primitive than the Thysanoptera. The anterior part of the head is produced forwards like a snout, and this is probably a derived character[209]. The clypeolabrum, genae (laminae mandibulares) and the basal sections of the maxillae (laminae maxillares) are all involved in this according to Poisson and Pesson (in Grassé, 1951). It is not clear whether the buccal pump is a further development of the cibarial pump of the Thysanoptera[210].

The only undoubted derived character in the wing is the single branch of the radial sector. According to Poisson and Pesson (in Grassé, 1951), the fore-wings are always slightly more sclerotized than the hind-wings. Moreover, the anal lobe is clearly demarcated from the rest of the wing by the anal furrow, along which the cubital vein (CuP) runs, and this lobe is called the clavus. In the ground-plan of the hind-wing, the anal lobe is folded beneath the rest of the wing, according to Poisson and Pesson.

Unfortunately, it is impossible to say whether these wing features are characteristic of the ground-plan of the Hemiptera or whether they were already present in the ground-plan of the Condylognatha. Von Kéler (1955) has stated that in the Thysanoptera the fore-wing is rather more strongly sclerotized than the hind-wing, which is always membraneous, and that the 'scale' in the anal corner of each wing is homologous with the clavus (anal lobe) of the Hemiptera. However, it is not possible to say anything about the form that the radial sector took in the stem-group of the Thysanoptera. This becomes disturbingly obvious when the fossils have to be interpreted. There are no difficulties after the Lower Permian, as Thysanoptera are known from this period, but older fossils with the ground-plan characters of the recent Hemiptera could belong to the Hemiptera or to the stem-group of the Condylognatha, from which both the Thysanoptera and the Hemiptera have descended. It would only be possible to determine the age of the Thysanoptera and Hemiptera (on the basis of fossil wings) if some Carboniferous wings were available with the derived characters of the Thysanoptera or with those of subgroups of the *Hemiptera. Some authors have actually attempted to deduce from fossil wings that subgroups of the *Hemiptera did exist in the lower Upper Carboniferous. Haupt (1944) asserted that one of the oldest insect wings, *Stygne roemeri* Handlirsch[211] from the Namurian, had 'indisputable fulgoroid features', and he also interpreted the wings of various Palaeodictyoptera and Blattodea (*Recula, Synarmoge, Blattinopsis*[212]) as 'Homoptera'. These interpretations are absurd and should not be taken seriously. Sharov (1966b) assigned *Stygne roemeri* to the family Paoliidae, in the stem-group of the Pterygota! From time to time, however, other authors have also assigned Palaeozoic wings or wing fragments to subgroups of the *Hemiptera.

Unfortunately, the relationships between the subgroups of the *Hemiptera are not very clear, and this is a serious drawback when attempting to interpret the fossils. Two suborders, the Homoptera and Heteroptera, are usually recognized. The Heteroptera are a well founded monophyletic group, but it is impossible to decide whether the Homoptera are also monophyletic. The diagnostic characters which are used to separate them from the Heteroptera are all primitive and must have been present in the common ancestors of the Homoptera and Heteroptera, the stem-group of the Hemiptera. This is particularly true of the wing venation. The question of whether the Coleorrhyncha are more closely related to the Heteroptera or to the Homoptera, or to a subgroup of the so-called Homoptera, has been hotly debated for a long time. I think it most likely that they are closest to the Heteroptera. Three monophyletic groups can be recognized within the Hemiptera: the Heteroptera (including the Coleorrhyncha), Sternorrhyncha, and Auchenorrhyncha.

This means that it will be necessary to examine the Palaeozoic fossils that are usually assigned to these three groups and to see if they can in fact be assigned to the stem-groups of these groups[213]. Extreme care is necessary if any attempt is to be made to resolve the considerable differences of opinion that still exist.

The most appropriate course is to begin with groups that have a particularly large number of derived characters to establish that they are monophyletic. The minimum age of such groups can be determined from the oldest fossils that can be definitely assigned to them, and from this it is possible to calculate the age of other groups which, according to their wings, had more primitive characters.

Revisionary notes

209. The morphology of the head capsule and its significance for the phylogeny of the Hemiptera were dealt with by Spooner (1938). [Willi Hennig.]

210. Parsons (1974) has discussed the morphology and possible origin of the hemipteran loral lobes. The significance of the pulvillus for the taxonomy of the land bugs has been dealt with by Goel and Schaefer (1970).

Langer and Schneider (1970) have drawn attention to the presence in *Gerris* of open rhabdomes and an arrangement of retinula cells similar to that in the Diptera. We still need to know whether this is true of all the Heteroptera or just of certain water-bugs; and, if of all the Heteroptera, then whether it is an autapomorphy or present in other Hemiptera. [Willi Hennig.]

211. For further analysis of *Stygne*, see Schwarzbach (1939). [Willi Hennig.]

212. In a recent study, Müller (1977a) has followed Kukalová (1959) in considering the Blattinopsidae to belong to the Protorthoptera. He has rejected suggestions that they are related to the fulgoroid cicadas or to the Protoblattoidea. [Dieter Schlee.]

213. Palaeozoic fossils of the 'Homoptera' were reviewed by Bekker-Migdisova (1960a). [Willi Hennig.]

2.2.2.2..3.2..2.2.1.   Heteropteroidea[214]

The Coleorrhyncha and the Heteroptera (bugs) have a number of derived characters in common which indicate that they form a monophyletic group of higher rank. There are only four antennal segments[215]. The body has a flattened dorsal area over which the wings are placed. The differentiated areas 'corium' and 'membrane', which are so characteristic of the true Heteroptera, are already weakly developed in the Coleorrhyncha.

The course of the anal veins is a particularly important feature. Wootton (1965) described how the apical sections of the two anal veins have fused together in all the Coleorrhyncha and true Heteroptera: they form a Y-vein (Fig. 71). The fused apical section crosses the claval furrow, which corresponds to the 'anal furrow' of the ground-plan of the Neoptera, and continues into the remigium where it fuses with CuA_2. I think that Wootton was correct in regarding this as a particularly convincing synapomorphic character of the Coleorrhyncha and Heteroptera. According to Schlee (1969d) a 'united anal cone'[216] is a further synapomorphic character of the Coleorrhyncha and

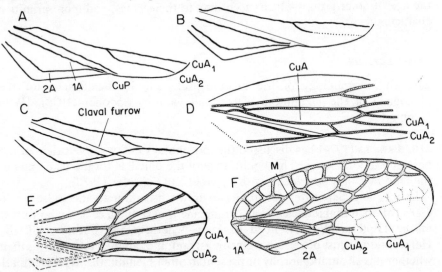

Fig. 71. Fore-wings of the Heteropteroidea. A–C, Hypothetical stages in the development of the cubito-anal area; D, an unnamed archescytinid, from the Lower Triassic of Kirgiziya; E, *Microscytinella radians* Wootton, from the Upper Triassic of Australia; F, *Peloridium hammoniorum* Breddin (Coleorrhyncha; recent). From Wootton (1965).

Heteroptera: the two segments that form the anal cone in the ground-plan of the Hemiptera have been fused.

There can only be a sister-group relationship between these two groups[217]. It must already have been in existence by the Upper Permian because the fossil *Paraknightia* (see below) is known from this epoch and can be assigned with some certainty to the Heteroptera. If this assignment is correct, the Coleorrhyncha must also have been in existence at the same time as an independent group. The Heteropteroidea may therefore have originated in the Lower Permian.

Revisionary notes

214. A comparative study of the mouth-parts of the Heteroptera has recently been published by Cobben (1978). [Dieter Schlee.] See note 210. [Willi Hennig.]

215. Schlee (1969d) listed additional synapomorphies of the Heteropteroidea (Heteroptera + Coleorrhyncha): the distal antennal segments are larger and more clavate than the proximal ones; in cross-section, the abdomen has flattened tergites, with the 'latero-tergites' protruding laterad and the spiracles directed ventrad; there is a characteristic horseshoe-shaped sclerite at the base of the aedeagus, etc. [Dieter Schlee.]

216. The anal cone is exposed, consists of one segment, and has a retractile distal flap. [Dieter Schlee.]

217. A diagram of the synapomorphies was given by Schlee (1969d, Fig. 4). [Dieter Schlee.]

2.2.2.2..3.2..2.2.1.1. Coleorrhyncha

The Coleorrhyncha consist of a single 'family', the Peloridiidae. There are only 19 recent species[218], and these are confined to New Zealand–Australia–Tasmania and southern Chile.

They are more primitive than the Heteroptera in lacking a gula and in certain other characters (see Schlee, 1969d). They also have a number of derived characters, such as the complete absence of ocelli and hind-wings, and only two tarsal segments[219].

Several authors have considered the lateral lobes of the pronotum to be a very important character. They are said to be homologous with the prothoracic paranota of the Palaeodictyoptera and the Coleorrhyncha have been regarded as the only recent group to have retained paranota (Pesson, in Grassé, 1951), but I think that this is very doubtful. I prefer to think of the lateral lobes of the pronotum as new acquisitions, and therefore as a further derived character of the Coleorrhyncha.

There are conflicting statements about the number of antennal segments. Heslop-Harrison (1952a) has stated that the Coleorrhyncha have four, like the Heteroptera, but Poisson and Pesson (in Grassé, 1951) gave the number as three. The coleorrhynchous antennae thus appear to be even more reduced than those of the Heteroptera.

The Peloridiidae used to be placed in the Heteroptera, but now they are usually assigned to the Homoptera[220]. Evans (1963) treated them as a third line of descent from the hypothetical 'Protohomoptera', alongside the Auchenorrhyncha and Sternorrhyncha.

I should point out that no characters are known that can be used to establish that the Homoptera are a monophyletic group[221] or to indicate that the Coleorrhyncha are closely related to the 'Homoptera'.

Poisson and Pesson[222] (in Grassé, 1951) believed that the mixture of 'homopterous' and 'heteropterous' characters indicated that 'the Coleorrhyncha undoubtedly belong to the Homoptera'. In fact, the characters common to these two groups are all primitive (absence of gula and stink-glands, etc.) or are the results of convergence (two tarsal segments in Coleorrhyncha and Sternorrhyncha). Agreements in undoubted derived characters, which can be regarded as synapomorphies, only exist between the Coleorrhyncha and Heteroptera[223].

Müller (1962) attempted to deduce the relationships of the Coleorrhyncha from their symbionts, but unfortunately he began with the assumption that they belong to the 'Homoptera'. He found that *Hemiodoecus fidelis* Evans, the only peloridiid that he examined, harbours a-symbionts and is 'the only homopteron without secondary or ancillary symbionts'. He concluded that the a-symbionts are 'the oldest and originally the most important symbionts, in other words the main symbionts', and he believed that his results proved that the Coleorrhyncha are 'at the base of at least all the Auchenorrhyncha'. But this phrase, 'at least all the Auchenorrhyncha', shows the weak link in his argument. If it cannot be shown that the a-symbionts are a characteristic new acquisition of the Auchenorrhyncha, and therefore a derived character of the group, it is possible that they are part of the ground-plan of the Hemiptera and must have been present in the stem-group of the Hemiptera. 'Phylogenetic trees' of the Homoptera (Evans, 1963) or of the 'Auchenorrhyncha' (Müller, 1962) which show the Coleorrhyncha but not the Heteroptera, or show the Heteroptera and the Sternorrhyncha with their component groups, are fundamentally misleading. Schlee (1969d) has recently discussed the soundness of Müller's conclusions. He found that '"a-symbionts of the fulgoroid type" are not a synapomorphic character of the Peloridiidae and Fulgoroidea but are part of the basic endosymbiotic fauna common to all Hemiptera'[224].

There are no Palaeozoic or Mesozoic fossils that can be definitely assigned to the Coleorrhyncha.

Wootton (1965) reported that the Actinoscytinidae[225] are the oldest fossils that agree with the Coleorrhyncha and Heteroptera in the form of the cubito-anal lobe. They are known from the Upper Permian to the Upper Jurassic.

Evans (1956) was the first to assign *Actinoscytina belmontensis* Tillyard (Fig. 72), from the Upper Permian of Australia, to the Heteroptera, although this was only 'provisional'. The most that can be deduced from the wing venation of this species is that it *could* belong to the stem-group from which the Heteroptera and Coleorrhyncha have descended. Unfortunately, the anal veins are absent. Subsequently, according to Evans (1963), Bekker-Migdisova expanded the Coleorrhyncha to include the Cicadocoridae, from

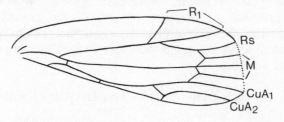

Fig. 72. *Actinoscytina belmontensis* Tillyard (Actinoscytinidae), from the Upper Permian of Australia. From Evans (1956).

the Triassic of Asia, and several other fossils that Handlirsch had previously called the 'Procercopidae'. Evans disagreed with this, and considered that all these species belong to the Actinoscytinidae. In the meantime, Bekker-Migdisova (in Rodendorf, 1962) had also come round to this view. However, little is gained by these assignments, and the position of the Actinoscytinidae still remains unclear. If *Paraknightia*, from the Upper Permian of Australia, really belongs to the stem-group of the Heteroptera (see below), the stem-group of the Coleorrhyncha must also have been in existence as a separate group at the same time. For determining the age of these groups it is of no practical importance whether the Actinoscytinidae are this stem-group or not. It is even possible that they are survivors from the pre-Upper Permian stem-group of the Heteropteroidea (Coleorrhyncha + Heteroptera). If this were so, it would have an important bearing on the origin of the recent Coleorrhyncha, which are of such zoogeographical interest, and also on the interpretation of the Mesozoic Actinoscytinidae. For example, are these Mesozoic Actinoscytinidae survivors from a stem-group of the Heteropteroidea, or are they an invalid stem-group containing the relatives of various subgroups of the Heteropteroidea? This problem can only be solved when a reliable analysis of the venation of the Heteropteroidea and of the course of its development has been made.

Revisionary notes

218. Woodward *et al.* (in CSIRO, 1970) gave a total of 20 species. Evans

(1972) has subsequently described a 21st (Woodward *et al.*, in CSIRO, 1974). [Dieter Schlee.]

219. It is doubtful whether these characters can be regarded as autapomorphies of the Coleorrhyncha: reduced ocelli also occur in the Heteroptera, and reduced tarsi in the Sternorrhyncha. According to Dolling (*in litt.*), China (1962) has figured the metathoracic wing of a fully macropterous specimen of *Peloridium hammoniorum* Breddin, but he omitted to mention in his description that the macropter also possessed ocelli on the front of the head adjacent to the eyes; thus neither the absence of metathoracic wings nor the absence of ocelli can be accepted as part of the ground-plan of the Coleorrhyncha. The presence of propleural antennal sheaths and the structure of the frontal lobes are autapomorphies of the Coleorrhyncha (Schlee, 1969d). [Dieter Schlee.]

220. A review has been given by Schlee (1969d). [Dieter Schlee.]

221. Seeger (1975), who studied the structure of the antennae in the Psocoptera and related groups, found that there is a reduction of the antennal intersegments in the Homoptera, which could be interpreted as an autapomorphy. However, reductive characters are generally rather suspect since they can easily arise convergently. In any case, this character cannot be regarded as a synapomorphy of the Homoptera + Coleorrhyncha, because antennal intersegments are present in both the Coleorrhyncha and Heteroptera. [Dieter Schlee.]

222. Poisson and Pesson (in Grassé, 1951), based their statements on the work of Myers and China (1929). This was analysed by Schlee (1969d), who showed that the characters thought to be common to the Coleorrhyncha and 'Homoptera' are either symplesiomorphies or are based on incorrect observations. [Dieter Schlee.]

223. Cobben's (1965, p. 62) conclusion that his work on the structure of the eggs would 'indicate rather a homopterous than a heteropterous kinship relation' for the Coleorrhyncha is meaningless in the sense of phylogenetic systematics. Furthermore, it is based on a series of convergences/parallelisms rather than on synapomorphies; nor are the 'Homoptera' or their subgroups characterized in a phylogenetically meaningful way (see Schlee, 1969d). [Dieter Schlee.]

224. Müller's (1962) work suffers from the defect that he has overlooked how frequently convergent developments of the 'characteristic' mycetomes and symbionts have taken place. This problem has been analysed by Schlee (1969d) who used Buchner's (1953) data, which Buchner originally presented in an unweighted form. From this it appears that the structure of the mycetomes and symbionts is directly dependent on the chemistry of their environment, such as contact metamorphoses between mycetome and ovarium, chemical effects during ontogeny, different population densities, etc.

Apart from this, the alleged agreement between the 'a-organs' of the Coleorrhyncha and Auchenorrhyncha–Fulgoromorpha is not a real one, even in a typological sense: in the one case they are enlarged mycetocytes, which produce enlarged infective forms, whereas in the other they are strongly reduced cells with reduced infective forms, etc.

Furthermore, the alleged agreements between the two 'phylogenetic trees' based on morphological and endosymbiotic characters are irrelevant in the sense of phylogenetic systematics. [Dieter Schlee.]

225. A discussion, with descriptions of new species from the Upper Triassic of Australia, has been given by Wootton (1963). [Dieter Schlee.]

2.2.2.2..3.2..2.2.1.2.   **Heteroptera (bugs)**[226]

In some characters the bugs are more primitive than their presumed sister-group, the Coleorrhyncha. The hind-wings are still present[227], and there are two ocelli[228]: the third ocellus is absent. The antennae may also be rather more primitive than those of the Coleorrhyncha. According to Muir, there are four segments in the ground-plan. Occasionally, the basal segment is reduced or the apical segment secondarily divided.

The bugs also have some derived ground-plan characters when compared with the Coleorrhyncha[229]. The occiput is closed by a gula (von Kéler's 'hypostomal bridge'). As in other insect groups, prognathy and a shift of the anterior surface of the head into a dorsal position are correlated with this. According to Ross (1965), the tentorium is absent. Stink-glands are present which open to the exterior in the adults on the metathorax above the hind coxae, and in the larvae on the 4th–6th abdominal tergites: it is not completely certain if this derived character is also part of the ground-plan of the Heteroptera.

For assessing the fossils, the structure of the fore-wings is more important than any of these characters. The fore-wings have been modified into hemelytra ('hemi-elytra') (Fig. 73). A basal hardened 'corium' is separated by the 'nodal line' from a softer distal 'membrane'. The 'clavus', which is

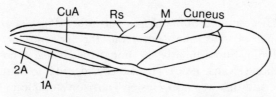

Fig. 73. Fore-wing of the Heteroptera. From Poisson (in Grassé, 1951).

separated from the corium by the claval suture along CuP, is not a special feature of the Heteroptera but is part of the ground-plan of the Hemiptera. In the Heteroptera, however, it is hardened like the corium. According to Evans (1963), the nodal line is a line of weakness which is found in many Hemiptera, Sternorrhyncha and Auchenorrhyncha. Evans also believed that it was characteristic of 'certain Protohemiptera'. Some stages in its development should therefore be part of the ground-plan of the Hemiptera(?). For this reason its presence in certain fossils cannot be used as evidence that there is any close relationship between these species and any particular subgroups of the *Hemiptera. The venation of the corium is greatly reduced, and in the membrane none of the veins reaches the wing margin.

According to Evans (1950, 1956), the most characteristic feature of the Heteroptera is not the nodal line but the costal fracture and the costal furrow that originates from it. This furrow is a line of weakness that divides the corium into 'cuneus' and 'embolium'. It is probable that the development of this second line of weakness in the corium is correlated with the hardening of this part of the wing and is a result of the technical requirements of flight(?).

There are no well founded ideas about the changes in the mode of life that may have led to the origin of the bugs (Heteroptera). China (1933) thought that the ancestors of the Heteroptera were terrestrial and phytophagous[230]. However, this must also have been true of the stem-group of all the Hemiptera[230]. The flattened body and leathery fore-wings point to a terrestrial and more or less cryptic mode of life for these 'beetles' of the Paraneoptera. However, this is also more or less true of the Coleorrhyncha and does not explain the characteristic features of the Heteroptera.

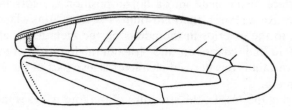

Fig. 74. Fore-wing of *Paraknightia magnifica* Evans (Heteroptera). Redrawn from Fig. 75.

Evans (1943) described *Paraknightia magnifica* (Figs. 74 and 75) from the Upper Permian of Australia (not from the Triassic, as was stated by Rodendorf, 1962). Subsequently, he noted that it had a costal furrow and reduced venation (Evans, 1950). He also considered it to be the oldest known heteropteron, and this was also Heslop-Harrison's opinion (1956, according to Evans, 1963). *Paraknightia* cannot be assigned to any of the recent subgroups of the Heteroptera.

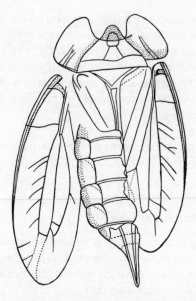

Fig. 75. *Paraknightia magnifica* Evans (Heteroptera), from the Upper Permian of Australia. From Evans (1943).

Revisionary notes

226. New comprehensive papers have been published on the eggs by Cobben (1968), the mouth-parts by Cobben (1978), and the comparative anatomy by Rieger (1976).

These papers are a rich source of information on anatomy and function, but both authors have taken diametrically opposing views on the matter of phylogeny. Cobben (1978, p. 191) attacked Rieger's thorough study as being based on 'insufficient evidence and . . . inconsistent and questionable evaluation of data', and he felt obliged 'to warn against a Hennigian analysis of the kind that Rieger had made'. However, I must emphasize that from the point of view of Hennigian phylogenetic systematics Cobben's own phylogenetic approach is inadequate. Furthermore, many of his arguments, and even his final conclusions (Cobben, 1978, Fig. 75) have nothing in common with the aims of phylogenetic systematics. For further details, see Schlee (1969d) and Seeger (1979). [Dieter Schlee.]

227. See note 219. [Dieter Schlee.]

228. See also note 219. [Dieter Schlee.]

229. Rieger (1976) has listed four further characters as part of the ground-plan of the *Heteroptera: (1) a maxillary lever to guide the piercing stylet, situated between the wing of the hypopharynx and the maxillary stylet; (2) a protractor and a retractor for the maxilla, which originate in the sac of the maxillary plate and at the cranium, respectively; (3) a flattened triangular mandibular lever, to which two protractors originating from the clypeus are attached; and (4) two retractors of the mandibular stylet, which originate on the cranium and are attached at the tip of the piercing stylet and subapically, respectively. [Dieter Schlee.]

230. The arguments for and against phytophagy have been given by Schlee (1969d), Rieger (1976, p. 181) and Cobben (1978, pp. 191 and 238). [Dieter Schlee.]

2.2.2.2..3.2..2.2.2. Sternorrhyncha

The Sternorrhyncha are probably a second monophyletic group of the Hemiptera, but this is not generally accepted. For example, Heslop-Harrison (1951, 1952a, 1958) has argued strongly against the division of the Homoptera into Sternorrhyncha and Auchenorrhyncha. He preferred to regard the Cicadomorpha + Psyllidomorpha + Jassidomorpha as one monophyletic group, and he based this mainly on the presence of a filter chamber in the alimentary canal in these groups. He thought it unlikely that such a specialized organ could have arisen on several occasions independently. He considered that the Aphidina and Fulgoroidea formed a second monophyletic group[231].

I am not convinced by Heslop-Harrison's arguments, because it is by no means certain that the filter chamber should be regarded as a synapomorphic character of the groups where it is present. So far as I know, there has been no adequate study of this character. Yet, according to Weber, its structure is very diverse, and it is undoubtedly an important feature for working out phylogenetic relationships.

The recent Sternorrhyncha have the following derived characters: the position of the base of the proboscis between or behind the fore coxae, the reduction of the tarsal segments to two, the presence of a saw-like egg-burster, the absence of the mesothoracic trochantin (Schlee, 1969b), and the reduction of the clavus, which has at most a single anal vein. Moreover, the subcosta is fused with the radius for most of its length and only the apical section is free. The basal sections of the radius, media, and cubitus are fused, and the media and cubitus are also fused for a short distance beyond this.

Only a few of these characters are visible in the fossils. Various authors have assigned Palaeozoic species to individual subgroups of the Sternorrhyncha, and if these assignments are accepted then it follows that some of the derived characters present in all the recent Sternorrhyncha must have arisen independently in several groups, by convergence. Until recently, all the Palaeozoic Hemiptera, except for *Paraknightia* and the Actinoscytinidae which have been transferred to the Heteroptera, were arranged into two families, the Archescytinidae[232] and Lithoscytinidae, which together formed the 'Palaeorrhyncha' (Carpenter, 1931b, 1933; Evans, 1943; Laurentiaux, in Piveteau, 1953). The Palaeorrhyncha were considered to be the stem-group of all the recent Homoptera, or at least of the Sternorrhyncha. Nowadays, however, most authors believe that a considerable number of subgroups of the Sternorrhyncha and Auchenorrhyncha can definitely be recognized amongst the Permian fossils.

For this reason it is necessary to discuss in more detail the characters of the Sternorrhyncha and the extent to which they are visible in fossils.

The labium is inserted between or behind the fore coxae, and according to Pesson and Poisson (in Grassé, 1951) this shift is a characteristic feature of the Sternorrhyncha. It is generally accepted that the orthognathy and open occiput of the Auchenorrhyncha have been taken over more or less un-changed from the ground-plan of the Hemiptera. In this case, the prognathy and closed occiput (gula, hypostomal bridge) of the Heteroptera and the hypognathy or opisthognathy of the Sternorrhyncha are both undoubted derived characters, representing two distinct lines of development from the ground-plan of the Hemiptera. It was only possible for the 'sternorrhynchy' of the Sternorrhyncha to develop because the occiput remained open. Obvious-ly, this is a character that can never be seen in fossils, even in rare cases where a complete impression of the body is preserved. In the Russian Textbook of Palaeontology (Rodendorf, 1962), Bekker-Migdisova assigned the 'Cicadop-syllidea' to the 'Psyllomorpha', a subgroup of the Sternorrhyncha, and stated that they were 'hypognathous or orthognathous'. It is not clear from the available figures on what she based this bold assertion.

The reduction of the 3rd tarsal segment is characteristic of all recent Sternorrhyncha. *Archescytina permiana* Tillyard had three tarsal segments, and Bekker-Migdisova assigned the Archescytinidae to the Aphidomorpha. If this is correct, then the reduction in the number of tarsal segments to two must have taken place independently in various subgroups of the Sternor-rhyncha. This view is not well founded. In any case, reduction in the number of tarsal segments is so common in the 'Polyneoptera' and Paraneoptera that it could easily have taken place through convergence in different groups of the Sternorrhyncha.

So far as I know, only Börner (1934) has pointed out that a saw-like egg-burster is a characteristic feature of the Sternorrhyncha. This is some-thing that is never visible in fossils and, like all characters of this type, is known only from a comparatively small number of random samples amongst recent species. It may well be an important character for indicating that the Sternorrhyncha are a monophyletic group, if it really does prove to be characteristic of the whole group, but it will never be of any assistance in interpreting the fossils.

It is the wing venation that is of prime importance for work on the fossils. Unfortunately, the venation of most recent Sternorrhyncha is so reduced that it can be 'derived' from almost any other venation, even though only slightly more primitive. This makes it extraordinarily difficult to interpret the fossils[233].

There are four monophyletic subgroups of the Sternorrhyncha, the Psyllina, Aleyrodina, Aphidina, and Coccina. However, their phylogenetic relationships have not been established with any certainty, and this is a source of further difficulty: almost every possible combination has been suggested at one time or another. Börner (1934) grouped together the Coccina + Aphidina and the Aleyrodina + Psyllina. On the whole this arrangement has been generally preferred, and does appear to be well founded (Schlee, 1969a, 1969b, 1969c, 1969e)[234]. In the sections that follow, I think it is best to begin with the recent species that have particularly striking derived characters in their wing venation, as these will be most helpful in interpreting the fossils.

Revisionary notes

231. Schlee (1969b) has given a critique of Heslop-Harrison's (1956 and subsequently) various statements about this. He concluded that the presence of a filter chamber does not prove that the groups in question constitute a single monophyletic group. [Dieter Schlee.]

232. A new study of the venation of the Archescytinidae has been made by Szelegiewicz and Popov (1978). [Dieter Schlee.]

233. Szelegiewicz (1971) has discussed the autapomorphies in the venation of recent subgroups of the Sternorrhyncha and their significance for the classification of the Palaeozoic fossils. [Willi Hennig.]

234. A diagram of the synapomorphies has been published by Schlee (1969b, Fig. 26). [Dieter Schlee.]

2.2.2.2..3.2..2.2.2.1. Aphidomorpha

It is almost impossible to find a name for the group containing the plant-lice and scale-insects that has not already been used and misused in various ways.

Börner (1904, 1934) and Heymons (1915) suggested that the plant-lice and scale-insects were closely related, and this has been re-examined by Theron (1958). He listed 10 characters common to the genera *Aphis* and *Margarodes* (which has probably diverged little from the ground-plan of the Coccina). However, it is difficult to decide whether some of these characters are primitive or derived.

Schlee (1969b) has made a careful study of these groups and has recognized the following synapomorphic characters of the Aphidina and Coccina[235]:

The antennal muscles are attached to the wall of the head capsule, and not to the tentorium as is the case in the ground-plan of the Sternorrhyncha. The prescutum is reduced anteriorly. There is an articulating joint between postalar and epimeron. A separate area called the 'lateropleurite' has become detached from the mesopleuron. The lobe-like appendages on the pretarsus have been reduced and are absent. There is a special wing-coupling apparatus which consists of a row of hooked bristles along the fore-margin of the hind-wing. These bristles grip the hind-margin of the fore-wing which is bare but thickened. The male claspers ('parameres') are reduced; the penis is thin-walled, retractile, and contained within a sheath. The female has no ovipositor: eggs are simply released. The anal tube is reduced in both sexes.

It is particularly important that Schlee (1969e) has been able to show that there are apparent synapomorphic agreements in the venation of the Coccina and Aphidina. The costal field is narrow and the fore-margin of the wing is straight. The stems of Sc, R, M, and Cu form a single, straight vein[236] and the free sections of these veins branch off the stem like the teeth of a comb. RS and M issue from the stem-vein close to the distended pterostigma[237]. Finally, the wing base is narrow in the area of the anal lobe[238].

White (1957) reported that post-reduction divisions take place during meiosis in both the Coccina and the Aphidina. However, it is unlikely that this is a well founded agreement because White's results are probably based only on a few random samples.

If the Coccina and Aphidina really form a monophyletic group, then the heterobathmy of their characters indicates that they must be sister-groups. It is not known when this relationship originated because the Upper Permian genus *Permaphidopsis*, which has usually been assigned to the Aphidina, could equally well belong to the common stem-group of the plant-lice and scale-insects[239].

Revisionary notes

235. The Aphidina + Coccina = Aphidiformes (sensu Schlee) or Aphidimorpha (sensu Hennig). I prefer the suffix -formes, for the reason given by Hennig in the first paragraph of this section on the Aphidomorpha. [Dieter Schlee.]

236. In cross-section, this vein appears to be a U-shaped fold, not a cylindrical tube. [Dieter Schlee.]

237. The pterostigma contains a large body of parenchyma. [Dieter Schlee.]

238. The narrow wing base should not be mentioned here because it is an autapomorphy of the Aphidina alone (Schlee, 1969e, Figs. 1 and 2), not a synapomorphy of the Aphidina + Coccina. [Dieter Schlee.]

239. Szelegiewicz and Popov (1978) have shown that the wing described as *Permaphidopsis* is actually the *hind*-wing of a species of Protopsyllidiidae. [Dieter Schlee.]

2.2.2.2..3.2..2.2.2.1.1. Aphidina (plant-lice, aphids)

The derived ground-plan characters of the Aphidina are the alternation of hosts and generations and the absence of Malpighian tubes. These indicate that the Aphidina are a well founded monophyletic group. However, it is obvious that none of these characters can be seen in the fossils. Schlee (1969e) has recently shown that there is a fundamental agreement in derived characters between the venation of the Aphidina and Coccina[240] (Fig. 76), which is the character most helpful for the interpretation of the fossils. As a result it is more or less impossible to assign older fossils to the Aphidina[240] because there are no autapomorphic ground-plan characters in the venation of this group.

Evans (1963) was correct in establishing that *Permaphidopsis sojanensis* Bekker-Migdisova, a fore-wing from the Upper Permian of the Arkhangelsk region (Fig. 77), agrees most closely with the recent Aphidina. The wing is only slightly more primitive than that of *Triassoaphis* (Fig. 79). Unfortunately, the fore-wing of *Kaltanaphis permiensis* Bekker-Migdisova (Fig. 78), from the Lower Permian of the Kuznetsk basin, is incomplete: the whole clavus, including CuP and any anal veins that may have been present, is missing. It differs fundamentally from *Permaphidopsis* and recent Aphidina in having M with only two branches. Furthermore, the origin of M, whether from R or CuA, cannot be seen. We could be more or less certain that the stem-group of the Aphidina (*Permaphidopsis*) existed during the *Upper* Permian, but not the Lower Permian, if we could be sure that *Permaphidopsis* did not belong to the common stem-group of the Aphidina + Coccina. However, this possibility can no longer be ruled out[241].

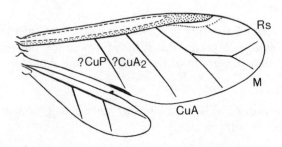

Fig. 76. Wings[242] of *Hyalopterus* sp. (Aphidina; recent). From Pesson (in Grassé, 1951).

I do not think that there is any real justification for the assignment of the Pincombeidae, from the Upper Permian of Australia, and of the genus *Tchecardaella* (*Tshekardaella*), from the Lower Permian[243] of the Urals, to the Aphidomorpha or even to the Aphidina. I shall discuss these groups again later (p. 263).

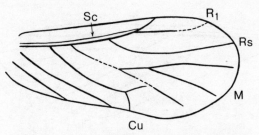

Fig. 77. Fore-wing of *Permaphidopsis sojanensis*
Bekker-Migdisova[244] (Aphidomorpha), from the
Upper Permian of the Arkhangelsk region. From
Rodendorf (1962).

Revisionary notes

240. This statement is not entirely correct, and was probably based on Hennig's incorrect interpretation of the narrow wing base, which he believed to be a synapomorphy of the Aphidina and Coccina (see note 238). Fossil *Aphidina can in fact be recognized by their narrow wing base. [Dieter Schlee.]

241. A recent paper by Szelegiewicz and Popov (1978) deals with the genera *Permaphidopsis, Kaltanaphis* and *Tchecardaella* (*Tshekardaella*), and has an important bearing on the discussion in this paragraph. The authors have assigned *Kaltanaphis* and *Tchecardaella* to the Archescytinidae, and *Permaphidopsis* to the Protopsyllidiidae. [Dieter Schlee.]

242. See Schlee (1969e) for notes on the nomenclature of the veins. The following interpretation of the two cubital veins can be suggested: Cu1a (labelled CuA in the figure) and Cu1b (labelled ?CuP/CuA$_2$), because both are convex veins. CuP should be a concave vein, and in fact there is a concave vein (claval fold/Cu2) alongside the narrow strip of the reduced clavus (not drawn in Fig. 76). [Dieter Schlee.]

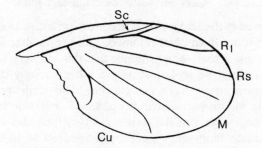

Fig. 78. Fore-wing of *Kaltanaphis permiensis*
Bekker-Migdisova[245] (Aphidomorpha), from
the Lower Permian of the Kuznetsk basin.
From Rodendorf (1962).

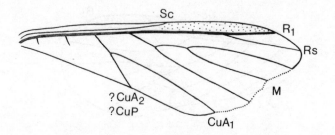

Fig. 79. Fore-wing[246] of *Triassoaphis cubitus* Evans (Aphidina), from the Upper Triassic of Australia (Mt. Crosby). From Evans (1956).

243. Szelegiewicz and Popov (1978) revised the nomenclature of *Tchecardaella* and assigned it to the Archescytinidae. They gave its age as late Lower Permian. [Dieter Schlee.]

244. Szelegiewicz and Popov (1978, Figs. 6 and 8) have given a new figure of this fossil, which is strikingly different in the details of the venation (especially the base and fork of Cu) from that given by Bekker-Migdisova and reproduced here by Hennig. Szelegiewicz and Popov have also revised the nomenclature of the veins, and have assigned it to the Protopsyllidiidae. [Dieter Schlee.]

245. In their new illustration and analysis of this fossil, Szelegiewicz and Popov (1978) have shown that Sc does not exist; the base of M is not interrupted but is actually obscured in the fossil; and all the veins reach the wing margin. They have interpreted it as the hind-wing of an archescytinid, and gave its age as early Upper Permian. [Dieter Schlee.]

246. For the venation, see note 242. [Dieter Schlee.]

2.2.2.2..3.2..2.2.2.1.2. Coccina (scale-insects, mealy-bugs)

In some characters the scale-insects are more primitive than the plant-lice: two Malpighian tubes have been retained; there is no alternation of generations and hosts, nor is there any evidence that this has been lost secondarily.

In other characters the scale-insects are more strongly derived than the Aphidina. The tarsi always have a single claw. The ocelli are absent (Evans, 1963). The male mouth-parts have atrophied. The female labium has been reduced to 1–2 segments, but the stylets are very long and withdrawn into a crumena. The male hind-wings have been reduced to haltere-like hooks. Females are completely wingless; they reach sexual maturity 1–2 moults before the males (neoteny!) and the moults are not associated with any fundamental morphological changes.

From this it is clear that fossils in which these characters are present at a more primitive stage of development, as in the recent Aphidina, could just as easily belong to the stem-group of the Aphidina + Coccina. This is even true of the wing venation (Fig. 80). In this character too the Coccina are more

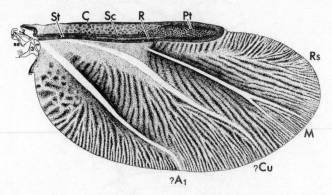

Fig. 80. Fore-wing of the male of *Sphaeraspis priaskaensis* Jakubski (Coccina; recent). From Schlee (1969e).

strongly derived[247] than the Aphidina: in their ground-plan, a section of the venation is very difficult to make out. I cannot see any reason for not deriving the venation of the Coccina as well as that of the Aphidina from *Permaphidopsis*! However, the venation of the Coccina could just as easily be derived from that of *Triassoaphis*!

This shows how dangerous it is to determine the age of the Aphidina using solely the formal agreement of fossils and recent species. Bekker-Migdisova described *Mesococcus asiaticus* (Fig. 81) from the Upper Triassic (Rhaetic) of Issyk-Kul', not the Permian as Evans (1963) inadvertently stated. This fossil is 1–2 mm in length, and she interpreted it as the female of a scale-insect. She may well be correct, but it is impossible to be certain[248]. On the other hand, Borkhsenius' (1958) suggestion that from the Carboniferous to the Permian the Coccina separated into no fewer than 11 subgroups is not based on any facts!

Bekker-Migdisova (in Rodendorf, 1962) assigned the Protopsyllidiidae (including the Permopsyllidae) to the stem-group of the Coccina: this is the only way I can interpret her arrangement of this family, together with *Mesococcus* and the recent Coccina, into a superfamily 'Coccidomorpha'. The venation of the Protopsyllidiidae is even more primitive than that of *Permaphidopsis*[249]. This means that the venation of the recent Coccina can be formally derived from that of the Protopsyllidiidae just as easily as from that of *Permaphidopsis*[250], but both derivations are equally unconvincing. In fact, I cannot see the slightest reason why the Protopsyllidiidae should belong to the stem-group of the Coccina. Their venation differs only slightly from that of the Archescytinidae, and I shall discuss these two families again below (p. 263).

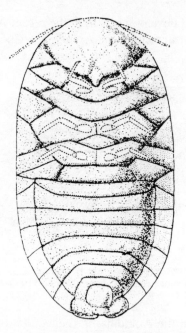

Fig. 81. *Mesococcus asiaticus* Bekker-Migdisova
(Coccina?), from the Upper Triassic of Issyk-Kul'.
From Rodendorf (1962).

Revisionary notes

247. The size of the clavus in the Coccina is not more derived than in the
Aphidina. [Dieter Schlee.]

248. The poor quality of the morphological details visible in a compression
fossil that is only 1–2 mm in length, and the general resemblance of what can
be seen to the young larvae of the Aleyrodina (see Weber, 1930, Fig. 279),
cast considerable doubt on the interpretation of '*Mesococcus*' as a species of
the Coccina. Incidentally, the drawing of '*Mesococcus asiaticus*' also looks
surprisingly similar to illustrations of the termitophilous bug *Termitaphis* (see
Poisson, in Grassé, 1951, Fig. 1496; Jordan, 1972, Fig. 61). Perhaps there are
also young larvae of the Heteroptera with a similar appearance? [Dieter
Schlee.]

249. A new study of *Permaphidopsis* has been published by Szelegiewicz and
Popov (1978). See also notes 239 and 244. [Dieter Schlee.]

250. Szelegiewicz and Popov (1978) assigned *Permaphidopsis* to the Protopsyllidiidae. [Dieter Schlee.]

2.2.2.2..3.2..2.2.2.2. Psyllomorpha[251]

Börner (1934) considered the Psyllina + Aleyrodina (series 'Psyllodea') to be the sister-group of the Aphidina + Coccina. As characters common to these two groups he mentioned the development of the hind-legs into jumping legs and the fact that the eggs 'are inserted into plant tissue by means of a short flagellum'.

Evans' (1963) views are not entirely clear. He published a phylogenetic tree in which he showed two sister-groups, the 'Psylloidea' and the rest of the Sternorrhyncha (his Aleyrodoidea + Aphidoidea + Coccoidea). In his text, however, he mentioned that Bekker-Migdisova (1960b) and Haupt (1935) had both placed the 'Psyllina' and 'Aleurodina' in a single group, and he considered that his own phylogenetic tree, in which the Psylloidea and Aleuroidea were placed next to each other, also conformed with this. I am not sure if he meant that his classification was a typological one, unlike his phylogenetic tree, or if he was saying that his phylogenetic tree did not express the actual relationships.

Schlee (1969a, 1969c) has recently made a careful study of this. He listed a number of synapomorphic characters of the Psyllina and Aleyrodina which clearly indicate that the Psyllomorpha are a monophyletic group: the presence of a sperm pump[252] at the base of the ductus ejaculatorius; the complete reduction of the ring muscles of the ductus ejaculatorius; the stalk-like abdomen, with the two basal segments reduced; the enlarged coxae, which are very close together. This last character is probably connected with the ability to jump, which in the Sternorrhyncha is restricted to the Psyllina and Aleyrodina[253].

These characters are very difficult or impossible to see in fossils. Consequently, it is virtually impossible to recognize any fossils as belonging to the stem-group of the Psyllomorpha[254].

The heterobathmy of characters suggests that there is a sister-group relationship between the Psyllina and Aleyrodina[255].

Revisionary notes

251. The evolution of the Psyllomorpha has been discussed by Bekker-Migdisova (1971). [Willi Hennig.]

252. It is not 'the presence of a sperm pump' that is important but rather its specialized structure, with the large number of specialized details such as an internal gland (see Schlee, 1969a). [Dieter Schlee.]

253. This sentence is not entirely accurate, because some Aphidina are also able to jump. However, the Aphidina use their fore-legs for jumping, whereas the Aleyrodina + Psyllina have special modifications in the structure of the hind coxae and the base of the abdomen. [Dieter Schlee.]

254. These characters *can* be seen in amber fossils. So far, species of the stem-group of the Aleyrodina have been found in Lower Cretaceous amber (Schlee, 1970). It is always possible that the stem-group of the Psyllomorpha will be found in older ambers and that it will be possible to recognize fossils as belonging to this stem-group. [Dieter Schlee.]

255. The evidence for this and a diagram of the synapomorphies has been given by Schlee (1969a). [Dieter Schlee.]

2.2.2.2..3.2..2.2.2.2.1. **Aleyrodina (white flies)**

The Aleyrodina differ from the Psyllina in having mobile hind coxae, which are not fused with the sternum. The wings do not have a coupling apparatus[256].

They also have a number of derived characters. The single anterior ocellus is absent. The female ovipositor is reduced, and there are only two Malpighian tubes[257] (four in the Psyllina). The tracheal system, spiracles, and wing venation are extensively reduced. Only the 2nd and 8th abdominal spiracles have been retained, according to Pesson (in Grassé, 1951). Moreover, there are no transverse connections between the longitudinal tracheal stems, nor is there any connection between the thoracic and abdominal parts of the tracheal system. Metamorphosis is rather complicated ('allometaboly'): the 1st instar larva is flat and active, but after the first moult it becomes

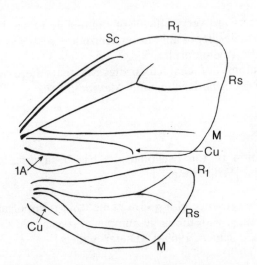

Fig. 82. Wings of *Udamoselis pigmentaria* Enderlein (Aleyrodina; recent). Nomenclature of the veins follows Pesson (in Grassé, 1951)[258].

apodous and stationary. The final instar has also been called a 'puparium' because the adult organs are formed inside it. However, it is not a resting stage, like the pupa of the Holometabola, but still continues to feed. According to White (1957), males of the Aleyrodina are haploid.

In all these characters the Psyllina are more primitive than the Aleyrodina. However, there are only a few characters in which the Aleyrodina are more primitive than the Psyllina (mobility of the hind coxae, absence of the wing-coupling apparatus) and these are seldom visible in fossils. Consequently, fossils with the characters of the Psyllina must really belong to the stem-group of the Psyllomorpha (Psyllina + Aleyrodina). The same is true of the venation. The venation of the Aleyrodina is very reduced (Fig. 82) and can be derived from something very similar to that of the recent Psyllina or only slightly more primitive. However, it is not necessary to make such a derivation.

The only known fossil[259] is *Permaleurodes rotundatus* Bekker-Migdisova, from the Upper Permian of the Kuznetsk basin (Fig. 83). This is 3 mm long, and Bekker-Migdisova (1959) considered it to be the nymphal stage of an aleyrodine. Evans (1963) thought that she could be right, though he was somewhat doubtful, and I share his scepticism.

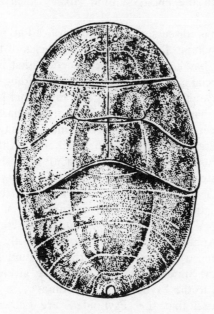

Fig. 83. *Permaleurodes rotundatus* Bekker-Migdisova (nymph, Aleyrodina?), from the Upper Permian of the Kuznetsk basin. From Rodendorf (1962).

Revisionary notes

256. There is a group of tiny bristles which have a sensory function and have nothing to do with mechanical wing coupling (Schlee, 1970). [Dieter Schlee.]

257. This may be the result of convergence, perhaps because of their reduced size: the Coccina also have only two Malpighian tubes (see p. 254). [Dieter Schlee.]

258. In my view, the nomenclature of the wing-veins should be different (Schlee, 1970): RS, M, and Cu should be called M, Cu, and claval fold (CuP), respectively. [Dieter Schlee.]

259. Fossils have been described from Burmese amber and Baltic amber (Tertiary), and from Lower Cretaceous Lebanese amber (Schlee, 1970). [Dieter Schlee.]

2.2.2.2..3.2..2.2.2.2.2. Psyllina (jumping plant-lice)

In many characters the Psyllina are more primitive than the Aleyrodina and other Sternorrhyncha, but they also have some derived characters. The most striking of these are the enlarged hind coxae which are fused with the sternum.

Schlee (1969a) has discussed a number of additional autapomorphic characters of the Psyllina. The dorsoventral muscles of the mesothorax are reduced. The penis has a movable apical section that is connected with a valve. There is a special mesosternal leg joint. According to Schlee, this joint is a new acquisition of the Psyllina and is a functional replacement for the mesotrochantin that has been lost in all Sternorrhyncha.

In the wings of recent Psyllina (Fig. 84) the media has two branches; one of the branches has therefore been reduced. Furthermore, the basal sections of R, M, and Cu are fused for some distance. CuP is the first vein to diverge from the common stem of R + M + Cu, and it is followed by M and CuA (Fig. 84A). M and CuA usually separate only after a further common section (Fig. 84B). This venation cannot be derived from that of *Permaphidopsis*. If we consider the Upper Permian *Permaphidopsis* to be correctly placed in the stem-group of the Aphidina, or possibly more correctly of the Aphidina + Coccina, then species with a more primitive venation must have been in existence at the same time and these could have given rise to the recent Psyllina, or possibly to the Psyllina + Aleyrodina[260].

Evans (1963) stated that the 'Psylloidea' are known from the Upper Permian, but unfortunately he did not mention any by name. Bekker-Migdisova (in Rodendorf, 1962) assigned the 'Aleurodidea', 'Psyllidea' and Cicadopsyllidea (Fig. 85) to her 'Psyllidomorpha', but it is not clear if she regarded the Cicadopsyllidea as the stem-group of the Psyllina or of the Psyllina + Aleyrodina. She assigned only two families to the Cicadopsyllidea,

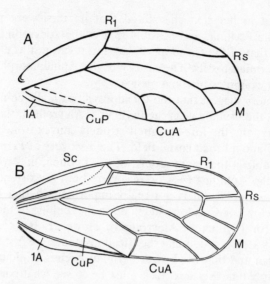

Fig. 84. Fore-wings of recent Psyllina. A, *Trioza urticae* Linnaeus; B, *Panisopelma quadrigibbiceps* Enderlein. From Haupt (1935) (A) and Enderlein (1910b) (B).

the Coleoscytidae and Cicadopsyllidae. These include about 15 species from the Upper and Lower Permian of the Soviet Union.

Almost all of these species have a three-branched media (four-branched only in the Lower Permian *Sojanopsylla*); two anal veins, whenever the clavus is preserved; and CuA not fused or fused for only a short distance with Sc + R or with M. In these characters they are more primitive than the recent Psyllina and Aleyrodina. Bekker-Migdisova (1960b and subsequently) has shown three tarsal segments in her reconstruction of *Scytoneura elliptica*

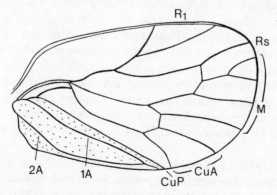

Fig. 85. Fore-wing of *Sojanopsylla brevipennis* Bekker-Migdisova (Cicadopsyllidea), from the Upper Permian of the Arkhangelsk region. From Rodendorf (1962).

Martynov, but in her text she stated that no tarsi were preserved! The venation of the Psyllina, and possibly that of the Aleyrodina, can easily be derived from that of the Cicadopsyllidae, but it can just as easily be derived from other Permian fossils. The history of the Aphidomorpha must provide many striking examples of convergence.

Two characters suggest that the Cicadopsyllidea really could belong to the stem-group of the Psyllina or of the Psyllina + Aleyrodina: the fore-wings are rather leathery; and the fore-margin is strongly convex, which gives the wing an oval shape and a broad costal field. The discovery of *Permaphidopsis* has made it probable[261] that stem-groups of the Psyllina, or Psyllina + Aleyrodina, and of the Aphidina (+ Coccina) were both in existence in the Upper Permian. Consequently, it is not really important whether the Protopsyllidiidae (see below) or Cicadopsyllidea do in fact belong to the stem-group of the Psyllina (or Psyllina + Aleyrodina). However, Bekker-Migdisova (in Rodendorf, 1962) has stated that the Cicadopsyllidea are first known from the Lower Permian and that the Lower Permian Archescytinidae are the ancestors of the Aphidina. My next task must be to see what conclusions should really be drawn from these Lower Permian fossils.

Revisionary notes

260. *Permaphidopsis* has been shown to be the hind-wing of a protopsyllidiid (Szelegiewicz and Popov, 1978). [Dieter Schlee.]

261. See the preceding note: no conclusions regarding the age of the Psyllina can be drawn from *Permaphidopsis*. [Dieter Schlee.]

Appendix: probable Sternorrhyncha from the Lower Permian

Bekker-Migdisova (in Rodendorf, 1962) grouped the Lower Permian Sternorrhyncha into five separate 'families': Archescytinidae, Permaphidopsidae[262], Protopsyllidiidae, Coleoscytidae, and Cicadopsyllidae. Two characters indicate that these groups are more primitive than the recent Sternorrhyncha: two anal veins are present, and CuA is connected to M by a short cross-vein (Fig. 86). More rarely, CuA touches the media or R + M at one point or is fused with it for a short distance only (Fig. 87). I must emphasize that *all* the recent Sternorrhyncha can be derived from these wing characters, and for this reason there have been many different assessments of these groups.

Carpenter (1931b and subsequently) called the Archescytinidae the 'Palaeorrhyncha' and asserted (Carpenter, 1933) that they did not belong to the Sternorrhyncha or the Auchenorrhyncha. Evans (1943) did the same. Haupt (1940) regarded the Archescytinidae as the direct ancestors of the Jassoidea (= Cicadelloidea), in other words as a subgroup of the Auchenorrhyncha. Evans (1948, 1956, 1958) rejected this, but still thought that the

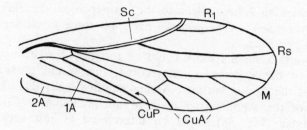

Fig. 86. Fore-wing of *Palaeoscytina brevistigma*
Carpenter (Archescytinidae), from the Lower Permian of Kansas. From Carpenter (1931b).

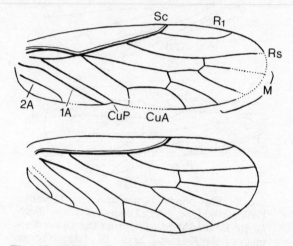

Fig. 87. Wings of *Archescytina permiana* Tillyard
(Archescytinidae), from the Lower Permian of Kansas. From Carpenter (1939)[263].

Archescytinidae and Jassoidea might be closely related and even suggested
(Evans, 1948) that the Archescytinidae were a specialized side-branch of one
of the jassoid families. Martynov (1933) had previously suggested that the
Archescytinidae were closely related to the Sternorrhyncha, and especially to
the 'Aphidoidea', but Carpenter had expressly rejected this. This view has
recently been put forward again by Bekker-Migdisova (1961). However, her
arguments are peculiarly unrealistic and are entirely typological. It now seems
certain that the Upper Permian *Permaphidopsis* (family Permaphidopsidae)
belongs to the stem-group of the Aphidina + Coccina (see above)[264]. The
only problem with the *Lower* Permian Archescytinidae, therefore, is whether
they belong to the stem-group of the plant-lice or Aphidomorpha. Bekker-
Migdisova also assigned the Lower Permian *Tchecardaella* (*Tshekardaella*)[265]
to the Permaphidopsidae[266], and so logically she should have rejected the
derivation of the plant-lice from the Archescytinidae instead of advocating it.
From the available descriptions and illustrations, I cannot find the slightest

264

reason for assigning *Tchecardaella* to the Permaphidopsidae rather than to the Archescytinidae[267]. *Tchecardaella* even has three tarsal segments like *Archescytina permiana*.

If any Lower Permian species could be recognized as belonging to the stem-group of the Psyllina + Aleyrodina, this would provide indirect evidence that the *Lower* Permian Archescytinidae really do belong to the stem-group of the Aphidina or even of the Aphidina + Coccina. The Protopsyllidiidae (Fig. 88) could be interpreted as just such a group.

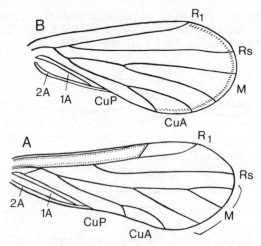

Fig. 88. Fore-wings of Palaeozoic and Mesozoic Protopsyllidiidae. A, *Tomiopsyllidium iljins-kiense* Bekker-Migdisova, from the Upper Permian of the Kuznetsk basin; B, *Asiopsyllidium unicum* Bekker-Migdisova, from the Upper Triassic of Issyk-Kul'. From Rodendorf (1962).

Bekker-Migdisova (in Rodendorf, 1962) considered the Protopsyllidiidae to be 'Coccidomorpha'. However, Evans (1956) and Heslop-Harrison (1951, 1952a, 1957) suggested that they belong to the Psyllina (or Psyllomorpha), and I find this far more convincing. Both groups have the two anal veins fused to form a 'Y-vein'. Moreover, the Protopsyllidiidae have not actually been found earlier than the *Upper* Permian. Bekker-Migdisova's statement that the Australian genera have been described from the 'Lower' Permian is incorrect: the Lower Permian *Permopsylla* is now generally placed in the Archescytinidae, even by Bekker-Migdisova herself.

Apart from the Archescytinidae, the only *Lower* Permian fossils are a few 'Cicadopsyllidea' consisting of the genera *Kaltanoscyta, Sojanopsylla, Scytoneura, Scytoneurella*, and *Cicadopsis*. The main difference between these genera (Fig. 85) and the Archescytinidae is that Sc does not run independently to the wing-tip alongside R + M and that the costal field (Bekker-Migdisova's 'subcostal field') is much broader. However, according to the available figures, these differences are not always very distinct. Evans first assigned *Kaltanoscyta* and *Sojanopsylla* to the Auchenorrhyncha, but later (Evans, 1963) he followed Bekker-Migdisova and placed them in the Sternorrhyncha. However, he said just as little about their relationships as he did about those of the other three cicadopsyllidean genera. He thought that *Scytoneura elliptica* Martynov, which is actually an *Upper* Permian species(!), could belong to the 'Protohomoptera', which is his hypothetical stem-group of all the 'Homoptera'. The published figures show that the fore-wings of the 'Cicadopsyllidae' (*Scytoneura, Scytoneurella*, and *Cicadopsis*) were strikingly similar to those of contemporary species that Bekker-Migdisova assigned to the Auchenorrhyncha. Bekker-Migdisova's (1960b) key even placed the Cicadopsyllidae close to 'families' that are said to belong to the Auchenorrhyncha.

I do not think that these fossils provide any evidence that two or more subgroups of the Sternorrhyncha were in existence during the Lower Permian and are still represented in the modern fauna. The Archescytinidae may be the stem-group of the Sternorrhyncha or may be an invalid stem-group containing species from several subgroups of the Sternorrhyncha, but it is impossible to come to any decision about this. The relationships of the other Lower Permian fossils are still uncertain.

Another problem is whether any Permian species survived into the Mesozoic, apart from the direct ancestors of the recent Sternorrhyncha. Bekker-Migdisova (in Rodendorf, 1962) assigned six species to the Protopsyllidiidae, which she considered to be the stem-group of the Coccina. Five of these are from the Upper Triassic (genera *Cicadellopsis, Cicadopsyllidium, Asiopsyllidium*, and *Triassothea*) and one is from the lower Lias (*Mesaleuropsis*). As discussed above, I think it more likely that the Protopsyllidiidae belong to the stem-group of the Psyllina, or of the Psyllina + Aleyrodina. According to Bekker-Migdisova, the oldest 'true Psyllina' is the Lower Jurassic *Liadopsylla*. I can see no reason why the so-called Mesozoic 'Protopsyllidiidae' (Fig. 88) should not have had the same early Mesozoic ancestors as the recent Psyllina. This idea is supported by the presence of a two-branched media in the Mesozoic 'Protopsyllidiidae', a character shared by the recent Psyllina but not by the Palaeozoic 'Protopsyllidiidae'. It is even possible that the so-called Mesozoic Protopsyllidiidae may still belong to the stem-group of the Psyllina + Aleyrodina, unless *Permaleurodes* really belongs to the Aleyrodina.

Revisionary notes

262. The 'Permaphidopsidae' have been synonymized with the Protopsyllidiidae by Szelegiewicz and Popov (1978). [Dieter Schlee.]

263. A new analysis of the archescytinid wing and the nomenclature of its venation has been given by Szelegiewicz and Popov (1978). [Dieter Schlee.]

264. See note 262. [Dieter Schlee.]

265. *Tchecardaella* is now assigned to the Archescytinidae (Szelegiewicz and Popov, 1978). [Dieter Schlee.]

266. See note 262. [Dieter Schlee.]

267. See note 265. [Dieter Schlee.]

2.2.2.2..3.2..2.2.3. **Auchenorrhyncha**

a. The monophyly and age of the Auchenorrhyncha as a whole

If we are to believe the results of Bekker-Migdisova (in Rodendorf, 1962) and Laurentiaux (in Piveteau, 1953), then some of the oldest insect fossils should be assigned to the Auchenorrhyncha and the group would be known as early as the lower Upper Carboniferous. However, the fact that all of the groups that might be the sister-group of the Auchenorrhyncha are not known before the Lower Permian suggests that the Carboniferous Hemiptera are only 'Auchenorrhyncha' in a typological sense. In fact, almost all the diagnostic characters that are used to distinguish the recent Sternorrhyncha and Auchenorrhyncha are present in the Auchenorrhyncha in a more primitive state. The common ancestors of all the Hemiptera must therefore have had virtually the same characters as the Auchenorrhyncha. Ross (1965) did not consider the Auchenorrhyncha to be a monophyletic group. His phylogenetic tree showed the 'Fulgoridae' (my Fulgoriformes) as the sister-group of the Cicadiformes + Sternorrhyncha.

There may in fact be a few derived characters that could be part of the ground-plan of the Auchenorrhyncha. Börner (1934) stated that the pronotum covers the entire width of the mesonotum, at least along the fore-margin, unlike the Sternorrhyncha. According to Ross (1965) this character is more strongly developed in the Cicadiformes than in the Fulgoriformes. Nevertheless, it appears to be part of the ground-plan of the Auchenorrhyncha.

It is rather doubtful whether stridulatory and auditory organs are part of the ground-plan of the group.

Until recently it was thought that sound production was restricted to the Cicadina. However, Ossianilsson (1949) showed that the ability to stridulate is much more widespread, and he suggested that it is present in all Auchenorrhyncha. In spite of the large number of species that Ossianilsson examined, his work can only be regarded as a preliminary sample. Evans (1963) attributed timbals to *both* sexes of the hypothetical stem-group of the Auchenorrhyncha. There has been no detailed study of the ground-plan of the timbals[268]. Sound is produced by a sort of 'tin-drum', the surface of which clicks in and out. The timbals themselves are fluted, and are formed from the sides of the 1st abdominal segment. They may be free (primitive species?) or covered by a reflector that arises from the 2nd tergite (derived species?). The entire area is protected by an operculum that arises from the epimeron of the 3rd thoracic segment. The resonator is an air-sac of unknown origin that is situated above the timbal muscles.

It is still not clear whether the ability to jump is a derived character of the Auchenorrhyncha. If we accept that this is so, then it must have arisen convergently in the Psyllomorpha (Sternorrhyncha), as a parallel development. An alternative suggestion would be that the ability to jump was in the ground-plan of at least the Homoptera. This would mean that it had been lost secondarily in the plant-lice and scale-insects. Consequently, its presence in the Psyllina and Aleyrodina could not be used as evidence that these two groups form a single monophyletic group of higher rank, as I suggested above (p. 257). It is essential that a careful comparative study should be made of the ability to jump in the Hemiptera.

Martynov (1933) suggested that the anojugal area of the wing has been secondarily enlarged in the Auchenorrhyncha, and Evans (1963) agreed with this. However, it is something that needs to be more carefully worked out before it can be used as evidence that the group is monophyletic.

The structure of the antennae in the ground-plan of the Auchenorrhyncha is still not clear. Handlirsch (1920–21) wrote: 'antennae with 2–3 larger basal segments and a segmented flagellum'. Pesson (in Grassé, 1951) referred to 1–3 enlarged, spherical basal segments. As a long antennal flagellum consisting of homonomous segments appears to be part of the ground-plan of the Hemiptera, the reduction of at least the fourth and following segments to a setiform flagellum must be one of the derived characters of the Auchenorrhyncha. Evans (1956) pointed out that many Cicadellidae have long antennae and that in some cases these appear to be multi-segmented: examples are *Ciccus latreillei* Distant, *Neocoelidia fuscodorsata* Walker, and *Oecinirrana arborea* Evans. However, the context of Evans' remarks shows that he was considering this character from a diagnostic point-of-view. The species that he mentioned occupy a very subordinate position in the Auchenorrhyncha and it seems most likely that their antennae have secondarily acquired a form that superficially resembles the primitive state.

Unfortunately, none of these characters is visible in fossils. Wings, and usually only fore-wings, are all that are preserved in Palaeozoic fossils and in most of the Mesozoic ones.

No definite derived characters are known in the venation of the Auchenorrhyncha, but this is of no importance as far as the Permian fossils are concerned. This is because the Lower Permian Archescytinidae can be definitely assigned to the Sternorrhyncha, and consequently species with the diagnostic wing characters of the recent Auchenorrhyncha can be positively assigned to this group. The only problem is that the original state of the venation of the Coleorrhyncha and Heteroptera is not known. This means that Lower Permian fossils with the wing characters of the 'Auchenorrhyncha'

268

could actually belong to the stem-group of the Heteroptera, of the Coleorrhyncha, or even of the Heteropteroidea (Coleorrhyncha + Heteroptera).

Only three alleged Hemiptera have been described from the Carboniferous. *Archeglyphis crassinervis* Martynov, from the lower Balakhonka Stage of the Kuznetsk basin, has even been assigned to a subgroup (Cicadidea) of the Auchenorrhyncha by Bekker-Migdisova (in Rodendorf, 1962). This is wrong. According to the published illustrations, *Archeglyphis* is only known from a few veins of the fore-wing but it is clear that the radial sector has two branches. However, in the ground-plan of the *Hemiptera, unlike that of the Psocodea, the radial sector is unforked! This means that *Archeglyphis* could equally well belong to the stem-group of the Psocodea or of the Hemiptera, but not to a subgroup of the *Hemiptera. Alternatively, it could be assessed differently. It shows all the ambiguities inherent in fossils with primitive characters: these have been enormously magnified in this particular case because only a few characters have actually been preserved. Evans (1963) was right to state that *Archeglyphis* could belong to the Archescytinidae but might not be a homopteron at all.

Bekker-Migdisova assigned *Blattoprosbole tomiensis* Bekker-Migdisova, also from the Balakhonka Stage of the Kuznetsk basin, and *Protoprosbole straeleni* Laurentiaux (Fig. 89), from the Namurian of Belgium, to a separate

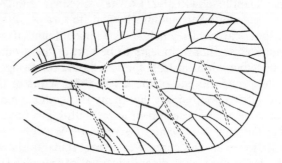

Fig. 89. Fore-wing of *Protoprosbole straeleni* Laurentiaux, from the Upper Carboniferous (Namurian) of Belgium. From Laurentiaux (1952).

'infraordo Blattoprosbolomorpha' in the Auchenorrhyncha. In these two species all of the veins are independent as far as the wing base. In this respect they are more primitive than all the later Hemiptera. There is not a single character that would support their assignment to a subgroup of the Hemiptera, either to the Auchenorrhyncha or one of the other groups. Furthermore, I am very doubtful whether they even belong to the Paraneoptera, because of the completely free subcosta, the presence of an archedictyon, and the convergence of all the main vein stems towards the centre of the wing-base (when this can be seen). Bekker-Migdisova herself alluded to a resemblance between these wings and those of the 'cockroaches' in her text and in the generic name she chose. Evans (1963) was right to point out that the ordinal position of all three Carboniferous species was uncertain. Sharov (1966b) thought that *Blattoprosbole* belonged to the Blattodea, and he was probably correct. He also suggested that *Protoprosbole* showed the first stage of development in the direction of the Paraneoptera. It must be obvious by now that there is no conclusive evidence that the Hemiptera existed during the Carboniferous. Bekker-Migdisova (quoted by Martynova, 1961a) may be correct in suggesting that *Protoprosbole* and *Blattoprosbole* descended from the Archimylacrididae. However, it is wrong to conclude that they link the Archimylacrididae with the Homoptera. Carboniferous Paurometabola can be easily confused with Hemiptera, as is shown by Heslop-Harrison's (1956) assignment of *Dictyocicada antiqua* Brongniart to the Hemiptera.

On the other hand, there are numerous Lower and Upper Permian fossils which, with the reservations discussed above, can be assigned to the Auchenorrhyncha. However, it is much more difficult to determine which subgroups were in existence during the Permian. Before dealing with this problem, I shall discuss the classification of the Auchenorrhyncha.

Additional note

268. According to Weidner's revision of Weber (1974), the timbals are on the sides of the 1st abdominal segment in the Cicadiformes, whereas in the Fulgoriformes they are on the dorsum of the abdomen in the area of the 1st and 2nd segments. [Willi Hennig.]

b. The monophyletic subgroups of the Auchenorrhyncha and their age

It now seems certain that there is a sister-group relationship between the Fulgoriformes and the Cicadiformes. Each group has a number of characteristic derived features that are not present in the other.

2.2.2.2..3.2..2.2.3.1. **Fulgoriformes**
2.2.2.2..3.2..2.2.3.2. **Cicadiformes**

In the Fulgoriformes the hind coxae are immobile. The meron is fused with the metathoracic epimeron. It is curious that Evans (1963) did not mention this character although it was also pointed out by Heslop-Harrison (1952a). The 2nd antennal segment (pedicel) is greatly enlarged and is provided with some striking sensory organs. Antennae and ocelli are situated below the eyes. I would like to regard this as a derived character, but Evans (1963, 1964) suggested that in the Homoptera the ocelli were originally situated below the eyes. 1A and 2A of the fore-wings always unite at their tips to form a Y-vein (Fig. 90). Müller (1962) found that secondary x-symbionts are always present, in addition to the main a-symbionts, whereas this is not the case in the Cicadiformes.

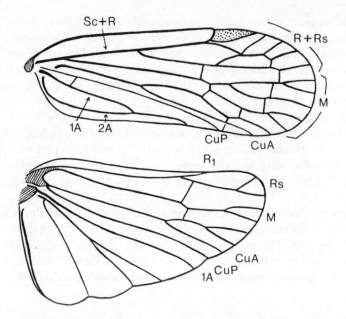

Fig. 90. Wings of *Cixius nervosus* Linnaeus (Auchenorrhyncha, Fulgoriformes; recent). (Original. Willi Hennig.)

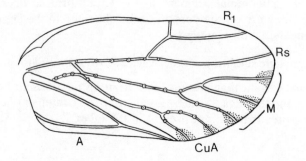

Fig. 91. Fore-wing of *Mundus nodosus* Bekker-Migdisova (Auchenorrhyncha, Fulgoriformes), from the Upper Permian of the Arkhangelsk region. From Rodendorf (1962).

In the Cicadiformes the tegula is absent. The hind-wing has a marginal vein (Figs. 92 and 93). According to Pesson (in Grassé, 1951), the claws are completely fused with the pulvilli. The alimentary canal has a filter chamber, into which the Malpighian tubes have also been absorbed. Evans (1963) also stated that the Cicadiformes have a special head structure ('type C'). However, it is not clear how far these characters have developed from the head of 'type B', which Evans ascribed to the Fulgoriformes (his 'Fulgoroidea'), or how far they have developed directly from the ground-plan head ('type A') of the Auchenorrhyncha.

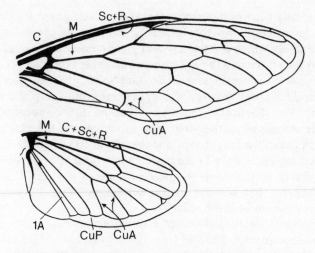

Fig. 92. Wings of *Cicada* sp. (Auchenorrhyncha, Cicadiformes; recent). From Pesson (in Grassé, 1951).

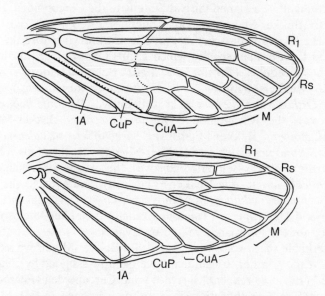

Fig. 93. Wings of *Tettiga, cta crinita* Distant (Auchenorrhyncha, Cicadiformes; recent). From Evans (1956).

The next problem to investigate is the age of the Fulgoriformes and
Cicadiformes. Almost all the Palaeozoic fossils consist of wings, usually only
fore-wings, and this poses almost insuperable difficulties because there are
virtually no ground-plan differences in the venation of these two groups.
Evans' (1963, 1964) papers are extremely valuable and provide an indispensable basis for any study of the fossils. He attempted to work out the
characteristic features of the venation in the various subgroups of the

'Homoptera' and then to determine the age of each group from the fossils. However, this attempt was too formal, at least in its presentation. Evans listed the features which are usually and characteristically present in recent species, and then discussed the differences and exceptions that occur amongst the recent and fossil species. His account gave no clear picture of the ground-plan differences between the individual groups, nor did he indicate whether the few diagnostic characters that he found were primitive or derived. He discussed neither the direction in which the characters have developed in the individual groups nor the extent to which parallel developments and convergence have taken place. It may well be impossible to obtain any clear picture of this from the present state of the classification of the Auchenorrhyncha, but without such a systematic basis it is impossible to make any assessment of the fossils.

Some of Evans' interpretations of Palaeozoic fossils, based on the wing venation, are completely different from those of Bekker-Migdisova (see also Evans, 1957). For example, he gave a list of selected Permian species of the Cercopoidea, a subgroup of the Cicadiformes, and presumably he thought that these could all be assigned fairly definitely to the Cercopoidea. However, according to Bekker-Migdisova (in Rodendorf, 1962) not one of them belongs to the 'Cercopidea', as this group is not known before the Triassic. She assigned the Permian species which Evans (1964) placed in the Cercopoidea to the 'Cicadidea' (*Permocicada nigrita* Bekker-Migdisova, *Kaltanopsis ornata* Bekker-Migdisova, *Prosboloneura kondomensis* Bekker-Migdisova, *Orthoscytina skokei* Bekker-Migdisova, *Prosbole kaltanica* Bekker-Migdisova), the 'Cicadellidea' (*Ingruo lanceolata* Bekker-Migdisova, *Kaltanospes elongata* Bekker-Migdisova, *Sarbaloptera sarbalensis* Bekker-Migdisova, *Scytinoptera picturata* Bekker-Migdisova), the 'Fulgoromorpha' (*Surijokocixius tomiensis* Bekker-Migdisova), the 'Auchenorrhyncha incertae sedis' (*Tychticola longipenna* Bekker-Migdisova), and even to the Sternorrhyncha (*Scytoneura elliptica* Martynov, *Cicadopsis rugosipenna* Bekker-Migdisova) and Heteroptera (*Tingiopsis reticulata* Bekker-Migdisova: from the Triassic, not the Permian as stated by Evans)!

It is remarkable that two of the species that Bekker-Migdisova assigned to the genus *Prosbole* in her 'Cicadidea' have been re-assigned by Evans to his 'Cicadoidea' (*Prosbole reducta* Martynov) and Cercopoidea (*Prosbole kaltanica*). Even before the publication of Evans' major review (1963), Martynova (1961a) had decided to follow Bekker-Migdisova's views: she considered that Evans' classification was not adequately founded and needed extensive revision. However, statements like this contribute little to progress. Bekker-Migdisova's own chapter in the Russian Textbook of Palaeontology (Rodendorf, 1962) had the obvious defect that the diagnostic characters given for individual groups were often drawn from recent species and are not visible in fossils. It is often not clear what characters she used to assign particular fossils to their groups. Her statements about differences in the venation have not led to any clearer understanding than did Evans'.

These differences of opinion do not inspire me with much confidence in the decisions of specialists who have assigned the Permian and other fossils to various subgroups of the *Auchenorrhyncha. The large numbers of fossil hemelytra often show considerable differences amongst themselves, and this has clearly encouraged specialists to suppose that they could recognize species from the stem-groups of very subordinate groups of the *Auchenorrhyncha. But so far it has proved impossible to say what groups these are.

We ought to be able to recognize the Fulgoriformes, because they have a clear derived ground-plan character in the fusion of the two anal veins to form a Y-vein (Fig. 90). However, this character frequently occurs in other Hemiptera. For example, in somewhat modified form it is characteristic of all the Coleorrhyncha and Heteroptera. Nevertheless, insofar as the wing is completely preserved, a Y-vein is visible in all the Upper Permian species that Bekker-Migdisova assigned to the Fulgoriformes (her 'Fulgoromorpha'): *Mundus* (Fig. 91), *Permopibrocha, Scytophara*, and probably *Scytocixius*. The Coleorrhyncha and Heteroptera can definitely be ruled out as close relatives of these fossils, and so it seems very probable that they belong to the Fulgoriformes. It is curious that Evans (1963, 1964) did not mention any of these genera in his discussion of the Fulgoriformes, although he did include *Permocixiella* in which the clavus has not been preserved. Bekker-Migdisova assigned this genus to the recent family Cixiidae, but this is unjustified. She also assigned the Upper Permian *Surijokocixius* to the Cixiidae, but according to Evans this actually belongs to the Cercopoidea!

Bekker-Migdisova assigned two Lower Permian genera to the Fulgoriformes, *Neuropibrachus* and *Kaltanopibrocha*. Unfortunately, in these fossils the basal part of the wing with the clavus is missing. The distal part of the wing has a highly reticulate venation, which is characteristic of very much more recent Fulgoriformes, but the significance of this character is not at all clear.

It is remarkable that none of the Permian fossils has been assigned to the stem-group of the Cicadiformes: they are all said to belong to subgroups of the Cicadiformes. This is not really so surprising, because no derived characters are known in the fore-wing venation of the Cicadiformes when compared with the ground-plan of the Auchenorrhyncha. It is impossible to see whether the tegula is present or absent in fossils. Nor have any Permian hind-wings been found with the marginal vein that is so characteristic of the Cicadiformes.

Before I can suggest my own conclusions about these fossils, I first need to investigate the monophyletic groups of the *Cicadiformes and their presumed relationships. Opinions about these are still very varied, but it seems clear that three monophyletic groups can be distinguished. Evans (1963) called them the Cicadoidea, Cicadelloidea, and Cercopoidea.

The Cicadoidea (= Cicadomorpha) occupy a special position ecologically. After hatching, the larvae burrow into the ground and their fore-legs are specially modified for digging. A marginal membrane is present in the fore-wing (Figs. 92 and 93). Evans (1963) listed two other characters in his phylogenetic diagram as characteristic of the Cicadoidea: reduction of the timbals in the female, and the presence of abdominal tympana in both sexes. However, in his text, he wrote that they are not present in *Tettigarcta*. They cannot therefore be part of the ground-plan of the Cicadoidea!

Bekker-Migdisova's 'Cicadomorpha' (in Rodendorf, 1962) is the same as Evans' Cicadoidea so far as recent species are concerned. She assigned no fewer than 85 Lower and Upper Permian species here, all of which she placed in the family Prosbolidae. She also included *Archeglyphis crassinervis* Martynov, the only Carboniferous species, but I have already discussed the doubtful position of this species. Virtually all the Permian species are known only from fore-wings, and in none of them is the marginal membrane evident, which is the characteristic derived feature of Mesozoic and recent Cicadoidea! A few hind-wings have been discovered: *Evanscicada rectimarginata* Bekker-Migdisova and *Pervestiga veteris* Bekker-Migdisova, from the Lower Permian of the Kuznetsk basin; and *Sojanoneura proxima* Martynov, *Orthoprosbole congesta* Martynov, and *Prosbole excisa* Martynov, from the Upper Permian of the Soviet Union. However, there is no mention in any of these descriptions of the marginal vein, which is the characteristic derived feature of the Cicadiformes! If the hind-wing really had no marginal vein, then this sheds an interesting light on *Orthoprosbole* and *Evanscicada* because both species have a conspicuous nodal line in the fore-wing: Evans (1963, 1964) believed that all fossils with a nodal line belong to the Cicadoidea. He also stated (1963) that the nodal line (Fig. 94) is a line of weakness that occurs in

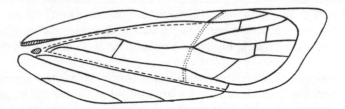

Fig. 94. Fore-wing of *Balata fulviventris* Walker (Auchenorrhyncha, Cicadiformes, family Hylicidae; recent), a homopteron with a nodal line. From Evans (1950).

many Hemiptera and is important for the mechanics of flight. He thought it was probably also present in certain 'Protohemiptera', which is evidently his name for the hypothetical stem-group of the Hemiptera. However, in the Cicadoidea the veins are broken and on the proximal side of this break (nodal line) they are thickened (Fig. 93). Hinton (1948) did not think that the nodal line had anything to do with flight but was connected with the subterranean life of the larvae. He thought that the larvae would be able to move backwards through the soil if the wing rudiments could flex in the middle. The nodal line would then have no function in the adults but would have been retained simply because at first the adults are contained within the last larval skin. This can hardly be a valid explanation of the origin of the nodal line in general because, as Evans pointed out, it also occurs in many Hemiptera that do not have subterranean larvae. It could, however, be the correct explanation for the special form that it has taken in the Cicadoidea and which, according to Evans, is characteristic of the whole group. It would be extremely valuable to establish that this morphological character is functionally correlated with a specific derived character of the Cicadoidea: the subterranean life of the larvae. It would mean, for example, that fossils with a nodal line present in the form that is characteristic of recent Cicadoidea could actually be assigned to this group, although with great caution.

Evans stated that a nodal line formed exactly as in the recent Cicadoidea was also present in the Permian family Prosbolidae. However, I cannot see this character in any of the published figures, nor is there any trace of it in Evans' own figure of *Prosbole reducta* Martynov (Evans, 1963, 1964). Moreover, he included a number of species of the 'genus *Prosbole*' in the Cercopoidea! In the light of this, it is by no means certain that the 'Cicadoidea' were in fact in existence during the Palaeozoic. Bekker-Migdisova's Prosbolidae, which contains almost 100 species, is probably an invalid stem-group.

Indirect evidence for the existence of the Cicadoidea in the Palaeozoic would be available if we could prove that the groups that might be their sister-group were also in existence in the Palaeozoic. This sister-group could be either the Cicadelloidea (= Jassidoidea) or the Cercopoidea. Unfortunately, it has not been established whether the Cicadelloidea are more closely related to the Cicadoidea or to the Cercopoidea. Even the symbionts fail to give any clue: Müller (1962) has shown that each of the three groups has its own set of secondary symbionts in addition to the main symbionts common to all Auchenorrhyncha.

The Cicadelloidea are a well founded monophyletic group. In his phylogenetic diagram, Evans (1963) listed the lack of any connection between anterior and posterior arms of the tentorium and the loss of the anterior ocellus as derived characters, and he considered the Jassoidea (Cicadelloidea) to be the sister-group of the Cicadoidea. He did not list any derived characters common to the two groups. The lack of any connection between the anterior and posterior arms of the tentorium is a peculiar character that is only found in the Cicadelloidea and is therefore an autapomorphy of this group. Consequently, it cannot be included in any discussion of the relationships of this group. The anterior ocellus is absent in the Cercopoidea as well as the Cicadelloidea and, unlike Evans, both Heslop-Harrison and Bekker-Migdisova placed these two groups together as the Jassidomorpha or Cicadellomorpha. However, the absence of the anterior ocellus is the only derived character used as evidence that the group is monophyletic, and a feature involving such simple atrophy could easily have arisen convergently.

No derived characters are known for the Cercopoidea, and so there is no justification for treating them as a monophyletic group. All the characters of the Cicadelloidea are derived when compared with the Cercopoidea. The absence of the connection between the anterior and posterior arms of the tentorium, the lateral dilation of the hind coxae, the absence of hind coxal spurs, and the absence of a free basal section of the subcosta are all derived characters.

There are very substantial differences of opinion over the assignment of Palaeozoic (Permian) fossils to the Cicadelloidea and Cercopoidea. Bekker-Migdisova recognized about 40 Lower and Upper Permian species as members of the 'Cicadellidea', and assigned them all to the family Scytinopteridae. She stated that no Cercopoidea were known before the Triassic. Evans (1963) also asserted that there were very many Palaeozoic and Mesozoic fossils that were probably correctly assigned to the Cicadelloidea yet, in a paper that was published almost simultaneously (Evans, 1964), he failed to mention any Palaeozoic Cicadelloidea. One of his figures shows the wing of *Homaloscytina plana* Tillyard, from the Upper Permian of Australia, together with other Triassic and recent Cicadelloidea, although he stressed that this wing was almost identical with that of *Prosbole reducta* Martynov, a species he assigned to the Cicadoidea, and that the main difference was the lack of a nodal line in *Homaloscytina*. At the same time he mentioned no less than 14 Lower and Upper Permian species as members of the Cercopoidea, a group that did not appear before the Triassic according to Bekker-Migdisova. Bekker-Migdisova herself assigned these 14 species to the 'Jassidea' (*Sarbaloptera sarbalensis* Bekker-Migdisova, *Scytinoptera picturata* Bekker-Migdisova, *Kaltanospes elongata* Bekker-Migdisova, *Ingruo lanceolata* Bekker-Migdisova), the Cicadidea (*Permocicada nigrita* Bekker-Migdisova, *Kaltanopsis ornata* Bekker-Migdisova, *Prosboloneura kondomensis* Bekker-Migdisova, *Orthoscytina skokei* Bekker-Migdisova, *Prosbole kaltanica* Bekker-Migdisova), the Fulgoromorpha (*Surijokocixius tomiensis* Bekker-Migdisova), the 'Auchenorrhyncha incertae sedis' (*Tychticola longipenna* Bekker-Migdisova), and even the Sternorrhyncha (*Scytoneura elliptica* Martynov, *Cicadopsis rugosipenna* Bekker-Migdisova) and Heteroptera (*Tingiopsis reticulata* Bekker-Migdisova: actually a Triassic species).

When two specialists differ so profoundly in their interpretation of the Permian fossils, the non-specialist can only conclude that a lot more work needs to be done to establish some basis for determining their systematic position.

In fact, neither the ground-plan nor the lines of development of the wing-venation have yet been worked out for the various groups, and yet the assessment of fossils is entirely dependent on this character. I should also point out that there is just as little sign of a marginal vein in the few poorly preserved hind-wings that Bekker-Migdisova has assigned to the Cicadellidea (= Cicadelloidea) (*Ivaia* sp., *Anaprosbole ivensis* Bekker-Migdisova and *Ingruo inopinata* Bekker-Migdisova) as there is in the supposed Permian Cicadoidea (see above). This vein appears to be in the ground-plan of the

Cicadiformes. It is possible that it has arisen independently and by convergence in the Cicadoidea, Cicadelloidea, and Cercopoidea, but no evidence has been found for this. I would say that on balance it is possible, even probable, that the subgroups of the Cicadiformes did actually originate during the Permian. However, there is no definite evidence for this. Bekker-Migdisova's Scytinopteridae, like the Prosbolidae, are probably an invalid stem-group.

2.2.2.2..4. **Holometabola**

a. Ground-plan and origin. The oldest fossils

There are no longer any reasonable grounds for doubting that the Holometabola are a monophyletic group. They include about 500 000 species, which is well over half the total of recent insects (88%, according to Carpenter, 1953, 1954a). If we measure the evolutionary success of a group by the number of its species, then the Holometabola are by far the most successful of the 8–9 groups of insects that are of the same age or even older. The first obvious line of enquiry must be to search for the reasons for this success.

The only derived characters in which the Holometabola differ from their sister-group, whichever group that may be, are the endopterygoty, holometaboly, and the presence of a 3rd, sternal, articulating joint in the coxa[269]. Furthermore, Sharov (1966b) stated that the anal furrow has been almost completely levelled out. The Holometabola have been called endopterygotes because the wing-buds develop as 'imaginal discs beneath the larval skin and are first extruded between the discarded larval cuticle and the epidermis at the end of the larval stage' (von Kéler, 1955, 1956). Externally the wing-buds are not visible until the pupal stage which is intercalated between the larval and imaginal stages (holometaboly) and is a special feature of the Holometabola. A great deal has been written about the origin and significance of the pupal stage. Weber (1949) emphasized the contrast between two different hormones: one stimulates larval growth and ecdysis, whilst the other triggers the final transformation into the pupa or adult. He considered that the most striking feature of holometaboly was that 'release of the "juvenile hormone" by the corpora allata was continued until the last two immature stages'. He concluded from this that 'holometaboly arose in the course of evolution as a result of a change in the activity of the corpora allata: this was genotypically conditioned, or might have been the result of a change in a complex gene'.

This explanation only deals with 'holometaboly' as an ontogenetic phenomenon and covers the last stage in a progressive series of causalities. It offers no explanation for the origin of holometaboly in the course of phylogeny or for the overwhelming success of the Holometabola. Handlirsch (1920–21) wrote: 'The profound climatic deteriorations of the Permian ice-age provided the insects with the first adverse conditions that they had experienced. The cold led to delays in the formation of wings in the larvae, and gave rise to the heterophyletic occurrence of complete metamorphosis'. Handlirsch's theory that the Holometabola arose polyphyletically ('heterophyletically') is no longer accepted. Nonetheless, it is interesting that he did not consider the acquisition of the pupal stage to be the first decisive step in the origin of the Holometabola but rather the 'delays in the formation of wings' during postembryonic development. In other words, he thought that the Holometabola were in the first place 'endopterygote'. Handlirsch expressed himself in Lamarckian terms, but even so it is not certain that delays in the formation of wings were initiated by the Permian ice-age. Whether or not his explanation can be accepted depends on the age of the individual subgroups of the *Holometabola. All recent Holometabola are endopterygote and have a pupal stage, and until there is proof to the contrary we must accept that these two features were present in the latest common ancestors of all the recent Holometabola. These must have lived during the Upper Carboniferous at the latest (see below). Unfortunately, the exact dating of the 'Permo-Carboniferous ice-ages is not definitely known' (Schwarzbach, 1950). However, it is interesting that they were restricted to the southern continents (Gondwanaland). If we were to follow Handlirsch in believing that the Permian ice-age was responsible for the 'delays in the formation of wings', we would also have to follow Jeannel (1950), who suggested that the Holometabola originated in the southern continents. There is no firm evidence for this[270].

Schwarzbach (1950) wrote that arid conditions were the most characteristic feature of the Permian. In such conditions the absence of external wing-buds could also have been a selective advantage. Hinton (1948) considered that this advantage lay in the ability of larvae without wing-buds to tunnel through firm substrates, such as soil and plant tissues. He stated that larvae with external wing-buds were unable to move backwards in firm substrates and he explained how exopterygote terrestrial species such as cicadid larvae have been able to get round this by using different means from the Holometabola. If Hinton is correct, it would be possible to explain the origin of the Holometabola as follows: reduction of the external wing-buds in the larvae enabled them to withdraw into damp media, such as soil and plant tissues, and thereby to avoid the arid conditions of the Permian (and even the late Upper Carboniferous). Lameere, as reported by Chen (1946), had previously 'attributed' the origin of holometaboly to the fact that larvae were able to bore into plant tissues. It is remarkable that thysanuroid or campodeiform larvae are found in the most primitive groups of the Holometabola, and various authors (e.g. Gilyarov, 1957; Martynova, 1957) have been right to regard these as the most primitive larval forms of the Holometabola. However, such thysanuroid or campodeiform larvae would hardly be able to burrow in firm soil or plant tissues as envisaged by Hinton or Lameere. For this reason, the delay in the appearance of external wing-buds can hardly be regarded as an adaptation to burrowing in firm substrates. It is a feature that must have arisen for other reasons, and as a 'preadaptation' it facilitated the origin of burrowing species. In this way it contributed to the subsequent evolutionary radiation of the Holometabola, but not to their origin.

Chen's (1946) suggestion that the larvae of the Holometabola originally lived in water must be rejected, unless we restrict 'in water' to mean 'in a moist environment'. Tillyard (1926e) was more likely to be correct in suggesting that they originally lived in vegetable debris and similar substances, as do the modern Choristidae (Mecoptera). It is the delay in the appearance of the wing-buds that shows most clearly the fundamental characteristic of the Holometabola, which is the divergence in form and habits of the larval and imaginal stages. The immature stages of the Paurometabola and Paraneoptera are very similar to the adults in their mode of life, and so they resemble the adults closely in their appearance ('imaginalization'). On the other hand, the larvae of the Holometabola have adopted completely different modes of life from the imagines, and as a result they differ considerably in appearance.

The origin of the pupal stage in the Holometabola appears to be connected with this. Hinton (1948, 1963) dealt with this problem in some detail and considered that the pupal stage arose during postembryonic development when the anatomical differences between larvae and imagines became so great that certain muscles used by the adults could no longer be formed in the larval body. He thought that its importance lay originally in providing a stage when certain muscles, probably some of those necessary for flight, could be formed.

In discussions on the origin of the pupal stage (e.g. Heslop-Harrison, 1958 and subsequently), considerable stress has been laid on the problem of whether the pupa of the Holometabola should be regarded as an imaginal stage (Poyarkov", 1914a, 1914b; Hinton in his early papers), or whether it corresponds to one of the nymphal stages of the 'Exopterygota' (Pérez, 1910; Handlirsch, 1920–21; Hinton, 1963; Sharov, 1966b) or even to all of them (Heslop-Harrison, 1958, 1961, following Poulton and Berlese).

In his early papers, Hinton followed Poyarkov"'s view that the pupa was an imaginal stage and should be compared with the 'subimago' of the Ephemeroptera, as Boas first suggested. This was also DuPorte's (1958a) view. Subsequently, Hinton (1963) changed his mind and suggested that the pupal stage of the Holometabola corresponded to the last larval stage of the 'Hemimetabola'. In this case the pupa of the Holometabola could only be compared with the subimago of the Ephemeroptera if this too corresponded to the last larval stage of the 'Hemimetabola'. However, no progress towards a solution of this problem has been made in any of the papers dealing with the interplay of hormones during postembryonic development (see Novák, 1955). No study of the metamorphosis of the Ephemeroptera has been made along these lines. Sharov (1966b) also followed Hinton's later views, which are said to have been put forward originally by Pérez and Handlirsch. He did not refer to his earlier statement, which would have been so important in this context, that subimaginal ecdysis had been found in a fossil species of the 'Paraplecoptera' (Neoptera!).

I shall now discuss briefly the interesting implications of the effect that these different views on the nature of the pupal stage have on evolutionary–biological explanations of holometaboly.

If the pupa is an *imaginal* stage and an original feature of the Pterygota, then we could follow DuPorte (1958a) who suggested that the pupa of the Holometabola did not *originate* because the larvae and imagines diverged in their morphology and mode of life, but rather that this divergent development was only possible because the Holometabola *retained* the subimaginal stage. In this way they were able to utilize the potential for an astonishingly successful radiation that was inherent in the division of their postembryonic development into two separate phases, each one able to take advantage of the most diverse ecological zones. On the other hand, the Paurometabola, Paraneoptera, and Plecoptera gave up the subimago and so lost this potential. The next step would then be to investigate why the 'hemimetabolous' groups gave up the subimago and presumptive pupa.

On the other hand, if the pupa developed from the last *larval* instar, it is no longer necessary to attribute at least part of the sucess of the Holometabola to their retention of a stage that was given up, either monophyletically or convergently, by other groups early on in their development.

Snodgrass (1961) pointed out that in primitive Holometabola the differences between larval and imaginal morphology are hardly as great as they are in groups like the Odonata. These differences increased in the course of phylogeny, and consequently one of the stages close to the division of the two phases in postembryonic development, either the last larval stage or a first imaginal stage (subimago), must already have become a resting stage before these two phases diverged. This resting stage became totally inactive, and this made possible the origin of the Holometabola. Snodgrass considered that this process was really 'retromorphosis', an ontogenetic reversion from the derived condition of the larva to the primitive structure of the adult.

Hinton (1949) wrote that in the ground-plan of the Holometabola the pupa is a 'pupa dectica', with strongly sclerotized mandibles that are used to break through the pupal case or cocoon. This is compatible with the view that at first the pupa of the Holometabola was a resting stage rather than a stage in postembryonic growth when morphological changes could take place. According to Hinton, 'pupae adecticae' without mobile mandibles are only found in derived groups[271]. It would be extraordinarily difficult to determine the sequence in which the characteristic derived features of the Holometabola originated.

The loss of the external wing-buds in the larvae and the origin of a resting stage (either final larval instar or subimago) are both said to be adaptations to circumvent adverse environmental conditions. These could have been the cold of the Permo-Carboniferous ice-age (Handlirsch), or extremes of drought and heat (Tillyard, 1926e). We have no idea of the sequence in which these and possibly other characters arose.

Carpenter (1947) reported on holometabolan larvae of unknown ordinal position, from the Lower Permian of Kansas and from the Permian of the Soviet Union (*Permosialis*: see below). These are of no practical importance because species that undoubtedly belong to subgroups of the *Holometabola, such as the beetles, have been described from the same epoch. The most important questions are *how many* of these subgroups can actually be recorded from the Permian and whether some of them or at least the stem-group of the Holometabola were already in existence during the Upper Carboniferous. These questions are extremely difficult to answer, because in their ground-plan the *Holometabola have an extraordinarily primitive wing venation. As I have just shown, none of the constitutive characters of the group is found in the wing venation and so is hardly likely to be visible in fossils.

Martynov noted that the presence of a single simple longitudinal vein in the neala[272] was a derived character of the Holometabola. However, this is a very trivial character and would also be very difficult to see in fossils[273].

For this reason, it is extraordinarily difficult to decide whether any of the Upper Carboniferous fossils do actually belong to the Holometabola. Previous authors have tried to 'derive' the Holometabola from Upper Carboniferous Palaeodictyoptera, Megasecoptera, Protoblattoidea, Protorthoptera, or even Paraplecoptera, but these attempts are worthless. They only show that it is possible to 'derive' particular characters of the Holometabola from particular characters of these fossil groups. The Upper Carboniferous *Fatjanoptera* (see p. 292: Raphidioptera) and *Metropator* (see p. 336: Mecoptera) have even been assigned to subgroups of the *Holometabola, but such assignments are not well founded.

Sharov (1966a) recently suggested that the Glosselytrodea and Caloneurodea belonged to the Holometabola ('Oligoneoptera'). Martynov (1938b) had previously considered that both groups were closely related to the Orthopteroidea (Saltatoria). The phylogenetic tree in the Russian Textbook of Palaeontology (Rodendorf, 1962, Fig. 27) followed Martynov, but in the text the Glosselytrodea were assigned to the Orthopteroidea and the Caloneurodea to the Plecopteroidea. Unfortunately, Sharov's methods for establishing that both groups belong to the Holometabola are rather dubious from the point of view of phylogenetic systematics. He did not work out the derived ground-plan characters in which these two groups agree with other groups, but merely pointed out that particular characters are present in a similar form and are widely distributed in the Holometabola ('Neuropteroidea') but not in the 'Orthopteroidea' or 'Polyneoptera'.

According to Sharov, the following characters indicate that the Glosselytrodea and Caloneurodea belong to the Holometabola: the shape and size of the compound eyes, the moniliform antennae (only in the Glosselytrodea), the hairs on the wing veins (only in the Glosselytrodea), the conformation of the pleural region of the thorax (only in the Glosselytrodea), the structure of the scutella in the mesothorax and metathorax (only in the Caloneurodea), the structure of the legs, and certain characters in the abdomen such as the presence of only nine externally visible segments, the shape and position of the tergites and sternites, and the one-segmented cerci. There are also some striking differences between these two groups: the Glosselytrodea have moniliform antennae, enlarged coxae, and generally short legs, whereas the Caloneurodea have filiform antennae, short coxae, and generally long legs.

There are particular problems in the wing venation. The Glosselytrodea have a broad precostal field, a character that is otherwise known only in the Orthopteroidea. On the other hand, the Caloneurodea generally have no precostal field, but Sharov wrote that he knew of a species discovered by Kukalová in the Lower Permian of Moravia that had an undoubted precostal field. This character is certainly part of the ground-plan of the Neoptera, as is shown by its presence in the Orthopteroidea. It could very well have been retained in the stem-group of the Holometabola and have been reduced later. However, it seems certain that the secondary enlargement of the precostal field in the Orthopteroidea is correlated with the lateral position of the wings: one result of this position is that the wing base was enlarged just before the basal articulation, and this was achieved mainly by the expansion of the precostal field. Something similar seems to have taken place in primitive Holometabola ('Neuropteroidea'), but in this case the wing base was enlarged by expansion of the costal field. This means that the precostal field must already have been reduced before this process began. This is hardly compatible with Sharov's assignment of the Glosselytrodea to the 'Neuropteroidea'[274].

The Glosselytrodea are only known from the Permian and the Triassic, and are therefore of no great importance for determining the age of the Holometabola or of any of their subgroups. This is because they are no older than fossils which undoubtedly do belong to other subgroups of the Holometabola, such as the Coleoptera. However, the position is different with the Caloneurodea, as these are known from the middle Upper Carboniferous of Commentry (Carpenter, 1962b). As I mentioned above, Sharov (1967) assigned both Caloneurodea and Glosselytrodea to the 'Neuropteroidea'. However, Sharov is a practitioner of 'typological systematics' and did not give any phylogenetic tree in his paper. For this reason, it is not clear whether his assignment of these two groups to the 'Neuropteroidea' means that he regarded them as members of a monophyletic group 'Neuropteroidea' or that he wished to assign a stem-group of the entire Holometabola to the 'Neuropteroidea' on the basis of their similarity in many characters to the Neuroptera. He did something like this in another phylogenetic tree (1965, 1967) when he assigned some fossils to the Ensifera for typological reasons although they were genealogically most closely related to the Caelifera. Sharov's (1967) discussion seems to me to lead to the same kind of unsatisfactory results with the Caloneurodea as were evident with the 'Protorthoptera' of older authors and the Paraplecoptera of Sharov, which he himself had previously placed close to the Caloneurodea (in Rodendorf, 1962).

Several subgroups of the *Palaeoptera and *Paurometabola have been shown to have been definitely in existence during the Upper Carboniferous. In view of this, it is all the more remarkable that efforts to find some reliable evidence for the assignment of certain Carboniferous fossils to the Paraneoptera or Holometabola have all failed, although both groups must have diverged widely from each other by the Lower Permian.

I have already pointed out (p. 167) that various explanations have been made to account for the different representation of the Palaeoptera + Paurometabola and the Paraneoptera + Holometabola in the fossil record. One of these suggestions is that the origin and early development of the Paraneoptera + Holometabola must have taken place in regions from which no fossils have yet been found. The discovery of Carboniferous Paraneoptera and Holometabola from localities where most of the known fossils consist of Palaeoptera and Paurometabola would be of vital importance for the confirmation or refutation of these suggestions, quite apart from their bearing on the vexed question of the assignment and relationships of the Plecoptera. For this reason, particularly strict criteria should be used when identifying Carboniferous fossils as Paraneoptera, Holometabola, or Plecoptera.

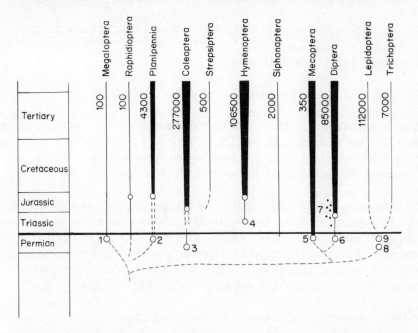

Fig. 95. Phylogenetic tree of the Holometabola. 1, *Permosialis*; 2, Palaeohemerobiidae and Permithonidae of Carpenter; 3, *Tshekardocoleus* and other genera; 4, *Archexyela*; 5, Mecoptera from the Upper Permian of Australia; 6 and 7, Paratrichoptera; 8, *Microptysma*; 9, *Microptysmodes*, *Cladochorista*. The numerous Permian 'Mecoptera' have not been included. They are probably an 'invalid stem-group' and include members of several orders from the Hymenoptera to the Siphonaptera and Diptera.

Revisionary notes

269. Cryptosterny is another derived character.

The 'laterosternites', which Weber (1933) reported in *Sialis* etc., are pleural rather than sternal elements.

Snodgrass (1954) noted that holometabolan larvae have short-sighted simple eyes, contrasting with the long-sighted compound eyes of the Ephemeroptera, Plecoptera, etc.

Snodgrass (1954) also stated that larvae of the Holometabola have only one claw on each leg (except for the Megaloptera, Raphidioptera, most Planipennia, and 'certain' Coleoptera), and he believed this to be a derived state. This suggests that the Hymenoptera may be more closely related to the Mecopteroidea. [Willi Hennig.]

270. Furthermore, it is clear that the epicentre of the distribution of the Holometabola is now in the tropics, whereas they are said to have arisen in response to a cold climate. Jeannel (1950) also wrote that the Holometabola (and Paraneoptera) were able to spread throughout the world as the climate became more subtropical and moderate. [Willi Hennig.]

271. According to Hinton (1949), the pupal mandible in decticous pupae is operated by an imaginal muscle. Surely this should be the derived state? See also Hinton (1971). [Willi Hennig.]

272. This vein is called the jugal bar. [Michael Achtelig.]

273. A jugal bar is also present in the hemipteroid orders (Hamilton, 1972b). Hamilton has described another characteristic of the 'neuropteroid orders' (= Holometabola): the ability to fold each wing along a jugal fold. However, this is very difficult to check in fossil wings. [Michael Achtelig.]

274. Hamilton (1972c) believed that the hemipteroid lineage probably began with the ancestors of the Glosselytrodea. His argument was that in these insects the plical vein (the common postcubitus) and the empusal vein (the common 1st anal vein) were close together, and that the two following anal veins (1A and 2A) were fused apically to form a 'Y-vein'—'the two features characteristic of the clavus of the Rhynchota'. [Michael Achtelig.]

b. The monophyletic subgroups of the Holometabola

The Neuropteroidea, Coleoptera, Strepsiptera, Hymenoptera, and Mecopteroidea[275] are the oldest and most comprehensive monophyletic groups of the Holometabola. There are undoubtedly close affinities between some of them, but these are still not clear. The only course open to me is to discuss these groups one at a time and, as the opportunity arises, to suggest what their relationships may be.

Additional note

275. According to Grassé (in Grassé, 1951, p. 116), Brauer proposed the name Petanoptera for the Mecoptera + Diptera + Trichoptera + Lepidoptera. [Willi Hennig.]

2.2.2.2..4.1. Neuropteroidea

Martynova (1957) called the Neuropteroidea 'the most archaic Holometabola'. This only means that in their most obvious characters they have diverged least from the ground-plan of the Holometabola, but it gives no clue as to whether they are a monophyletic group. A few derived characters indicate that this may in fact be so. The adults and larvae have a gula, and in the adults the labium has no paraglossae[276]. The 'gonarcus' of the male genitalia is a further character that has been mentioned. Tjeder (in Tuxen, 1956) said nothing about the morphology and function of the gonarcus. He reported that it is very differently developed in individual cases but is generally an arcuate structure with a downwardly or inwardly directed shaft. However, Acker (1960) did not think that Tjeder's gonarcus was a homologous structure. Tjeder's other statement, that the female lacks the 8th abdominal sternite, is equally ambiguous, because he himself pointed out that an 8th sternite has arisen secondarily in the Corydalidae and Coniopterygidae. It is difficult to see how this statement is to be understood[277].

The simplification of the ovipositor is not an important character[278]. Ross (1965) reported that the 1st and 2nd valvulae are reduced[279]; the 2nd valvifer and 3rd valvula have been retained. In the Coleoptera (and Strepsiptera) and Mecopteroidea, the ovipositor is even more extensively reduced or completely absent. Consequently, however simplified the ovipositor of the Neuropteroidea may be, it must be regarded as a relatively primitive character.

Even if we take no account of the doubtful characters in the male and female abdomen, it is still true that the other two derived characters (presence of a gula, and absence of paraglossae in the adult labium) are not found together in any of the other groups of the Holometabola that might be the sister-group of the Neuropteroidea. The suggestion that the Neuropteroidea are a monophyletic group appears to be an adequately founded working hypothesis[280].

It is clear that all the derived characters used to establish that the Neuropteroidea are a monophyletic group are cryptic and cannot be seen in fossils. Nor are any derived ground-plan characters known in the wing venation of the group. Consequently, Palaeozoic fossils that agree with relatively primitive recent Neuropteroidea in their venation should not be automatically assigned to this group. On the other hand, fossils which agree with one of the subgroups of the *Neuropteroidea with undoubted derived characters in their venation are quite another matter.

The Megaloptera, Raphidioptera, and Planipennia (Neuroptera sensu stricto) have long been recognized as monophyletic groups of the Neuropteroidea. However, Achtelig (1967) has recently suggested that the Raphidioptera are most closely related to the megalopterous family Corydalidae, and this appears to be well founded[281]. The Megaloptera without the 'Raphidioptera' would then be a paraphyletic group. Unfortunately, it has not been possible for me to discuss Achtelig's views and the considerable bearing they have on the interpretation of fossils in any more detail here.

Revisionary notes

276. It could be suggested that these three characters are not actually synapomorphies. A larval gula is only present in the Raphidioptera and the Corydalidae. The gular region, though not the true 'gula', is always membraneous in the larvae of several families of the Planipennia (Killington, 1936). According to Kristensen (1975), the ventral closure of the head capsule in the Planipennia is not effectuated by a true gula, as it is in the Raphidioptera and Megaloptera, but rather by a sclerotized bridge lying anterior (ventral) to the metatentorial pits. Paraglossae may be present in some Megaloptera, and presumably are also present in the Planipennia as an essential component of the ligula (Snodgrass, 1935). [Michael Achtelig.]

277. According to Aspöck and Aspöck (1971) and Achtelig (1977), most Inocelliidae and Raphidiidae also have an 8th sternite. [Michael Achtelig.]

278. Mickoleit (1973) has described two very important synapomorphies in the ovipositor of the Neuropteroidea: the 3rd valvulae are fused dorsally and are equipped with an internal musculature. [Michael Achtelig.]

279. In the Raphidioptera, the 1st valvulae are fused and elongate, and the 2nd valvulae are absorbed into the 3rd valvulae during morphogenesis (Achtelig, 1978). [Michael Achtelig.]

280. Riek (1967b: 337) found that the Planipennia, Raphidioptera, and Megaloptera possess 'two associated structures on the wings and body . . . with a possible stridulatory function'. These structures do not occur in any other Holometabola and appear to be an apomorphic character which indicates that the Neuropteroidea are monophyletic—with the proviso that they have not been lost in the Coleoptera or any other group! [Willi Hennig.]

There are some characters in the metathorax and basal abdominal segments which, according to Achtelig (1975), may add further support to this hypothesis. They are as follows: (1) the notum of the 1st abdominal segment has a $\curlywedge$-shaped caudally-bifid sagittal suture; (2) the postnotum of the metathorax is always divided medially; (3) the 2nd abdominal sternum has a transverse sternal suture (or an inner ridge), a character which is repeated on each sternite up to and including the 6th in some Planipennia, such as the Hemerobiidae; (4) there is a connection between the metafurca and the lateral antecosta of the 1st abdominal spiracle which originates at the tip of the metafurca. Panov and Davydova (1976) described some common characters in the medial neurosecretory cells in the brain of the Neuropteroidea. [Michael Achtelig.]

281. This view has subsequently been rejected by Achtelig and Kristensen (1973) because of the striking synapomorphies in the derived aquatic larvae of the Megaloptera, particularly the Corydalidae and Chauliodidae, as well as in the adults. The characters which were previously thought to be synapomorphies are now considered to be plesiomorphies or the results of convergence. [Michael Achtelig.]

2.2.2.2..4.1..1. Megaloptera (alder flies)

It is unfortunate that the most important derived characters used to show that the Megaloptera are a monophyletic group are found only in the larvae. The larvae are aquatic and, as an adaptation to this mode of life, they have segmented tracheal gills on the first seven or eight abdominal segments, each consisting of six sections. Snodgrass thought that these tracheal gills were homologous with the abdominal tracheal gills of the Ephemeroptera and the styli of the 'Thysanura'. He even homologized some vesicles that are present at the base of the first seven tracheal gills in the larva of *Corydalis*, and that have tufts of filaments, with the coxal glands of the 'Thysanura'. He may well have been right to homologize the gills with the rudiments of the abdominal appendages present in the insect ground-plan. However, this would only show that the Holometabola have retained in their ground-plan some extraordinarily primitive pterygotan features, and there are in fact other reasons for believing this to be so. We would then have to suppose that the rudiments of the abdominal appendages had been retained as primordia and were then brought into use secondarily as respiratory organs when the larvae of the Megaloptera adopted an aquatic mode of life. Bradley (1947) considered that these larvae were primarily aquatic and attempted to derive them directly from the aquatic larvae of the Plecoptera and even of the Ephemeroptera. I do not think that this suggestion is well founded. It would compel us to the conclusion that the larvae of the Raphidioptera, Planipennia and other hemimetabolous groups, which breathe through an open tracheal system with a normal complement of spiracles, acquired this system secondarily (derived). There is no evidence that spiracles can be lost and then be acquired again in their original position: on the contrary, experience shows that larvae that are believed to have given up an aquatic in favour of a terrestrial life, such as terrestrial larvae of the Chironomidae and Ceratopogonidae (Diptera), never re-acquire the spiracles that their ancestors lost when they first became aquatic. For this reason, I agree with Ross (1955) who suggested that the larvae of the Neuropteroidea were primarily terrestrial, although they may have lived in moist environments close to water margins.

The larva of *Corydalis*, which Chen (1946) described as having two thoracic and eight abdominal spiracles, may provide the key to the solution of this problem.

The aquatic mode of life of the larvae (apoecy!) and the associated morphological adaptations (apomorphies!) show that the Megaloptera are a monophyletic group.

Unfortunately, it is almost impossible to recognize any characters in the ground-plan of the adults as being definitely derived, from the point of view of the ground-plan of the Neuropteroidea. The only character that appears to be derived is the reduced female ovipositor, for the Raphidioptera seem to show that the ovipositor in the ground-plan of the Neuropteroidea is much better developed than it is in females of the Megaloptera. Further possible apomorphies are the telotrophy of the meroistic ovarioles[282] and the absence of the pterostigma[283] (see p. 291). Martynova (in Rodendorf, 1962) stated that the adults are nocturnal and live close to water; they do not feed. This is also a derived state[284].

The oldest fossils to be assigned to the Megaloptera have been described from the Permian. The most convincing of these is a larva from the Upper Permian of the Urals (Fig. 96), which Sharov (1953) assigned to the genus

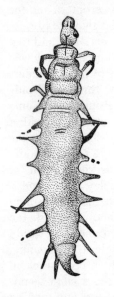

Fig. 96. Larva of *Permosialis* sp. (Megaloptera)[285], from the Permian of the Urals. From Rodendorf (1962).

Permosialis. This larva appears to be prognathous and to have large mandibles. It also has nine pairs of appendages that appear to correspond to the tracheal gills of megalopterous larvae. However, recent larvae have no more than eight pairs of tracheal gills, and in this respect the Permian larva is more primitive[285].

Sixteen species of the genus *Permosialis* have been described from wings found in the Permian of Arkhangelsk, the Urals region, and the Kuznetsk basin. It is striking that in all these wings Sc is rather widely separated from R_1 even at the costa (Fig. 97), whereas in all recent Megaloptera the apical part of Sc is fused with R_1 for a short distance (Fig. 98). However, there are good reasons for believing that in the ground-plan of the Holometabola, and possibly of a much larger group (p. 165), a branch of Sc (Sc_2) is fused with R_1. In recent Megaloptera the fusion has taken place at the site of this branch, and so it is surprising that Sc and R_1 are completely independent in *Permosialis*. This does not rule out the possibility that *Permosialis* belongs to the stem-group of the Megaloptera, perhaps as a 'side-branch', but there is nothing to prove that it does belong to the Megaloptera. The discovery of the larvae and other lines of reasoning provide much better evidence for the existence of the Megaloptera in the Upper Permian than do the fossil wings.

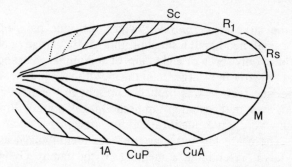

Fig. 97. Fore-wing of *Permosialis bifasciata* Martynov (Megaloptera), from the Upper Permian of the Arkhangelsk region. From Rodendorf (1962).

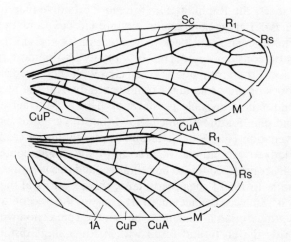

Fig. 98. Wings of *Sialis lutaria* Linnaeus (Megaloptera; recent). From Tillyard (1919a).

Revisionary notes

282. The telotrophic ovarioles of the Megaloptera and Raphidioptera are very similar. In addition, they have secondary nurse chambers for the formation of large micropyles (Achtelig, 1978). [Michael Achtelig.]

283. The blood sinus of the pterostigma may be unpigmented, but it is generally present in the Pterygota (Arnold, 1963), including the Megaloptera. [Michael Achtelig.]

284. There are several reports in the literature of *Sialis* feeding on pollen (Achtelig and Kristensen, 1973). [Michael Achtelig.] Also by Schlee (1977a). [Dieter Schlee.]

285. Unlike the larva of *Permosialis*, which has nine pairs of abdominal appendages, a small head, and a small prothorax, the larvae of all the recent Megaloptera have only seven or eight pairs of abdominal gills, and a head and prothorax that are strikingly larger than the mesothorax and metathorax. The latter characteristic feature was probably present in the stem-group of the Megaloptera or even in the stem-group of the Megaloptera + Raphidioptera. There is really very little justification for assigning *Permosialis* to the Megaloptera. There are far better grounds for identifying it as an aquatic coleopterous larva, resembling a member of the family Gyrinidae where larvae are known to have a small head, small prothorax, and nine segmental appendages. [Michael Achtelig.]

2.2.2.2..4.1..2. Raphidioptera (snake flies)[286]

In many respects the Raphidioptera are more primitive than the Megaloptera (alder flies). Larvae and adults are predaceous, and the larvae are terrestrial. The elongate and neck-like extension of the prothorax, which gives the head a considerable degree of mobility, is the most characteristic derived feature of the group and is connected with their predaceous habits. The lateral lobes of the pronotum extend ventrally over the pleura down to the narrow and almost invisible prosternum. This is a further characteristic derived ground-plan feature[287] which shows that the group is monophyletic. In the wing venation (Fig. 99), Sc terminates freely in the costa and this is probably a derived ground-plan character of the Raphidioptera. Tillyard (1919a) actually considered this to be primitive, and this view has been shared by most authors. In 1953 I suggested that in the ancestors of the Raphidioptera the tip of Sc was connected to R_1, as in the recent Megaloptera. Subsequently, the section immediately before the connection with R_1 moved forward towards the costa and fused with it for a short distance, but the original course of Sc is shown by the proximal 'cross-vein' delimiting the pterostigma. This suggestion has been completely ignored, probably because

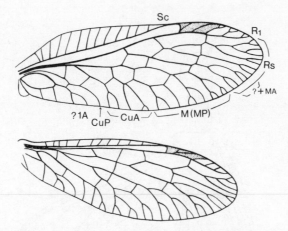

Fig. 99. Wings of *Raphidia notata* Linnaeus (Raphidioptera; recent). From Tillyard (source not located).

it has never come to the attention of Raphidioptera specialists, but I still consider it to be plausible even though it has not yet been proved. It could be a very important character for assessing fossil wings. A further derived character of the Raphidioptera is the partial fusion of CuA and M[288].

On the other hand, the Raphidioptera have a well developed ovipositor and in this respect are probably more primitive than the Megaloptera[289]. The same may also be true of the presence of a pterostigma in the wing[290]. Martynov (1932) published some interesting ideas about the significance of this structure, but his paper has been almost completely overlooked. He found that in the pterostigma the hypodermis is preserved between the two layers of wing integument, and sometimes it can be found to contain blood cells, oenocytes, and fat-body cells. He was right to emphasize that such a pterostigma shows the original structure of the wing. Naturally, the pterostigma is frequently a dead structure, without any living hypodermal cells. It is nonetheless clear that species with a pterostigma cannot be derived from those without. If this is all correct, then it must have a considerable influence on the interpretation of the Palaeozoic fossils. A pterostigma is present in the Upper Jurassic Mesoraphidiidae (Fig. 100) and Baissopteridae. So far as is known, both groups are more primitive than the recent Raphidioptera in having the prothorax shorter than the mesothorax and CuA only touching M at one point. However, the pterostigma is absent in the Palaeozoic fossils that Martynova (1961b; in Rodendorf, 1962) assigned to the Raphidioptera. She particularly emphasized this character, though elsewhere she included the *presence* of the pterostigma in her characterization of the Raphidioptera.

Martynova stated that the Letopalopteridae, consisting of the genus *Letopaloptera* with only two species from the Upper Permian of the Arkhangelsk region, had jumping legs(!) and a peculiar wing venation (very

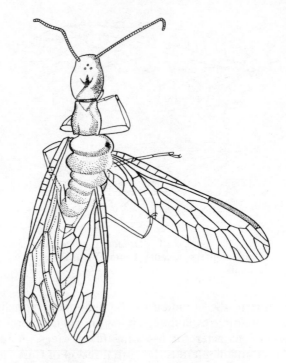

Fig. 100. *Mesoraphidia pterostigmalis* Martynova (Raphidioptera), from the Upper Jurassic of Kazakhstan. From Rodendorf (1962).

short Sc, etc.). Several characters of the Sojanoraphidiidae (*Sojanoraphidia rossica* Martynova, also from the Upper Permian of Arkangelsk) indicate that they too belong to the Raphidioptera: the habitus, the long ovipositor, the 'flattened' and apparently prognathous head, and, according to Martynova (1961b), the elongate prothorax. However, apart from the absence of the pterostigma, the wing shows no sign of an enlarged costal field which appears to be part of the ground-plan of the Neuropteroidea and also of the Raphidioptera (see below).

No matter how these fossils are interpreted, the existence of the Megaloptera and Planipennia during the Upper Permian provides strong evidence that the Raphidioptera too must have been in existence at the same time[291]. However, this conclusion would be untenable if Achtelig's (1967) interpretation of the Raphidioptera, briefly mentioned above, is correct.

Martynova also assigned a wing from the Upper Carboniferous of the Tunguska basin (*Fatjanoptera mnemonica* Martynova) to the Raphidioptera. This species also lacks the pterostigma and a broad costal field. I cannot see the slightest reason for assigning it to the stem-group of the Neuropteroidea, let alone to the Raphidioptera[292].

Martynova apparently regarded the supposed partial fusion of RS and MA as the criterion for assigning all these Palaeozoic fossils to the Raphidioptera, because she mentioned it in her diagnosis of this group[293]. In this she followed Tillyard (1932b), but it is a highly problematic character. 'MA' is probably attached to RS in all the Holometabola, and also in the Plecoptera and Paraneoptera(!). It is not at all clear how far the Raphidioptera show a derived state in comparison with other Neuropteroidea.

The Permoraphidiidae, from the Lower Permian of Kansas, were assigned by Laurentiaux (in Piveteau, 1953) and Carpenter (1943a) to the 'Raphidiodes' or Raphidioptera. They are now regarded as Orthopteroidea, and the precostal field of the fore-wing shows that this assignment is correct.

It is very unfortunate that the sister-group of the Raphidioptera is not known.† If the Neuropteroidea are considered to be a monophyletic group, then this sister-group must be either the Megaloptera or the Planipennia or the Megaloptera + Planipennia. There is no sound evidence for the third of these possibilities. Nor do I know of any derived characters common to the Raphidioptera and Megaloptera. Previously (Hennig, 1953) I considered that the partial fusion of CuA and M was just such a character, but I was wrong: the Sialoidea are the only Megaloptera with this character. It cannot be in the ground-plan of the Megaloptera and must therefore have arisen convergently in the Sialoidea and Raphidioptera. The only other hypothesis is that the Raphidioptera and Planipennia are most closely related, as was suggested by Klingstedt (1937) and Martynova (1961b). However, there is no clear support for this view either.

Brues and Melander (1954) stated that the Raphidioptera and Planipennia lack a folded anal fan in the hind-wing. A careful study may show this to be a synapomorphy of these two groups. It has also been suggested that the larvae of the Raphidioptera and Planipennia have become completely independent of the aquatic habitat and have lost their abdominal appendages. However, a serious objection to this is that the larvae of the Sisyridae (Planipennia) have tracheal gills that are derived from abdominal appendages[294]!

Revisionary notes

286. A new monograph on the Raphidioptera has been published by Aspöck and Aspöck (1971). [Dieter Schlee.]

287. This condition is not found in the Inocelliidae (Lauterbach, 1972c) and therefore cannot yet be part of the ground-plan of the Raphidioptera. [Willi Hennig.]

288. The partial fusion of CuA and M takes place during ontogenesis, as does the development of the anal cell, which is formed by 1A (which is arched anteriorly) and the fused 2A + 3A. In recent Raphidiidae, the anal cell of the

---

† Compare this paragraph with Achtelig's (1967) recent paper.

hind-wing is secondarily reduced by the further fusion of CuP + 1A. [Michael Achtelig.]

289. Smith (1969) and Mickoleit (1973) considered the sliding interlock (olistheter) in the ovipositor of the Raphidioptera to be a secondary structure. From this it would follow that the elongate condition must be derived. However, on the basis of the ontogeny and comparative anatomy Achtelig (1978) has refuted this view. [Michael Achtelig.]

290. See note 283. [Michael Achtelig.]

291. See notes 281, 285, and 299. [Michael Achtelig.]

292. Subsequently, the Letopalopteridae, Sojanoraphidiidae, and Fatjanopteridae have also been excluded from the stem-group of the Raphidioptera (Aspöck and Aspöck, 1971; Achtelig, 1976). [Michael Achtelig.]

293. The morphogenesis of the wing shows that M and R never fuse in the Raphidioptera. The vein in the hind-wing of recent Raphidioptera that is often called MA or simply 'x' is actually nothing more than a secondary cross-vein that is formed at the same time as all the other cross-veins, during the last stage of the development of the wing. [Michael Achtelig.]

294. In the Megaloptera, the hind-wing is folded in exactly the same way as in the Raphidioptera, Planipennia, and lower Mecopteroidea: it only flexes along the plica jugalis (Achtelig and Kristensen, 1973). Some of the other characters that have previously been thought to provide evidence of a sister-group relationship between the Raphidioptera and the Planipennia are also known to occur in other insect orders: a spermatheca with a ductus seminalis in addition to a ductus receptaculi is present in some Coleoptera (see note 309 on p. 307); 'distance segregation' of the sex chromosomes is known in some Coleoptera, Mecoptera, Heteroptera, and Phasmatodea (White, 1954) and in the Coccidia (Hughes-Schrader, 1975); a characteristic epimeral suture is known in some primitive Diptera, Hymenoptera, and Homoptera (Achtelig, 1975). Other similarities, which cannot be accepted as proof that the Raphidioptera and Planipennia constitute a single monophyletic group, have been discussed by Achtelig and Kristensen (1973).

On the other hand, there are several characters common to the Raphidioptera and Megaloptera that can hardly be considered to have arisen convergently. (1) When compared with the Planipennia, they have several derived characters in their ovaries: the racemose shape of the ovaries, the high number of ovarioles, the telotrophic ovarioles, and the development of special nurse chambers for the formation of large micropyles (or aeropyles) (Achtelig, 1978). (2) The metathoracic epimeron is firmly connected to a caudal 'postepimeron', which is a derivate from the laterotergite of the 1st abdominal segment; it bears an 'epimeral apophysis', an enlargement of the 1st antecosta, which is connected to the metafurca by a ligament or a skeletal bridge, which in turn is derived from the furco-antecostal muscle of the Neuropteroidea. (3) The characteristically modified tergite of the 2nd abdominal segment has a reinforced acrotergite which is the point of origin for a powerful polyintersegmental muscle between the 2nd and 5th tergites in the Raphidioptera and Sialidae (Achtelig, 1975). Additional characters, which are probably apomorphies, are (4) the derived cleaning behaviour (Jander, 1966), and (5) the imaginal gula (see also notes 276 and 309). Panov and Davydova (1976) found that species of the Megaloptera and Raphidioptera are characterized by multiplication of the neurosecretory cells during the larval stages. However, the phylogenetic significance of their results is not clear at present. [Michael Achtelig.]

2.2.2.2..4.1..3. Planipennia (lacewings, ant-lions)

A number of derived larval characters show that the Planipennia are a monophyletic group.

The tarsi are one-segmented. The mouth orifice is reduced to a narrow cleft. The mandibles and the lacinia of the maxillae (Berland and Grassé, in Grassé, 1951; galea of Weber, 1949) are greatly enlarged and are placed close together to form a suctorial tube. The maxillary palpi are absent. The mid and hind intestines are not connected. The tips of the Malpighian tubes are also connected to the rectum (cryptonephry).

The mouth-parts have clearly been adapted to a specialized form of predatory activity, the seizing and draining of much larger species of prey. The peculiar characters of the alimentary canal are probably also correlated with this. The evolutionary success of the Planipennia, when compared with the Megaloptera and Raphidioptera, is probably a direct result of these specialized larval adaptations.

Unfortunately, it is much more difficult to evaluate the adult characters. The orthognathy of the head may be a derived character. In general, orthognathy is regarded as a primitive character in the insects, and this appears to be correct. For this reason, it would be tempting to treat the prognathy of the Megaloptera and Raphidioptera as evidence (a synapomorphy?) that these two groups are closely related. However, it is probable that there was originally a close connection between prognathy and the closure of the occiput by a gula. This gula is present in all the Neuropteroidea and adult Coleoptera. It could be that it is part of the ground-plan of a monophyletic group of higher rank which would include the Neuropteroidea and the beetles (see below). Beetles are prognathous in their ground-plan, as are the Megaloptera and Raphidioptera, but orthognathy and even opisthognathy occur secondarily in some subgroups of the beetles. It is therefore possible that orthognathy in the Planipennia has arisen secondarily, as it has in some of the beetles, and that a prognathous head is part of the ground-plan of the

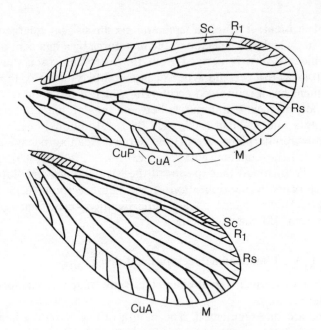

Fig. 101. Wings of *Sisyra fuscata* Fabricius (Planipennia; recent). From Berland and Grassé (in Grassé, 1951).

Neuropteroidea as well as of the beetles. Some evidence for this is provided by the fact that orthognathy is found only in the adults of the Planipennia, whereas the larvae of all three neuropteroid groups are prognathous. A detailed study should be made of this problem[295].

A further difficulty concerns the genital openings of the Planipennia. Female Megaloptera and Raphidioptera have two separate genital openings: a sexual opening (the tip of the bursa copulatrix) and a reproductive opening (oviduct)[296]. However, in the Planipennia, bursa copulatrix and oviduct have a common aperture. The Mantispidae are the only group with two separate genital openings in the female, as in the Megaloptera and Raphidioptera. In the Lepidoptera the presence of two separate openings ('ditrysy') is undoubtedly a derived character, and this could be true in the Neuropteroidea too, particularly as a single opening in the female genitalia is part of the ground-plan of the Holometabola. On the other hand, it is also possible that the separation of the sexual and reproductive openings has been secondarily suppressed in the Planipennia, except for the Mantispidae. This is another problem that needs urgent study.

The recognition of derived characters in the wing venation is particularly important for assessing the fossils. One such character is probably the comb-like proliferation of branches of the radial sector (Fig. 101) (Martynova, in Rodendorf, 1962)[297]. Two further unmistakable trends are an increase in the number of cross-veins and an increase in the number of secondary branches of the longitudinal veins, and both of these are correlated with the enlargement of the wing. However, in the Coniopterygidae the venation has been drastically reduced. In the ground-plan of the fore-wing the costal field appears to be enlarged.

Both Laurentiaux (in Piveteau, 1953) and Carpenter (1954b) considered the Permoberothidae (*Permoberotha villosa* Tillyard), from the Lower Permian of Kansas (Fig. 102), to be the oldest species of the Planipennia.

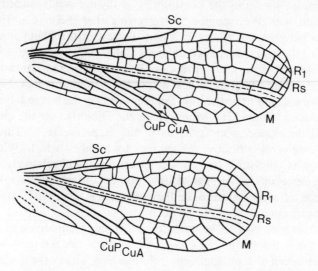

Fig. 102. Wings of *Permoberotha villosa* Tillyard (Plani-pennia), from the Lower Permian of Kansas. From Carpenter (1943a).

Martynova (in Rodendorf, 1962) assigned *Permoberotha* to the Glosselytrodea, in other words to the grasshopper-like insects (Paurometabola). Sharov considered that all the Glosselytrodea, including *Permoberotha*, belong to the Neuropteroidea (see pp. 281–282)[298].

Upper Permian fossils that definitely belong to the Planipennia are quite numerous. They all have the pectinate branches of the radial sector, the enlarged costal field, and the secondarily increased number of longitudinal veins and cross-veins[299]. Unfortunately, it is not known whether they actually belong to the stem-group of the Planipennia or to subgroups of the *Planipennia. This is because the monophyletic groups of the *Planipennia and their relationships have not yet been clearly worked out.

Withycombe (1924) published a phylogenetic tree of this group (Fig. 130). In his discussion he distinguished the superfamilies Ithonidoidea, Coniopterygoidea, Osmyloidea, Hemerobioidea, and Myrmeleonoidea (see Berland and Grassé, in Grassé, 1951), and these do actually correspond to the monophyletic groups shown in his phylogenetic tree. However, as is so often the case in systematics, the classification does not reflect all the phylogenetic relationships shown in the phylogenetic tree. For example, it is not evident from

Withycombe's classification that there is a sister-group relationship between the Myrmeleonoidea and Hemerobioidea, or that the Myrmeleonoidea + Hemerobioidea have a sister-group relationship with the Osmyloidea, etc. His phylogenetic classification is incomplete, but nonetheless it is absolutely correct from the point of view of phylogenetic systematics because it does not contain any paraphyletic or polyphyletic groups. The only question is whether the relationships shown in the dendrogram really are well founded. Tillyard (1916) agreed with Withycombe in suggesting that there is a sister-group relationship between the Ithonoidea and the rest of the Planipennia. However, Tillyard thought that the larvae of the Polystoechotidae are closer to the 'archetype' larvae of the Neuroptera (= Planipennia) than are those of any other recent group. On the other hand, the Coniopterygidae[300] and the Psychopsidae are said to be the only families in which the larvae have retained a well developed labium. The Mantispidae and Dilaridae are the only groups with a well developed ovipositor, as in the Raphidiidae[301]. Tillyard thus attributed ostensibly primitive characters to families which Withycombe showed in his phylogenetic tree as widely separated and relatively subordinate groups along different lines of development. Martynova's (1958) phylogenetic tree differed in certain fundamental details from Withycombe's, but was no better founded.

The most urgent task to be undertaken in the Planipennia is a precise analysis of the heterobathmy of larval and imaginal characters[302]. This is the only way in which the phylogenetic relationships within the group can be worked out, and no definite assessment of the fossils can be made until this has been done.

Martynova (in Rodendorf, 1962) included four superfamilies in the Planipennia: Myrmeleontidea, Coniopterygoidea, Polystoechotidea, and Hemerobiidea. The last two were said to have been found as early as the Upper Permian. Some of the earlier authors thought that the Permian fossils had affinities with *different* families in the modern fauna (see Laurentiaux, in Piveteau, 1953) and these opinions were often reflected in the family names that they chose, such as Permithonidae, Palaemerobiidae, and Permosisyridae. Apart from the Lower Permian Permoberothidae, which probably do not belong to the Planipennia (see above), all these families have been grouped by Carpenter (1954b) into the two 'families' Palaemerobiidae and Permithonidae, which he distinguished by the number of branches of the radial sector. These two groups correspond to Martynova's Polystoechotidea and Hemerobiidea. She separated her groups mainly by the width of the fore-wing and by the course of Sc: Sc terminates in the radius in the Polystoechotidea, but in the Hemerobiidea it reaches the costa. However, not all of Martynova's figures show this distinction: some of the species which she placed in the Hemerobiidea clearly show Sc terminating in the radius! Martynova also stated that all recent and Mesozoic Polystoechotidea and Hemerobiidea have the base of MA fused with RS or R, whereas it is said to be free in all the Palaeozoic fossils of these groups. Once again, not all her figures agree with this, for example her Fig. 864 of the Upper Cretaceous *Grammopsychops* (in Rodendorf, 1962). These contradictions within the same work do not inspire me with any confidence in the author's phylogenetic 'derivations'. Consequently, it is still not clear whether any of the direct ancestors of the various monophyletic subgroups of the *Planipennia were in existence during the Palaeozoic. About 30 species have been described from the Upper Permian of Arkhangelsk, the Urals, and the Kuznetsk basin, and 6 species from the Upper Permian of Australia.

Wille (1960) placed the Neuropteroidea, Coleoptera, Strepsiptera, and Hymenoptera in a single group. He used Crampton's (1924) name for them, the Panneuroptera, and regarded them as the sister-group of the rest of the Holometabola (Panmecoptera). However, he was unable to find any un- doubted derived characters for the 'Panneuroptera'. He only stated that the mesothoracic coxal mera, which he considered to be part of the ground-plan of the Holometabola, had been lost in the Coleoptera, Strepsiptera, and Hymenoptera. However, this has also taken place in the Siphonaptera ('Panmecoptera'), and coxal mera are present in the Neuropteroidea (Pan- neuroptera). He also stated that males of the 'Panneuroptera' tend to develop processes on the 'gonocoxopodites', which are the basal segments of Snod- grass' parameres, and these are usually called volsellae. However, this is much too vague a character.

At first Ross (1955) came to the same conclusions as Wille and showed the Hymenoptera, Coleoptera, and Neuropteroidea as a single branch in his phylogenetic tree. More recently (Ross, 1965), he considered that the Neuropteroidea alone are the sister-group of the Mecopteroidea, and he showed the Hymenoptera and Coleoptera arising as separate branches from the main stem of the Holometabola *before* the Neuropteroidea and Mecop- teroidea.

The best founded suggestion may be that the Coleoptera (and Strepsip- tera?) are the sister-group of the Neuropteroidea[303].

Revisionary notes

295. The 'ligula' of the labium is a striking derived character of adult Planipennia. Further synapomorphies are to be found in the subdivision of the 1st sternite and in the articulation of the thorax and abdomen: there is one joint between the laterotergite (which has the condyle) and the epimeron, and another between the epimeron (which has the condyle) and the 1st sternite (Achtelig, 1975). [Michael Achtelig.]

296. There is a single genital opening in the Raphidioptera and Megaloptera, as well as in the Planipennia (Mickoleit, 1973; Achtelig, 1978). [Michael Achtelig.]

297. Surely the pectinate branching of the radial sector cannot be a derived

character. It is clearly present in the Corydalidae, and also in the Raphidiop-
tera in a reduced form (Fig. 99). It was also present in the Palaeodictyoptera
(Kukalová, 1969a, 1969b, 1970; Kukalová-Peck, 1972a) and in the
Megasecoptera (Kukalová-Peck, 1972b). [Michael Achtelig.]

298. On the other hand, the non-pectinate radial sector, the fusion of the
bases of M and Cu, and the distal fusion of the anal veins (Fig. 102) are not
characters of the Neuropteroidea, let alone of the Planipennia (see note 274).
[Michael Achtelig.]

299. The most striking similarity between the wings of several of the Permian
Planipennia and a number of recent families is the presence of trichosors
(small thickenings of the wing margin between the tips of the veins and
veinlets). [Michael Achtelig.]

300. A new monograph of the Coniopterygidae has been published by
Meinander (1972). [Dieter Schlee.]

301. However, the Mantispidae and Dilaridae differ from the Raphidioptera
in the absence of the 1st valvulae and of the styli. [Michael Achtelig.]

302. Meinander (1972) has given a review of previous classifications. His own
results, which are focused on the systematic position of the Coniopterygidae,
are summarized in his Fig. 4: the closest relatives of the Coniopterygidae are
the 'higher Planipennia' (i.e. all the Planipennia except for the Ithonidae and
Coniopterygidae) and the Coniopterygidae + 'higher Planipennia' are the
sister-group of the Ithonidae. [Dieter Schlee.]

303. See note 309. [Michael Achtelig.]

2.2.2.2..4.2.   Coleoptera (beetles)[304]

The Coleoptera are as well founded a monophyletic group as we could ever
hope to find. They have the following ground-plan characters. There is a
prothoracic connection between notum, pleura, and sternum. The fore-wings
have been modified into rigid elytra and do not have any real venation, and
their sides are often reflexed to form the 'epipleura'. The articulation of the
hind-wing is protected by a 'humeral callus'. The hind-wings are the true
organs of flight, and when at rest they are folded beneath the elytra in a
complicated manner that varies from group to group (Forbes, 1926)[305]. The
venation is highly derived and is influenced by the way in which the wing is
folded[306]. Where the abdominal segments are covered by the elytra, the
tergites are membraneous. The tergites of the 7th abdominal segment
('propygidium') and of the 8th ('pygidium') are free and strongly sclerotized.
In both sexes, the segments after the 8th are withdrawn into the abdomen and

are not visible externally. The cerci and ovipositor have been reduced[307]. Bradley (1947) mentioned that the appendages of the 9th abdominal segment of the larvae have been modified into 'pseudocerci' (= 'urogomphi') and considered this to be a further derived character. However, Sharov (1966b) thought that the urogomphi were true cerci, in which case their presence in beetle larvae would be a primitive character.

The Coleoptera are virtually the only group of the Holometabola that have modified the fore-wings into protective covers for the hind-wings, the true organs of flight, and to a certain extent for the abdomen too. In this respect they have followed a line of development that was also taken by some important groups in the Paurometabola and Paraneoptera. It is certainly no coincidence that all these groups are markedly terrestrial and seldom use their wings.

Rodendorf (in Rodendorf, 1962) thought that the acquisition of the elytra was responsible for the extraordinary evolutionary success of the Coleoptera: not only are they by far the largest order of insects, but as adults and larvae they have colonized the most diverse habitats. The rigid connection between pronotum, propleura, and prosternum and the enlargement of the abdominal sternites are also closely associated with the origin of the elytra. These three features have converted the entire upper body surface, at least in the ground-plan of the Coleoptera, into an armoured plate with virtually no gaps or membraneous areas between the constituent sections[308].

I shall digress for a moment to discuss whether the assertion that the Coleoptera are the 'most successful' of all the insect orders or even of all the 'insect groups' has any real meaning. Before asking this question, let alone answering it, it is first necessary to understand what is meant by a 'successful' group. The simple assertion that the largest group is the most successful is certainly no answer. Saying that one group is more successful than another involves a comparison, and so a further preliminary step is needed to clarify whether and why the groups in question should actually be compared at all. It could be maintained that only groups which represent one particular closed functional type should be compared, and in this case it would have to be shown that the Coleoptera and the 'less successful' groups which are being compared do actually meet this definition. This could be true of adult beetles but not of their larvae. It is even possible that the beetles have been so successful because they have *not* restricted any of their postembryonic phases to one particular closed 'functional type'. However, this takes no account of the time factor, and this cannot be ignored. In this sense the only valid comparisons that can be made should be between sister-groups which began their evolutionary development with the same holomorphological and ecological endowment (characters present in their common stem-species) and which subsequently continued to have the same intervals of time and the same opportunities available for their development. This definition means that the Coleoptera, perhaps with the Neuropteroidea (see below), can only be compared with the rest of the holometabolan orders *in toto*, because it seems most probable that there is a sister-group relationship between the Coleoptera and almost all the other Holometabola. Such a comparison shows clearly that in numbers of species and in ecological valency, if these are the features to be considered as the criterion of evolutionary success, the Coleoptera do not substantially overshadow their sister-group.

There is no definite evidence yet to support the idea of a sister-group relationship between the beetles and all the other holometabolan orders, but this is immaterial for the comparison I have just been suggesting. The Neuropteroidea, with some 4500 species, and the Coleoptera should probably be considered together because they may form a monophyletic group of higher rank. However, no entirely satisfactory evidence has yet been put forward for this hypothesis, nor has it been generally accepted.

This hypothesis could be correct because there are several possible synapomorphies common to the Neuropteroidea and Coleoptera. The most important of these is the gula. Admittedly, a gula has arisen convergently in a number of groups. Weber (1933) considered that the gula originated through division of the submentum in the beetles, Neuropteroidea, and a few other groups that are not closely related to the beetles. On the other hand, DuPorte (1960) did not think that the origin of the gula in adult beetles had anything to do with the submentum. He believed that it developed through sclerotization of the membraneous area behind the submentum. However, he did not examine any Neuropteroidea, and so it is not absolutely certain that the gula of this group really is the same as the gula of the Coleoptera (see Achtelig, 1967).

In the larvae of the Neuropteroidea and Coleoptera, the 'gula' has arisen independently and convergently because it is absent from the larval ground-plan in both groups.

Prognathy is also connected with this character. DuPorte was right to suggest that prognathy is part of the ground-plan of the Coleoptera, at least in the imaginal stage. Species with orthognathous (hypognathous) and even opisthognathous heads do occur, but they have all retained the gula and it is likely that their orthognathy or opisthognathy has developed secondarily from prognathy. It is not clear how the various head forms of beetle larvae should be assessed.

There may be a second synapomorphy of the Neuropteroidea and Coleoptera in the fore-wings. In most Holometabola the wings are folded flat over the abdomen, but in the Neuropteroidea they are held more obliquely so that they also cover the sides of the abdomen. The enlargement of the costal field, which is so characteristic of the Neuropteroidea (but see the Dilaridae in Ross, 1965), is probably correlated with this. It is probably a parallel development to the enlargement of the precostal field in the Orthopteroidea: unlike the cockroach-like Paurometabola, the Orthopteroidea hold their wings like the Neuropteroidea.

In the Coleoptera, the elytra are arched and cover the sides of the body as well as the dorsum. The position of the wings in the Neuropteroidea may be a forerunner of this. A possible deduction from this is that the epipleura of the beetles could have developed from the enlarged costal field of the Neuropteroidea. This is undoubtedly a daring idea, but one day it may have a certain heuristic value[309].

Various authors have considered individual subgroups of the *Neuropteroidea, rather than the Neuropteroidea as a whole, to be the closest relatives of the Coleoptera.

According to Théodoridès (1952), Lameere has been the only author to derive the beetles from the 'Planipennia'.

Martynov (1931), Bradley (1947), and Jeannel (in Grassé, 1949) thought that the Megaloptera and Coleoptera had common ancestors. There are several characters common to these two groups, especially in the larvae, and these are supposed to rule out the possibility of any alternative view. However, they are all symplesiomorphies. The 'common ancestors' which had the characters retained by recent Megaloptera and Coleoptera could just as well have been the common ancestors of all Neuropteroidea and Coleoptera. Bradley and Jeannel have fallen into the usual trap of overlooking a fundamental distinction: the important feature is not whether these groups had 'common ancestors' but whether they had ancestors that were 'common only to them'. There is no evidence for this[310].

Ross (1955) considered the Raphidioptera to be the closest relatives of the Coleoptera. His reasons for this were the condensation of the wing venation and the development of a retractile male genital capsule. He was right to discard this idea in his later book (Ross, 1965).

If Bradley (1947) was correct in deriving the pseudocerci (urogomphi) of beetle larvae from rudiments of the appendages of the 9th abdominal segment, the larvae of the ancestors of the beetles must have been even more primitive than those of all the recent Neuropteroidea. This is because the Neuropteroidea, including the Megaloptera, have abdominal appendages present on no more than the first eight abdominal segments, as tracheal gills. It is therefore highly significant that a larva has been discovered in the Upper Permian (*Permosialis*, p. 288) which is supposed to belong to the Megaloptera and which has processes on the 9th abdominal segment as well. The larvae of recent Neuropteroidea and Coleoptera can easily be derived from a larva such as this. According to Chen (1946) the larvae of the Gyrinidae are supposed to have ten pairs of 'abdominal styli', but this statement is not entirely clear[311]. However, if Sharov (1966b) was correct in suggesting that the urogomphi of beetle larvae are true cerci, then the ideas I have put forward in this paragraph would be untenable.

The oldest fossils that undoubtedly belong to the Coleoptera are from the 'Lower' Permian of southern Siberia and the Urals (Fig. 103). About 15 species in 7 different families have been described from the Lower Permian of the Kuznetsk basin and the Perm region. Other species are known from the Upper Permian of the Arkhangelsk region, the Kuznetsk basin, the Urals, and Australia. Some of them have been assigned to the same 'families' as the

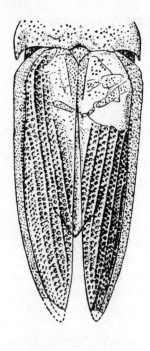

Fig. 103. *Tshekardocoleus magnus* Rodendorf (Coleoptera), from the Lower Permian of Chekarda (Urals). From Rodendorf (1962).

Lower Permian fossils. Consequently, it hardly matters that the Permian 'Protelytroptera', which were originally described as Coleoptera, actually belong to the Paurometabola (see p. 187). The same is true of the Permofulgoridae.

Laurentiaux (in Piveteau, 1953) removed the families Permophilidae, Permosynidae, Permarrhaphidae, and Sojanocoleidae from the true 'Coleoptera' and assigned them to a separate order 'Paracoleoptera'. Jeannel (in Grassé, 1949) had already used the name Archicoleoptera for this group. Some of them are even more primitive than the other fossils, including the Lower Permian ones, and the recent Coleoptera. Their most striking characters are the membraneous elytra, which are scarcely or only weakly arched, and the absence of the humeral callus. Jeannel thought that they were still able to use the fore-wings as active organs of flight. However, some 'true' beetles are known from the same epoch. Consequently, these Archicoleoptera can only be survivors from an even older stem-group, and they also provide an indication that the Coleoptera were in existence during the Upper Carboniferous.

An equally important question is whether any affinities can be discerned between the Permian fossils and individual subgroups of the recent Coleoptera. Rodendorf and Ponomarenko (in Rodendorf, 1962) assigned eight Lower Permian and seven Upper Permian species from the Kuznetsk basin to the recent family Cupedidae (= Cupidae). This family has usually been placed in the Adephaga, but there is some evidence that there is a sister-group relationship between the Cupedidae and the rest of the Coleoptera (Machatschke, 1962). For this reason the Cupedidae have been removed from the Adephaga + Polyphaga and called the 'Archostemata'. Machatschke even wanted to restrict the name Coleoptera to the Adephaga + Polyphaga without the Archostemata, but no one is likely to follow him in this. However, he was right to remove the Micromalthidae from the Archostemata, an assignment that was made by Crowson[312].

Machatschke noted two characters in which the Cupedidae (Archostemata) are more primitive than all the other Coleoptera: the elytra are not completely sclerotized and the original longitudinal veins and cross-veins can still be seen; the hind-wings still have numerous cross-veins, and have the distal part spirally coiled in repose instead of folded beneath the elytra as in the other Coleoptera. He also found two characters in the Cupedidae that are derived when compared with the rest of the Coleoptera: the scales on the elytra and the invagination of the 7th abdominal segment[313]. These characters show that the Cupedidae are a monophyletic group. There are a few other derived characters which add further support to this view but which are less convincing because they are also found in some of the other groups of the Coleoptera (see Machatschke, 1962).

Other authors have treated the Cupedidae as the sister-group of the Polyphaga rather than the rest of the Coleoptera (see p. 388).

If Permian fossils really do belong to the Cupedidae, then the stem-group of the rest of the Coleoptera (Adephaga + Polyphaga) must have been in existence in the Lower Permian; or, if the Cupedidae actually belong to the Polyphaga, the Adephaga and the rest of the Polyphaga must both have existed as separate groups at that time. None of this can be proved, because none of the derived characters that establish the recent Cupedidae as a monophyletic group are visible in the Permian fossils. The invagination of the 7th abdominal segment[313], and some other derived characters found by Machatschke, such as the cryptogastrous abdomen and the presence of only four Malpighian tubes, are hardly likely to be visible in fossils. Nor have I been able to find any mention of scales on the elytra or of the absence of an alula in the Russian descriptions of fossils. Admittedly, the alula is a rather insignificant character and would not mean a great deal on its own. Until there is some evidence to the contrary, there is no alternative but to regard the characters common to all the Permian fossils, including the so-called Permian 'Cupedidae', and the recent Cupedidae as symplesiomorphies. The Permian fossils have only been assigned to the Cupedidae because these are the only recent beetles that have retained the relatively primitive structure of the elytra which is also visible in the Permian fossils. However, this structure must have been present in the direct common ancestors of all the recent beetles.

The only possible course is to assign all the Permian fossils, including the so-called Cupedidae, to the stem-group of the Coleoptera and to admit that no close relationship with any of the subgroups of the *Coleoptera can be

306

detected. On the other hand, it is possible that some of the subgroups of the *Coleoptera were actually in existence during the Permian.

As I mentioned above, the sister-group of the Coleoptera is still not known. This is largely because it has not been possible to establish that the Neuropteroidea are a monophyletic group, although it seems most likely that they are the sister-group of the Coleoptera. It is possible that one of the subgroups of the Neuropteroidea is most closely related to the Coleoptera. Further difficulties are caused by the uncertain relationships of the Strepsiptera[314].

Revisionary notes

304. Abdullah (1975) has given a list of the families of the Coleoptera, including an inventory of the fossil records. [Willi Hennig.]

305. Chen and T'an (1973) have described a new family, the Umenocoleidae, from the Lower Cretaceous of Kansu. Amongst other things, this is characterized by the well developed and mostly non-parallel veins of the elytra and by the hind-wings apparently being unfolded when at rest. [Michael Achtelig.]

306. The venation of the hind-wing has been discussed by Wallace and Fox (1975). [Willi Hennig.]

307. Abul-Nasr (1954) reported that in all adult beetles the gonocoxites (gonopods) are reduced. The paramere lobes have divided: the inner pair has fused with the intromittent organ, whilst the outer pair has been retained to form the true parameres. [Willi Hennig.]

308. Hlavac (1972) has described just such a hypothetical ground-plan for the beetle prothorax. However, there are a number of objections to this view (Baehr, 1976, 1979). All the Adephaga and the Cupedidae (Archostemata) have a thorax resembling an armoured plate, which is best exemplified by the prothorax. For this reason, movement of the procoxa is restricted. On the other hand, many of the Polyphaga have the prothorax fragmented and with extensive membraneous areas. In this case, the procoxa is freely mobile. This is the situation found in various primitive groups and in families which are the most primitive members of particular family-groups, such as the Cantharoidea, Silphidae, Staphylinidae, Lymexylonidae, Pyrochroidae, and Meloidae.

In addition, Baehr (1979) has described some apparently highly primitive characters in the prothorax of *Atractocerus* (Lymexylonidae) which suggest that in its ground-plan the prothorax of the Coleoptera is very similar to the ground-plan of the Planipennia. These characters are the presence of the anapleural suture in the so-called precoxal bridge; presence of the pleurosternal suture in the propleura; a very simple articulation between the propleura and pronotum; the presence of several post-pleural sclerites, a large spinasternum, and two lateral cervical sclerites and one dorsal cervical sclerite on each side.

In various families of the Polyphaga, which are considered to be derived on the basis of other characters, this fragmented prothorax has been secondarily modified into the compact adephagan prothorax (e.g. the families Oedemeridae, Alleculidae, and Tenebrionidae of the Heteromera). For this reason, Baehr considered it possible that a compact prothorax was not originally present in the ground-plan of the Coleoptera but arose subsequently and possibly independently in the Cupedidae, Adephaga, and various groups of the Polyphaga. [Martin Baehr.]

309. A most striking synapomorphy has been described by Mickoleit (1973) in his study of the female genitalia of the Coleoptera and Neuropteroidea. He found that in their ground-plan both groups have a simplified ovipositor which consists of a pair of vaginal palps. These have developed from the 3rd valvulae and the gonocoxites of the 9th segment, through coalescence. They are medially fused in the Neuropteroidea but remain separate in the Coleoptera. They have a subapical division which is considered to be the last annulus of the style (= lamnium of Smith, 1969) or the whole style (Achtelig, 1978). The vaginal palps have an external 'lateral sclerite', which articulates at its base with the anterolateral corner of the 9th tergite, indicating that the gonangulum has been absorbed into this tergite. The function of the cerci has probably been taken over by the 3rd valvulae which have a number of sensilla on the apical segment. The 1st valvulae are very short, perhaps through convergence, and have probably been absorbed into the inner wall of the 3rd valvulae. The 2nd valvulae have atrophied and have probably been absorbed into the vaginal palps. This absorption can be observed clearly during morphogenesis of the raphidiopteran ovipositor (Achtelig, 1978). After describing some other synapomorphies in the ovipositor of the Neuropteroidea, Mickoleit concluded that the only conceivable relationship between the Coleoptera and the Neuropteroidea is that of sister-groups. [Michael Achtelig, Martin Baehr.]

Further evidence has been provided by Chadwick (1959) who drew attention to the absence of the supposedly primitive cruciate cervicle muscle in the Neuropteroidea and Coleoptera, which is retained in the Hymenoptera and some Mecopteroidea. For further discussion of this character, see Kristensen (1975). [Michael Achtelig, Martin Baehr.]

The Raphidioptera, several Planipennia, and several Coleoptera have a specialized spermatheca, which has one connection to the bursa copulatrix for sperm uptake and another to the vagina for egg fertilization. The morphogenesis of this character is strikingly similar in the Raphidioptera and Coleoptera (Achtelig, 1978). Different conditions in the Coleoptera and Megaloptera can be explained by the incomplete development of the inner genital ducts. [Michael Achtelig.]

310. Hamilton (1972c) has recently discussed this view in the light of certain characters in the venation of both groups. Kristensen (1975) discussed Hamilton's arguments and challenged the validity of his results. [Martin Baehr.]

311. See note 285. [Dieter Schlee.]

312. There are hardly any grounds for Crowson's view. The only possible evidence is the presence of a so-called sclerome in the hypopharynx of the Cupedidae and Micromalthidae, but it seems most likely that Klausnitzer (1975) was right in suggesting that this character arose convergently because the larvae of both groups have very similar habits. The Micromalthidae lack the characteristic elongate notopleural suture present in the Cupedidae (see the illustration of the prothorax of *Micromalthus* by Hlavac, 1975: Fig. 5), and the only characters that support the assignment of the Micromalthidae to the Archostemata are symplesiomorphies. In fact, the Micromalthidae may be rather simplified members of the Cantharoidea or Lymexylonidea (Polyphaga). [Martin Baehr.]

313. This is an error, as the Cupedidae have seven distinctly separate sternites. The 1st sternite is very thin and is situated in the membrane between the metacoxa and the 2nd sternite. As a result it is not visible externally and was overlooked by Machatschke (1962). However, it was first described by Edwards (1953) and has been recently illustrated by Baehr (1975). Elytral scales are also present in other beetles, but may have arisen convergently. At present, it is still very difficult to establish that the Cupedidae are a monophyletic group. [Martin Baehr.]

314. See note 309. [Michael Achtelig.]

2.2.2.2..4.3. **Strepsiptera (stylops)**[315]

There can be no doubt that the Strepsiptera are a monophyletic group. Many of their derived characters are connected with the endoparasitic habits of their larvae. Most females also remain within the host, but in their ground-plan they are free-living. Males have the fore-wings reduced to 'halteres', whilst the hind-wings are greatly enlarged, can be plaited, and have the venation very reduced. Females are apterous. Male antennae are short, flabellate, and have seven segments in their ground-plan. Both sexes have the mouth-parts reduced. The tentorium is absent (Kinzelbach, 1967), and so are the ocelli and the prothoracic spiracle. The alimentary canal is rudimentary in both sexes, and is closed behind. The Malpighian tubes are absent. So far as is known, the nervous system is highly concentrated with only one thoracic and one abdominal ganglionic mass. The female genital opening is behind the 7th abdominal sternite. Males lack the gonopods (parameres) and females the ovipositor. During embryonic development there is only one embryonic envelope, and this is considered to be the amnion. The larvae have very reduced thoracic legs and after the 1st stage are endoparasites of other insects[316].

Whilst it is easy to show that the Strepsiptera are a monophyletic group, it is extremely difficult to work out their affinities. Earlier authors such as Lameere regarded them as a subgroup of the Coleoptera. Even Handlirsch thought that they were no more than a parasitic family of the Malacodermata, and Ross (1965) followed Crowson by treating them as a superfamily ('Stylopoidea') of the beetles. He thought that the triungulin larvae, the reduced wing venation, and the parasitic mode of life indicated that they were closely related to the Meloidae.

Crowson (1960) put forward the most convincing case for not regarding the Strepsiptera as more than a group 'close to the Coleoptera and arising from the same lineage' (Ulrich, 1943), and he considered that they belong to a relatively subordinate group of the *Coleoptera[317].

According to Crowson, the assignment of the Strepsiptera to the Polyphaga is supported by the absence of a notopleural suture on the prothorax of the adult and by the absence of one of the leg segments in the 1st larval instar: this has taken place by reduction or, according to Crowson, by fusion of the tarsus and pretarsus. Within the Polyphaga, the adults agree with the Cucujiformia in the absence of the 8th abdominal spiracle and in the structure of the metendosternite[317]. Unfortunately, two other important derived characters of the Cucujiformia or Polyphaga are not present in the Strepsiptera: cryptonephry (Cucujiformia) and telotrophy of the ovaries (Polyphaga). The Malpighian tubes are completely reduced in the Strepsiptera, and according to Lauterbach (1954) the females have no true reproductive organs: the abdomen is completely filled with a random arrangement of eggs and fat-body cells. Both characters are autapomorphies of the Strepsiptera, and it is impossible to determine whether the Malpighian tubes and ovaries were like those of the Cucujiformia before they were lost.

The derived characters that Crowson found to be common to the Strepsiptera and Cucujiformia could be either synapomorphies or the result of convergence. Their interpretation depends on whether the Strepsiptera also have any character in which they are more primitive than all the Coleoptera. If they do have such primitive characters, then the characters that they share with the Cucujiformia could not be regarded as synapomorphies. The lack of a gula in the Strepsiptera could be such a character, if it is in fact primitive. The presence of a gula, and of the prognathous head associated with it, appears to be one of the derived ground-plan characters of adult Coleoptera, or even of a group of higher rank consisting of the Coleoptera and Neuropteroidea. The ancestors of the Strepsiptera must have had both of these characters if they really do belong to the Cucujiformia. Unfortunately, it is not certain whether the gula is primitively absent in the Strepsiptera or whether it has been secondarily lost as a result of the reduction of the mouth-parts. There are various beetles with secondarily hypognathous heads in which the gula appears to have been retained. A careful morphological study may be able to shed some light on this problem[318].

Pierce (1964) challenged the assignment of the Strepsiptera to the Coleoptera, but his results were fundamentally unsound[319]. He did no more than list the peculiarities of the Strepsiptera, all of which appear to be autapomorphies of the group and are not found in the beetles. He stated that these characters are found in a similar stage of expression in groups like the Paraneoptera and certain Diptera, but there are other reasons for not regarding any of these as closely related to the Strepsiptera. He was right to point out that the tarsal isomery is a strong argument against the commonly held view that there is a close relationship between the Strepsiptera and the heteromerous families Rhipiphoridae and Meloidae, unless further unfounded hypotheses are brought forward to add support[319].

Unlike the authors who attempted to associate the Strepsiptera with the beetles, Jeannel (1945) has tried to show that they are most closely related to the Hymenoptera[319]. At first sight, his most appealing argument was that the Hymenoptera and Strepsiptera are the only holometabolous groups which have a single embryonic envelope. However, Weber (1949) pointed out that only the Hymenoptera Aculeata have a simple embryonic envelope and that it is usually interpreted as the serosa; in the Strepsiptera, on the other hand, the single embryonic envelope is homologous with the amnion. According to Jeannel, a truly derived character common to the Strepsiptera and the Hymenoptera is the fusion of the 1st abdominal segment with the thorax. The morphology of the basal abdominal segment is also used for defining morphological subgroups of the beetles, and so it may provide a valuable feature for more detailed morphological studies.

Kinzelbach (1967) has recently used the head morphology to work out the relationships of the Strepsiptera. He found 'many characters that suggest a close relationship with the mecopteroid complex, particularly with the Trichoptera, Zeugloptera, and Lepidoptera'. However, he has done no more than produce a list of strepsipteran characters and show that they are present in the same stage of expression in various species of the Mecopteroidea that he examined. Inevitably, individual characters of the Strepsiptera were found to be present in diverse species. Kinzelbach's assertion that 'most of the characters that the Strepsiptera share with species of the mecopteroid complex, particularly the Trichoptera, Zeugloptera, and Lepidoptera, should be regarded as synapomorphies' cannot be accepted as it is premature and methodologically incorrect[319]. If the concept of a synapomorphy is to mean anything at all, it can only be used when a particular character can be shown to be present in an undoubtedly derived state in the *ground-plan* of two groups. Kinzelbach's paper may well be of value as a stimulus for further investigations but his conclusions about the relationships of the Strepsiptera are no better founded than were those of earlier authors.

The earliest fossil Strepsiptera have been found in Baltic amber[320].

Revisionary notes

315. Recent revisions have been published by Kinzelbach (1971a, 1971b, 1978). [Ragnar Kinzelbach.]

316. There can be no doubt that the Strepsiptera are a monophyletic group. The endoparasitic mode of life of the secondary larvae and, in the Stylopidia, of the adult females, have made the most significant contribution to their complement of highly derived characters, that is to the large number of their autapomorphies. The tiny primary larvae, which are produced in enormous numbers, are the infective stage. Adult males are short-lived, and their activities are confined to seeking, finding, and fertilizing the females.

The antennae of the adult male are flabellate and consist at most of seven segments. Mandibles are absent only in the Corioxenidae. The maxillae consist of a single segment, with an unsegmented palp, or are partly fused with the head capsule (Halictophagidae). The labium, the free hypopharynx, and at least part of the labrum form a membraneous or weakly sclerotized mouth area which contains the more or less prognathous mouth. There is no trace of a sclerotized gula: the submentum closes the head ventrally. The site of the tentorium can be located by the tentorial grooves, but it is no longer used in the adult as a point of origin for the muscles. The compound eyes, which have individual facets of an ocellar type, are a special apomorphy. Vestiges of the median ocelli are present beneath the cuticle.

The metathorax is very much larger than either prothorax or mesothorax. The hind-wings are large and supported only by longitudinal veins. They can be plaited and held backwards over the body. The fore-wings are short, without any supportive surface, and function like halteres. The fore and mid coxae are free and well developed, but the hind coxae have been absorbed in varying degrees into the structure of the metathorax. The fore and mid trochanters are fused with the femora, but the hind trochanters are free. All the tibiae have a more or less conspicuous, well defined, proximal 'pseudo-patella'. In primitive species the tarsi are 5-segmented and have claws, but in most families the apical 1, 2, or 3 segments have been lost on all legs.

The middle intestine is closed behind in both sexes. Two or three Malpighian tubes are present in the hind intestine. The ganglion pharyngeum inferius (oesophageal ganglion) forms a common ganglionic mass with the thoracic ganglia and some or all of the abdominal ganglia; the remaining abdominal ganglia may coalesce to form a separate mass in the abdomen. The Mengenillidae have spiracles in the mesothorax, metathorax, and abdominal segments 1–7; in all other families only the spiracles of the 1st abdominal segment are functional in both sexes. Gonopods and parameres are absent.

Females are always wingless. In the Mengenillidae fertilized females are free-living, but so far as is known they live permanently within the host in all the other families. The anterior part of the body ('cephalothorax') protrudes out of the host as far as the 1st abdominal segment, enclosed in the cuticle of the final larval instar and of the pupa. Either the ventral surface (Corioxenidae) or the dorsal surface (other families) is directed towards the host. In primitive species the genital opening is behind the 7th abdominal sternite; there is no ovipositor.

Primary larvae are free-living and have only one tarsal segment. Secondary larvae pass through at least five stages (often with hypermetamorphosis), and hosts range through at least nine orders of insects. [Ragnar Kinzelbach.]

317. Crowson (1960) has laid some stress on a number of characters common to the Strepsiptera and plesiomorphic Coleoptera, but not all of his findings were correct: Strepsiptera have no evident gula; a notopleural suture is present on the prothorax; two or three Malpighian tubes are present in most families, particularly the most plesiomorphic ones. Other characters mentioned by Crowson, such as the one-segmented tarsus of the primary larvae, the absence of the 8th abdominal spiracle, and the structure of the metendosternite, are too vague to be used as evidence of a synapomorphy. As they occur so frequently, they should be regarded as parallel reductions or as symplesiomorphies.

Abdullah's (1974) proposal to group all the various entomophagous Coleoptera and the Strepsiptera into a suborder 'Entomophaga' is not worth discussing. [Ragnar Kinzelbach.]

318. Crowson (1960) was also unable to find any convincing synapomorphies shared by the Coleoptera and Strepsiptera. The only characters in common that have any weight are the opisthomotoria and prognathy. However, several other orders have a tendency towards these two conditions. I should also point out that prognathy in the Strepsiptera only affects certain parts of the head.

Hennig has clearly formulated the most important question: can it be shown that the Strepsiptera have characters that are more primitive than those of the Coleoptera? The following four characters come into this category: (1) the fore-wing has several distinctly separate veins which are serially homologous with those of the hind-wing; it is impossible to envisage such a structure developing from an elytron; (2) like the hind-wing, the fore-wing is attached to the side of the thorax; unlike any known elytron, it can be rapidly whirred; (3) the structure of the thorax is plesiomorphic, as is shown by the position of the subalar and basalar, and by the basal articulation of the wing; apart from the opisthomotoria, there are no evident synapomorphic specializations common to Strepsiptera and Coleoptera; (4) there are ten abdominal segments. Several additional characters could be mentioned. However, it is not always possible to decide whether an apparently plesiomorphic character is in fact an original plesiomorphy or has resulted from the retention of a larval character ('paedomorphosis'). As a result these characters do not carry much weight. Further comparative studies need to be made.

In addition to the morphological facts, there is some further evidence that has received very little attention.

Crowson (1960) suggested that the Strepsiptera arose as a subgroup of the Coleoptera during the Cretaceous. However, the order must be considerably older than this, even though the earliest fossils are from the Baltic and Dominican ambers of the Tertiary. There are several reasons for this: (1) the large number and complexity of the morphological autapomorphies of the *Strepsiptera, and their biological specializations; (2) uniformity of structure and paucity of characters in the *Strepsiptera; (3) many taxa of higher rank, with relatively few species apart from the large number of 'subjective' species that still have to be rationalized; (4) cosmopolitan distribution; (5) a wide spectrum of hosts, but at the same time a high host specificity; the origin of this presupposes a considerable period of time (the beetle families Meloidae and Rhipiphoridae are restricted to a few groups of hosts; (6) a well balanced parasitism; (7) it is possible to find phylogenetic parallels between the Strepsiptera that live in Hymenoptera and their groups of hosts; this must place the origin of the Stylopidia at least as early as the Upper Jurassic; the more plesiomorphic stem-groups from other hosts must be older still; it is therefore likely that the order originated during the Permian. This means that there can be no question of the Strepsiptera being a derived subgroup of the *Coleoptera. They may possibly have 'shared part of their evolutionary history with the ancestors of the Coleoptera' (Ulrich), in other words with the stem-group of the Coleoptera or with a number of holometabolous orders but not with the *Coleoptera. However, they must have diverged before the *Coleoptera acquired their full complement of characters. They have remained a relict group, archaic in some respects yet exhibiting a specialized form of parasitism. [Ragnar Kinzelbach.]

319. None of Pierce's conclusions are worth any attention except those in which he argued against the assignment of the Strepsiptera to the Heteromera. There is no sound evidence for Jeannel's proposed assignment of the Strepsiptera to the Hymenoptera. Hennig was also right to criticize my own earlier assignment of the Strepsiptera to the 'Mecoptera, Zeugloptera, Trichoptera' (Kinzelbach, 1967), which I myself have completely retracted (Kinzelbach, 1971b, 1978): my arguments were based on plesiomorphies. [Ragnar Kinzelbach.]

320. Until recently, the only known fossil Strepsiptera were six males of a single species, *Mengea tertiaria* Menge, from Baltic amber. This species could not be assigned to any recent family and was therefore placed in its own family, the Mengeidae (Pierce, 1908). Kinzelbach (1978) has provisionally treated it as a separate family pending the discovery of the female. In the light of present knowledge, the Mengeidae could be the sister-group of the Mengenillidae or could belong to the stem-group of the Mengenillidae.

The original series of this species is thought to be lost. Kinzelbach (1978) based his detailed description on a newly discovered specimen from a private collector.

In 1977 I discovered two new strepsipterous fossils, which considerably extend our knowledge of the order since they belong to two families not previously known from fossils. One is from Baltic amber and the other from Dominican amber (Schlee and Glöckner, 1978), which is the first record of a fossil strepsipteron from America. These fossils are being studied by Kinzelbach, and form part of the amber collection in the Department of Phylogenetic Research, State Museum of Natural History, Stuttgart. [Dieter Schlee.]

2.2.2.2..4.4. Hymenoptera[321]

The relationships of the Hymenoptera are not clear. As a monophyletic group they are just as well founded as the Coleoptera and Strepsiptera but, unlike these groups, it is almost impossible to find any correlation between their numerous derived ground-plan characters and the peculiarities in their mode of life. I would suggest with some confidence that the phytophagous habits of the larvae, in other words their restriction to living plants, is an apoecous ground-plan character of the Hymenoptera. However, there is no apparent correlation between the morphological peculiarities of the larvae and their changed mode of life. Eruciform larvae are part of the ground-plan of the Hymenoptera, and in comparison with thysanuroid or campodeiform larvae they probably represent a derived larval type, particularly as the femur and tibia are fused according to Hinton (1958). On the other hand, it is possible that the Hymenoptera share the eruciform larval type with their sister-group and that it was inherited from their non-phytophagous ancestors. Unfortunately, it is not known which group is their sister-group, and it is also clear that the largest subgroups of the Hymenoptera have moved away from phytophagy. Nevertheless, it seems safe to say that as a group the Hymenoptera owe their origin to the acquisition of the phytophagous habit by their larvae. However, their evolutionary success did not result from this, but apparently came about because they were able to take advantage of other possibilities made available by their newly acquired mode of life.

Apart perhaps from the fusion of femur and tibia, no undoubted derived ground-plan characters are known in the larvae, although their adoption of phytophagous feeding was the seminal change that marked the point of origin of the Hymenoptera. On the other hand, the adults have numerous derived characters but it has not been possible to correlate them with any of the peculiarities in their mode of life.

The wing venation is very difficult to interpret (Ross, 1936, 1937) and has many derived characters which I do not need to list here. The hind-wings are small and are joined to the fore-wings by hamuli; their venation is even more reduced than that of the fore-wings. Berland and Bernard (in Grassé, 1951) considered their function to be for steering rather than for active flight. The 'cenchri' are also a derived ground-plan character of the Hymenoptera since they are found in the most primitive groups of the order (all the so-called Symphyta, except for the Cephidae). The 'cenchri' are 'two raised bosses protruding from the metathoracic membrane between the scutum and scutellum'. Their upper surface is 'roughened by a layer of microscopic fine lamellae of varying structure. There is an area of spinules on the lower surface of the anal lobe of the fore-wing: when the wing is at rest this fits on to the cenchri and anchors the wing over the abdomen' (von Kéler, 1955, 1956).

In the ground-plan of the Hymenoptera the 1st abdominal segment ('propodeum') is firmly fused with the metathorax. The 1st sternite is very reduced and is usually absent. The abdominal tergites and sternites overlap. Berland and Bernard (in Grassé, 1951) reported that all Hymenoptera have poison glands, which are accessory glands of the female reproductive organs. They are therefore part of the ground-plan of this group although nothing appears to be known of their function amongst the primitive phytophagous species ('Symphyta').

A further ground-plan character is that all the males seem to be genetically haploid and always arise from unfertilized eggs (Berland and Bernard in Grassé, 1951). White (1957) noted that as a consequence of these cytological facts all genes in the Hymenoptera behave as if sex-linked[322].

As with other groups that are characterized by a large number of special derived characters ('autapomorphies'), it is difficult to work out the relationships between the Hymenoptera and other groups of the Holometabola[323]. The most divergent views have been put forward.

Ross (1955) suggested that similarities in the male genitalia indicate that there is a close relationship between the Hymenoptera and Coleoptera (including Strepsiptera). In his phylogenetic tree he showed these two as sister-groups within a single monophyletic group of higher rank. Hinton (1949) found that the Coleoptera, Strepsiptera, and Hymenoptera have developed a unique method of escaping from the pupal case or cocoon that is not found in any other insects: the adult remains in the cocoon until the mandibles are hard enough to be used for biting a way out. This could well be regarded as a derived character (as a synapomorphy), supporting the assignment of these three groups to a single monophyletic group. However, I doubt if this interpretation is correct. The interpretation of the structure of the coxae is equally difficult. Wille (1960) attributed coxal mera to the ground-plan of the Holometabola. Their absence in the Hymenoptera, Coleoptera and Strepsiptera would then be a derived character (a synapomorphy?) of these three groups. However, coxal mera are also absent in the Siphonaptera. On the other hand, if the absence of mera is actually a primitive character, the opposite of Wille's view, then it would be difficult to place the Neuropteroidea, which do have these mera. The mera would have had to have originated independently in the Neuropteroidea and Mecopteroidea, or would indicate that these two groups are closely related. The

Hymenoptera agree with other groups of the Holometabola in other un-doubted derived characters, but it is not certain which of these are synapo-morphies and which are the result of convergence.

Martynov (1937) based his views on similarities in the wing venation between the Hymenoptera, the Raphidioptera, and in part the Sialidae (Megaloptera). He concluded that the Raphidioptera might be a conservative side-branch in the phylogenetic tree of the Holometabola that developed from ancestors that were closely related to the Hymenoptera. His phylogene-tic tree of the insect orders (Martynov, 1938b) showed the Hymenoptera and Raphidioptera as sister-groups. This view was followed in the Russian Textbook of Palaeontology (Rodendorf, 1962, Fig. 718 on p. 240) and by Ross (1955) (in his text, but not in his phylogenetic tree).

However, the suggestion that there is a sister-group relationship between the Hymenoptera and the Raphidioptera conflicts with all the evidence indicating that the Neuropteroidea are a monophyletic group (see p. 285). It is so difficult to interpret the venation of the Hymenoptera that it should not be used for working out the phylogenetic relationships of the group. It is possible that most of the similarities that Martynov noted in the venation of the Hymenoptera, Raphidioptera, and Sialidae are symplesiomorphies. This is certainly the case with the enlarged pronotum, another character which Martynov emphasized. The Hymenoptera have no trace of the enlarged costal field which is so characteristic of the Raphidioptera and the rest of the Neuropteroidea.

The Hymenoptera differ from all other Holometabola by having a fully formed orthopteroid ovipositor. This is apparently the character that encour-aged Ross (1965) to reject his earlier views and to suggest that there is a sister-group relationship between the Hymenoptera and the rest of the Holometabola. The reduction of sections of the orthopteroid ovipositor would then have to be regarded as a constitutive synapomorphic character of 'the rest of the Holometabola'. However, reductions in the ovipositor have taken place so frequently in the insects through convergence that they cannot be used on their own to establish that a group is monophyletic. There appears to be no other evidence for suggesting that there is a sister-group relationship between the Hymenoptera and the rest of the Holometabola.

Finally, it has also been suggested that the Hymenoptera are phylogeneti-cally close to the 'Mecopteroidea' (usually including the Siphonaptera). Two characters appear to support this: the firm attachment of the pronotum to the mesonotum, although the prosternum remains mobile; and the pupal cocoon constructed by the larva with a silk-like secretion produced by one of the labial glands. If any of the Neuropteroidea and Coleoptera form a cocoon, it is made with material from the Malpighian tubes[324]. However, Mickoleit's (1967) fundamental study of *Merope* has shown that the pronotal character is not yet part of the ground-plan of the Mecoptera. It is possible that the presence of a single pretarsal claw in larval Hymenoptera and Mecopteroidea (Snodgrass, 1961) indicates that these two groups are closely related. The larvae of the Neuropteroidea and Coleoptera have retained two claws in their ground-plan.

It may be possible to show that the Hymenoptera are most closely related to the Mecopteroidea (or Mecopteroidea + Siphonaptera?), in which case the two groups can only have a sister-group relationship[325]. This is because the Mecopteroidea have a number of derived characters (see p. 322) which are not present in the Hymenoptera.

In the Russian Textbook of Palaeontology (Rodendorf, 1962), as well as in her earlier papers (1957, 1959), Martynova suggested that the late Palaeozoic and early Mesozoic 'Paratrichoptera' should be regarded as the stem-group of the Hymenoptera and Diptera. She did not state whether the Siphonaptera should also be included. This conflicts with Rodendorf's own view that there is not likely to be any close relationship between the Hymenoptera and Mecopteroidea, and with the phylogenetic tree given in his Textbook (Rodendorf, 1962, Fig. 718).

Martynova's main reason for this suggestion seems to have been the trend towards reduction of the hind-wing in these two groups and in the fossil Paratrichoptera. However, this view is in direct conflict with all the evidence which indicates that the Mecopteroidea are a monophyletic group, and it cannot be accepted. Martynova has made a mistake that is very common amongst palaeoentomologists: she has found that it is *possible* to make a formal derivation of particular characters present in *different* recent groups from the *same* fossils, but has then overlooked all the morphological evidence making it impossible for these particular groups to have originated from the same stem-group.

It is not certain how far the larvae of the Hymenoptera agree with those of the Mecoptera in the fusion of femur and tibia which, according to Hinton (1958), is one of their ground-plan characters. Even if this proves to be a character common to these groups, it is hardly likely to be a synapomorphy.

The earliest fossils that can be reliably assigned to the Hymenoptera are from the Mesozoic (Triassic)[326]. All the groups that might be the sister-group of the Hymenoptera are known to have been in existence during the Palaeozoic, and so there is little doubt that the Hymenoptera must have arisen by the Permian at the latest. Because there are so many difficulties in interpreting the venation of this group, it is extraordinarily difficult to decide whether particular Permian fossils could belong to the stem-group of the Hymenoptera or not.

The old view put forward by Tillyard (1927), that the Megasecoptera or one of their subgroups was the stem-group of the Hymenoptera, has now been completely rejected.

It is most likely that some of the Palaeozoic fossils assigned to the 'Mecoptera' actually belong to the stem-group of the Hymenoptera.

There is one derived character in the venation that could be of importance for assessing the fossils: the anal veins, and sometimes others, no longer reach the hind-margin of the wing. Their apical sections have been reduced so that

318

each anal vein appears to terminate in the one in front of it. Martynova (1959) thought that the Paramecoptera, from the Upper Permian of Australia, were an extinct side-branch of the 'Eumecoptera', but Riek (1953a) considered them to be the stem-group of the Hymenoptera (Fig. 104).

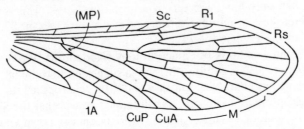

Fig. 104. Fore-wing of *Belmontia mitchelli* Tillyard (Paramecoptera), from the Upper Permian of Australia. From Tillyard (1919d).

All that can be definitely established at present is that it is not possible to trace any of the subgroups of the *Hymenoptera back into the Palaeozoic. However, it is very likely that the group as a whole originated during the Palaeozoic, but no fossils have yet been found that will prove this.

Revisionary notes†

321. Königsmann (1976, 1977, 1978a, 1978b) has published a Hennigian phylogenetic analysis of the Hymenoptera, with a discussion of all the relevant literature. [Eberhard Königsmann.]

322. At present the ground-plan of the *Hymenoptera includes 19 tentative characters (i.e. autapomorphies of the *Hymenoptera). These have been listed by Königsmann (1976). [Eberhard Königsmann.]

323. The sister-group of the Hymenoptera is believed to be the Mecopteroidea (Königsmann, 1976). [Eberhard Königsmann.]

324. In *Rhynchaenus* (Coleoptera, Curculionidae), the cocoon is made with a silk-like secretion produced from the peritrophic membrane of the midgut, not from the Malpighian tubes (Streng, 1973). [Willi Hennig.]

325. See note 323 above. [Eberhard Königsmann.]

326. It can be confirmed that the oldest fossils are from the Triassic, and they can be assigned to the Xyeloidea (Königsmann, 1976). [Eberhard Königsmann.]

---

† These notes have been prepared by Dieter Schlee from Königsmann (1976) and are published here with the approval of Dr Eberhard Königsmann.

2.2.2.2..4.5. Siphonaptera (fleas)

The Siphonaptera are one of the few orders to form a really well founded monophyletic group. All the derived ground-plan characters of the adults are the direct or indirect result of their mode of life as temporary ectoparasites of birds or mammals.

The body is laterally compressed. The tergites and sternites overlap, rather as in the Hymenoptera, and form a closed protective covering. Fleas are completely apterous. The hind-legs are adapted for jumping. The pretarsus on all three pairs of legs lacks the pulvilli and empodium, but it is not clear how far this is a derived character. The ocelli are completely absent. The compound eyes have been replaced by a pair of atypical eye facets. The antennae have an annulated club and are lodged in antennal grooves. The antennal sockets are joined by an interantennal groove which may be only weakly developed, and Wagner (1939) considered that this was also part of the ground-plan of the Siphonaptera. The same is true of the ctenidia on the head, which function as grasping organs. 'It is possible that the oral, genal, and antennal ctenidia are all that is left of an original ctenidium which extended horizontally along the entire lower margin of the head, over the genal lobes, and vertically up the edge of the antennal grooves. *Macropsylla* has this type of ctenidium and a number of other primitive characters' (Wagner, 1939, p. 17). The proboscis is an absolutely unique structure. The labrum (or epipharynx) and the maxillary laciniae have been modified into stylets. The mandibles are absent. The labial palps act as a sheath for the proboscis, and the comparatively large number of segments (usually 5, but as many as 17) is probably correlated with this. The female has no trace of an ovipositor. In both sexes the 10th abdominal tergite has been modified into a pygidial plate which always has some 'trichobothria'. The larvae have no eyes or legs.

Although it is easy to show that the Siphonaptera are a monophyletic group, it is difficult to work out their relationships.

The old idea, held by Brauer and Lameere, that the fleas are closely related to the beetles, is not supported by the presence of any derived characters in common. For the moment it is not clear how the absence of coxal mera should be interpreted. The production of a secretion from one of the labial glands to form a pupal cocoon could indicate that the fleas are closely related to the Hymenoptera and Mecopteroidea.

The most widely accepted idea used to be that the Siphonaptera were most closely related to the Diptera. Indeed, individual families such as the Phoridae or Mycetophilidae were thought to be the closest relatives of the fleas. There are definite echoes of this in Rodendorf's (1957a) suggestion that the fleas originated from the 'Bibionomorpha', specifically from the 'Fungivoroidea'. However, there is absolutely no evidence for this. If there is a close relationship between the Diptera and the Siphonaptera, then it can only be that the two are sister-groups. However, there are substantial objections to this view too. The only derived characters common to both orders are two larval features, the absence of thoracic legs and the one-segmented labial palps (Hinton, 1958), but these have arisen very commonly amongst the insects through convergence: on their own they provide no argument against the strong evidence suggesting that there is no particularly close relationship between the Diptera and the Siphonaptera.

More recently it has been suggested that the Siphonaptera are most closely related to the Mecoptera. Ross (1965) drew attention to a remarkable agreement in the internal structure of the proventriculus in these two groups, and in his phylogenetic tree he showed them as sister-groups[327].

The same arguments can be used to refute this view as were used against the suggestion that the Siphonaptera and Diptera are closely related. The absence of spermatophores in the Siphonaptera, Diptera, and Mecoptera could provide some evidence that the Siphonaptera are more closely related to these two groups than to the rest of the Holometabola. All three groups transfer liquid sperm by means of a sperm pump. However, it is not known whether the sperm pump is actually homologous in all three groups. If it were to prove to be homologous, then the only possible sister-group relationship would be between the Siphonaptera and the Mecoptera + Diptera. According to Ross (1965) the Mecoptera and Diptera have the lower margin of the mesopleuron, beneath the spiracle, fused with the lower margin of the pronotum. The Siphonaptera do not have this character, and Ross considered the connection to have been secondarily lost. However, there is no evidence for this.

There are also some synapomorphic agreements in the structure of the proboscis between the Mecoptera and the Diptera which are not shared by the fleas. For example, the labial palp is two-segmented in the Mecoptera and Diptera, and its modification into the labella which are so characteristic of the Diptera has its direct antecedent in the Mecoptera. The proboscis of the fleas is not derived from anything like this. Their labial palps have been modified into a protective sheath for the stylets, and in their ground-plan they apparently have four or five segments, which is a larger number than in the Mecoptera and Diptera and is probably a primitive character.

None of these characters indicates that the fleas are more closely related to the Diptera or the Mecoptera, but they would be compatible with the hypothesis that there is a sister-group relationship between the Siphonaptera and the Diptera + Mecoptera.

On the other hand, there are some characters that make it doubtful whether the Siphonaptera do actually belong to the Mecopteroidea (Mecoptera + Diptera + Trichoptera + Lepidoptera). Larsén (1945a, 1945b) found that they have retained the tergal leg remotor muscle which is absent in all the Mecopteroidea, as has been confirmed by Mickoleit (1967). According to Matsuda (1965a), the larvae of the Mecopteroidea have a tergo-galeal muscle, which has been derived from the stipito-galeal muscle by changing its point of origin. This change, which is so characteristic of the Mecopteroidea, does not appear to have taken place in the Siphonaptera.

Extreme care needs to be taken when using characters of the internal anatomy because they have usually been discovered by taking a few random samples. There is no guarantee that they will apply throughout a large group. Nevertheless, the fact remains that there is no definite evidence that the Siphonaptera belong to the Mecopteroidea or are most closely related to one of their subgroups.

No fossils that might belong to the Siphonaptera are known before the Tertiary (Baltic amber)[328]! However, the age of the groups that have to be considered as possible sister-groups of the Siphonaptera makes it virtually certain that the Siphonaptera must have arisen in the Upper Permian at the latest. It is unlikely that it will ever be possible to recognize any fossils as belonging to the stem-group of the fleas.

Revisionary notes

327. Richards (1965, p. 256) wrote: 'In conclusion, the Mecoptera, Neomecoptera [i.e. the Boreidae], and Siphonaptera all have a ring of spine-like projections in the swollen proventriculus. No other insects whose proventriculi are known to me show this characteristic'. The significance of this agreement is still thoroughly puzzling. Both the tergal leg remotor and the structure of the proboscis argue against its being a synapomorphy. Yet, if it is a ground-plan character of the *Mecopteroidea, it must have been lost independently in the Diptera and Amphiesmenoptera, and this does not seem likely. [Willi Hennig.]

328. Riek (1970) has recorded two fleas from the Lower Cretaceous freshwater siltstones of Koonwarra (Australia, Victoria). One of these is of normal pulicid form and size. The other, according to Riek, combines a typical flea abdomen and genitalia with a very primitive head structure; the hind femur is not enlarged for jumping. Moreover, the body length is 7 mm, and this is large when compared with recent species. It is doubtful whether it really is a flea.

Ponomarenko (1976) has described the wingless species *Saurophthirus longipes* from the Lower Cretaceous of Transbaikalia (Buryat Mongolian Republic, Sosnovo-Ozerskiy district, banks of the River Vitim below the mouth of the River Baysa; Tsatsa suite). He provisionally assigned it to the Siphonaptera, adding that the host was probably one of the Pterosaurians. He concluded (1976, p. 339) that 'it seems most probable that we are here concerned with a form that originated from the same mecopterans from which the fleas also originated and that possesses some features in common with fleas, but is adapted to an attached rather than a leaping mode of life'. In fact, this insect not only has long legs but also has a body that is not laterally compressed, and its systematic position is still completely uncertain. [Rainer Willmann.]

2.2.2.2..4.6. Mecopteroidea[329]

Most entomologists and palaeontologists now agree that the Trichoptera, Lepidoptera, Mecoptera, and Diptera form a single monophyletic group, but it is not easy to find reliable evidence for this. Martynov (1937) pointed out that the pronotum is small when compared with the Neuropteroidea, Coleoptera, and Hymenoptera, but Mickoleit (1967) found that this was not so in *Merope* and that it was therefore not part of the ground-plan of the Mecopteroidea. According to Larsén (1945a, 1945b) and Mickoleit (1966), the tergal leg remotor muscle is absent. Matsuda (1965a) reported that the larvae have a tergo-galeal muscle which has been derived from the stipito-galeal muscle by changing its point of origin. Ross (1965) stated that in the hind-wing the bases of 1A and CuP are fused for a short distance. The ovipositor is always completely reduced. According to Mickoleit (1966), there is a muscle extending from the first pteralium to the pleural ridge and this is also characteristic of the group. It is homologous with part of the posterior tergo-pleural muscle in the ground-plan of the Pterygota. This is not a very impressive list of derived ground-plan characters, but it is enough. None of them has any obvious functional connection with features in the mode of life.

Hardly any of these characters are of any value for interpreting fossils, and so it is not surprising that no fossils have been definitely assigned to the stem-group of the Mecopteroidea.

The oldest and most primitive members of the Mecopteroidea are thought to be the Mecoptera, but this is only meaningful in a typological sense. Hinton (1958) was right to point out that this is because the recent Mecoptera, which are probably a monophyletic group (see below), have two characters in which they are more primitive than the other three orders: the wing venation and the presence of compound eyes in the larvae. Most of the fossils are represented only by fore-wings, and so it is not surprising that they have been considered to be Mecoptera even though they may have been primitive species of other orders, or of the stem-group of the Mecopteroidea or even of the entire Holometabola.

In the venation of the fore-wing, the primitive Mecoptera are very similar to the most primitive Neuropteroidea. For this reason these two groups have frequently been thought of as 'closely related'. According to their particular points of view, individual authors have regarded the 'Mecopteroidea' as descendants of the 'Neuropteroidea' (Rodendorf's introduction to the Holometabola, in Rodendorf, 1962; Martynova, 1957, 1959) or the 'Neuropteroidea' as descendants of the 'Mecopteroidea' (Tillyard, 1926b; Sharov, 1966b). As I explained above, there is some good evidence for believing that the Coleoptera (+ Strepsiptera) are more closely related to the Neuropteroidea, and the Hymenoptera and Siphonaptera to the Mecopteroidea. However, this is by no means certain. The Neuropteroidea and Mecopteroidea are the only two groups of the Holometabola with coxal mera. If this is a derived character, rather than a ground-plan character of the Holometabola, it would be highly probable that the Neuropteroidea and Mecopteroidea are sister-groups. But, as I stated above, this is still an unresolved question.

The Mecopteroidea are probably more primitive than the Neuropteroidea in not having an enlarged costal field in the fore-wing. In the Mecopteroidea and the primitive Neuropteroidea (Raphidioptera and Megaloptera) dichotomous branching of the radial sector has taken place. Martynova (1957) thought that the pectinate branching of the radial sector was 'primitive', but it is most probably a derived character restricted to the Planipennia (see p. 296).

If the Mecopteroidea have characters in their fore-wings that are even more primitive than those of the Neuropteroidea, it will not be possible to decide whether fossil fore-wings that are formally identical with those of the Mecoptera do actually belong to the Mecopteroidea or even to the Mecoptera. This is because the characters in common are symplesiomorphies. It is therefore impossible to interpret the so-called Upper Carboniferous Mecoptera. I shall discus this problem and the interpretation of the Permian fossils in the section on the Mecoptera (pp. 334–339).

There are most probably two monophyletic groups (sister-groups) in the Mecopteroidea: the Amphiesmenoptera (caddis flies and butterflies + moths) and the Antliophora (scorpion flies and true flies).

Additional note

329. Grassé (in Grassé, 1951, p. 116) pointed out that Brauer proposed the name Petanoptera for the Mecoptera + Diptera + Trichoptera + Lepidoptera. [Willi Hennig.]

2.2.2.2..4.6..1. Amphiesmenoptera[330.]

The Amphiesmenoptera (Trichoptera + Lepidoptera) are a very well founded monophyletic group of the Mecopteroidea. In one respect they are more primitive than the rest of the Mecopteroidea: they have no sperm pump. They have retained, or at least retained initially, the old method from the ground-plan of the insects of transferring sperm by means of a spermatophore.

The most striking character which indicates that the Amphiesmenoptera are a monophyletic group is that they are the only insects with heterogametic females. White (1957) believed that the replacement of male heterogamety by female heterogamety only took place once in the history of the insects, and that all the Trichoptera and Lepidoptera have descended from this single species. Some cases of female heterogamety are known in the more derived Diptera.

The looped anal veins of the fore-wing are a further derived character of the Amphiesmenoptera. Martynova (1957) suggested that the anal veins have lost their apical sections, although the veins themselves have remained connected by the cross-veins that originally joined them together (Fig. 107). She also stated that the anal loop lies in front of a fluted postanal lobe which is covered with hairs and is an adaptation for holding a layer of oxygen when the wing is immersed in water. She believed that the dense covering of hairs on the wings was connected with the occasional entry of the adults into water.

This is an attractive explanation and is the only one which attempts to correlate the ecological and morphological ground-plan characters of the Amphiesmenoptera. However, it can only be accepted with some reservations, because strictly speaking the Amphiesmenoptera are not aquatic. In the ground-plan of the group the larvae had an open tracheal system.

It seems that fossil Amphiesmenoptera are known as early as the Lower Permian. According to Riek (1953a), Tillyard's *Cladochorista belmontensis* (Fig. 105), from the Upper Permian of Australia, belongs to the 'Trichop-

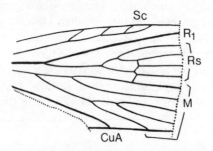

Fig. 105. Fore-wing of *Cladochorista belmontensis* Tillyard (Amphiesmenoptera), from the Upper Permian of Australia. From Tillyard (1926b).

tera'. Martynova (1958) described a second species from the Upper Permian (*Microptysmodes uralicus*, from Tikhiye Gory in the Urals region) and another from the Lower Permian (Kuznetsk Stage) of the Kuznetsk basin (*Microptysma sibiricum*, Fig. 106). The assignment of these three species to the 'Trichoptera' is probably incorrect. In comparison with the Lepidoptera, the ground-plan venation of the Trichoptera only has primitive characters,

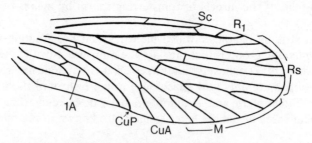

Fig. 106. Fore-wing *Microptysma sibiricum* Martynova (Amphiesmenoptera), from the Lower Permian of the Kuznetsk basin. From Rodendorf (1962).

and these must also have been present in the common ancestors of the Trichoptera and Lepidoptera. The Permian fossils can at most belong to the stem-group of the Amphiesmenoptera.

The existence of the Amphiesmenoptera (but not the Trichoptera!) at least as early as the Upper Permian is indirectly proved by the existence at that time of their sister-group, the Antliophora.

The Trichoptera and Lepidoptera are both well founded monophyletic groups of the Amphiesmenoptera and are sister-groups.

Revisionary note

330. Several additional amphiesmenopteran autapomorphies have recently been discovered and have been reviewed by Kristensen (1975).

In addition to the female heterogamety, a number of caryological specializations in the Amphiesmenoptera were listed by Suomalainen (1966): achiasmatic oogenesis; the presence of a Feulgen-negative elimination plate between the plates of the 1st anaphase; probably holocentric chromosomes; the frequent occurrence of apyrene sperm; high and almost identical basic chromosome number (Trichoptera $n = 30$, Lepidoptera $n = 31$). Baccetti *et al.* (1970) have shown that the amphiesmenopteran sperm is unique among the known insect types in having the outer accessory filaments particularly stout and filled with a glycogen/protein material. The hypopharynx is fused with the prelabium in both adults and larvae of the Amphiesmenoptera (see also Denis and Bitsch, in Grassé, 1973), whereas it is retained as a free lobe in the ground-plan of the Mecoptera. In adult Amphiesmenoptera the posterior arms of the pterothoracic furca are fused with the hind margin of the adjacent epimera (Brock, 1971; Matsuda, 1970). The ventral diaphragm muscles of adult Amphiesmenoptera insert directly on to the ventral nerve cord, a condition not encountered elsewhere among the insects (Kristensen and Nielsen, in press a). The presence of a setiferous process ('pseudempodium' of Debaizieux, 1935) above the claw base of the adult pretarsus is perhaps another derived ground-plan character of the Amphiesmenoptera (Kristensen, unpublished work). The presence of large paired glands opening anteriorly on the 5th abdominal sternite is apparently a further synapomorphy of adult Trichoptera and Lepidoptera. These glands are sometimes present in both sexes, and for this reason a defensive rather than a scent-producing function has tentatively been ascribed to them (Kristensen, 1972; see also Davis, 1975); recently, this suggestion has been confirmed in the case of a limnephilid caddis fly (Duffield *et al.*, 1977). However, in other cases they are present in one sex only or are differently developed in the two sexes, and here they are probably scent organs after all. They have been lost independently in numerous lineages within both the Trichoptera and Lepidoptera. The presence of goblet cells in the midgut of Trichoptera-Annulipalpia (Gibbs, 1966) and Lepidoptera indicates that this specialized type of cell is part of the ground-plan of the Amphiesmenoptera. [Niels Kristensen.]

2.2.2.2..4.6..1.1.   Trichoptera (caddis flies)[331]

Ecologically, the Trichoptera (caddis flies) are most clearly defined by their strictly aquatic larvae. Their tracheal system has been adapted to an aquatic life and is closed (apneustic). According to Ross (1967), tracheal gills are not yet a derived ground-plan character of trichopterous larvae. However, a gula appears to be a derived ground-plan character of the larvae, and according to Weber it originated in the same way as the gula of the Neuropteroidea and Coleoptera. Because DuPorte has recently given a different explanation of the origin of the gula in the Coleoptera, this particular character in the Trichoptera needs to be re-examined. Ross (1967) did not mention it[332].

It is not certain whether anal prolegs are primarily or secondarily absent in the larvae of the Trichoptera. Ross (1967) did not mention them. Martynova (1957) wrote that eruciform larvae are part of the ground-plan of the Mecopteroidea (panorpoid complex). However, like all other workers, she distinguished both campodeiform and eruciform larvae in the Trichoptera (in Rodendorf, 1962). In general, the campodeiform larva is considered to be the more primitive type, but it is not certain whether this is true in the Trichoptera. If there really is a sister-group relationship between the Hymenoptera and the Mecopteroidea, this would be very strong evidence that the eruciform type of larva is in the ground-plan of the Hymenoptera + Mecopteroidea. On the other hand, it is possible that the typological concepts eruciform and campodeiform are too schematic and superficial to be of any value in the study of the Trichoptera.

In the larval head, the bases of the labium and maxillae are fused to form a single labio-maxillary complex. This is also a derived character, and appears to have been an adaptation by primitive trichopterous larvae to a specialized mode of life[333]. Ross (1967) did not mention this character, but he did describe the larvae as having one-segmented antennae with one or two apical papillae. In lepidopterous larvae, the antennae consist of one or two segments.

The larvae of recent Trichoptera have developed very diverse feeding habits: they may be predators, plankton feeders, or plant feeders. As the presence of a gula is usually associated with prognathy and with carnivory, it seems likely that carnivory must have been the original feeding habit of Trichoptera larvae[334].

The derived characters of the adults are also obvious adaptations to their feeding habits: they lap up nectar. The mandibles have been reduced and are non-functional, 'but they are usually visible on each side of the labrum, as papillae or more rarely as tiny pointed hooks' (Weber, 1933, p. 66). Ross (1967) did not mention this character. The maxillary lacinia is absent. 'The haustellum is very characteristic and consists of the labium and its fused extensible lobes. Its basal dorsal part is lined with the weakly grooved hypopharynx, in which the elongate labrum is inserted. The labrum and hypopharynx form a short flattened tube which imbibes the liquid mopped up by the haustellum' (von Kéler, 1955)[335].

These characters suggest that the Trichoptera are a monophyletic group, but they are all present in organs or stages that are never preserved as fossils. For this reason the only clue as to when the Trichoptera first originated is provided by fossils that appear with some certainty to belong to the Lepidoptera or to one of the subgroups of the *Trichoptera with derived venation. The earliest of such fossils are known from the Mesozoic, and the Trichoptera are not known at all from the Palaeozoic.

Revisionary notes

331. Two centuries of Trichoptera systematics were reviewed by Fischer (1965). [Willi Hennig.] A new monograph on the Trichoptera has been published by Malicky (1973). [Dieter Schlee.]

332. Apart from the extreme reduction of the antennae (p. 326), the cephalic specializations of trichopterous larvae mentioned in this paragraph are not genuine apomorphies of the group. It is unlikely that a true gula is ever present in caddis fly larvae. In the prognathous larvae of the Annulipalpia and primitive Integripalpia (of Ross, 1967), as well as in various unrelated species of the hypognathous higher Integripalpia, the posteroventral closure of the head capsule is effectuated by a hypostomal (post-genal) bridge (Denis and Bitsch, in Grassé, 1973). This is also the case in the Mecoptera and some of the lowest Lepidoptera such as *Agathiphaga* (Kristensen, unpublished work) and probably *Micropterix* (according to Lorenz, 1961). [Niels Kristensen.]

333. The labio-maxillary complex in larval Trichoptera is basically similar to that in lepidopterous caterpillars. On the other hand, the greatly reduced larval tentorium seems to be a genuine autapomorphy of the Trichoptera. Interestingly, this reduction of the tentorium has been paralleled in most of the Lepidoptera, but a strong tentorium has been retained in the Zeugloptera (Hinton, 1958; Yasuda, 1962). [Niels Kristensen.]

334. See note 332. [Niels Kristensen.]

335. The ground-plan of adult Trichoptera is similar in most details to that of the Amphiesmenoptera. It is doubtful whether the maxillary lacinia is really absent in the ground-plan of the Trichoptera: for a review of the controversies surrounding this character, see Denis and Bitsch (in Grassé, 1973). True autapomorphies include the peculiar haustellum mentioned by Hennig; the loss of the primitive proximal articulation of the labrum, and, associated with this, of the extrinsic labral musculature (Klemm, 1966; Kristensen, 1968; Denis and Bitsch, in Grassé, 1973); the loss of a true mandibular articulation, which is paralleled in the Lepidoptera-Glossata; and probably the elongation of the pterothoracic prealar arms, which was first mentioned by Weber (1924). [Niels Kristensen.]

2.2.2.2..4.6..1.2.   Lepidoptera (butterflies, moths)

It may seem curious, but the Lepidoptera are the section of the Amphiesmenoptera that initially diverged least from the common stem-group of the Lepidoptera + Trichoptera in their mode of life and morphology. Their larvae have remained terrestrial and have retained an open tracheal system. Their larvae are all of the eruciform type, and this is probably primitive like all the other larval characters. Probably their only derived character is the reduction of the metathoracic spiracle. However, Hinton (1958) found that this spiracle is present and fully functional in the Micropterigidae, and so no derived ground-plan character is known for the larvae. Unfortunately, nothing definite is yet known about the mode of life of the original lepidopterous larvae. Martynova (1957) suggested that they lived on mosses near water, and this appears to be true of some of the very primitive recent Micropterigidae. This is a habitat from which the Trichoptera too could have developed. This hypothesis seems preferable to considering root-feeding larvae like the Hepialoidea to represent the original mode of life.

Adult Lepidoptera were probably flower visitors like the Trichoptera. This may be a derived (apoecous!) ground-plan character of the Amphiesmenoptera, because in their ground-plan adult Mecopteroidea, like the first Holometabola, must have been predaceous. Unlike the most primitive Trichoptera, however, the most primitive Lepidoptera have no suctorial proboscis. They feed on pollen, and have functional mandibles and rather primitive maxillae. During the progressive development of the Lepidoptera, a suctorial proboscis was formed from the maxillary galea and this is different from what took place in the Trichoptera.

The Lepidoptera only have the following derived characters: anterior ocellus absent; wings with scales (hairs modified into scales!); M_4 of the fore-wing fused with CuA_1 (Fig. 107); female cerci absent and no lateral suture present to delineate the prescutum (Ross, 1967)[336].

It is impossible to find a plausible correlation between any of these characters and the mode of life of the Lepidoptera. It was Martynova (1957) who suggested that wing hairs were useful only when insects entered the water: as the Lepidoptera became independent of water, the wing hairs were no longer needed and were modified into scales. This negative explanation is, of course, no explanation at all, and the functional significance of the derived ground-plan characters of the Lepidoptera is still not known.

Hinton (1958) recognized the Zeugloptera, with the single family Micropterigidae, as a third major group alongside the Lepidoptera and Trichoptera.

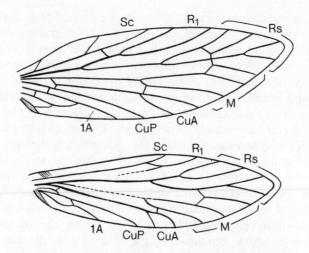

Fig. 107. Wings of *Sabatinca incongruella* Walker (Lepidoptera, Micropterigidae; recent). From Tillyard (1935a).

He was only able to adduce similarities to support this, by saying that in many respects the Zeugloptera resemble the Trichoptera more closely than the Lepidoptera to which they are usually assigned. Probably without realizing it, Hinton has slipped into the methodology and argumentation of typological systematics, although as a theoretician he always defended phylogenetic systematics. Mickoleit (1969) has recently put forward new arguments challenging Hinton's views and supporting the assignment of the 'Zeugloptera' to the Lepidoptera.

No Palaeozoic fossils are known that could belong to the Lepidoptera, nor is there any indirect evidence that the Lepidoptera were in existence as a separate group during the Palaeozoic.

Revisionary note

336. Although no sound autapomorphies have yet been found for the immature stages, several additional autapomorphies for adult Lepidoptera (including the Zeugloptera) have been recognized during recent studies of the lowest moths (Kristensen, 1971, 1975, and unpublished work; Common, 1975). Further revisions of the suggested lepidopterous ground-plan are likely to appear shortly. In particular, the very recent availability of adequately preserved material of the Australo-Pacific genus *Agathiphaga* is of considerable importance: this genus was described as recently as 1952, but for a long time remained known only from very few specimens (review by Robinson and Tuck, 1976). It is important because it has a number of characters in a more plesiomorphic condition than any other Lepidoptera I have examined (Kristensen, unpublished).

330

The recently recognized apomorphies in the ground-plan of the Lepidoptera include the following characters: the protibial epiphysis, which was first described in detail by Philpott (1924) and whose function is for cleaning the antennae (Jander, 1966); a posteromedian process on the corporotentorium; the folding of the maxillary palp, with sharp bends between segments 1–2 and 3–4, and simplification of its musculature; an invaginated group of chemoreceptors, situated apically on the distal segment of the labial palp; the loss of the dorsal longitudinal muscle of the salivarium; the presence of an apodeme, directed anterodorsally and invaginated from the upper part of the mesopleural suture; the separation of the labral nerve and the frontal ganglion connective immediately after their origin from the tritocerebrum. Some of the other characters that were discussed in the papers mentioned above as alleged lepidopterous autapomorphies need to be re-examined, such as the cranio-stipital muscle, the lateral labral seta-bundle, the absence of the pterothoracic prescutal demarcation, the bilobed male dorsum 10, and the unpaired abdominal interganglionic connectives. A few of them are now known to be positively erroneous. Particularly important is the fact that vein M_4 is completely retained in *Agathiphaga* (Fig. IX) and that the ground-plan

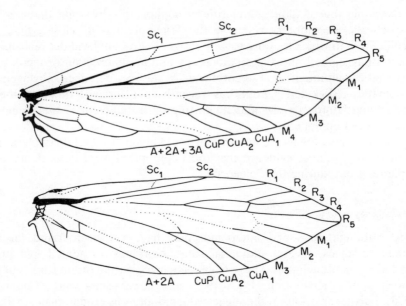

Fig. IX. Wings of *Agathiphaga queenslandensis* Dumbleton (Lepidoptera; recent). From Common (1973).

venation of the Lepidoptera is therefore indistinguishable from that of the Trichoptera. This will obviously make it very difficult to recognize the earliest fossil Lepidoptera. [Niels Kristensen.]

2.2.2.2..4.6..2. **Antliophora**

The probable sister-group of the Amphiesmenoptera is the Mecoptera + Diptera. This group has never been named, and I propose to call it the Antliophora.

The characteristic feature of the Antliophora is that they no longer transfer sperm by means of spermatophores, as most other insects do: they have developed a sperm pump[337]. The sperm pump is part of the ground-plan of the Diptera, but it is not certain whether it is also present in all Mecoptera or whether it could even be part of the ground-plan of a monophyletic group of higher rank that includes the Siphonaptera as well as the Mecoptera + Diptera[338]. Moreover, it is not certain that it is homologous in these three groups. Hinton (1958) stated that the absence of retractor muscles in the hypopharynx and of ventral muscles in the common duct of the labial glands in the larvae were further derived characters of the Antliophora. He also stated that the larval legs are much reduced. They have usually only been retained in the Mecoptera, but even here there are only three segments beyond the coxa, the most distal of which is considered to be the pretarsus. Hinton was unable to decide whether the other two segments are a tibio-tarsus and a trochantero-femur, or are a tarsus and a trochanter + femur + tibia. The pretarsus has no muscles[339].

These derived characters also appear to be present in the Siphonaptera, and it would be reasonable to assign this group also to the Antliophora. There are however a number of characters that appear to exclude the fleas from the Mecopteroidea (p. 320). Nevertheless, the possibility remains that the Siphonaptera might really belong to the Antliophora in addition to the Mecoptera and Diptera.

Mickoleit (1967) published a paper that was characterized by the clear formulation of a number of problems to be investigated. He found three derived ground-plan characters in the adults: the structure of the posterior tergal articulatory process in the wing base, the lower pleural arm, and probably the rigid connection between pleura and postnotum. Subsequently (Mickoleit, 1969, p. 172) he added the 'displacement of the pleural-ridge-scutum muscle from the postero-lateral margin of the scutum to the tip of the tergal arm or to the 4th pteralium . . . this displacement led to a specific modification of the posterior tergal articulatory process'. Finally, two further characters should be mentioned, the reduction of the galea of the maxilla and the presence of only two segments in the labial palpi[340].

Most of the primitive characters of the Antliophora have been retained by the Mecoptera. This is particularly true of the wing venation. Consequently, many of the Palaeozoic fossils whose venation agrees formally with that of the recent Mecoptera definitely do not belong to this particular monophyletic group. They belong either to the stem-group of the Mecopteroidea, or to the stem-group of the Antliophora, or are very primitive species of various subgroups of the Antliophora. Hinton was right to point out that these fossils, which have been assigned to the Mecoptera because only wings were available, would certainly be placed in other groups if other structures or the larvae had been preserved.

Leaving aside the Siphonaptera, whose phylogenetic relationships are not entirely clear, the Antliophora include two monophyletic sister-groups, the Mecoptera and the Diptera.

Revisionary notes

337. For discussion of sperm pumps and spermatophores, see note 338. [Gerhard Mickoleit.]

338. Mickoleit (1971) described the plesiomorphic genitalia of *Notiothauma reedi* McLachlan, and this was the first step towards understanding the origin of the sperm pump during the phylogenetic development of the Mecoptera. He was able to establish that a sperm pump is present in the following recent families of the Mecoptera: Nannochoristidae, Bittacidae, Eomeropidae (= Notiothaumidae), Meropeidae, Choristidae, Apteropanorpidae, Panorpodidae, and Panorpidae. The sperm pump is absent in all the species of Boreidae that have been examined so far. Mickoleit has also shown that the sperm pump of the Mecoptera differs from that of the Diptera in certain fundamental features. There are also differences in the mode of operation: in the Mecoptera the pistil and the main body of the pump close together like a pair of pincers, whereas in the Diptera the sperm pump operates like a piston-pump. Mickoleit concluded that in each group the sperm pump must have arisen from homologous elements in the genitalia and that the latest common stem-species of the Mecoptera + Diptera must have had at least a primitive arrangement for the expression of liquid sperm. It is astonishing that in *Boreus westwoodi* Hagen the sperm is transferred by means of a spermatophore, which is presumably the case in the rest of the genus *Boreus*. If this outline of the evolution of the sperm pump is correct, and if the Boreidae really are true Mecoptera, then the absence of a sperm pump in the Boreidae must be regarded as a reduction. This would imply that the spermatophore of the Boreidae must have arisen convergently in the stem-group of this family. In fact, it is possible for spermatophores to arise secondarily as new structures, for there are examples in the Diptera, the group most closely related to the Mecoptera: individual species of the Ceratopogonidae, Chironomidae, and Simuliidae are known to have developed spermatophore production secondarily (Pomerantsev, 1932; Davies, 1965; Wenk, 1965). There is further evidence that the spermatophore of *Boreus* has arisen convergently when compared with the spermatophore of the ground-plan of the Pterygota, because there are plausible grounds for believing that the sperm pump of *Boreus* has been reduced. It seems likely, therefore, that the stem-group of the Mecoptera still retained the ability to produce a sac of sperm enveloped in a secretion. It is conceivable that sacs of sperm where the enveloping secretion had not hardened were expressed by a primitive sperm pump. It is possible that this or a similar mode of sperm transfer was present in the stem-group of the Boreidae and formed the starting point for the new formation of a firm spermatophore. [Gerhard Mickoleit.]

339. According to Mickoleit (1971), the Mecoptera and the Diptera are the only groups to have lamellate, criss-crossing mandibles in which the anterior points of articulation have been reduced whilst the posterior ones articulate with a prop-like subgenal process. It seems most likely that the common stem-species of the Amphiesmenoptera had an orthopteroid biting mandible, and so this mandibular structure could be a further synapomorphy of the Antliophora. [Gerhard Mickoleit.]

340. The common stem-species of the Amphiesmenoptera (the sister-group of the Antliophora) had a labium that was close to the ground-plan of the Pterygota in its external conformation as well as in its musculature. Particularly notable characters are the five-segmented labial palpi and the presence of premental muscles in the glossa and paraglossa. In the Antliophora, on the other hand, the labium is strikingly derived. It differs from that of the Amphiesmenoptera by the two-segmented labial palps, the median surface of which is extensively membraneous. Both the Mecoptera and the Diptera use the labial palpi to anchor the rostrum on to the food surface. Glossae and paraglossae have atrophied. Further synapomorphies are the reduction of the labral muscles to the following: M. tentorio-praementalis dorsalis, M. praemento-palpalis externus, M. praemento-palpalis internus, and M. palpo-palpalis. [Gerhard Mickoleit.]

2.2.2.2..4.6..2.1. Mecoptera (scorpion flies)[341]

There are a few derived characters which suggest that the Mecoptera are probably a monophyletic group. In the adults the hypopharynx is completely absent[342]. However, Arora (1956) identified it with a conical membraneous lobe hanging freely in the mouth chamber and with the salivary glands opening at its base. The 1st abdominal tergite is transversely divided, and the anterior part is fused with the metanotum. The 1st abdominal sternite is free but has been reduced to two small sclerites. The genital opening of the female is situated behind the 9th sternite, in a genital chamber formed from the invaginated membrane between segments 9 and 10. The accessory glands, spermatheca, and oviduct all have separate openings here, and the chamber acts as a bursa copulatrix[343].

Unfortunately, I am not sure whether these characters really are present in all Mecoptera or whether they have only been worked out from study of the comparatively derived species of the northern hemisphere. Tillyard (1935a) divided the Mecoptera into two separate groups, 'Protomecoptera' and

'Eumecoptera', both of which he thought were already in existence by the Palaeozoic. The descendants of the Protomecoptera consist of no more than a few relict genera (*Merope, Austromerope, Notiothauma*). On the other hand, the descendants of the Eumecoptera are said to include the rest of the recent Mecoptera (which only consist of about 350 species), and all the Diptera[344]. Using this division as her basis, Martynova (1959) developed some highly fanciful ideas and asserted that even the recent families of the Mecoptera could be traced back as separate lineages far back into the Carboniferous. Her phylogenetic trees in this paper show differences that are extremely puzzling.

For this reason, it is very important to know for certain whether the recent 'Protomecoptera' and the recent 'Eumecoptera' form a monophyletic group, which is my view, or whether the Eumecoptera are more closely related to the Diptera, which is the view expressed by Tillyard and Martynova[345].

The oldest species of the 'Protomecoptera' is *Platychorista venosa* Tillyard (Fig. 108), from the Lower Permian of Kansas. Tillyard originally described

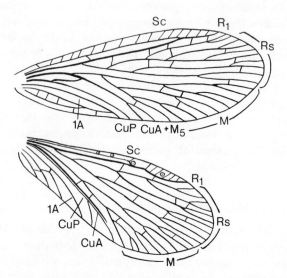

Fig. 108. Wings of *Platychorista venosa* Tillyard (Mecoptera), from the Lower Permian of Kansas. From Carpenter (1930c).

the fore-wing of this species as *Protomerope permiana*[346]. Since then, a further 22 species have been described, from the Permian of the Urals region and in particular from the Kuznetsk basin, by Martynova (1958) and Zalesskiy (1946). These 'Protomecoptera' are all characterized by an independent fork of CuA and the presence of rather numerous cross-veins in the comparatively broad costal field between Sc and C. They are therefore a precursor from which not only the Mecoptera but also all the recent Diptera,

Trichoptera, and Lepidoptera could be formally derived. There is no good reason for not regarding them as members of the stem-group of the entire Mecopteroidea, perhaps even including the Siphonaptera and Hymenoptera(!).

The Lower Permian of the Kuznetsk basin, which is slightly younger than the Lower Permian of Kansas in which *Platychorista* was found, has recently produced *Microptysma sibiricum* Martynova (Fig. 106) which has the looped anal veins that are characteristic of the Amphiesmenoptera. For this reason it is not necessary to regard the Palaeozoic 'Protomecoptera' as members of the stem-group of all the Mecopteroidea.

Further Lower Permian fossils are known (see below) which could be justifiably assigned to the stem-group of the Diptera and the majority of the recent Mecoptera. They have CuA unforked, and the number of cross-veins in the costal field is very reduced. This means that there are in fact only two possible interpretations of the Lower and Upper Permian 'Protomecoptera': they could be survivors of an older, possibly Upper Carboniferous, stem-group of all the Mecopteroidea; or they could belong to a stem-group from which only the three recent 'Protomecoptera' have descended—*Merope* in North America, *Austromerope* in Western Australia, and *Notiothauma* in Chile. If the second of these hypotheses is correct, we should then enquire whether there is a sister-group relationship between these Protomecoptera and the Eumecoptera + Diptera, or whether the recent and fossil Protomecoptera and Eumecoptera form a monophyletic group which would then be known to have existed during the Lower Permian. Again, the second of these hypotheses would imply that the ancestors of the Diptera must also have been in existence at the same time as a separate group.

I think it most unlikely that these problems will ever be solved by studying the wings alone. Nevertheless, the history of the Nannochoristidae ('Eumecoptera') appears to indicate that when the line of descent of the recent 'Protomecoptera' is traced back into the past it remains distinct from that of the rest of the Mecoptera and only merges with it in the Permian. Martynova (1959, Fig. 5; and, with clearer captions, in Rodendorf *et al.*, 1961, Fig. 435) thought that the Lower Permian Platychoristidae from North America and the Urals were the stem-group of the recent Meropeidae (one Nearctic species) and Notiothaumidae (one Chilean species), whilst the recent Austromeropidae (one Western Australian species) were most closely related to the Upper Permian Permomeropidae, containing two species from Australia. The separation of these two lineages is supposed to have taken place in the Carboniferous. If this hypothesis is well founded, then it could provide an extremely interesting explanation of a distribution pattern that is not uncommon amongst the insects. Unfortunately, this does not seem to be the case: the North American *Merope* and the Western Australian *Austromerope* have usually been assigned to the same family, as for example by Byers (1965), and they do seem to have some derived characters in common, such as the absence of the ocelli[347].

According to Martynova, the Permian 'Eumecoptera' differ from the 'Protomecoptera' by the small number of 'cross-veins' in the costal field and the simple ('unforked') CuA (Fig. 102). CuA appears as a simple vein because its anterior branch (CuA_1) has fused with M_4, the posterior branch of the media. This has also taken place in all the recent Diptera, which are supposed to have descended from the Eumecoptera, in all the Lepidoptera, and in many Trichoptera, although in these groups it has undoubtedly arisen independently. I do not know if this explanation will be valid for all the fossil Eumecoptera. However, it may be important for defining the Eumecoptera if the group is to be restricted to include only the fossils that are more closely related to the recent Eumecoptera than they are to any other recent group of insects: there is no doubt that species have often been assigned to the Eumecoptera that do not belong there at all.

Metropator must also be removed from the Mecoptera. *Metropator pusillus* Handlirsch was described from the middle Upper Carboniferous (lower Pottsville = lower Westphalian) of Pennsylvania in North America. Tillyard (1926b, p. 265) described it as 'definitely mecopterous in structure'. Subsequently (Tillyard, 1935a) he still considered it to belong to the stem-group of the entire Mecopteroidea.

Carpenter (1930c), following Crampton's investigations, doubted whether *Metropator pusillus* belonged to the Mecoptera. On the other hand, Martynova (1958; in Rodendorf *et al.*, 1961) thought that *Metropator* belonged to the Eumecoptera and that it proved that the separation of the Mecoptera(!) into the two suborders Protomecoptera and Eumecoptera had already taken place by the middle Upper Carboniferous. In fact there is absolutely no evidence for this. The published illustration of *Metropator* does not show that CuA_1 is fused with M_4. Even if it were, this is a character that has arisen independently in the Pterygota on so many occasions that Martynova would be wrong to use it as the basis for a hypothesis with such far-reaching implications. Carpenter (1965) thought that *Metropator pusillus* was probably the hind-wing of a species of the order Miomoptera which 'is close to' the Protorthoptera. However, this is not much of a step forward because the relationships of this 'order' and of the Protorthoptera are not at all clear. Nonetheless, it does show that *Metropator* cannot be assigned to the Mecoptera, or even to the Holometabola, without a great deal more evidence.

Other Carboniferous 'Mecoptera' are even more doubtful. Schmidt (1962) was right to say that *Metropatorites kassenbergensis* Keller, from the Namurian of the Ruhr valley Coal Measures, could not be interpreted. However, the same is true of the hind-wing described by Schmidt himself (1962) as *Eopanorpella ernsti*, from the Carboniferous (Westphalian D) of Ibbenbüren. I do not see why this hind-wing should not belong, for example, to the Archimylacrididae, like the hind-wing of *Parapanorpa ungensis* Zalesskiy which Zalesskiy originally described as a mecopteran.

The assignment of *Permopanorpa formosa* Tillyard, from the Lower

Permian of Kansas (cf. Fig. 109), and of a few other Permian fossils to the Eumecoptera seems to be more justified. However, it is possible to make a formal derivation of the Diptera from them, and this diminishes their importance considerably. The wing of the genus *Ctenostematopteryx*, from the Rothliegende of the Thüringer Wald (Haupt, 1951), has either been badly drawn or does not belong to the Mecoptera at all.

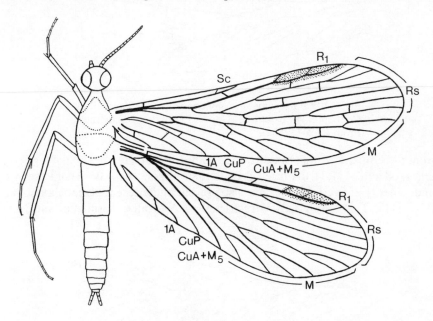

Fig. 109. *Permopanorpa inaequalis* Tillyard (Mecoptera), from the Lower Permian of Kansas. From Tillyard (1935a).

Thus for the present it is not possible to say how far the Antliophora as a whole and the Mecoptera in particular had separated into subgroups by the Lower Permian.

There are many Upper Permian species, but they are of no particular importance. This is because they could belong to the stem-groups of several recent orders but there is other evidence to show that all these orders were already in existence during the Upper Permian. Two species from the Upper Permian of Australia, *Belmontia mitchelli* Tillyard and *Parabelmontia permiana* Tillyard, have often been grouped together as the 'Paramecoptera' and, according to Riek (1953a) and Laurentiaux (in Piveteau, 1953), they have CuA forked as in the Protomecoptera. Tillyard (1919b; see Laurentiaux) considered that they could be the ancestors of the Lepidoptera and Trichoptera, whereas Riek (1953a) thought that it was possible to derive the Hymenoptera from them. Martynova (1959; and in Rodendorf, 1962) treated the Paramecoptera as an extinct side-branch of the Eumecoptera, in which CuA is unforked[348].

338

The differences between most of the Upper Permian species (or wings!) of the 'Eumecoptera' are very slight. By far the most interesting species are *Nannochoristella reducta* Riek (Fig. 110) and *Neochoristella optata* Riek,

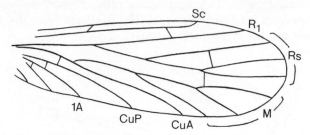

Fig. 110. Fore-wing of *Nannochoristella reducta* Riek (Mecoptera), from the Upper Permian of Australia. From Riek (1953a).

from the Upper Permian of Australia. They are characterized by having R_{2+3} unforked and M and CuA fused close to their origin. These two characters are undoubtedly derived and they are also present in the recent Nannochoristidae (Fig. 111). This family is now restricted to Australia, Tasmania, and New

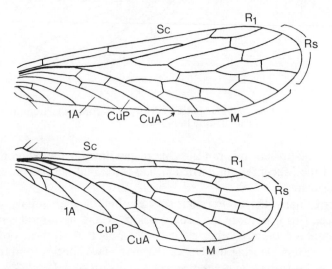

Fig. 111. Wings of *Nannochorista dipteroides* Tillyard (Mecoptera; recent). From Tillyard (1935a).

Zealand (four or five species), and to Chile and Argentina (two species). It would seem perverse not to accept that the Australian Upper Permian species and the recent Australian species are closely related, as they are so very similar. In any case, such a situation provides valuable insights into the distribution of the recent species. There is a parallel case in the Plecoptera (p. 172).

It is particularly interesting that the small recent family Choristidae (Fig. 140), containing only three species, is also restricted to Australia and also appears to be closely related to the Nannochoristidae[349]. However, the Choristidae are even more primitive than the Upper Permian genera that Riek (1953a) assigned to the Nannochoristidae: R_{2+3} is forked and CuA touches M at one point only. For a revision of the Australian Mecoptera, see Riek (1954b). There seems to be no course but to accept that the Choristidae and Nannochoristidae had separated by the Upper Permian. This would mean that we would have to distinguish at least three separate lineages in the Mecoptera, without counting the Protomecoptera. Moreover, if the Antliophora are in fact a monophyletic group, then these fossils also provide indirect evidence that the Diptera too must have arisen in the Upper Permian, if not earlier.

There are about 140 other Permian 'Eumecoptera', which have been described from all the known Lower and Upper Permian localities. Carpenter (1954b) assigned most of them to the family Orthophlebiidae, which contains many Mesozoic species as well as these Palaeozoic ones. This is certainly not correct. It seems that the Mesozoic fossils can be associated with the two main groups of the recent Mecoptera, of which the most important families are the Panorpidae and Bittacidae. It is not possible to do this with the Palaeozoic fossils, even the Upper Permian ones. None of the subgroups of the *Mecoptera can be traced back into the Palaeozoic except possibly the 'Protomecoptera' and the Nannochoristidae, and maybe indirectly the Choristidae. As I have mentioned before, there is further uncertainty over the interpretation of the Permian fossils as it is impossible to establish whether some of them may not actually belong to the stem-group of the Siphonaptera or to that of the entire Antliophora.

Revisionary notes

341. A new monograph on the Mecoptera has been published by Kaltenbach (1978). Schlee and Schlee (1976) have given a bibliography of recent and fossil Mecoptera, and Willmann (1978) a catalogue of the fossil Mecoptera. [Dieter Schlee.]

342. Hepburn's (1969b) work on the skeleto-muscular system of the head has shown that the hypopharynx is present in all the families of the Mecoptera. The hypopharynx of *Panorpa communis* Linnaeus was first described and illustrated by Steiner (1930) and Grell (1942). [Gerhard Mickoleit.]

Hepburn's (1969b) discussion of the musculature of the head has revealed a number of further derived ground-plan characters in the Mecoptera (see note 344). Further derived ground-plan characters are the absence of the clypeo-labral suture, the elongation of the subgena along the lateral margin of the clypeus, and the absence of the posterior tentorial pits. [Willi Hennig.]

For a discussion of the proventriculus, see note 327 and Richards (1965); also Hepburn (1969a). [Willi Hennig.]

343. This must be discarded as a derived ground-plan character of the Mecoptera. According to Mickoleit (1975; see also Byers, 1954), the opening of the genital chamber in females of the Nannochoristidae is between the 8th and 9th abdominal segments, and not between the 9th and 10th segments as has generally been thought. The genital chamber, together with the openings of the ductus receptaculi and the oviduct, is entirely contained within the 8th abdominal segment.

In all the other recent families of the Mecoptera, the genital segments of the female have been fundamentally modified. In the course of phylogeny, the floor of the genital chamber, which consisted of the fused genital appendages of the 8th segment, became elongated. As a result of this, the posterior end of the genital chamber together with the opening was displaced to the posterior edge of the 9th abdominal segment. A further consequence of this modification was the displacement of the ductus receptaculi and the opening of the appendicular gland to the venter of the 9th segment. [Gerhard Mickoleit.]

344. There is no longer any doubt as to the monophyletic origin of the Mecoptera. Mickoleit (1971) has listed the following synapomorphies of the Mecoptera. (1) The labrum is fused to the clypeus, to form a uniform appendage of the head-capsule. (2) The labral muscles have been entirely reduced except for the epipharyngeal compressor. Compared with this, the labrum of the Diptera is far more plesiomorphic: the Diptera have a completely isolated labrum and at its base there is a single clypeal retractor which Matsuda (1965b) considered to represent the fusion of the fronto-labral muscles. (3) Reduction of the tentorial adductor of the mandible. (4) Reduction of all the hypopharyngeal muscles (Hepburn, 1969b). In the Diptera, on the other hand, the tentorial adductor of the mandible, the lateral suspensor of the mouth-edge, and the dorsal dilator of the salivary opening are all present (Matsuda, 1965b). See also note 342. These all provide decisive arguments against Tillyard's (1935a) suggestion that the Mecoptera are paraphyletic. [Gerhard Mickoleit.]

345. The only characters that have been used to define the Protomecoptera are in the wing-venation: (1) the subcosta has a large number of branches; (2) as pointed out by Riek (1953a) and Martynova (in Rodendorf, 1962), the fore-wing has Cu_1 forked, unlike the Eumecoptera; (3) Riek pointed out that the fore-wing has looped anal veins, as in the Trichoptera and Lepidoptera. It is not difficult to see that the first two of these are symplesiomorphies, since a pectinate subcosta and forked Cu_1 are part of the ground-plan of the entire Mecopteroidea. The looped anal veins that are said to be present in fossil Protomecoptera could at best be regarded as a synapomorphy of the

Amphiesmenoptera + Antliophora. However, it seems far more likely that the species with looped anal veins are not Mecoptera at all but belong to the stem-group of the Amphiesmenoptera: everything that is known about the phylogeny of the Mecopteroidea suggests that this stem-group must have had a forked Cu_1, pectinate subcosta, and looped anal veins. So far as I know there are no characters that can be used to establish that the Protomecoptera are a monophyletic group.

The Eumecoptera are characterized by the reduced number of cross-veins and the simple Cu_1. However, it seems highly probable that these characters have arisen through convergence: similar reductions are known to have taken place convergently in the most diverse insect orders and even in those most closely related to the Mecoptera, such as the Diptera and the Amphiesmenoptera. Consequently, they cannot be used as evidence that the Eumecoptera are a monophyletic group.

Mickoleit (1975, 1976, 1978) has been able to show that the Eomeropidae (= Notiothaumidae), Choristidae, Apteropanorpidae, Panorpodidae, and Panorpidae probably form a monophyletic group within the Mecoptera to which the Nannochoristidae and Bittacidae do not belong. The following characters of the eomeropid group of families provide support for this division: the opening of the genital chamber has been displaced to the hind-margin of the 9th abdominal segment; the division of the gonocoxosternite of the 8th abdominal segment into a cranial and a caudal section which is connected with the evolution of a telescopic abdominal tip; the fusion of the basal segments of the cerci with the dorsum of the 11th abdominal segment, with the result that the marginal areas of the tergum are retained as so-called circumapical sclerites. It is probable that the Meropeidae also belong to the same phylogenetic lineage as the Eomeropidae, Choristidae, Apteropanorpidae, Panorpodidae, and Panorpidae. However, the phylogenetic relationship of the Bittacidae and Nannochoristidae to this group of families is still not clear.

Penny (1975) also considered the Protomecoptera to be a poorly founded group. He regarded the Meropeidae as the sister-group of all the other families of the Mecoptera, which share two synapomorphies: the reduced number of costal cross-veins and the presence of bulbous basistyles. He placed the Eomeropidae (Notiothaumidae) close to the Panorpidae and Panorpodidae, on the basis of the wing-clasping organ (notal organ), the elongate rostrum, and the strongly developed 9th sternum and tergum. He thus had to interpret the numerous cross-veins of *Notiothauma* as a new acquisition. Like Mickoleit, Penny considered the Nannochoristidae to be a comparatively primitive branch of the Mecoptera, and he pointed out that the Nannochoristidae have retained a primitively short 9th tergum. [Gerhard Mickoleit.]

346. Riek (in CSIRO, 1970; see also Kristensen, 1975) transferred the Platychoristidae to the Trichoptera but subsequently (in CSIRO, 1974) returned them to the Mecoptera 'sensu lato'. Sukacheva (1976) assigned *Platychorista* to the Permotrichoptera, which she considered to belong to the Trichoptera. In fact, the Permotrichoptera should be regarded as members of the stem-group of the Amphiesmenoptera (Trichoptera + Lepidoptera) (Willmann, 1978). Willmann followed previous authors in excluding *Platychorista* from the Mecoptera because of the looped anal veins of the fore-wing. This character is now considered to be a synapomorphy shared by *Platychorista* and the Amphiesmenoptera, though in *Platychorista* there are two veinlets that connect A_2 and A_3 with the hind-margin of the wing. [Rainer Willmann.]

347. Several characters suggest that *Austromerope* and *Merope* undoubtedly form a single monophyletic group: the reduced ocelli; the characteristically elongate gonobases and gonostyli of the male genital forceps; and above all a sclerotized, plate-like process at the base of the hind-margin of the fore-wing, which attaches the fore-wing to the mesonotum when at rest. [Gerhard Mickoleit.]

Re-examination of *Austromerope poultoni* Killington has shown that the striking characters which it shares with *Merope* are synapomorphies and cannot be interpreted as the results of convergence (Willmann, 1979). [Rainer Willmann.]

348. According to Willmann (1978), CuA has no true fork in *Belmontia*. It seems most likely that what has been called the first branch of CuA is actually the last branch of M (M_{4b}). The 'cross-vein' between M_4 (now believed to be M_{4a}) and CuA should therefore be interpreted as the basal part of M_{4b}. Riek (1953a: 58) identified a wing as belonging to *Belmontia* in which 'CuA$_1$ appears as a fork on M_4 with CuA appearing simple'. Riek's interpretation agrees with my own. [Rainer Willmann.]

349. See note 430. [Gerhard Mickoleit.]

2.2.2.2..4.6..2.2. Diptera (flies)[350]

Adult Diptera are characterized in both sexes by the modification of the hind-wings into halteres. In the fore-wing, CuP (Cu_2) is reduced: its basal part is developed as a vein, and the rest is only visible as a fold (Fig. 112). Males lack the 8th abdominal spiracle. The only character in which the mouth-parts are more primitive than those of the Mecoptera is the presence of a hypopharynx. However, Hoyt (1952) and Ferris both thought that the so-called hypopharynx of the Diptera was a new structure, the labial lonchus. The Mecoptera and Diptera resemble each other in that of the two maxillary plates only the lacinia has been retained. The labial proboscis is a strikingly derived ground-plan character of the Diptera. The labial plates are very reduced; the apical segment of the labial palpi, which are two-segmented as in

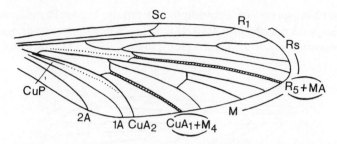

Fig. 112. Ground-plan of the venation of the Diptera. The interpretation and nomenclature of veins R_5 + MA and CuA_1 + M_4 are still controversial, and these two veins have been specially emphasized in the figure. Based on Hennig (1954).

the Mecoptera, has been modified into the labellum. In their ground-plan the 'pseudotracheae' of the labella are very simple: in *Tipula* there are two simple channels in each labellum. The setiform mandibles and the loss of the anterior mandibular articulation are two further derived characters[351].

Larvae of the Diptera, like those of the Siphonaptera, are characterized by the absence of thoracic legs. The spiracles have no closing apparatus (Hinton, 1958).

It is obvious that the Diptera have been vastly more successful than their presumed sister-group, the Mecoptera. However, it is difficult to see precisely what factors have been responsible for this success. It seems clear that the derived ground-plan characters of larval and adult Diptera are connected with their mode of life, but it is not easy to see exactly how they differ from the Mecoptera. Martynova (1957) gave an 'evolutionary–biological' explanation of the origin of the Diptera, but this is of no value whatsoever. Some authors have thought that the blood-sucking habit was part of the ground-plan of the Diptera, but this is certainly not correct: the most primitive Diptera are not blood-suckers but, like the Mecoptera, are predators of other insects (see Downes, 1958)[352].

The oldest fossils that can be reliably assigned to the Diptera are from the Mesozoic. Martynova (1957, 1959) stated that their Palaeozoic stem-group was the Paratrichoptera, which she thought had also given rise to the Hymenoptera at the same time. As I have already explained (p. 317), this is certainly incorrect.

Most of the species that Martynova (in Rodendorf, 1962) assigned to the order 'Paratrichoptera' are of no interest because they are from the Mesozoic. True Diptera are known from the same epoch, and so these 'Paratrichoptera' can be no more than survivors from the stem-group. Martynova stated that the 'Paratrichoptera' differ from all the other 'suborders' of the 'Mecoptera' by having the basal part of the radius straight, CuA and CuP close together, and only one or two anal veins. These characters do indeed suggest the Diptera. Riek (1953a) mentioned two, or more rarely three, anal veins. He described three species of 'Protodiptera' from the Upper Permian of Australia: *Permotanyderus ableptus* Riek (Fig. 113), *Choristotanyderus nanus* Riek, and *Permotipula patricia* Tillyard (see Tillyard, 1929). He also reported that

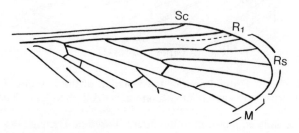

Fig. 113. Wing of *Permotanyderus ableptus* Riek ('Protodiptera'), from the Upper Permian of Australia. From Riek (1953a).

Martynova (1948) had given the name *Robinjohnia tillyardi* to a specimen figured by Tillyard (1937a), also from the Upper Permian of Australia[353]. *Permotipula borealis* Martynova, from the Upper Permian of the Kuznetsk basin, has been described as a fifth species. Some of these species definitely do not belong to the ancestral line of the Diptera because the anterior branch of the radial sector is unforked. The probable existence of the Mecoptera (Nannochoristidae) in the Upper Permian provides the best (indirect) evidence that the Diptera must have been in existence at the same time. It is impossible to trace any of the subgroups of the *Diptera back into the Palaeozoic.

Revisionary notes

350. For additional comments on the Diptera, see notes 434–452 on pp. 437–439. [Dieter Schlee.]

351. According to information given by Crowson (1970), the numbers of chromosomes are no greater in the Coleoptera than in the Diptera. Downes (1958) also drew attention to the loss of the cuticular hairs that act as a filter in the proventriculus; see also note 327. [Willi Hennig.]

352. Both Rupprecht (*in litt.*) and Hepburn (1969a, 1969b) believed that in their ground-plan the Mecoptera are detritivorous or phytophagous rather than predaceous. [Willi Hennig.]

353. Riek (1968a) has subsequently assigned *Robinjohnia* to the Nannochoristidae (Mecoptera). [Willi Hennig.] This assignment has been followed by Willmann (1978). [Adrian Pont.]

C. Review of the history of the Insecta during the Mesozoic

a. Introduction

In the preceding section of this chapter I have shown that almost all the monophyletic groups of insects that have traditionally been called 'orders' can be recorded from the Palaeozoic. Only in a few cases is there any evidence to show that the latest common ancestors of various 'orders' (Phthiraptera and Psocoptera) did not arise until the Mesozoic. In a very few cases, it seems likely that subgroups of so-called orders were already in existence during the Palaeozoic.

The recent insects are therefore derived from comparatively few stem-species, which lived in the early Mesozoic as descendants of the Palaeozoic insect fauna. If we stick firmly to demonstrable facts, there can have been no more than 40–50 such stem-species. My next task will be to trace the history of these stem-species and their descendants through the Mesozoic, and I shall pay particular attention to what took place during the Cretaceous. Amongst the insects, as amongst other animals, it seems that a rather limited number of early Mesozoic species produced descendants that were able to survive the Upper Cretaceous and reach the present. The Cretaceous seems to have been a critical period for the insects, as was the transition from the Palaeozoic to the Mesozoic, and many groups were unable to survive it.

Unfortunately, it is much more difficult to work out insect phylogeny during the Mesozoic than during the Palaeozoic. There are several reasons for this. The first and most important is that much less is known about the phylogenetic relationships of the monophyletic groups within the so-called 'orders' than about the relationships of the orders themselves. Until there is an improvement in this situation, interpretation of the fossils will always be uncertain. This is because the task of phylogenetic research is to work out the phylogenetic classification of the recent fauna, and the role of palaeontology, as I have already explained (p. 3), is to locate the fossil species as accurately as possible within the framework of this classification.

A second reason, which is no less important, is the extraordinary scarcity of Cretaceous fossils[354]. Most Mesozoic insects are from the Jurassic, with a smaller number from the Upper Triassic. As a result, any attempt to analyse the history of the insects during the Mesozoic must be restricted to a few conclusions derived from the limited number of well founded monophyletic subgroups of the so-called 'orders' which can be recognized in the Jurassic or Upper Triassic.

Finally, there is a third difficulty that obtrudes more and more in the study of insect phylogeny as we approach the present. This is that the constitutive characters of monophyletic groups consist of morphological details that become increasingly difficult to detect as we come to deal with more and more subordinate groups in the phylogenetic tree. This makes it extremely difficult to interpret the fossils. The beetles are a good example of the kinds of difficulty that are encountered. Some of the constitutive characters of the monophyletic subgroups of the Coleoptera are to be found in the structure of the basal abdominal segment, the male genitalia, the position of the Malpighian tubes (cryptonephry), the larval morphology, etc. However, most fossils consist only of elytra or of impressions of the outline of the body. It is not even possible to see the structure of the tarsi and antennae, which also provide important characters for recognizing some of the monophyletic groups of the beetles. The relatively large number of Mesozoic beetle fossils thus contributes a disproportionately small amount to our understanding of the phylogenetic development of this important order.

As a result, I shall have to restrict myself in this chapter to a critical examination of the evidence on which some of the more recent phylogenetic trees or 'phylogenetic classifications' of various orders have been based, so far as this can be done by a non-specialist. I shall take particular note of how far the systematic position accorded to Mesozoic fossils within these classifications is justified or leads to any worthwhile conclusions.

I have been unable to give a detailed account of the recent review of Jurassic insects from Karatau (Rodendorf, 1968).

Additional note

354. See note 92 on p. 82. [Adrian Pont.]

b. The individual orders during the Mesozoic

1. to 2.2.1.   **The primarily wingless groups ('Apterygota')**

There are five monophyletic groups of primarily wingless insects ('Apterygota'), Diplura, Protura, Collembola, Archaeognatha, and Zygentoma, and these must have separated during the Devonian. It is impossible to determine how many of their subgroups go back as far as the early Mesozoic or even the Palaeozoic because virtually no fossils are known. What is more, it is unlikely that future discoveries will shed any light on this because a satisfactory analysis of characters would be possible only with exceptionally well preserved fossils.

The only early Mesozoic species is *Triassomachilis uralensis* Sharov (Fig. 114), from the Upper Triassic of the Bashkirskaya Republic. Sharov (in Rodendorf, 1962) assigned it to the paraphyletic group 'Thysanura'.

There is some evidence that *Triassomachilis* belongs to the Archaeognatha and not to the Zygentoma (see p. 117). If this assignment is correct, then *Triassomachilis* would be more primitive than all the recent species in having the compound eyes separated, and it could only belong to the stem-group of the Archaeognatha rather than to one of the subgroups. Sharov (1948) reported that the Triassomachilidae have segmented abdominal styli. These also occur sporadically in recent species. Janetschek (1957, p. 9) thought it

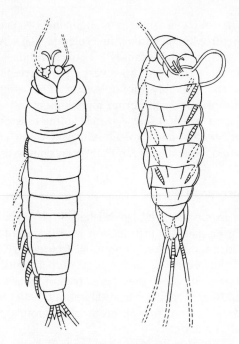

Fig. 114. *Triassomachilis uralensis* Sharov (Archaeognatha), from the Upper Triassic of the Bashkirskaya Republic. From Sharov (1948).

possible that 'segmentation of the abdominal styli was completely suppressed in the species that immediately followed the Triassomachilidae, the ancestors of the recent Machilidae'. He thought 'that the occurrence of aberrantly segmented styli in the bristle-tails could be interpreted as phyletic reversion, as a case of the re-acquisition of a character that had been lost from time to time in the course of phylogeny'.

The Collembola, with about 2000 species, are the largest group of primarily wingless insects in the recent fauna. Consequently, it is likely that some of their subgroups arose in the Mesozoic, or even in the Palaeozoic.

In a discussant's comment on Handschin's (1958) lecture, Delamare Deboutteville stated that the Middle Devonian *Rhyniella* (Fig. 19, p. 106) belonged to the tribe Pseudachorutini of the Collembola. From this he concluded that the recent subgroups of the *Collembola separated during the Devonian. This suggestion would be acceptable only if three conditions could be met: that the alleged assignment of *Rhyniella* to the 'Pseudachorutini' is not based on shared primitive characters (symplesiomorphies); that the recent subgroups of the Collembola, including the Pseudachorutini, are monophyletic; and that the genealogical interrelationships of these subgroups are known. However, none of these conditions can be met, and so there is no support for Delamare Deboutteville's conclusions.

Most authors now follow Börner in recognizing two 'suborders' in the Collembola, the Arthropleona and Symphypleona. The Arthropleona include the Pseudachorutini (Stach's family Pseudachorutidae) and most of the other recent Collembola. They are, however, a paraphyletic group, whereas the Symphypleona are probably monophyletic. The assignment of fossil species to the 'Arthropleona' is therefore meaningless.

I have already shown (p. 107) that it is misleading to assign the Devonian *Rhyniella* and *Rhyniognatha*, and *Protentomobrya* (from Canadian amber of the Upper Cretaceous), to a single 'family Protentomobryidae'.

2.2.2.1..1. Ephemeroptera[355]

A few years ago Edmunds (1962) published a 'phylogenetic tree' of the Ephemeroptera, but he gave no justification for his views from the point of view of phylogenetic systematics. Demoulin (particularly in 1958b) had previously grouped the recent species into six 'superfamilies'. Some of the families which he placed close together are not genealogically related according to Edmunds' phylogenetic tree. Demoulin believed that his superfamilies could be derived from two different 'stem-groups' that had already separated by the late Palaeozoic.

Demoulin (1955a, 1955b, 1955c) called one of these stem-groups the Mesephemeridae, and its oldest species is *Palingeniopsis praecox* Martynov (Fig. 116), from the Upper Permian (Kazanian Stage) of northern Russia. He regarded the recent Palingenioidea as the descendants of this stem-group, and included the families Behningiidae and Caenidae, which Edmunds showed as widely separated in his phylogenetic tree. Only two Mesozoic species of this group are known, *Mesephemera prisca* Germar and *M. litophila* Germar, from the Malm of Solnhofen[356].

Demoulin's second stem-group was the hypothetical Praepaedephemeridae (Demoulin, 1955d). This group is supposed to have existed during the Upper Permian alongside the Mesephemeridae (*Palingeniopsis*!), but no actual fossils have ever been found. Their recent descendants are supposed to be the Oligoneuroidea, Siphlonuroidea, and Ephemeroidea. Here too Demoulin has grouped a number of families together which do not have any close genealogical relationship according to Edmunds' phylogenetic tree. The oldest fossils that he assigned to this complex are from the Malm of Solnhofen. More recently, he has assigned the Mesozoic genera *Hexagenites* and *Ephemeropsis* to the family Hexagenitidae, together with the recent East Asian genus *Chromarcys* (= *Pseudoligoneuria*) (Demoulin, 1967b; see also Demoulin, 1953). However, he thought that the relationships of the Hexagenitidae (whether to the Oligoneuridae or to the Siphlonuridae) were still not clear. He also thought that the phylogenetic relationships of the Jurassic genus *Paedephemera*, the only genus in the 'family Paedephemeridae', needed to be re-examined (Demoulin, 1967a).

Demoulin said nothing about the relationships of two of his six superfami-

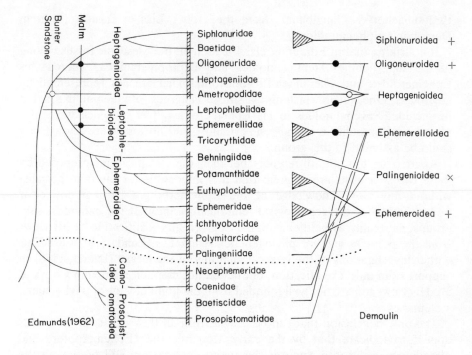

Fig. 115. Phylogenetic tree of the Ephemeroptera. A comparison of the views of Demoulin and Edmunds. On the right I have marked with + the groups that Demoulin derived from his hypothetical late Palaeozoic 'Praepaedephemeridae', and with × the groups which he derived from the Mesephemeridae.

lies, the Heptagenioidea and Ephemerelloidea. However, they are said to be known from Mesozoic fossils. The Heptagenioidea are represented by *Mesoplectopteron longipes* Handlirsch, a larva from the Lower Triassic of the Vosges (in Demoulin's Ametropodidae), and *Mesobaetis sibirica* Brauer, Redtenbacher & Ganglbauer, a larva from the Upper Lias of the Irkutsk region on Lake Baikal (in Demoulin's Leptophlebiidae). The Ephemerelloidea are represented by *Turfanella tingi* Ping, a larva from the Upper Jurassic of the Sinkiang-Uighur region, and *Mesoneta antiqua* Brauer,

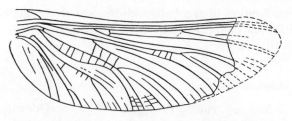

Fig. 116. Hind-wing of *Palingeniopsis praecox* Martynov (Ephemeroptera), from the Upper Permian of the Arkhangelsk region. From Rodendorf (1962).

Redtenbacher & Ganglbauer, from the Upper Lias of Irkutsk (both in Demoulin's Ephemerellidae).

It is striking that in Edmunds' classification all the Mesozoic fossils belong to the superfamilies Heptagenioidea and Leptophlebioidea, whereas his Ephemeroidea and Caenoidea-Prosopistomatoidea are not represented by any fossils. One way in which his classification agrees with Demoulin's is that he included seven families in his Ephemeroidea, five of which were also placed by Demoulin in a single superfamily, and no fossils are known that could be assigned to this group.

Apart from this, the differences between the classifications proposed by these two specialists are so great that it is impossible for the non-specialist without any close knowledge of the Ephemeroptera to gain any clear understanding of the genealogical relationships amongst the individual sub-groups, especially as neither author has given any justification of his views from the point of view of phylogenetic systematics. Landa (1959) used his earlier investigations of the internal anatomy of the larvae (Landa, 1948) to support Edmunds' classification, but he did not mention Demoulin's work at all. His views too are not well founded from the point of view of phylogenetic systematics.

The only conclusion that can be drawn from all this is that certain fossils appear to indicate that by the early Mesozoic the *Ephemeroptera had already separated into some of the subgroups that are still present in the recent fauna.

Additional notes

355. Demoulin (1969–71) has published a number of papers dealing with Mesozoic fossils of the Ephemeroptera, which are supplementary to his earlier review of the fossil history of this order (Demoulin, 1954c). The biogeography and evolution of the order have been discussed by Edmunds (1972). Further to the paper by Edmunds and Traver (1954) on flight mechanics, an important paper on the evolution of the mayfly wing has been published by Brodskiy (1974). The eggs and their bearing on mayfly phylogeny have been dealt with by Koss and Edmunds (1974). [Willi Hennig.]

356. For a redefinition of the Palingeniidae and their significance for the higher classification of the Ephemeroptera, see McCafferty and Edmunds (1976). [Willi Hennig.]

2.2.2.1..2. Odonata[357]

Three suborders are usually distinguished in the Odonata: Zygoptera, Anisozygoptera, and Anisoptera. There is no doubt that the Anisoptera are a monophyletic group, and they also have a derived character in their venation: the discoidal cell, which is known as the 'quadrilateral' in the Zygoptera and

Anisozygoptera, is divided by a single vein into the so-called 'triangle' and 'supratriangle' (Fig. 118). There are further derived characters but they are rarely visible in fossils.

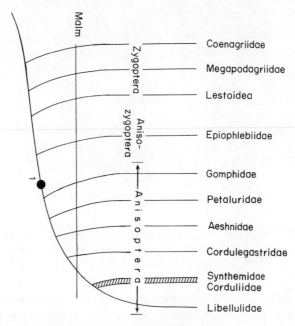

Fig. 117. Phylogenetic tree of the Odonata. 1, *Liasso-gomphus*. Re-drawn from Fraser (1954).

The oldest fossil that may belong to the Anisoptera is *Liassogomphus brodiei* Buckman, from the European Lias (Fig. 119). In the Russian Textbook of Palaeontology (Rodendorf, 1962), this species was assigned to the 'Aeshnidea', a subgroup of the *Anisoptera, but there is little justification for this. Tillyard and Fraser (1938–40) considered it to be a transitional form between the Anisozygoptera and the Gomphidae. As these authors (and Fraser, 1954) regarded the Gomphidae as the most primitive family of the *Anisoptera and as the Anisozygoptera are correctly considered to be the sister-group of the Anisoptera, I would express this by saying that Liassogomphus belongs to the stem-group of the Anisoptera.

Several undoubted Anisoptera are known from the Upper Jurassic (Malm) of Solnhofen. Some of them have been assigned to recent families: *Mesuropetala* and *Libellulium* (= *Cymatophlebia*) to the Petaluridae; *Nannogomphus*, *Protolindenia*, *Necrogomphus*, and *Phengothemis* to the Gomphidae. However, these assignments do not mean a great deal because, as Fraser (1954) pointed out, the Gomphidae and Petaluridae are the most primitive families of the Anisoptera and were the first to separate from the main stem of the group. The characters common to the Upper Jurassic fossils and these two

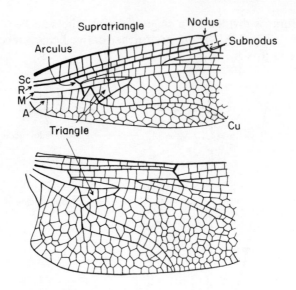

Fig. 118. Proximal half of the wings of *Aeshna juncea* Linnaeus (Odonata, Anisoptera; recent). From Chopard (in Grassé, 1949).

families could therefore be symplesiomorphies. In the Russian Textbook of Palaeontology (Rodendorf, 1962), the genera *Aeschnidium* and *Urogomphus*, from the Malm of Solnhofen, were included in a separate family Aeschnidiidae, together with the genera *Aeschnidiella* and *Aeschnidiopsis*, from the Cretaceous of Australia and the Soviet Union. This family was then placed in the superfamily 'Aeshnidea', which is the Aeshnoidea of other authors and which also includes a number of other recent families. These assignments also mean very little because the phylogenetic trees of Fraser (1954) and Watson (1956) show that the Aeshnidea (Aeshnoidea) are paraphyletic.

It is possible that some of the subgroups of the *Anisoptera were in

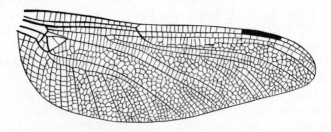

Fig. 119. Hind-wing of *Liassogomphus brodiei* Buckman (Odonata, Anisoptera), from the Lower Jurassic of England. From Tillyard (1925a).

existence during the Upper Jurassic but there is no evidence for this in the fossil record.

The Anisozygoptera are correctly regarded as the sister-group of the Anisoptera. Only two recent species belong here: *Epiophlebia superstes* Sélys (Fig. 120) and *E. laidlawi* Tillyard, from Japan and the Himalayas. They

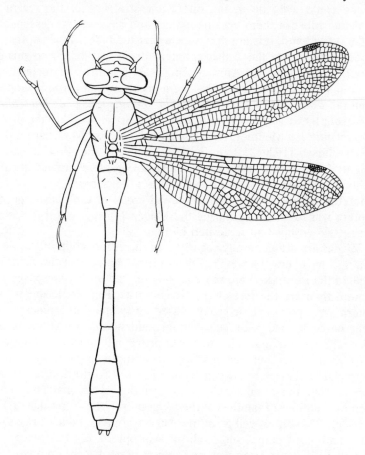

Fig. 120. *Epiophlebia superstes* Sélys (Odonata, Anisozygoptera; recent). From Handlirsch (1920–21).

agree with the Anisoptera in the enlargement of the eyes, which are rather close together on the frons; in the structure of the anal appendages at the tip of the male abdomen, which consist of three arms (see below); and in the enlargement of the anal lobe of the hind-wing. These all appear to be derived characters (synapomorphies of the Anisoptera and Anisozygoptera). The Anisozygoptera are more primitive than the Anisoptera in having the discoidal cell complete and not divided into the triangle and supratriangle. Unfortunately, I have not been able to read Asahina's (1954) paper on *A. superstes.*

If it is true that the Anisoptera were already in existence in the Upper Jurassic, and probably the Lower Jurassic too (*Liassogomphus*), then their sister-group the Anisozygoptera must have been as well. In fact, numerous fossils have been described since the Lias and assigned to the Anisozygoptera. Most of them really could belong to the stem-group of the Anisozygoptera and some could even be in the direct ancestral line of the recent genus *Epiophlebia*. Most of them became extinct and have left no descendants.

On the other hand, it is not certain whether the older, Triassic, fossils that have been assigned to the Anisozygoptera do in fact belong to this group. They only agree with the Anisozygoptera in primitive characters, and so could equally well belong to the stem-group of the Anisozygoptera + Anisoptera. This applies to *Mesophlebia, Periassophlebia, Triassophlebia*, and *Triassolestes*, from the Upper Triassic of Australia, and to the problematic *Piroutetia liasina* Meunier, from the Rhaetic of Fort Mouchard in France, which Handlirsch (1920–21) continued to assign to the 'Protodonata'. It is possible that some of the other Jurassic 'Anisozygoptera' are in fact survivors from the stem-group of the Anisozygoptera + Anisoptera.

It therefore seems definite that the stem-group of the Anisozygoptera + Anisoptera was in existence during the Upper Triassic, but it is not certain that these two groups had separated by then.

The Zygoptera are generally regarded as the sister-group of the Anisozygoptera + Anisoptera. However, Fraser (1954) stated that the Agrionoidea (also called the Lestoidea) are more closely related to the Anisozygoptera + Anisoptera than are the rest of the 'Zygoptera'. This would mean that the Zygoptera are a paraphyletic group, based on symplesiomorphies.

Fraser based his views on the general tendency in the Odonata for the origins of the longitudinal veins to travel proximally towards the base of the wing. He pointed out that a change appears abruptly in the Lestidae in which the intercalary vein (IR_3) between R_3 and R_4, and R_4 itself, originate from a point nearer to the arculus than to the nodus. This character has been retained in a number of families of the 'Zygoptera' (the Agrioidea of Tillyard and Fraser, 1938–40) as well as in the Anisozygoptera, which Fraser did not actually mention by name, and in all the Anisoptera. For this reason, Fraser thought that all these groups shared ancestors that were common only to them, whereas the Coenagriidae and Megapodagriidae separated from the main lineage of the *Odonata earlier. If he was correct, then several characters will have to be re-assessed. It is generally believed (e.g. by Chopard, in Grassé, 1949) that the median inferior anal appendage at the tip of the male abdomen in the Anisozygoptera and Anisoptera is homologous with the paracercus, and that the cerci are absent. On the other hand, the paired inferior anal appendages of the Zygoptera are supposed to have developed from the cerci, whereas the paracercus is absent. The cerci and the paracercus are both part of the ground-plan of the Pterygota, and this can only mean that the anal appendages of the Zygoptera and Anisozygoptera + Anisoptera have developed from a precursor with these characters more primitive than they now are in these groups. The Zygoptera and the Anisozygoptera + Anisoptera have each developed in a different direction from this precursor, and the resultant alternating apomorphies show that they are sister-groups. On the other hand, if Fraser was right, then the 'Zygoptera' must have retained the ground-plan characters of the Odonata and only the Anisozygoptera + Anisoptera modified them into fundamentally more derived states. We would then have to conclude that the median inferior anal appendage of the Anisozygoptera + Anisoptera is homologous with the fused cerci rather than with the paracercus, or that the paracercus was retained in rudimentary form in the Zygoptera and was then secondarily enlarged in the Anisozygoptera + Anisoptera. It will undoubtedly be possible to work out a solution to these difficulties. At all events, Fraser's views are very important because they are based on the venation, which is a character readily visible in the fossils, and they should provide a stimulus for further morphological investigations. These should take account of characters such as the secondary copulatory organs at the base of the male abdomen[358] and the larval morphology.

It used to be thought that the broad wing base of the Anisozygoptera + Anisoptera was a primitive character and matched very closely the conditions in Palaeozoic Odonata, but, as mentioned above (p. 353), Fraser found this view to be untenable. He considered that the broad wing base of the Anisozygoptera + Anisoptera had developed from a petiolate wing base, a character which has been retained in the Zygoptera and which is primitive in the *Odonata though not in the Odonata(!).

The existence of Upper Triassic fossils that probably belong to the stem-group of the Anisozygoptera + Anisoptera shows that species of the sister-group of the Anisozygoptera + Anisoptera must also have been in existence at the same time. However, it is curious that no fossils belonging to this group have been found before the Upper Jurassic. According to Rodendorf (1962), these fossils belong to the 'Agrionidea', which is the group that Fraser (1954) considered to be more closely related to the Anisozygoptera + Anisoptera than are the other 'Zygoptera'. For the moment, it does not matter whether the assignment of these fossils to recent families is correct or not: *Steleopteron deichmuelleri* Handlirsch has been assigned to the Amphipterygidae, and *Euphaeopsis* and *Pseudoeuphaea* to the Epallagidae. It is curious that no Mesozoic species of the 'Coenagrionidea' are known, although Fraser considered them to be an even more primitive group.

The Australian *Hemiphlebia mirabilis* Sélys occupies a special position amongst recent Odonata. As the only species of the 'Hemiphlebioidea', it has usually been placed at the beginning of the 'Zygoptera', but unfortunately Fraser (1954) made no mention of it in his phylogenetic tree. There may be a sister-group relationship between *Hemiphlebia* and the rest of the recent Odonata or, to be more precise, of the recent *Odonata. This is because it is rather more primitive than the rest of the *Odonata in a number of characters, according to Chopard (in Grassé, 1949): in the fore-wing at least, there is no cross-vein separating the discoidal and medial cells; the so-called penis at the base of the abdomen is less complex than in the rest of the Zygoptera; in the larvae the hypopharynx has three lobes, and in the labium the glossae and paraglossae are separate.

Wings with the open discal cell of the Palaeozoic Odonata are also known from the Triassic of Australia, the Lias of Europe, and the Malm of Central Asia (Karatau). In the Russian Textbook of Palaeontology (Rodendorf, 1962), they were all assigned to the family Protomyrmeleontidae: genera *Protomyrmeleon, Triassagrion*, and *Tillyardagrion*. The absence of the nodus, which is present in *Hemiphlebia* and all other recent Odonata, is a primitive character and would not exclude the Protomyrmeleontidae from the stem-group of the Odonata. On the other hand, the absence of ante-nodal cross-veins is a derived character according to Fraser (1954). Ante-nodal cross-veins (as a primitive character) and the forerunner of a nodus were present in certain Permian species (Permagrionidae), and so it seems probable that both characters were present in the latest common ancestors of all the *Odonata. These could have been in existence during the Lower Triassic, but have not been found in the fossil record. There is certainly no justification for tracing various subgroups of the *Odonata back into the Permian.

However, a number of subgroups that are still represented in the recent fauna must already have been in existence in the Upper Triassic and Jurassic. It is not possible to say how many there were, mainly because there is still no satisfactory phylogenetic classification of the *Odonata. It seems certain that the Protomyrmeleontidae, like some of the Protodonata, survived into the Mesozoic independently of the direct ancestors of the *Odonata and then became extinct. The extent to which the *Odonata underwent renewed expansion in the Cretaceous is not known.

Additional notes

357. Pfau (1971) has discussed the structure and function of the secondary copulatory organs. Heymer (1973) has dealt with the ethology and evolution of the Calopterygidae, with discussion of the evolution of the Odonata generally. Kiauta (1968) has given the chromosome counts of 236 species of Odonata, but his conclusions were challenged by Crowson (1970). [Willi Hennig.]

358. See the paper by Pfau (1971) mentioned above. [Willi Hennig.]

2.2.2.2..1. Plecoptera

It is extraordinarily difficult to work out the history of the Plecoptera during the Mesozoic because it is not known how many subgroups of the *Plecoptera originated during the Palaeozoic. If Illies' (1965) phylogenetic tree is to be believed, there must have been at least five groups, but this is open to dispute (see pp. 171–172)[359].

The Australian 'genus *Stenoperlidium*' is the only Upper Permian genus that is also known from the Upper Triassic. It is probable (see p. 172) that these fossils do actually belong to an archaic group of the *Plecoptera which still exists in Australia—the Archiperlaria, though possibly not in such a restricted sense as that of Illies (1965)[360].

Seven species, two of which are only larvae, have been described from the Jurassic (Lias and Dogger) of Central Asia and Siberia–China, and in the text of his paper Illies (1965) assigned them to the recent family 'Taeniopterygidae', together with the few known Cretaceous fossils. However, in his phylogenetic tree (Illies, 1965, Fig. 4), he assigned at least the Jurassic fossils to the stem-group of the Taeniopterygoidea, a superfamily that includes three other recent families as well as the Taeniopterygidae. It is not clear whether the characters common to the Jurassic fossils and the 'Taeniopterygoidea' are symplesiomorphies. It will be impossible to see how the order developed during the Mesozoic until this problem has been solved and the phylogenetic classification of the *Plecoptera itself more clearly worked out[361].

Revisionary notes

359. See note 174 on p. 175. [Peter Zwick.]

360. See note 172 on p. 173. [Peter Zwick.]

361. See notes 172 and 175 on pp. 173 and 175. [Peter Zwick.]

2.2.2.2..2. **Paurometabola**

No Embioptera are known from the Mesozoic, and so it is impossible to say when the recent subgroups originated. However, this is not a particularly important matter as there are only about 150 recent species[362].

The same is true of the Notoptera, which only include 12 recent species.

A few Mesozoic Dermaptera are known from the Upper Jurassic, and I have already discussed the importance of these fossils (p. 187). It is almost impossible to make any assessment of them because there is no well founded classification of the Dermaptera. Until very recently the two suborders Protodermaptera and Eudermaptera have usually been recognized. Martynova (in Rodendorf, 1962) also adopted this division but used the name Forficuloidea for the Eudermaptera and used the name Eudermaptera in a different sense.

Popham (1965) replaced these two groups with four superfamilies. Two of these, the Forficuloidea and Labioidea, combined groups of the old Protodermaptera with various groups of the old Eudermaptera[363].

It is not clear from Popham's paper whether his superfamilies really are monophyletic groups. In any case, he only used characters of the male genitalia and so even if they are monophyletic it is not known whether they could also be recognized by other characters that might be visible in fossils. It is also impossible to decide whether the characters found by Martynova to be common to the Upper Jurassic fossils and the recent Labiidae and Forficulidae are only symplesiomorphies.

It is not known at all how many subgroups of the *Dermaptera separated during the Mesozoic or which groups these may have been.

The Mantodea include over 1500 recent species, and only a few Mesozoic fossils can definitely be assigned here, as was the case with the Palaeozoic fossils. Nothing at all is known about when the separation of the *Mantodea into subgroups began. The relationships between the groups traditionally recognized within the order are also very unclear.

No Mesozoic fossils of the Isoptera are known[364] although the termites must have existed during this epoch if, as seems likely, the group originated during the Permian or even earlier (p. 207). All workers seem to agree that there is a sister-group relationship between the Mastotermitidae (one recent species in Australia) and the rest of the Isoptera[365], although no names have been proposed to formalize these higher categories. However, it is not known when this sister-group relationship arose.

The Blattariae are well known from Mesozoic fossils. It is always possible that some of these may in fact be survivors from the stem-group of the Blattodea or may belong to the stem-group of the Mantodea. None of them can be assigned to any of the subgroups of the *Blattariae. Unfortunately, the main sister-group relationships within the *Blattariae are still not clear (Fig. 121). Princis (1960) and McKittrick (1964) have both discussed the classification of the Blattariae in recent papers and each has proposed two major divisions which, according to their 'phylogenetic trees', must be sister-groups. However, their suggested sister-groups are not the same: Princis recognized the Polyphagoidea + Blaberoidea and Blattoidea + Epilamproidea, and McKittrick the Blattoidea and the Blaberoidea. In view of this lack of agreement, it is pointless to enquire when the main sister-group relationships may have arisen[366].

In this connection it should be noted that Martynova (in Rodendorf, 1962) assigned all the recent Blattariae to the single family Blattidae, in which she also included all the Mesozoic fossils and even a number of Palaeozoic ones, from the Upper Carboniferous onwards. On the other hand, Princis (1960) recognized 23 families of recent Blattariae.

Revisionary notes

362. There are about 800 species of Embioptera (Ross, 1970). [Willi Hennig.]

363. The suprageneric classification of the Dermaptera has also been dealt with by Steinmann (1975). [Willi Hennig.]

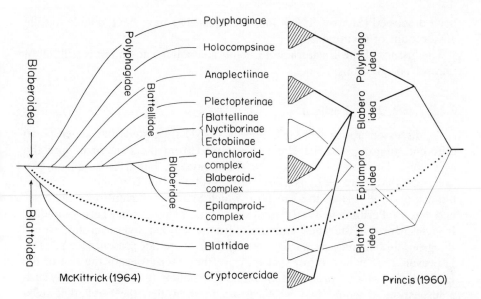

Fig. 121. Phylogenetic tree of the Blattariae. A comparison of the views of McKittrick and Princis.

364. In the last decade, Isoptera have been found in Lower Cretaceous amber from the Lebanon (no detailed analysis has yet been made). They have also been found in Upper Cretaceous amber from Siberia (Zherikhin and Sukacheva, 1973) and Middle Cretaceous amber from France (Schlüter, 1978). [Dieter Schlee.]

365. The Mastotermitidae are defined by symplesiomorphies rather than by synapomorphies, and so it is not possible to establish a correct sister-group relationship based on a scheme of synapomorphies. [Dieter Schlee.]

366. See Roth (1970), who has also included some phylogenetic trees, and Huber (1974) who constructed several dendrograms from a phenetic numerical taxonomic study of 37 species of cockroach. [Willi Hennig.]

2.2.2.2..2.2..2. **Orthopteroidea**

The Phasmatodea are one of the three monophyletic groups into which the Orthopteroidea can be divided. Their history during the Mesozoic is virtually unknown. In the Russian Textbook of Palaeontology (Rodendorf, 1962), a number of Mesozoic fossils has been grouped together under the name Chresmododea, but it is not known whether they really belong to the Phasmatodea (see p. 222). The fossils which Sharov (1967) mentioned from the Lower Triassic of Central Asia and which are supposed to connect the Palaeozoic Tcholmanvissiidae with the *Phasmatodea, have only recently been described (Sharov, 1968) and consequently I have been unable to take any account of them here.

The history of the Ensifera and Caelifera is better known but is still full of unresolved questions.

2.2.2.2..2.2..2.1. **Ensifera**

As discussed above (p. 218), Sharov (1967) suggested that at least two different subgroups of the Ensifera survived from the Palaeozoic into the Mesozoic. This suggestion may be well founded, but it is based on two premises: that there really is a sister-group relationship between the Tettigonioidea and the rest of the Ensifera, which Sharov believed to be the case, and that the Permian genus *Tettavus* really does belong to the Tettigonioidea. It seems probable that the *Tettigonioidea are a monophyletic group, even if the ground-plan characters suggested by Sharov are not present (see p. 216). However, it does not follow that they are the sister-group of the rest of the Ensifera, for which I proposed the name 'Gryllodea' (p. 216). To justify such a suggestion, it would first be necessary to show that the Gryllodea are a monophyletic group, and this is not an easy task.

The Grylloidea, with the families Gryllidae, Gryllotalpidae, and Oecanthidae, are undoubtedly monophyletic. They have several derived ground-plan characters of which the most peculiar is the method of folding the fore-wing. According to Ragge (1955a) the fore-wing forms a box-like cover for the proximal part of the flexed hind-wings and for the dorsal and lateral parts of

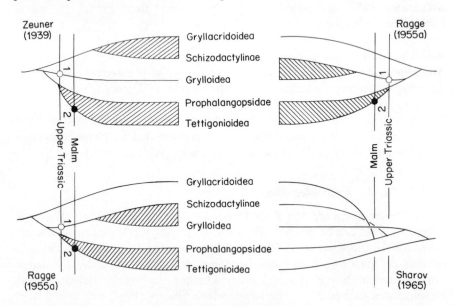

Fig. 122. Phylogenetic tree of the Ensifera. A comparison of the views of Ragge, Sharov, and Zeuner. 1, *Archaegryllodes*; 2, *Termitidium*.

the abdomen. The division between the dorsal field and the lateral field runs roughly along the median longitudinal axis of the wing where Sc, R, M, and CuA are closely parallel. Additional folding occurs in the median fan, an area formed by the divergent distal sections of R and M, and this is a very characteristic feature of the Gryllidae. However, it does not appear to be part of the ground-plan of the Grylloidea because it is absent in the Gryllotalpidae. The Gryllotalpidae are also more primitive in other respects.

On the other hand, the enlarged costal field is a derived ground-plan character and it has arisen through the displacement of Sc along the median longitudinal axis of the wing. The costal field contains numerous anterior accessory branches of the subcosta. On the other hand, the costa is very reduced: it is no longer an independent vein, but appears as the anterior branch of the subcosta.

According to Ragge (1955a), the hind-wings of the Grylloidea fold up in a similar way to those of the Tettigonioidea but the folding is more complete and extends up to the anterior margin of the wing. In this respect the Grylloidea are undoubtedly more derived than the Tettigonioidea; these in turn are more derived than the Caelifera, for example, because part of the remigium is included in the folding area of the wing which, in the ground-plan of the Orthopteroidea (and Caelifera), is confined to the anal lobe. When folded the hind-wings form a characteristic spike that generally protrudes beyond the fore-wings. Ragge (1955a) pointed out that the division of CuA into two free branches, which enclose a rather sclerotized area of wing surface, is correlated with this.

In male Grylloidea the harp and mirror, which are accessory structures of the stridulatory organ, are much larger than in the Tettigonioidea. They occupy almost the entire posterior half of the fore-wing. As Sharov suggested that the stridulatory ribs of the Tettigonioidea and Grylloidea are not homologous, it would be particularly valuable to work out the manner in which the stridulatory organ has developed from its common forerunner in the Tettigonioidea and also in the Grylloidea.

The next fundamental problem in the phylogenetic classification of the Ensifera concerns the position of the Gryllacridoidea, Schizodactyloidea, and Prophalangopsoidea.

The Gryllacridoidea and Schizodactyloidea agree in having no trace of any stridulatory organ in the fore-wing. A great deal has been written about whether this is primary (primitive) or secondary (derived), and this is a matter of particular importance for the interpretation of the fossils.

Karny and Zeuner believed that the stridulatory organs were primarily absent in the Gryllacridoidea and Schizodactyloidea, whereas Handlirsch and Sharov thought it more likely that they had been secondarily lost. Zeuner (1939) was not convinced by Karny's arguments that this absence is primary, but on the basis of his own investigations of the prothoracic spiracle he continued to advocate this hypothesis. On the other hand, Sharov (1967) claimed that he was following Martynov (1938b) in suggesting that the ostensibly primitive venation of the Gryllacridoidea is secondary and is a result of the cryptic and nocturnal mode of life adopted by this group. The ability to fly was lost and the wings only functioned as protective covers for the body.

If Karny and Zeuner are correct in considering the primary absence of stridulatory organs in the Gryllacridoidea and Schizodactyloidea to be primitive, then we would have to follow Zeuner and Ragge (1955a) in suggesting that there is a sister-group relationship between the Tettigonioidea + Grylloidea and the Gryllacridoidea + Schizodactyloidea (if the last two are monophyletic groups). This relationship must have arisen before the Lower Permian because Sharov has reported that fossils with a stridulatory organ on the fore-wing are known from this epoch.

However, other characters do not confirm this sister-group relationship. For the present, therefore, the best course is to leave the stridulatory organ out of consideration and to search for characters that might be synapomorphies of the Gryllacridoidea, Schizodactyloidea, and Grylloidea.

Zeuner (1939) stated that the Gryllacridoidea and Schizodactyloidea agree in having fore-wings that 'wrap' the dorsal and lateral surfaces of the abdomen. In this respect they resemble the Grylloidea rather than the Tettigonioidea, but differ from the Grylloidea by having no median longitudinal fold. However, Zeuner noted that in the Schizodactyloidea the line dividing the fore-wing into the dorsal and lateral parts is in the same position as in the Grylloidea. Ragge (1955a) also found some particularly significant characters common to the Schizodactyloidea and Grylloidea. He listed the following undoubtedly derived characters that only occur in these two groups: the characteristic folding of the fore-wing along the basal sections of R, M, and Cu; the development of a median fan in the fore-wing; the extension of the folding area of the hind wing as far as the most anterior branch of the media; the spike formed by the distal end of the folded hind-wing; and the division of CuA into two branches in the hind-wing.

Apparently, Ragge considered these to be synapomorphies and he showed the Schizodactyloidea and Grylloidea as sister-groups in his phylogenetic tree. However, there are problems with this view too. For example, the median fan appears to be absent in the ground-plan of the Grylloidea (see above). Moreover, there are several characters shared by the Gryllacridoidea and Schizodactyloidea that would have to be treated as the results of convergence if the Schizodactyloidea are in fact most closely related to the Grylloidea. It is indeed possible that some of these shared characters have arisen convergently, and this is a problem that still needs to be clarified. Zeuner (1939) pointed out that the longitudinal veins of the Schizodactyloidea and Gryllacridoidea radiate out from the base of the fore-wings and develop secondary branches in their distal sections, and he stressed that this is correlated with the special way in which these groups wrap their wings round the abdomen. He was right to regard this as a derived character, and he also thought it possible that the loss of the stridulatory organs and of the fore-tibial tympanal organs was connected with this. If we also take into account the fact that the wing of the Schizodactyloidea and Gryllacridoidea agrees best with that of female Grylloidea in many respects and that all three groups fold the fore-wing more or less along the median longitudinal axis then the following working hypothesis may be of some use:

The development of a stridulatory organ first began in the fore-wing of the Ensifera when the wings were still carried more or less flat over the body. This is the only way to explain why it was CuP, one of the veins in front of the anal lobe, that was modified into the stridulatory rib. Moreoever, the reports by Zeuner (1940b) and Sharov (1967) that the wings of the primitive 'Haglidae' from the early Mesozoic were still carried flat over the abdomen would also lend some support to this. As a later development, part of the wing surface was wrapped round the sides of the body. The same development took place in the Caelifera, where the anal furrow was used as the bending fold. However, the Ensifera could not make use of this furrow because the stridulatory rib had already been developed just in front of it and so they had to fold their wings just in front of the stridulatory rib. The comparatively long and narrow fore-wings of the Oedischiidae (and Caelifera) show that this wing shape is primitive: on the whole, it was retained in the Tettigonioidea or led to even longer and narrower wings. By contrast, the wings of the Gryllodea appear to have become shorter and broader. This appears to be correlated in some way with the enlargement of the stridulatory organ and the expansion of the costal field.

The only way in which the wings could be folded and wrapped round the sides of the abdomen was along a line more or less following the median longitudinal axis of the wing. Sharov (1967) thought that the wings folded along this line because the body of the Grylloidea had been compressed, and this in turn is supposed to result from the cryptic life in the soil and under stones that this group adopted. However, it seems more likely that this line of development began at the base of the wing in the stem-group of the Gryllodea: these still enjoyed a free and open mode of life, as is shown by their coloured wings, and they were still rather grasshopper-like in form, as is clear from Sharov's (1967) illustration of *Aboilus*. It may therefore be more correct to correlate the start of this development with the width of the stridulatory organ. The Gryllacridoidea and Schizodactyloidea must have lost the stridulatory organ early on but the wing had already begun to develop a line where it could be folded along the median longitudinal axis. The wings of these two groups may appear to be primitive, but it is not nearly so difficult to explain this condition as many authors have thought: it is only necessary to suppose that the wing structure of the female has been transferred to the male. The transfer of a character from one sex to the other is not unusual. It has also taken place in the Ensifera in the opposite direction: the stridulatory organ undoubtedly originated in males, but in some genera it is also found in the females.

It is not clear whether the loss of the stridulatory organ is a synapomorphy of the Gryllacridoidea and Schizodactyloidea or whether it is a convergent feature in these groups, as Sharov (1967) believed.

The trend towards short and broad fore-wings that is characteristic of the Gryllacridoidea and Schizodactyloidea is also evident in the Prophalangopsoidea.

According to Chopard (in Grassé, 1949) the bending fold in the fore-wing of *Prophalangopsis*, and perhaps of other genera, is at the middle of the wing and close to the radius, as it is in the Grylloidea. The structure of the cerci may lend some support to the view suggested by the characters of the fore-wings that the Gryllodea (Prophalangopsoidea, Gryllacridoidea, Schizodactyloidea, and Grylloidea) and the Tettigonioidea are monophyletic groups. The cerci are long and flexible in the Gryllacridoidea, Schizodactyloidea, and Grylloidea: this may be a derived character, and in this case it could be a synapomorphy of these groups. The cerci are short in all the other Orthopteroidea, including the Prophalangopsoidea. One genus of the Gryllacridoidea (*Lezina*) even has multi-segmented cerci (Zeuner, 1939). Many authors have been inclined to treat this as a primitive character. However, it seems far more likely to be an autapomorphy of *Lezina* because there is absolutely no evidence that the cerci were long and multi-segmented in the primitive Orthopteroidea, Ensifera, or Gryllodea.

There is another potentially important character which was noted by Chopard (in Grassé, 1949). In the Grylloidea and in the Stenopelmatidae, which is considered to be the most primitive family of the Gryllacridoidea, there is a peculiar 'pseudo-tympanal organ' of unknown significance at the base of the abdomen. It is possible that this organ is a derived ground-plan character of the Gryllodea and has been secondarily lost in the groups where it is now absent.

On the whole, there seems to be some basis for the view that the Gryllodea are a monophyletic group. However, the sister-group relationships within the Gryllodea are still completely unclear. In his phylogenetic tree, Sharov (1965, 1967) showed two sister-groups: the Prophalangopsoidea (which he called Haglidae) + Gryllacridoidea + Schizodactyloidea, and the Grylloidea (Gryllotalpidae, Gryllidae, and Oecanthidae). However, he gave no evidence for this classification. It is just as likely that there is a sister-group relationship between the Prophalangopsoidea and the rest of the Gryllodea. The structure of the cerci might lend some support to this suggestion.

Sharov appears to have based his views mainly on his study of the fossils. The oldest fossils that undoubtedly belong to the Gryllodea are from the Triassic, the Lower Triassic according to Sharov (1967, 1968). Some have been assigned to the Haglidae and others to the Gryllidae. According to Sharov, the fossil 'Gryllidae' differ from the Haglidae in having the median longitudinal axis of the fore-wing with a right-angled bend. Zeuner (1939) assigned the species from the lower Lias and Upper Triassic of South Africa to a separate subfamily 'Protogryllinae' of the Gryllidae (Fig. 123): I know of no species older than *Archaegryllodes stormbergensis* Haughton (Upper

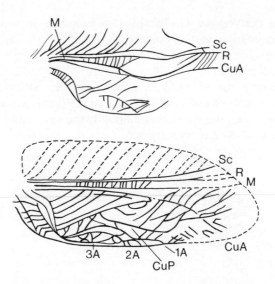

Fig. 123. Fore-wings of *Protogryllus* (*Archaegryllodes*) *stormbergensis* Haughton (above), from the Upper Triassic of South Africa, and *Protogryllus dobbertinensis* Geinitz (below), from the Lower Jurassic of Mecklenburg. From Zeuner (1939).

Triassic), and none are mentioned in the Russian Textbook of Palaeontology (Rodendorf, 1962). However, the 'Protogryllinae' probably belong to the stem-group from which the recent Gryllotalpidae, Gryllidae, Oecanthidae, and possibly other families have descended. It is not known whether the rest of the recent Gryllodea (Prophalangopsoidea, Gryllacridoidea, and Schizodactyloidea) could also have descended from this group. I think that it is possible. As I mentioned above, Sharov (1967) considered the Haglidae (Fig. 124) to be the stem-group of the other recent Gryllodea, except for the Grylloidea, but he gave no evidence for this. I know of no derived characters common to the 'Haglidae' and individual subgroups of the *Gryllodea. According to Zeuner, females of *Hagla* have extremely primitive fore-wings from which it is possible to derive the fore-wings of all the other Mesozoic and subsequent 'Saltatoria' (and not just the Gryllodea!). However, this type of statement is of no use for establishing relationships. Sharov (1967) asserted that it was impossible to derive the venation of the Schizodactyloidea from that of the Gryllodea although these had the more primitive venation, and for this reason he derived the Schizodactyloidea from some unspecified Jurassic 'Haglidae' which had independently acquired some of the peculiarities of the 'Grylloidea'. This conflicts with the work of Ragge (1955a) who, as I mentioned above, was able to deduce from the venation that there is an extremely close relationship between the Schizodactyloidea and the Grylloidea.

Sharov also derived the Gryllacridoidea from the Jurassic 'Haglidae' in which a re-orientation in the branching of MP + CuA had already taken place. This has been retained virtually unchanged by the most primitive recent Gryllacridoidea.

Sharov included the recent Prophalangopsoidea, which consist of only three species, and the fossil Haglidae in a single family which he called the 'Haglidae'. However, the characters shared by the recent and fossil species of this 'family' are all symplesiomorphies.

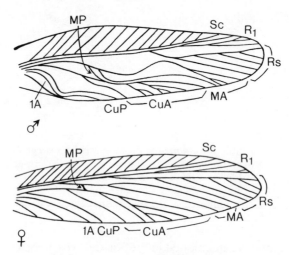

Fig. 124. Fore-wings of *Hagla gracilis* Giebel (Ensifera), from the lower Lias of England. Above, male; below, female. From Ragge (1955a).

It is always possible that some of Sharov's views will prove to be correct, but at present none of them appears to be well founded. Sharov, as he himself has frequently repeated, is a disciple of typological systematics, and so it is not surprising that he has made use of the rather convenient methods which are appropriate to this kind of systematics. His views may have a certain heuristic value but they can never provide any binding results.

The oldest fossils that have been assigned to the Haglidae are as old as, or slightly older than, the oldest 'Protogryllinae': *Notopamphagopsis bolivari* Cabrera (1928), from the Rhaetic of Argentina (Mendoza); *Prohagla superba* Riek (Fig. 125) and *Neohagla*, from the Middle and Upper Triassic of Australia (Riek, 1953b, 1955). They are of no greater use than the 'Protogryllinae' for determining the age of the Gryllodea.

It is still uncertain when the *Gryllodea separated into subgroups. It is likely that the oldest subgroups arose in the Jurassic or even in the Triassic, but it is not obvious exactly which ones these may have been.

According to Sharov (1967), the 'Haglidae' were the most abundant group of Mesozoic Orthopteroidea. However, it is not at all clear how far the

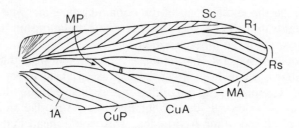

Fig. 125. Fore-wing of *Prohagla superba* Riek (Ensifera), from the Triassic of Australia. From Riek (1953b).

Middle and Late Mesozoic 'Haglidae' belonged to surviving side-branches of the stem-group of all the Gryllodea, which subsequently became extinct, and how far they belonged to subgroups of the *Gryllodea.

2.2.2.2..2.2..2.2. Caelifera

Since Ander's papers (1939a, 1939b), it has been usual to divide the Caelifera into the Tridactyloidea, which include only about 50 recent species, and the Acridoidea. Within the Acridoidea, Ragge (1955a) considered the

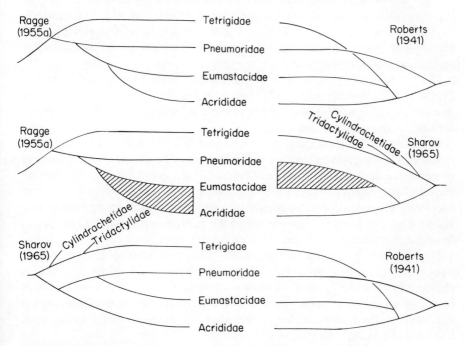

Fig. 126. Phylogenetic tree of the Acridoidea (Caelifera). A comparison of the views of Ragge, Roberts, and Sharov.

Tetrigidae to be the sister-group of the remaining families, whereas Roberts (1941) thought that there was a sister-group relationship between the Pneumoridae and all the other Acridoidea[367].

Sharov's views (1965, 1967) were not so very different from those of Ragge. Insofar as Sharov's conclusions can be expressed at all in the language of phylogenetic systematics, the only difference seems to be that he considered the Tetrigidae + Tridactyloidea rather than the Tetrigidae alone to be the sister-group of all the other Acridoidea. This would mean that the conventional division of the Caelifera into Tridactyloidea and Acridoidea does not reflect the true relationships within the group. As discussed earlier (p. 221), Sharov began with the assumption that the development of a large arolium is the derived ground-plan character of a monophyletic group consisting of the Caelifera + Phasmatodea. He believed that this character originated when the common ancestors of these two groups adopted a life on plants: at first these must have been bushes and trees as there was no herbaceous vegetation. They were what Bey-Bienko and Mishchenko called 'tamnobionts'.

Sharov thought that the Tetrigidae gave up this mode of life and that the secondary loss of the arolium was a result of this. He called them 'herpetobionts', which feed on plant detritus, fungal mycelia, lichens, etc. He also thought that the Tridactyloidea arose from 'early Tetrigidae'. In addition to the loss of the arolium, he considered the reduction of the fore-wing, certain changes in the hind-wing, and probably the backwardly directed process of the pronotum to be derived ground-plan characters of the Tetrigidae + Tridactyloidea. The recent Tridactyloidea do not have this process, but Sharov said that it was still present in Lower Cretaceous species, which he has apparently not yet described.

The origin of the sister-group relationship between the Tetrigidae + Tridactyloidea and the rest of the Acridoidea is therefore the earliest event that can be recognized in the history of the Caelifera. Sharov placed it in the Jurassic, without giving any evidence.

In describing the subsequent history of the Caelifera, Sharov reverted to his typological methodology and as a result his conclusions are imprecise and obscure. He derived the other families from the Locustopsidae, although he provided the Pamphagidae, Pyrgomorphidae, and Acrididae with common ancestors. His Locustopsidae appear to be an invalid stem-group to which he assigned species from the stem-group of the entire Caelifera as well as species from the stem-groups of many individual subgroups.

For example, he stated that some of the Upper Jurassic 'Locustopsidae' had teeth on the dorsal surface of the first hind tarsal segment, which is characteristic of the Eumastacidae according to Dirsh (1961). On the other hand, in his 'phylogenetic tree', which has all the traits of a typological dendrogram, the Eumastacidae are not shown to derive from the Locustopsidae until the Middle Cretaceous.

Sharov was correct to reject the view put forward by Smart (1951) and Ragge (1955a) that the venation of the Pneumoridae is primitive. However,

Sharov's own work has not led to any clearer understanding of the relationships of this family.

Sharov's (1967) discussion included a great deal that should be of some heuristic value, but nothing more precise can be gathered from his work beyond saying that certain subgroups of the Caelifera probably originated during the Jurassic.

In addition to the fossils that can be reliably assigned to the Ensifera, Caelifera, and, according to Sharov, the Phasmatodea, Sharov listed another three families which he considered to be survivors from the stem-group of the Orthopteroidea: he thought that they became extinct during the Jurassic (Bintoniellidae) or Lower Cretaceous (Vitimiidae and Elcanidae). Here too, however, his views are very typological. His 'phylogenetic tree' shows that all three groups could be more closely related to the *Ensifera than to any other subgroup of the Orthopteroidea.

Additional note

367. The classification of the 'Acridomorpha' has been dealt with by Dirsh (1975). [Willi Hennig.]

2.2.2.2..3. Paraneoptera

The history of the Zoraptera, Psocodea, and Thysanoptera during the Mesozoic is not at all clear. There are two reasons for this: very few fossils are available which can be satisfactorily interpreted, and there is no well founded phylogenetic classification of these groups[368].

As there are only about 27 recent species of Zoraptera, this is not a particularly important matter.

The Thysanoptera[369] are represented by *Liassothrips crassicornis* Martynov, from the Upper Jurassic (Malm) of Kazakhstan. The name of this genus was based on an incorrect appraisal of the geological age of the localities. Like *Permothrips longipennis* Martynov, it has been assigned to the 'Terebrantia'. However, this is a meaningless assignment because the Terebrantia are not a monophyletic group whereas the other suborder of the Thysanoptera, the Tubulifera, which is first known from the early Tertiary, is monophyletic. The assignment of *Liassothrips* to the Terebrantia only provides evidence that the Thysanoptera were in existence during the Jurassic, but this had already been deduced from the existence of the Permian *Permothrips*[370].

Various Mesozoic fossils are known that should belong to the Psocodea. They have usually been assigned to the Psocoptera, and in the Russian Textbook of Palaeontology they were even placed in the 'Parapsocida'. However, like the thysanopterous group Terebrantia, this is probably a paraphyletic group. Moreover, it is not certain that the 'Psocoptera' of contemporary classifications are a monophyletic group[371]. In a careful and painstaking study, Königsmann (1960) came to the conclusion that the Phthiraptera are probably a monophyletic group, a conclusion that was supported by White's (1957) cytological results, but he was unable to identify their sister-group[372]. It is possible that one of the subgroups of the 'Psocoptera' is most closely related to the Phthiraptera. No assessment of the Mesozoic fossils can be made until this question has been resolved. It is even possible that the early Mesozoic fossils still belong to the stem-group of the Psocodea[373].

Revisionary notes

368. See notes 370 and 373. [Wolfgang Seeger.]

369. See also the catalogue of world species by Jacot-Guillarmot (1970, 1971). Schliephake (1975) has published a phylogenetic tree of the Thysanoptera. [Willi Hennig.] A re-classification of the families has been proposed by Mound, *et al.* (1980). [Adrian Pont.]

370. The same applies in principle to Zur Strassen's (1973) study of fossil Thysanoptera from Lebanese amber (early Lower Cretaceous). Zur Strassen assigned his seven new species to the suborder 'Terebrantia'. The most pressing problem is whether the five new families that he erected (Jezzinothripidae, Neocomothripidae, Rhetinothripidae, Scaphothripidae, and Scudderothripidae) and their inclusion in the Heterothripoidea are justified from the point of view of phylogenetic systematics. Even Zur Strassen had some doubt about the propriety of his decisions (1973, pp. 5 and 43), since neither his new families nor the Heterothripoidea themselves are defined by derived characters.

As an example, the family Rhetinothripidae obviously lacks any derived characters. The two new species that Zur Strassen assigned here (*Rhetinothrips elegans* and *Progonothrips horridus*) have 15 or even 16 antennal segments. In this respect they differ from the ground-plan of the *Thysanoptera, where 6–10 antennal segments are present. The same is apparently true of *Archankothrips pugionifer* Priesner, from Baltic amber, which has 13 antennal segments (Priesner, 1924). It appears that a reduction in the number of antennal segments has taken place during the phylogenetic development of the Paraneoptera. Consequently, *Rhetinothrips* and *Progonothrips* could belong to the stem-group of the Thysanoptera in which reduction in the number of antennal segments had not progressed very far. If this hypothesis is correct, then these two species should be located at the base of the dendrogram of the Thysanoptera: the family Rhetinothripidae is probably paraphyletic and cannot be retained within a phylogenetic classification. [Wolfgang Seeger.]

371. Hennig's doubts have been resolved by the work of Seeger (see note 203 on p. 234). [Wolfgang Seeger.]

372. An important paper on the Amblycera was published by Clay (1970). Haub (1973) suggested that the Trichodectoidea are the sister-group of the rest of the Phthiraptera (but without the Rhynchophthirina).

New proposals for the classification of the Psocoptera have been put forward by Wong (1970) and Smithers (1972). [Willi Hennig.]

373. Fossil Psocoptera have been found in Lower Cretaceous amber from the Lebanon (Schlee and Dietrich, 1970; Schlee and Glöckner, 1978), but it was not known until recently whether these species belonged to the stem-group of the Psocoptera or even to the *Psocoptera. A study of these fossils is in progress (Seeger, in preparation).

A number of well preserved fossil Psocoptera have been found in Upper Cretaceous amber (Coniacian to Santonian) from Yantardakh in the Khatanga depression of Siberia, and they have been analysed by Vishnyakova (1975). She found that the suborders Trogiomorpha, Troctomorpha and Psocomorpha, particularly the families Trogiidae, Amphientomidae, and Lachesillidae, were represented. Consequently, there is no longer any doubt that the *Psocoptera appeared towards the end of the Mesozoic. On the other hand, the systematic position of two of the 14 species described by Vishnyakova has not been adequately resolved: *Khatangia inclusa* has been assigned to the Psyllipsocidae (Trogiomorpha) and *Cretapsocus capillatus* to the Elipsocidae (Psocomorpha). The problem arises because no convincing synapomorphies are known for either family, and because both fossils have retained some important archaic characters.

Khatangia inclusa has no retinaculum (wing-coupling mechanism) and the female hypopygium still has all three pairs of valves. These two characters suggest that *Khatangia* is more primitive that the Psyllipsocidae and is indeed more primitive than all the *Trogiomorpha. Such a conclusion would still be valid if we were to accept the unproven possibility that the retinaculum has arisen convergently within the different suborders and is not a synapomorphy of the Psocodea or of the Psocoptera (see note 203 on p. 234).

The existence of the family Elipsocidae in the Upper Cretaceous is also doubtful. *Cretapsocus capillatus*, which Vishnyakova assigned to this family, appears to have CuP and 1A reaching the margin of the fore-wing independently, in other words there is no nodulus, and in this respect it is more primitive than any other Psocomorpha. On the other hand, the two-branched media of *Cretapsocus* appears to be a derived character, as the usual state in the Psocoptera is for the media to have three branches: reduction of the media has taken place in 4 of the more than 20 recent genera of Elipsocidae, but in these genera it is associated with a tendency towards reduction in wing size, which affects other veins too and even varies in extent from species to species. As the occurrence of this character is undoubtedly a convergent development, it cannot be used to support the inclusion of *Cretapsocus* in the Elipsocidae, especially as the wings of *Cretapsocus* are fully developed.

Nevertheless, the fact remains that all three suborders and three of the 35 recent families of the Psocoptera are known from the Upper Cretaceous. Most of the taxa described by Vishnyakova can be assigned to recent families, but the systematic position of at least two species is doubtful because they have characters of the stem-group. This provides a clue as to when the three main lineages within the Psocoptera must have arisen. It also indicates that the main diversification of the order probably took place during the Cretaceous.

Since the Phthiraptera are a monophyletic group according to White (1957) and Königsmann (1960), and since the Psocodea and Psocoptera also appear to be monophyletic (see notes 201 and 203 on pp. 233–234), it seems highly likely that there is a sister-group relationship between the Phthiraptera and the Psocoptera. The existence of the *Psocoptera in the Upper Cretaceous thus implies that the Phthiraptera had separated from the Psocodea by the end of the Mesozoic.

It is most unlikely that George's (1969) record of a species of *Liposcelis* from the Upper Cretaceous intertrappean beds of Nagpur, Maharashtra State, India, is actually a true fossil. The insect described and illustrated by George evidently belongs to the Liposcelidae, and is most probably a species of *Liposcelis* because the metafemora are very thickened, the apical abdominal segments are fused, and the pronotum is tripartite. However, it seems extraordinary that this minute and extremely weakly chitinized insect should have survived as a compression fossil in such a perfect state of preservation that even internal structures of the tiny head are visible, such as details of the hypopharynx and the lacinia (5–10 μm broad!). It should be borne in mind that this might be a modern book louse accidentally contaminating rock specimens. Moreover, the intertrappean beds extend from the Upper Cretaceous (Danian/Maestrichtian) to the Lower Eocene (Holland et al., 1957), and George did not give a precise age for his fossil. [Wolfgang Seeger.]

2.2.2.2..3.2..2.2.1.2. **Heteroptera**

The Heteroptera, with about 25 000 recent species, are one of the few orders where we can build up some idea of their development during the Mesozoic. These ideas are not as precise as they might be, partly because the constitutive characters of many monophyletic groups are not visible in fossils and partly because, as is so often the case, the phylogenetic classification of the Heteroptera has not been entirely adequately worked out. If this were so, then the fossils that can be reliably interpreted would provide the basis for many more far reaching conclusions than are warranted at present.

There appears to be good evidence that the sister-group relationship between the water bugs (Hydrocorisae = Cryptocerata) and the land bugs (Geocorisae = Gymnocerata)[374] must have arisen before the Lias, probably during the Triassic and possibly even earlier. Not all authors have given clear expression to this sister-group relationship[375].

The Gelastocoridae and Ochteridae have often been assigned to the

Geocorisae because they are terrestrial, although they live on water margins[376]. However, China (1955) has shown convincingly that these two families are more closely related to the Hydrocorisae, and Parsons (1964) found some derived characters in the facial sclerite to confirm this. Some doubt has also been cast on the assignment of the Corixidae to the Hydrocorisae[377], but China (1955) used the studies and results of earlier authors to show that these doubts were unjustified. The Hydrocorisae thus appear to be a well founded monophyletic group[378], even though China himself did not show them to be so in his earlier phylogenetic tree (1933).

It is particularly valuable to have shown that the Hydrocorisae are a monophyletic group[378] because most of their subgroups, whether treated as 'families' or 'superfamilies', have a very characteristic appearance and fossils can be assigned to them with some certainty.

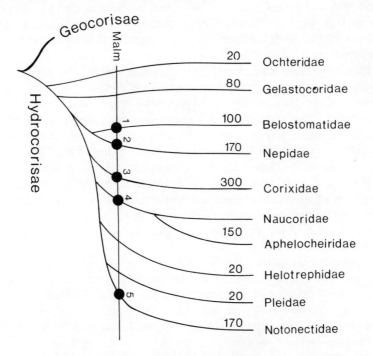

Fig. 127. Phylogenetic tree of the Hydrocorisae (Heteroptera)[379]. 1, *Mesobelostomum*; 2, *Mesonepa*; 3, *Karataviella*; 4, *Palaeoheteroptera*, *Nepidium*; 5, *Notonectites*, *Asionecta*. From China (1955).

Bekker-Migdisova (in Rodendorf, 1962) reported that the following families are known from the Upper Jurassic (Malm): Belostomatidae (*Mesobelostomum deperditum* Germar, from Bavaria), Nepidae (two species of the genus *Mesonepa*, from Bavaria), Corixidae (*Karataviella brachyptera* Bekker-Migdisova, from Karatau; a corixid, according to a footnote by Popov),

Naucoridae (*Palaeoheteroptera lapidaria* Weyenbergh, from Bavaria, and *Nepidium stolones* Westwood, from England), and Notonectidae (*Notonectites elterleini* Deichmüller, from Germany, and *Asionecta curtipes* Popov, from Kazakhstan)[380].

If these assignments and China's (1955) phylogenetic tree are correct[381], it would mean that at least nine of the subgroups of the Hydrocorisae were already in existence during the *Upper* Jurassic (Malm)[382], although they only account for about 1000 recent species (Fig. 127). However, assignments of the *Lower* Jurassic (Lias) fossils are far less certain. Bekker-Migdisova distinguished several families whose relationship to the recent species, or even suborders, is not at all clear. The family 'Apopnidae' (*Apopnus magniclavus* Handlirsch, from the upper Lias of Mecklenburg) is an example of how cautious we need to be. Bekker-Migdisova thought that this family might be related to the 'Notonectidae'. If this really is so and China's (1955) phylogenetic tree is correct, it would mean that approximately the same number of groups of the Hydrocorisae were in existence during the upper Lias as during the Malm.

Bekker-Migdisova considered two 'families' from the upper Lias of Mecklenburg to be respectively the 'apparent relatives' and the 'apparent ancestors' of the Naucoridae: the Aphlebocoridae (*Aphlebocoris*, with two species) and Probascanionidae (*Probascanion megacephalum* Handlirsch). China (1955) suggested that all the Hydrocorisae, except for the terrestrial Ochteridae and Gelastocoridae, had descended from the 'Protonaucoridae'[383]. This can only mean that they must have been most similar to the recent Naucoridae. As a result, the *most* that should be read into Bekker-Migdisova's interpretation of the Aphlebocoridae and Probascanionidae is that they could have belonged to the stem-group of the majority of the Hydrocorisae, from which only the Ochteridae and Gelastocoridae may already have separated[383].

The same is true of the 'family Triassocoridae' (genus *Triassocoris*, from the Upper Triassic of Australia. Bekker-Migdisova considered these to be 'the ancestors of a large proportion of the recent Hydrocorisae', whilst Poisson (in Grassé, 1951) thought that they 'probably' belonged to the Notonectoidea. It therefore seems quite probable that they do at least belong to the Hydrocorisae[384].

In spite of these uncertainties in interpreting the Liassic and Triassic fossils, it is still remarkable that such a large number of subgroups of the Hydrocorisae must have been in existence during the Upper Jurassic, particularly as this group comprises such a small percentage of the recent heteropterous fauna.

If it is accepted that the Triassocoridae are the stem-group of the Hydrocorisae, then it follows that their sister-group, the Geocorisae, must have been in existence at the same time[385]. Unfortunately, there are no Triassic fossils that can be definitely assigned to this group. The interpretation of the Upper Triassic Actinoscytinidae, listed in the Russian Textbook of

Palaeontology (Rodendorf, 1962), is also not certain: some have thought that they consist of survivors from the stem-group of the Heteropteroidea, others that they include some species from the stem-group of the Coleorrhyncha. *Tingiopsis reticulata* Bekker-Migdisova from Central Asia was assigned at different times in the same Textbook to the Upper Triassic and Lower Jurassic! Bekker-Migdisova assigned it to the recent family Tingidae, but Evans thought that it belonged to the Cercopoidea (Auchenorrhyncha, see p. 272). This species is best left out of consideration here.

The Lower Jurassic fossils that can be reliably assigned to the Geocorisae have all been placed in their own families. The relationships of these families are not clear, and all that is usually said of them is that they 'are reminiscent' of particular recent families, such as the Nabidae. Poisson (in Grassé, 1951) stated that the 'families' described by Handlirsch from the Lias of Mecklenburg have some characters of the Lygaeoidea and others of the Pentatomoidea. This would mean that the Geocorisae too had already separated into a number of subgroups by the Lias, and this definitely appears to be the case by the Upper Jurassic. Unfortunately, there is still no sound basis for a complete appraisal of these fossils. China (1933) published a phylogenetic tree of all the Heteroptera, but since then opinions on both the Geocorisae and the Hydrocorisae have undergone considerable change. Leston *et al.* (1954) divided the Geocorisae into two groups, the Cimicomorpha and Pentatomomorpha. Ross's (1965) simplified phylogenetic tree appears to be based on this division, but he used the names 'bed bug group' and 'stink bug group'. Leston *et al.* listed a large number of characteristic features for each of their groups, but unfortunately they did not distinguish between what is primitive and what is derived. Nevertheless, each group does appear to have several derived characters that are present in the other in a more primitive state. In spite of this, they should only be provisionally treated as sistergroups because the position of some families is still uncertain. This is particularly true of the Enicocephalidae and the Saldoidea. Leston *et al.* also failed to mention the pond skaters (Amphibiocorisae). Ross (1965) considered the pond skaters + Saldoidea to be the sister-group of the other Cimicomorpha. If he was correct, then the sister-group of the Pentatomomorpha would be the Amphibiocorisae + Cimicomorpha and not just the Cimicomorpha. According to Ross's phylogenetic tree, the Saldoidea have the same relationship to the pond skaters (Amphibiocorisae) as the Gelastocoridae do to the true water bugs (Hydrocorisae in the strict sense). However, Leston *et al.* assigned the Saldoidea to the Pentatomomorpha. Gupta's (1963a, 1963b) investigations have not helped to clarify these relationships[386].

In these circumstances, the Upper Jurassic fossils will not shed any light on the number of subgroups into which the Geocorisae had separated at that time. Bekker-Migdisova (in Rodendorf, 1962) mentioned Upper Jurassic fossils which are said to belong to the Coreidae (in the Pentatomomorpha-Coreoidea), Miridae and Nabidae (both in the Cimicomorpha-Cimicoidea). However, it is very doubtful whether these assignments to individual recent families are well founded, nor is it known for certain whether these families are monophyletic or possibly paraphyletic.

Revisionary notes

374. According to Rieger (1976), there is still no definite evidence for this sister-group relationship. On the one hand, the Cryptocerata are undoubtedly a monophyletic group with synapomorphies that include the elongate female subgenital plate, etc. On the other hand, the Amphibiocorisae + 'Saldoidea' could be a monophyletic group, and their sister-group could be the Cryptocerata or the 'rest of the Gymnocerata' (= the traditional 'Gymnocerata' without the Saldoidea and Amphibiocorisae) or both. See Fig. X. [Dieter Schlee.]

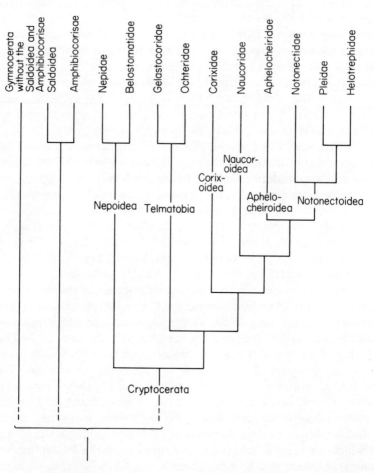

Fig. X. Phylogenetic diagram of the Heteroptera. From Rieger (1976).

375. This paragraph and the next two should be read together with Rieger's (1976) important analysis of the Heteroptera. [Dieter Schlee.]

376. The main purpose of Rieger's investigation was to determine the systematic position of the Ochteridae. His results were summarized in a scheme of synapomorphies (see Fig. X) in which the Ochteridae and Gelastocoridae are shown as sister-groups (Rieger, 1976, synapomorphies nos. 14–25 on pp. 184–185). He called these two families the Telmatobia, and considered the Telmatobia to be the sister-group of the group Corixoidea + Naucoroidea + Aphelocheiroidea + Notonectoidea. [Dieter Schlee.]

Popov (1971) has discussed the classification and historical development of the 'Nepomorpha' (i.e. the Cryptocerata). [Willi Hennig.]

377. According to Rieger (1976), the Corixidae should be assigned to the Cryptocerata (see note 376). [Dieter Schlee.]

378. Rieger (1976) has listed several characters which show that the Hydrocorisae = Cryptocerata are a monophyletic group: reduced antennae; weak antennal muscles; complete absence of apodemes in the head-capsule for the antennal muscles; the caudal elongation of the female subgenital plate, which covers the ovipositor; and possibly some embryological similarities (from Cobben, 1968). [Dieter Schlee.]

379. Rieger's (1976) scheme of synapomorphies (see Fig. X) differs from China's dendrogram in showing several different sister-group relationships. The only agreement between Rieger and China is that the Nepidae and Belostomatidae are sister-groups, and that the Corixidae are the sister-group of the Naucoridae + Aphelocheiridae + Helotrephidae + Pleidae + Notonectidae. [Dieter Schlee.]

380. Bradbury and Kirkland (1972) have discussed a species of bug occurring abundantly in the Upper Jurassic Todilto formation of New Mexico (Ojo del Espiritu Santo Grant). They considered that it might be ancestral to the modern families Notonectidae and Belostomatidae. This was the first record of a North American Jurassic insect. [Willi Hennig.]

381. Rieger (1976) has shown that the dendrogram in Fig. 127 is not entirely correct. [Dieter Schlee.]

382. Although China's dendrogram is partly incorrect, there is direct or indirect evidence that almost all the branches were in existence during the Upper Jurassic.

Hennig's Fig. 127 shows that all the groups included must have been in existence in the Malm, except for the branch Naucoridae + Aphelocheiridae. The only question here is whether the Naucoridae and Aphelocheiridae had already separated during the Malm or whether they were still represented by their common stem-group. [Note that Popov (1970) treated the Aphelocheiridae as no more than a tribe of the Naucoridae. Willi Hennig.]

When Rieger's (1976) new results are taken into account, the following questions are still posed by Hennig's Fig. 127. (1) Had the Pleidae and Helotrephidae already separated, or were they still represented by their common stem-group? (2) The same question for the Gelastocoridae and Ochteridae. (3) What is the true systematic position of fossil no. 4 of Hennig's diagram, and what are the consequences to be drawn from its new assignment? Hennig considered it to belong to the stem-group of the Naucoridae + Aphelocheiridae, but it is now known that these two families are not sister-groups. [Dieter Schlee.]

383. These views are no longer valid, because the Ochteridae and Gelastocoridae should be placed 'in the middle' of the phylogenetic system of the Cryptocerata (Rieger, 1976). [Dieter Schlee.]

384. There is no evidence that the Triassocoridae belong to the stem-group of the Hydrocorisae. [Dieter Schlee.]

385. See the preceding note. It is not clear whether the Amphibiocorisae + Saldoidea (which may form a single monophyletic group) belong to the Gymnocerata (Geocorisae), or whether they are a separate group and possibly the sister-group of 'the rest of the Gymnocerata' or of the Cryptocerata: for this reason nothing of consequence can yet be deduced from the existence of the Triassocoridae. [Dieter Schlee.]

386. Rieger (1976) considered the Amphibiocorisae + 'Saldoidea' to be a monophyletic group of uncertain phylogenetic affinity: he thought that they could be the sister-group of 'the rest of the Gymnocerata', or of the Cryptocerata, or of the Cryptocerata + 'the rest of the Gymnocerata'. Consequently, neither the Amphibiocorisae + 'Saldoidea' nor the Amphibiocorisae alone can be the sister-group of the Cimicomorpha, nor can the Amphibiocorisae + Cimicomorpha be the sister-group of the Pentatomomorpha. Moreover, the Gelastocoridae are the sister-group of the Ochteridae, according to Rieger (1976), and not of the 'Hydrocorisae *sensu stricto*'. [Dieter Schlee.]

2.2.2.2..3.2..2.2.2. Sternorrhyncha

Hardly anything definite can be said about the history of the Sternorrhyncha in the Mesozoic[387].

A few fossil Aphidomorpha are known from the Upper Triassic and Jurassic.

Triassoaphis cubitus Evans, from the Upper Triassic of Australia (Mt.

Crosby), and *Genaphis valdensis* Brodie, from the Upper Jurassic (Purbeck) of England, have both been assigned to the Aphidina[388]. However, it is not difficult to make a formal derivation of the Coccina too from this type of fore-wing venation. *Mesococcus asiaticus* Bekker-Migdisova, from the Upper Triassic of Issyk-Kul', is the only available evidence that the Coccina and Aphidina had separated by the Upper Triassic, but this presupposes that *Mesococcus* really does belong to the Coccina[389].

It is impossible to show that there is any close relationship between these fossils and any particular subgroups of the *Aphidina or *Coccina.

Unfortunately, the classification of both groups still presents problems.

The Aphidina have been discussed in several interesting recent papers (Mackauer, 1965; Richards, 1966; Heie, 1967; Steffan, 1968)[390]. I have shown in Fig. 128 how Mackauer (1965) and Heie (1967) differ in their ideas about the relationships of the individual subgroups. They agree that there is a sister-group relationship between the Adelgidae + Phylloxeridae (Phylloxeroidea of Steffan, 1968 = Aphidina ovipara of older authors) and all the other Aphidina (Aphidina vivipara of older authors). At present it is not possible to say when this sister-group relationship arose. Heie assigned *Triassoaphis cubitus* Evans, from the Upper Triassic of Australia, to the Phylloxeroidea. However, this species is only known from one wing, and it is extremely doubtful whether reliable conclusions about the relationships of fossils in this group can be drawn from the venation alone. This is because the ground-plan venation of the Aphidina is very derived and reduced, as is clear from Richards' (1966) illustrations. *Triassoaphis* has a three-branched media, and in this character at least it is more primitive than all the *Phylloxeroidea. No synapomorphies have been found.

Heie (1976) assigned *Genaphis valdensis* Brodie, from the Upper Jurassic (Malm) of England, to one of the subgroups of the Aphidoidea, a name available for the 'Aphidina vivipara' of earlier authors. It is possible and even probable that by the Upper Jurassic the sister-group relationship between the Aphidoidea and Phylloxeroidea had already arisen and that each group had already separated into a number of subgroups. However, the reservations that I felt in dealing with the wing of *Triassoaphis* also apply to the wing of *Genaphis*.

Richards (1966) has described six species from Upper Cretaceous Canadian amber. The two species of the 'Palaeoaphidinae' have seven-segmented antennae and in this character at least they appear to be more primitive than all the recent Aphidina. Steffan (1968) wrote: 'The Upper Cretaceous Palaeoaphididae have a number of plesiomorphic characters, and with these both the Phylloxeroidea and the Pemphigidae + Thelaxidae subfamily Hormaphidinae can be derived from this group'. This statement is not accurate. Literally, it implies that the currently accepted sister-group relationship between the Phylloxeroidea and Aphidoidea (p. 380) is not correct. However, the fact that it is possible to 'derive' several recent groups from a fossil group with plesiomorphic characters is not an adequate basis for

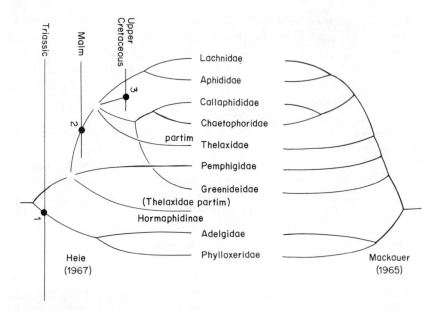

Fig. 128. Phylogenetic tree of the Aphidina. A comparison of the views of Heie and Mackauer. 1, *Triassoaphis*; 2, *Genaphis*; 3, *Canadaphis*.

establishing that these groups are closely related or, to put it more precisely, that these groups form a single monophyletic group.

Richards said nothing about the relationships of the 'Canadaphidinae', which contain three species. Heie (1967) who only knew one of them, assigned it to a group of families in the Aphidoidea which he considered to be a monophyletic group. On the other hand, Mackauer did not believe that this group was monophyletic.

In these circumstances, *Aniferella bostoni* Richards is probably the most interesting species. Richards assigned it to the subfamily Neophyllaphidinae, which had previously been ranked as a tribe of the Callaphididae and appeared as such in the phylogenetic trees of Heie and Mackauer (Fig. 128). If the 'Neophyllaphididae' are closely related to the 'Callaphididae' and if the assignment of *Aniferella* to this group is correct, then several conclusions could be inferred from the phylogenetic trees of Heie (1967) and Mackauer (1965): this single fossil from Canadian amber would mean that by the Upper Cretaceous (Campanian = upper Senonian) the sister-group relationship between the Aphidoidea and Phylloxeroidea had already arisen and that the Aphidoidea had already separated into several subgroups[391].

In the Phylloxeroidea, Steffan (1968) believed that the Adelgidae and Phylloxeridae arose 'before the Oligocene and probably in the Cretaceous'. He did not think that the Elektraphididae, which consist of two species from Baltic amber, were more closely related to either the Adelgidae or the Phylloxeridae. He thought that they represented a 'surviving side-branch' of the stem-group of the Phylloxeroidea. In his opinion the reduced ovipositor was the only character that might be a synapomorphy. 'Unlike the Adelgidae, the Elektraphididae and Phylloxeridae have the ovipositor reduced. It is always possible that this reduction has taken place independently in these two families, that in each case an autapomorphy has arisen convergently'. However, he appears to have overlooked an undoubtedly apomorphic character in the reduction of the radial sector (radial branch): the Elektraphididae ('fore-wing with the radial branch very reduced') and the Phylloxeridae ('fore-wing without any trace of a radial branch') both agree in this character. The radial sector has been retained complete in the Adelgidae, and the Elektraphididae and Adelgidae really do not appear to have any apomorphic characters in common.

It is certainly not 'impossible to decide which of the other two families of recent Aphidina ovipara is most closely related to the Elektraphididae', as Steffan asserted. The Elektraphididae and Phylloxeridae have two clearly derived and independent characters in common, in the venation and the ovipositor, and this makes it very probable that they are closely related, that the Elektraphididae belong to the stem-group of the Phylloxeridae. Steffan did not appear to notice that this conclusion provides further support for his suggestion that the Phylloxeridae and Adelgidae both arose before the Baltic amber era, perhaps in the Cretaceous.

Balachowsky (1942) divided the Coccina into three superfamilies, the Margarodoidea, Lecanioidea, and Diaspidoidea, and other authors such as Theron (1958) supported this classification[392]. However, Balachowsky's diagram and list of characters show that the Lecanioidea are more closely related to the Diaspidoidea. In fact, Borkhsenius (1958) suggested that there is a sister-group relationship between the Archaeococcoidea (which consist basically of the Margarodoidea) and the Neococcoidea (= Lecanioidea + Diaspidoidea). He assigned the Phenacoleachiidae, which only includes a single species from New Zealand, to the Archaeococcoidea, even though Balachowsky's work indicated that this family should be regarded as the most primitive of the Neococcoidea. Without the Phenacoleachiidae, the Neococcoidea do indeed appear to have several derived characters, such as the absence of abdominal spiracles, which show that they are a well founded monophyletic group. This is not true of the Archaeococcoidea, which may be a paraphyletic group based on symplesiomorphies. However, in view of the relatively small numbers of species involved, this uncertainty is not a particularly important matter. There is every justification for starting with the working hypothesis that there is a sister-group relationship between the Archaeococcoidea and Neococcoidea, and the question that needs to be investigated first is the one that is most important for phylogenetic research: when did this sister-group relationship arise? Borkhsenius (1958) suggested that the Archaeococcoidea and Neococcoidea separated as early as the Lower Carboniferous and that no fewer than 11 subgroups of the Coccina survived from the Palaeozoic into the Triassic. However, there is no evidence for this.

It has been suggested that the Psyllina[393] and Aleyrodina had separated before the end of the Palaeozoic, and this was based on the discovery of *Permaleurodes rotundatus* Bekker-Migdisova in the Upper Permian of the Kuznetsk basin. However, it is not certain that this species does actually belong to the Aleyrodina[394].

Several Mesozoic fossils have been assigned to the Protopsyllidiidae: *Cicadopsyllidium elongatum* Bekker-Migdisova, *Asiopsyllidium unicum* Bekker-Migdisova, and three species of *Cicadellopsis*, from the Rhaetic (Upper Triassic) of Issyk-Kul'; and *Mesaleuropsis venosa* Martynov and *Cicadellopsis incerta* Martynov, from the Lias of Central Asia. It is impossible to decide whether these do in fact belong to the stem-group of the Psyllina + Aleyrodina (Psyllidomorpha) or to one of these two subgroups. The same is true of the genus *Liadopsylla*, described from the Lower Jurassic of Germany and Central Asia, although Bekker-Migdisova assigned it to its own subfamily (Liadopsyllinae) of the Psyllidae.

Revisionary notes

387. Schlee (1970) has published an analysis of fossils from Lower Cretaceous Lebanese amber. These proved to belong to the stem-group of the Aleyrodina, and Schlee also found fossils in Tertiary Baltic amber that belong to the *Aleyrodina. The evolution of the stem-group of the Aleyrodina into the *Aleyrodina thus parallelled the evolution of the flowering plants, which are now almost the sole hosts of the *Aleyrodina. [Dieter Schlee.]

388. Evans (1971) described *Crosaphis anomala* into the 'Aphidoidea', noting that it resembled *Kaltanaphis* in wing-shape and *Triassoaphis* in venation. [Willi Hennig.]

389. There is some doubt about this: see note 248 on p. 256. [Dieter Schlee.]

390. Trends in the evolution of the Aphidina have been outlined by Shaposhnikov (1971). [Willi Hennig.]

391. Kononova (1975, 1976) has described eleven new species, six new genera and two new families of Aphidina from Cretaceous amber from the Taymyr (Siberia). [Adrian Pont.]

392. The phylogeny of the Coccina, based on the mouth-parts, has been discussed by Koteja (1974). The taxonomy of male scale-insects (Diaspididae) has been dealt with by Ghauri (1962). [Willi Hennig.]

393. A phylogenetic tree of the Psyllina has been published by Bekker-Migdisova (1973). [Willi Hennig.]

394. The Lower Cretaceous Aleyrodina (see note 387 above) prove that the Psyllina and Aleyrodina must have separated by the end of the Mesozoic at the latest, but they do not provide any hint as to when this separation actually took place, which must have been some time during the early Mesozoic. [Dieter Schlee.]

2.2.2.2..3.2..2.2.3. Auchenorrhyncha

The sister-group relationship between the Fulgoriformes and Cicadiformes is supposed to have originated in the late Palaeozoic, but this is not easy to prove. In the Mesozoic, both groups are definitely known from the Upper Triassic onwards. Of the 18 families of the Fulgoriformes recognized by Metcalfe, Bekker-Migdisova (in Rodendorf, 1962) recorded two from the Upper Triassic, the Cixiidae and Ricaniidae. The Cixiidae are even supposed to have been found in the Lower Triassic of Siberia (*Boreocixius rotundatus* Bekker-Migdisova and *B. sibiricus* Bekker-Migdisova). Bekker-Migdisova recorded two further families, the Issidae and Lophopidae, from the Lower Jurassic. However, these claims mean very little because the relationships amongst the recent families of the Fulgoriformes are not at all clear. In Mesozoic fossils it is usually the fore-wings that are preserved, and even if these are identical with the fore-wings of recent families it is still possible that the characters in common are all symplesiomorphies. The fossils mentioned above could well belong to the stem-group of several recent families. Fennah (1961) has even described a species as *Cixius petrinus*, from the Upper Jurassic (Wealden clay) of England (Surrey), assigning it to the recent genus *Cixius*. Nevertheless, it is very likely that the Fulgoriformes, which contain about 6500 recent species, according to Pesson (in Grassé, 1951), had separated into a number of subgroups by the Jurassic and possibly even by the Upper Triassic. However, this is no more than a vague guess at present.

The Cicadiformes present similar problems. There is no well founded phylogenetic tree for this group which could be used for appraising the phylogenetic relationships of the fossils. The Cicadoidea are a well founded monophyletic group (see p. 273) and, according to Pesson (in Grassé, 1951), they include about 1500 recent species. There was some doubt about the assignment of Palaeozoic fossils to this group, but there is no doubt about its existence in the Upper Triassic. Evans (1964) mentioned *Fletcheriana triassica* Evans and *Mesogereon superbum* Tillyard, from the Upper Triassic of Australia, and *Cicadoprosbole sogutensis* Bekker-Migdisova (Fig. 129), from the Upper Triassic of Tien Shan (inadvertently stated by Evans to be Jurassic). Bekker-Migdisova (in Rodendorf, 1962) assigned *Cicadoprosbole* and various other Jurassic species to the recent family Tettigarctidae. This is another case where the assignments appear to be based clearly on symplesiomorphies, for Evans (1964) pointed out that the Tettigarctidae are the most primitive family of the recent Cicadoidea. It is possible that the early Mesozoic species belong to the stem-group of the Cicadoidea. Tillyard

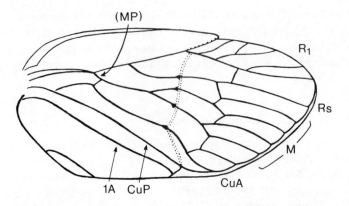

Fig. 129. Fore-wing of *Cicadoprosbole sogutensis* Bekker-Migdisova (Auchenorrhyncha, Cicadiformes), from the Upper Triassic of Issyk-Kul'. From Bekker-Migdisova (1947). Compare the vein labelled 'MP' with Figs. 36 and 37 ('arculus'), Figs. 39, 57, 58, 61 and 104 ('MP'), and the many other wings illustrated. The similar form of this vein throughout this series of wings is one reason why it has been suggested that at least in all the Neoptera (i.e. in the ground-plan of the Neoptera) a posterior branch of the media could be fused with CuA.

(1921b) established that the genus *Mesogereon*, which consists of four species from the Upper Triassic of Ipswich (Australia), agrees most closely with the recent genus *Tettigarcta* which is restricted to South Australia and Tasmania. It would also be extremely interesting if *Eotettigarcta scotica* Zeuner, from the Eocene of Scotland, were to be confirmed as a member of the otherwise exclusively Australian Tettigarctidae. This would provide a parallel with the Isoptera (termites): the most primitive family of this group, the Mastotermitidae, is also now restricted to Australia but has been recorded from the Tertiary of Europe.

In the other Cicadiformes there is too much uncertainty over the relationships of the individual subgroups for any definite assessment of the fossils to be made. As I showed in the section on Palaeozoic Auchenorrhyncha, it is not certain whether the Cicadelloidea are more closely related to the Cicadoidea or to the Cercopoidea. This affects fossil studies because we do not know which groups had a common stem-group or what this stem-group might have looked like.

The Cicadelloidea contain the most species (about 7000) and are also the most likely to be a monophyletic group (see p. 275). Bekker-Migdisova (in Rodendorf, 1962) thought that she could recognize various recent 'families' as early as the Upper Triassic: Biturritidae (genus *Absoluta*, with two species from Issyk-Kul'), Cicadellidae (= Jassidae: *Triassojassus* and *Mesoledra*, from the Upper Triassic of Australia), Eurymelidae (*Mesojassus*, with 25 species from the Upper Triassic of Australia) and Membracidae (even two of the recent subfamilies, Centrotinae and Membracinae, from the Upper Triassic of Issyk-Kul')[395]. The relationships of the various 'families' are still largely unclear but, even if this problem is left out of consideration for the present, to establish the existence of several subgroups of the Cicadelloidea in the Upper Triassic would be extremely important. However, it is very doubtful whether Bekker-Migdisova's assignment of early Mesozoic fossils to individual 'families' is well founded. Evans (1964) limited himself to stating that certain Triassic, Jurassic, and even Cretaceous fossils belong to the Cicadelloidea, without assigning them to any subgroups. For the present, this is all that is justified.

If the Cicadelloidea are definitely known from the Upper Triassic, it follows that the Cercopoidea must also have been in existence at the same time, no matter whether the Cicadelloidea are most closely related to the Cercopoidea or the Cicadoidea. Bekker-Migdisova (in Rodendorf, 1962) and Evans (1964) both listed a number of Upper Triassic and Jurassic fossils which are supposed to belong to the Cercopoidea, but they both differed widely in their interpretations of the individual fossils. So far the fossils have given no information about the separation of the Cercopoidea into subgroups during the Mesozoic.

The conclusion to be drawn from all this is that the Fulgoriformes, Cicadoidea, Cicadelloidea (= Jassoidea), and Cercopoidea probably existed from the Upper Triassic onwards, but the extent to which they had already separated into subgroups is not at all clear at present. The main reasons for this are that there is no well founded classification of these groups, and that the fundamental constitutive characters of each monophyletic group have not been worked out clearly enough to give an indication of the systematic position of fossils. We are forced to rely on the decisions of specialists whose opinions are frequently contradictory.

Additional note

395. The phylogeny, evolution, and zoogeography of the Membracidae has been discussed by Strümpel (1972). [Willi Hennig.]

2.2.2.2..4. Holometabola

It is still not known for certain how many groups of the Holometabola originated during the Palaeozoic. In some cases this affects the interpretation of Mesozoic fossils.

2.2.2.2..4.1. Neuropteroidea

Amongst the Neuropteroidea, it has been established that the Planipennia were in existence during the Permian, and this means that the Megaloptera and Raphidioptera must also have originated by the Permian at the latest. However, the subsequent history of these two small groups is largely unknown.

So far as the Megaloptera are concerned, no Mesozoic fossils appear to have been found. Ponomarenko (in Rodendorf, 1962) reported that the larva described by Ping (1928) as *Coptoclava longipoda*, from the Upper Jurassic of China (originally thought to be Lower Cretaceous), does not belong to the Megaloptera as Ping thought but to the Coleoptera. Sharov (also in Rodendorf, 1962) thought that Handlirsch's Chaulioditidae from the Triassic might belong to the Paraplecoptera.

The Raphidioptera are represented by several fossils from the Upper Jurassic (Malm) of Kazakhstan, Transbaikalia and western Europe. These have been assigned to the families Mesoraphidiidae and Baissopteridae, but they do not appear to be closely related to individual subgroups of the *Raphidioptera. Consequently, they must belong to the stem-group of this order, and they do not shed any light on its history[396].

A large number of fossils belonging to the Planipennia is known, and for this reason it would be very interesting to clarify the history of this group. However, this task is hampered by the lack of any well founded phylogenetic classification. Withycombe (1924) published a phylogenetic tree in which he showed the presumed sister-group relationships within the order, and his classification was followed by Berland and Grassé (in Grassé, 1951). However, not all the sister-group relationships suggested by Withycombe are well founded. The classification adopted by Martynova (in Rodendorf, 1962), and that illustrated in her earlier phylogenetic tree (Martynova, 1952), both differ substantially from Withycombe's views (Fig. 130). For example, the families that Martynova included in her 'Hemerobiidea' were placed along different branches of the phylogenetic tree by Withycombe, and the same applies to her 'Polystoechotidea'. In these circumstances, the fossils cannot provide very much information about the age of various subgroups of the *Planipennia.

Amongst the fossils whose relationships seem to be most definite are the Upper Jurassic Mesochrysopidae: two species of *Mesochrysopa*, from Bavaria, and *Mesypochrysa latipennis* Martynov, from Kazakhstan. There can be no doubt that they belong to the stem-group of the recent Chrysopidae (including the Apochrysidae): as these are a relatively subordinate group of the *Planipennia, this means that the *Planipennia must already have separated into a number of subgroups by the Upper Jurassic. Enderlein (1909) stated that *Archiconiopteryx liasina* Handlirsch, a wing from the Lias of Dobbertin that Handlirsch considered to be a hind-wing of the Archipsyllidae (Psocodea), actually belonged to the Coniopterygidae, a family that is characterized by its very reduced venation[397].

The Kalligrammatidae, from the Malm of Solnhofen, are a celebrated group of butterfly-like insects, in particular the giant species *Kalligramma haeckeli* Walther with fore-wings measuring up to 90–120 mm. Unfortunate-

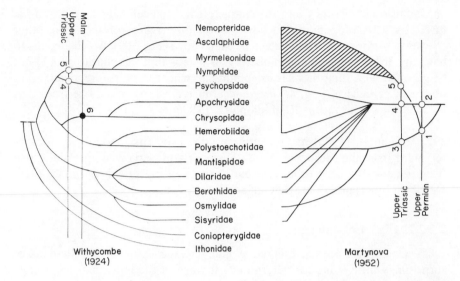

Fig. 130. Phylogenetic tree of the Planipennia. A comparison of the views of Martynova and Withycombe. 1, Permithonidae (including Permegalomidae) and Archeosmylidae, from the Upper Permian of Australia and the Soviet Union; 2, Palaemerobiidae, from the Upper Permian of the Soviet Union; 3, *Petrushevskia borisi* Martynova, from the Upper Triassic of Issyk-Kul'; 4, *Triassopsychops superba* Tillyard, from the Upper Triassic of Ipswich (Australia); 5, *Sogjuta speciosa* Martynova, from the Upper Triassic of Issyk-kul'; 6, *Mesypochrysa latipennis* Martynov, from the Upper Jurassic of Karatau (Kazakhstan).

ly, the exact position of this group in the phylogenetic classification of the Planipennia cannot be determined at present.

Revisionary notes

396. Carpenter (1967) has described a new family, the Alloraphidiidae, from the Upper Cretaceous of Labrador. This has extremely elongate wings, with a very narrow costal field, and may be most closely related to the Mesoraphidiidae.

The modern Raphidiidae and Inocelliidae should be regarded as sister-groups. Their most striking synapomorphies are the elongate prothorax; the so-called 'MA' (or 'x') of the hind-wing, which originates from MP rather than from R as in the most archaic species; the strongly bilobed 3rd tarsal segment and very short 4th tarsal segment. Only two of the Mesozoic species appear to have any of these characters. One only consists of a head and an elongate prothorax, and was identified by Martynova (1961c) as a species of *Baissoptera*. It is unfortunate that no wings have been preserved because other Baissopteridae have a primitive venation, and this makes Martynova's assignment rather implausible: baissopterids do not have the characteristic partial fusion of M and Cu or the characteristic anal cell between 1A and the fused 2A + 3A (depending on how the venation is interpreted). The other species is *Mesoraphidia pterostigmalis* Martynova (Fig. 100), which appears to have the 3rd tarsal segment bilobed. This means that this species should not be grouped with the other Mesoraphidiidae, where the tarsal segments were obviously cylindrical (e.g. *M. inaequalis* Martynov), but should be treated as an ancestor of the recent Raphidioptera. [Michael Achtelig.]

397. Meinander (1975) has recently described the first undoubted Coniopterygidae from the Mesozoic: one rather poorly preserved compression fossil from the Jurassic of Kazakhstan, and three well preserved specimens (2♂, 1♀) from Siberian amber. [Dieter Schlee.]

2.2.2.2..4.2. **Coleoptera**

The Coleoptera are represented by many Mesozoic fossils, but these are of surprisingly little value for determining the age of the individual subgroups. This is mainly because the constitutive characters of these groups are only rarely visible in fossils. Very little progress can be made using the only characters that are usually preserved, namely the elytra and the general shape.

What is more, there is still no satisfactory phylogenetic classification into which the fossils can be fitted, and this is a serious limitation for the interpretation of fossils. However, some important principles for a phylogenetic classification do appear to have been established.

For a long time the Coleoptera have been divided into the Adephaga and Polyphaga, and these appear to be well founded monophyletic groups. However, the position of the 'Archostemata', with the single family Cupedidae, is controversial and this has had some effect on the interpretation of the Permian fossils. It is also important for the study of Mesozoic fossils. The old idea that the Cupedidae belong to the Adephaga, which was still accepted by Bradley (1947) and Schilder (1949), is untenable[398]. The only decision to be taken is whether the Archostemata are the sister-group of the rest of the Coleoptera (Adephaga + Polyphaga) (Machatschke's view) or of the Polyphaga alone (Crowson's view). Crowson based his conclusions on the fact that the final instar larvae of the Cupedidae have the tarsus and pretarsus fused into a 'tarsungulus', as in the Polyphaga, whereas these two segments are distinct in the earlier larval instars. However, the significance of this character is diminished by Crowson's own statement (1962) that the larvae of only two species of *Cupes* are known and that these belong to a relatively derived species group. For this reason it is not certain whether the fused tarsus and pretarsus of the final larval instar is part of the ground-plan of the Archostemata. According to Chen (1946), the larvae of the staphylinid genera *Philonthus* and *Bledius* have retained a divided apical tarsal segment[399]. It is therefore possible that a single 'tarsungulus' has arisen

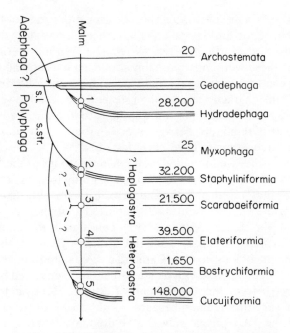

Fig. 131. Phylogenetic tree of the Coleoptera. An attempt at a graphical representation of Crowson's (1955, 1960) classification. 1, *Actea* and other genera; 2, Hydrophilidae, 3, *Proteroscarabaeus*; 4, *Tersus*, *Elaterina*; 5, *Praemordella* and other genera (see p. 394).

several times in the Coleoptera through convergence and is not part of the ground-plan of the Polyphaga at all.

No one is likely to disagree with the statement that of all the *Coleoptera the Archostemata (Cupedidae) have the most primitive elytral structure. For this reason I was unable to decide whether the Permian fossils belong to the Archostemata or to the stem-group of all the Coleoptera. This difficulty also applies to the Mesozoic fossils. However, Zeuner (1961) described *Pseudosilphites natalensis*, from the Middle Triassic (Molteno Beds) of South Africa (Natal), and assigned it to the Silphidae, a subgroup of the Polyphaga(!). If he was correct, it follows that the Archostemata must have been in existence as a separate group at the same time, as well as the Adephaga and Polyphaga, no matter whether we follow Machatschke in treating the Archostemata as the sister-group of the rest of the Coleoptera or Crowson in treating them as the sister-group of the Polyphaga alone. The problem of whether the early Mesozoic 'Cupedidae' [*Moltenocupes townrowni* Zeuner (1961), from the Middle Triassic of Natal; *Metacupes harrisi* Gardiner (1961), from the Rhaetic of South Wales] really do belong to the Archostemata or might be survivors from the Palaeozoic stem-group of the Coleoptera is therefore a comparatively subsidiary one and has no bearing on the age of these groups. Crowson (1960) reported that galleries have been found in fossilized timber from the Triassic of North America (Walker, 1938) and Germany (Linck, 1949) and that these were most probably produced by 'cupedid-like' beetles. This would mean that at that time the Cupedidae were following a mode of life which is still characteristic of their recent species. However, the life histories of only two species are known and, according to Crowson (1962), these belong to a relatively derived species-group. Only about 20 recent species of the Archostemata have been described[400].

There are two other well founded monophyletic groups of the *Coleoptera. The Adephaga are shown to be a monophyletic group by having the first three abdominal sternites fused and the hind coxae immovably fixed to the 1st abdominal sternite[401]. Crowson (1960) reported that they are known with certainty from the Lias. Zeuner (1961) considered the Middle Triassic *Umkoomasia depressa* Zeuner, from the Molteno Beds of Natal, to be a species that might belong to the 'Carabidae'. As stated above, it is highly likely that the Adephaga and Polyphaga had already separated by the Middle Triassic. Nevertheless, the assignment of Middle Triassic fossils to the 'Carabidae' is very problematic because, as shown by Crowson, of the two subgroups of the *Adephaga only the Hydradephaga can be regarded as monophyletic. The Geodephaga, to which the Carabidae belong, are probably a paraphyletic group[402], and early Mesozoic fossils with characters of the Geodephaga could well belong to the stem-group of all the Adephaga. It would therefore be particularly important to find some undoubted Mesozoic Hydradephaga. Crowson (1960) stated that the Upper Jurassic *Actea* is very similar to the recent Dytiscidae, and according to Rodendorf (in Rodendorf, 1962) the Gyrinidae are also supposed to be recorded from this epoch. As Crowson has shown that the Dytiscidae + Gyrinidae + Noteridae form a well founded monophyletic group[403] within the Hydradephaga, this would mean that other, more primitive subgroups of the Hydradephaga and Geodephaga must also have been in existence during the Upper Jurassic. This seems very likely, on the basis of what is known of other holometabolous orders and the Heteroptera. However, for the present nothing more precise than this can be said.

The Adephaga include 28 182 species, according to Schilder (1949), which is only 12% of the recent beetle fauna, and they are a much less significant group than the Polyphaga. They are entirely predaceous, whereas Crowson (1955) wrote that there is hardly any mode of life open to the insects that has not been adopted by at least some Polyphaga.

Although most of the Polyphaga have some derived characters which indicate that the group as a whole is monophyletic, it is not easy to find even a single derived ground-plan character when the 'Myxophaga' are also taken into account. Crowson mentioned the fusion of the tarsus and pretarsus to form a 'tarsungulus', but, as I mentioned above, the value of this character is seriously weakened by Chen's (1946) claim that this fusion has not taken place in certain genera of the Staphylinidae[404]. Further derived characters present in most Polyphaga are the telotrophic ovarioles[405], the absence of the cell called the 'oblongum' in the hind-wing, and the absence of the suture dividing the propleura and the pronotum. However, these characters are absent in four families (Calyptomeridae[406], Lepiceridae, Sphaeriidae, and Hydroscaphidae) which together include a mere 25 species and which Crowson (1955) 'provisionally' grouped together as the 'Myxophaga'[407].

These characters suggest that there is a sister-group relationship between the Myxophaga and the rest of the Polyphaga, and this has been shown in the phylogenetic trees that Günther (1962) and Ross (1965) constructed from Crowson's results. No Mesozoic fossils are known that might belong to the 'Myxophaga'. However, if Crowson's classification is correct and if the Silphidae (Polyphaga in the restricted sense) were in existence during the Middle Triassic, as Zeuner (1961) believed, it follows that the Myxophaga must also have been in existence at the same time[408].

In all the other Polyphaga (Polyphaga s. str.), the wing cell called the 'oblongum' is reduced, the suture between the pronotum and the propleura is absent[409], and, according to Crowson but not Paulian (in Grassé, 1949), the ovarioles are telotrophic (acrotrophic).

The most soundly based monophyletic group of the Polyphaga s. str. is the Cucujiformia, which contains more than 50% of all beetles, according to Crowson. In addition to the Cucujiformia, Crowson (1960) recognized the Staphyliniformia, Scarabaeiformia, Elateriformia, and Bostrychiformia. Crowson (1955, 1960) was unable to give any unambiguous derived ground-plan characters for these groups. He was forced to use words such as 'usually', 'rarely', 'generally', etc. Nevertheless, it should be pointed out that each of these groups has an extensive 'monophyletic core'. Most of the uncertainties are caused by a few small 'families' whose relationships are still unclear, but there is also some uncertainty over the relationship between the Staphyliniformia, Scarabaeiformia, Elateriformia, Bostrychiformia, and Cucujiformia themselves.

Crowson (1960) attempted to divide the Polyphaga s.str. into three major groups, which he thought originated during the Triassic: Staphyliniformia, Scarabaeiformia + Elateriformia, and Bostrychiformia + Cucujiformia. At least, this is how I would interpret his conclusions in the language of phylogenetic systematics. He rejected the old division into two sister-groups, Haplogastra and Symphiogastra (or Heterogastra), which Paulian (in Grassé, 1949) had advocated and which he himself had followed in his earlier table (Crowson, 1955).

This repudiation of his earlier views may be justified insofar as the Haplogastra (Crowson's Staphyliniformia + Scarabaeiformia) appear to have only primitive characters when compared with the Heterogastra (= Symphiogastra). Paulian (in Grassé, 1949) reported that the 2nd abdominal sternite is invaginated medially and only visible at the sides, and he thought that this was a derived character. However, Crowson (1955) stated that this sternite was occasionally though rarely completely developed (free). Paulian also stated that the ovarioles are telotrophic, and this could be a derived character as he said that they are polytrophic in the Heterogastra. However, according to Crowson, telotrophic (acrotrophic) ovarioles are characteristic of all the Polyphaga s.str.! Bradley (1947) listed clavate (clubbed) antennae as a character of the branch in his phylogenetic tree that corresponds to the Haplogastra, but neither Crowson nor Paulian mentioned this.

Urogomphi with a basal articulation are found only in larvae of the Haplogastra. Crowson (1960) attempted to show that this might be a derived character, but it was not a very productive line of enquiry because he was discussing the relationships of the Scarabaeiformia which do not have urogomphi. In any case, I think it much more likely that articulating urogomphi are a primitive character, as Crowson himself suggested in his earlier paper (1955), and that they have been retained in the Adephaga and Staphyliniformia.

At present it is impossible to find any definite evidence that the Haplogastra (Schilder's Protopolyphaga) are a monophyletic group. It is possible, as Crowson (1960) suggested, that the Scarabaeiformia are more closely related to the Heterogastra than to the Staphyliniformia. Naturally, they have not yet acquired all the derived characters of the Heterogastra.

Mesozoic fossils that can be assigned to the Staphyliniformia are rare. The most interesting of them is *Pseudosilphites natalensis* Zeuner, from the Middle Triassic (Molteno Beds) of Natal, and Zeuner (1961) was convinced that it belonged to the Silphidae. This would mean that by the Middle Triassic the Staphyliniformia had separated into a large number of subgroups. It therefore seems most pertinent to suggest that the most that should be said of *Pseudosilphites natalensis* is that it belongs to the Polyphaga, or perhaps to the stem-group of the Polyphaga.

Théodoridès (1952) apparently followed Handlirsch and stated that the Hydrophilidae were known with some certainty from the Lower and Upper Jurassic. In the Russian Textbook of Palaeontology (Rodendorf, 1962) this family was not recorded until the Tertiary.

The only record of the Scarabaeiformia (= Lamellicornia) appears to be *Proteroscarabaeus yeni* Grabau, from the Upper Jurassic of the Shantung Peninsula (originally thought to be Lower Cretaceous).

In addition, several subgroups of the Heterogastra are definitely known from the Jurassic, and they provide much more reliable evidence of the simultaneous existence of the Haplogastra, or Staphyliniformia + Scarabaeiformia (unless the latter are more closely related to the Heterogastra), than could any fossils that might be definitely assigned to the Haplogastra.

Crowson (1955) listed two derived ground-plan characters for the Heterogastra (his Symphiogastra): the larvae have fixed and unjointed urogomphi, and the adults have the pleurite of the 2nd abdominal segment 'almost always' fused with that of the 3rd segment.

Recent classifications of the Heterogastra have suggested different groupings: Ross (1965) apparently followed Crowson (1955) and suggested that there is a sister-group relationship between the Elateriformia and Bostrychiformia + Cucujiformia, whereas Günther (1962) apparently followed Crowson (1960) and in his phylogenetic tree included the Elateriformia as the sister-group of the Scarabaeiformia (a section of the Haplogastra).

Crowson's (1960) Elateriformia are essentially the same as Schilder's Deuteropolyphaga and correspond to Crowson's (1955) Dascilliformia without the Dascillidae. Crowson did not mention any derived ground-plan characters for this group. It is therefore possible that it is paraphyletic and that one of its subgroups is more closely related to the Bostrychiformia + Cucujiformia than are the others. Crowson (1960) listed these groups as the Byrrhoidea, Dryopoidea, Buprestoidea, Rhipiceroidea, Elateroidea, and Cantharoidea (= Malacodermata). Earlier authors, such as Bradley (1947), did not consider the families that Crowson later included in his Elateriformia to form a monophyletic group.

Of the families in Crowson's Elateriformia, Théodoridès (1952) recorded the Buprestidae from the Lower Jurassic, a statement apparently taken from Handlirsch. However, in the Russian Textbook of Palaeontology (Rodendorf, 1962), Buprestidae were first recorded from the early Tertiary.

Martynov (1926) assigned *Tersus crassicornis* Martynov, from the Malm of Karatau, to the Elateridae, and Gardiner (1961) described *Elaterina liassica*, from the Lower Jurassic of England (Lyme Regis) in the same family[410]. On the other hand, Théodoridès (1952) stated that *Elateridopsis permiensis*, which Zalesskiy (1932) described from the Permian of Tikhiye Gory, is probably a plant fragment.

Crowson did not mention any derived ground-plan characters for the group to which the Bostrychiformia and Cucujiformia belong (Schilder's Tritopolyphaga). Ross (1965) stated that there are fewer than seven visible abdominal segments, but he too failed to mention any derived characters for the Bostrychiformia. It is therefore possible that this is a paraphyletic group.

It seems that no Mesozoic fossils have been assigned to any of the families that Crowson included in his Bostrychiformia. On the other hand, several subgroups of the Cucujiformia are definitely represented amongst the Jurassic fossils. There is no doubt that this group is monophyletic. Crowson (1955) listed two derived ground-plan characters in the adults: the absence of functional spiracles in the 8th abdominal segment, and the cryptonephric condition in which tips of the Malpighian tubes terminate in the peritoneal membrane of the rectum.

Crowson stated that more than 50% of all the recent described beetles belong to the Cucujiformia, and Schilder (1949) gave the figure of 66%. On its own this group is larger than any other insect order! Crowson also assigned the Strepsiptera here, as the family-group Stylopoidea[411].

Théodoridès (1952) did not mention any authentic records of this group from the Mesozoic. However, he overlooked two of Martynov's papers (1926, 1935). According to the Russian Textbook of Palaeontology (Rodendorf,

1962), the following families of Cucujiformia are known from the Malm of Karatau: Ostomatidae (*Lithostoma expansum* Martynov), Nitidulidae (*Nitidulina exclavata* Martynov), Praemordellidae (*Praemordella martynovi* Shchegoleva-Barovskaya), Oedemeridae (*Necromera baeckmani* Martynov), Cerambycidae (*Parandrexis parvula* Martynov), Chrysomelidae (*Mesopleurites jurassicus* Martynov), and Curculionidae (*Archaeorrhynchus tenuicorne* Martynov). Some of these species, such as *Archaeorrhynchus* and *Praemordella*, are so characteristic that it is hardly possible to mistake their relationships. On their own they could prove that the *Cucujiformia had separated into a large number of subgroups by the Upper Jurassic. Unfortunately, the phylogenetic classification of the Cucujiformia, and of the Coleoptera in general(!), has not been worked out in sufficient detail for any reliable conclusions to be drawn from the few fossils whose relationships are beyond dispute.

Martynov (1935) has also described a species *Mesosagrites multipunctatus*, from the lower Lias of Chelyabinsk, which he assigned to the Chrysomelidae. This is known from only a single damaged elytron, and I doubt very much if it is really possible to recognize the 'family Chrysomelidae' from such a fragmentary fossil.

Revisionary notes

398. The complete invagination of the coxosternum into sternite 7 (Bils, 1976) and a series of possibly synapomorphic similarities in the structure of the prothorax (Baehr, 1975, 1979) may provide evidence that the Cupedidae are most closely related to the Adephaga. However, the possibility that these derived characters have arisen convergently cannot be ruled out yet. [Martin Baehr.]

399. Emden (1934) has denied that any larvae of the Polyphaga have a separate apical pretarsus. Consequently, the 'tarsungulus' is probably part of the ground-plan of the Polyphaga. [Martin Baehr.]

400. The Archostemata have been discussed by Atkins (1958) and Ponomarenko (1969a). The same author (Ponomarenko, 1969b) has described a new family, the Labradorocoleidae, from the Cretaceous of northern Labrador, Canada. [Willi Hennig.]

401. Further evidence supporting the monophyletic origin of the Adephaga has been provided by Bils (1976) from characters in the female genitalia and by Baehr (1979) from the prothorax. [Martin Baehr.]

402. Burmeister (1976) and Baehr (1979) have provided further evidence that the Geodephaga and the so-called Carabidae are paraphyletic. Bell (1964, 1966, 1967), Crowson (1968), Lindroth (1969), Bils (1976), Burmeister

(1976) and Baehr (1979) all agree that the Hydradephaga and Trachypachini together form a monophyletic group. [Martin Baehr.]

The higher classification of the Carabidae has also been dealt with by Ali (1969), Lindroth (1969), and Kryzhanovskiy (1976). [Willi Hennig.]

403. Burmeister (1976) has pointed out that some presumably synapomorphic peculiarities in the musculature of the female genitalia of the Haliplidae and Noteridae suggest that these families are closely related. According to Burmeister, these families and the Gyrinidae are the sister-group of the Amphizoidae + Hygrobiidae + Dytiscidae.

Baehr (1979) has also stressed that the Haliplidae and Noteridae are closely related, and has mentioned the highly synapomorphic prosternal–mesosternal articulation in support of this. However, he considered it possible that the Gyrinidae are the sister-group of the rest of the Hydradephaga. [Martin Baehr.]

404. See note 399. [Martin Baehr.]

405. Bonhag (1958) appeared to confirm that the Adephaga have polytrophic ovarioles and the Polyphaga telotrophic ovarioles. [Willi Hennig.]

406. The Calyptomeridae certainly belong to the Polyphaga. The Myxophaga also include the Torridincolidae, which were recently described by Steffan (1964). [Martin Baehr.]

407. The systematic position of these four families is still not settled. Barlet (1972, 1974), who studied the propleura and the sternal apophysis in three of them, did not think that they could all be grouped together into a single separate suborder. He tentatively assigned the Sphaeriidae and Torridincolidae to the Polyphaga. Because of their adephagan propleura, he assigned the Hydroscaphidae to the Adephaga.

Reichardt (1973a, 1976) and Reichardt and Vanin (1977) have described a number of new genera and species of the Torridincolidae and Hydroscaphidae. Reichardt (1973b, 1974) also concluded from his studies of the immature stages and mouth-parts that these families are closely related. Britton (1966) emphasized the similarities between larvae of the Sphaeriidae and those of the Torridincolidae and Hydroscaphidae, and he too concluded that these three families are closely related. [Martin Baehr.]

408. Klausnitzer (1975) has recently suggested a similar classification. He divided the Coleoptera into two groups, the Archostemata (= Cupedidae) and the rest of the Coleoptera, for which he proposed the name Pantophaga. He further divided the Pantophaga into two sister-groups (Adephaga and Heterophaga) and the Heterophaga into two sister-groups (Myxophaga and Polyphaga). [Martin Baehr.]

409. This is not entirely correct. The notopleural suture is not absent, since the propleura and pronotum are actually connected. Normally this connection takes the form of an articulating joint. In the Polyphaga it is short and is not visible externally because it is usually concealed beneath the lateral flange of the pronotum. [Martin Baehr.]

410. The systematics of the Mesozoic Elateridae have been discussed by Dolin (1975). [Willi Hennig.]

411. For additional comments on the Strepsiptera, see notes 315–320 on pp. 311–314. [Dieter Schlee.]

2.2.2.2..4.4. **Hymenoptera**[412]

There are no Palaeozoic fossil: that can be reliably assigned to the Hymenoptera. The earliest fossils are from the Triassic, and by the Upper Jurassic several subgroups of the *Hymenoptera appear to be represented[413]. Unfortunately, the phylogenetic classification of the *Hymenoptera has not been worked out sufficiently well for us to deduce from the few Mesozoic fossils the number of subgroups into which the order had separated by this epoch. Moreover, many of the characters that are necessary for making a reliable appraisal of phylogenetic relationships are not visible in the fossils.

It is now almost certain that the old division of the *Hymenoptera into Symphyta and Apocrita does not reflect the actual relationships. The 'Symphyta' appear to be a paraphyletic group[414]. Some of them are more closely related than others to the Apocrita, which are very probably monophyletic. This situation is shown very clearly in the phylogenetic trees published by Börner (1919) and Ross (1937, 1965), even though there are considerable differences in detail. It is possible that there is a sister-group relationship between the Tenthredinoidea and the rest of the Hymenoptera, as was suggested by Börner (1919) and Ross (1937) in their phylogenetic trees. The name Strophandria Crampton (= Ectropoda Börner) could be used for the Tenthredinoidea. For the rest of the Hymenoptera, and not just the Symphyta without the Tenthredinoidea(!), the name Orthandria Crampton (= Anectropoda Börner) is available.

According to Ross (1937), there are two characters that show that the Strophandria are a monophyletic group: the male genitalia are inverted and the pronotum is reduced to a narrow collar. Ross also stated that the trapezoidal mesosternum and the absence of larvapods (prolegs) on the 1st and 9th abdominal segments of the larvae are probably further derived characters.

Ross (1937) listed two derived characters for the Orthandria: the sternopleural sutures meet along the mid-line, and the metepisterna form a prosternal bridge. Börner (1919, p. 147) considered the presence 'of a single (or double) row of characteristic flattened or spatulate bristles on the fore

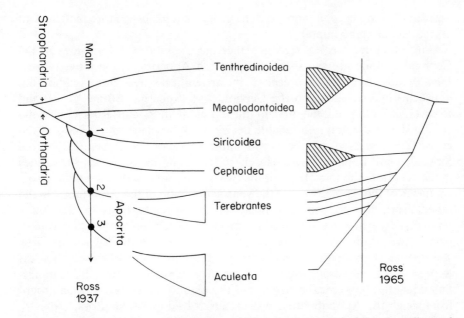

Fig. 132. Phylogenetic tree of the Hymenoptera. A comparison of Ross's earlier and later views (1937, 1965). 1, *Pseudosirex*; 2 and 3, *Mesohelorus*. For determining the age of these groups, it does not matter to which one *Mesohelorus* actually belongs, because the two available alternatives are sister-groups.

tibia and metatarsus, which appear to be a forerunner of the preening organ of the Apocrita', to be a derived character. 'The true Tenthredinidae completely lack these flattened or spatulate bristles on the fore legs, and this indicates that they are the most primitive members of the Symphyta'. Börner's account is not entirely free from ambiguities because it is not clear whether the 'flattened or spatulate bristles' are actually absent in the ground-plan of the Strophandria: this would mean that they really are a derived ground-plan character of the Orthandria.

Both Ross (1937) and Börner (1919) suggested that in the Orthandria there is a sister-group relationship between the Megalodontoidea (Xyelidae, Pamphilidae, Megalodontidae) and the rest of the Orthandria. Ross did not mention any derived ground-plan characters for the Megalodontoidea. More recently (Ross, 1965), he suggested that this group was most closely related to the Tenthredinoidea. He recognized the Tenthredinoidea + Megalodontoidea, and not the Tenthredinoidea alone, as the sister-group of the rest of the Hymenoptera. It is not clear what considerations induced him to change his mind. Possibly he thought that the derived characters which he had previously (Ross, 1937) listed for the Orthandria (including the Megalodontoidea) had in fact arisen convergently, but his reasons for this new interpretation are not known.

In the rest of the Orthandria the larvae lack segmented thoracic legs. The

Apocrita belong here, a monophyletic group that includes more than 90% of all the recent Hymenoptera.

Unfortunately, it is not clear whether the Cephoidea + Siricoidea or the Cephoidea alone are the sister-group of the Apocrita. If the latter is true, then it must have been preceded by an earlier sister-group relationship between the Siricoidea and the Cephoidea + Apocrita. Börner (1919) and Ross (1965) followed the first of these possibilities, and Ross (1937) the second. Unfortunately, no reliable decision can be made on the basis of the characters discussed by these two authors. Ross (1937) stated that the Siricoidea and Cephoidea are 'closely related', and based this on similarities in the venation, the loss of one of the two apical spurs on the fore tibia (also a character of the Apocrita), and the ventral direction of the tentorium. On the other hand, he stressed that many of the specializations of the lower Apocrita, by which he appeared to mean characters present in the ground-plan of the group, are present at least in some form in the Cephoidea. This was why he suggested that the Cephidae, the only family of the Cephoidea, 'represent the ancestral form of the primitive Apocrita' (Ross, 1937, p. 42). Subsequently, however, he wrote that the identity of the 'symphytan group' from which the Apocrita arose was an unresolved question (Ross, 1965).

There is an impressively close agreement in undoubtedly derived characters between the Cephoidea (without the Siricoidea) and the Apocrita, which suggests that there is a sister-group relationship between these two groups. It would therefore be an important task to investigate how far the characters common to the Siricoidea and Cephoidea are symplesiomorphies, in terms of the Siricoidea + Cephoidea + Apocrita, or have arisen convergently.

Börner (1919) listed several derived characters for the Apocrita[415]. The larvae have no anus. In the adults the labial parts can be retracted into the mouth opening. There is always a preening organ on the fore-legs, 'consisting of a row of differentiated bristles on the metatarsus which fits against the single tibial spur' (Börner, 1919: 147). As mentioned above, Börner believed that this preening organ had developed from the 'flattened or spatulate bristles' that are present in certain 'Symphyta'. 'It is only necessary to suppose that the flattened or spatulate bristles of the metatarsus were retained in the Apocrita and were completely adapted to a new function' (Börner l.c.). There is an articulating joint between the 1st abdominal tergite, which is attached to the thorax, and the 2nd tergite, and Ross (1937) reported that it is present in a more primitive form in the Cephoidea where there is the suggestion of a joint (an indentation) between these tergites. According to Ross (1965), the mesothoracic scutum has a flexible suture which facilitates flight in this unusually robust group, and this character is also present in the Cephoidea. The way in which the head capsule is closed ventrally may be a further character common to the Cephoidea and the Apocrita. An open head is part of the ground-plan of the Hymenoptera, and Ross (1937) showed that different groups are characterized by the manner in which the head capsule is closed. He found that the maxapontal type is characteristic of the Cephoidea, and that the Siricoidea belong to another type: in the Cephoidea the ventral parts of the maxillaries are fused along the mid-line and form a bridge ('maxaponta') between the postgenae. Ross (1937) showed that in the Braconidae, which he apparently only introduced into his discussion as they are relatively primitive members of the Apocrita, the type of head closure and the structure of the tentorium can be derived from the conditions found in the Cephoidea. This argument is not weakened by Benson's (1938) discovery that the 'maxapontal' head type is also present in a number of the Tenthredinoidea, although Ross stated that it was only found in the Cephoidea and, in a derived form, in the Apocrita. Finally, again according to Ross, the structure of the male genitalia in the Cephoidea, in particular the almost complete fusion of the gonostipes and harpes, is a precursor of the condition found in the Apocrita. According to the Russian Textbook of Palaeontology (Rodendorf, 1962) the Apocrita also agree with the Cephoidea in having the subcosta and radius of the fore-wing completely fused.

Several recent authors have dropped the old division of the Apocrita into Terebrantes and Aculeata[416]. Ross's (1965) phylogenetic tree showed the Aculeata but not the Terebrantes as a monophyletic group. Ross broke up the Terebrantes into a number of genealogical units some of which are more closely related to the Aculeata than others.

Oeser (1961) found that the female ovipositor provided clear evidence that the Aculeata really are a monophyletic group and that the Terebrantes are not. His studies showed that the Chrysidoidea (= Bethyloidea) also belong to the Aculeata, a conclusion that had been rejected by Börner (1919) but accepted by a number of other authors, such as Berland (in Grassé, 1951). Oeser suggested that there is a sister-group relationship between the Chrysidoidea and the rest of the Aculeata (Aculeata s.str.). He also showed that the Formicidae belong to the Aculeata s.str., not to the Chrysidoidea (Bethylidae s.lat., Cleptidae, Chrysididae) as Ross (1965) indicated in his phylogenetic tree.

Using the female ovipositor, Oeser was unable to show that individual subgroups of the Terebrantes are closely related to the Aculeata. In a discussant's comment on Bradley (1958), Townes suggested that the Aculeata could not be 'derived' from the parasitic Hymenoptera (Terebrantes), and Bradley himself confirmed that the most that can be said is that the two groups had common ancestors. This would mean that there is a sister-group relationship between these two groups.

Bradley considered the absence of the posterior lobe of the hind-wing to be a derived character of the Terebrantes. This appears to be the same as the anal lobe (= vannus), and has been retained in the Aculeata and in the ground-plan of the Hymenoptera[417]. Bradley thought that it had been lost once and for all in the Terebrantes. In this case, its absence really could be regarded as constitutive derived character of this group. However, it is also present in the Evaniidae, a small family whose larvae are parasitoids of cockroach eggs, and Bradley (1958) wanted to treat this group as a separate superfamily at the base of the two main lineages of the Apocrita. In short, he appeared to be suggesting that there is a sister-group relationship between the Evaniidae and the rest of the Apocrita. On the other hand, Emden (1957) and Short (1952) both thought that the Evaniidae were more closely related to the Aculeata. It is important to work out the position of this interesting group in the phylogenetic classification of the Hymenoptera as this will also have some bearing on the interpretation of the fossils[418].

There are difficulties in the interpretation of Mesozoic fossils: firstly, the constitutive characters of the monophyletic subgroups are not found in the venation (Fig. 133), at least amongst the more primitive Hymenoptera, and

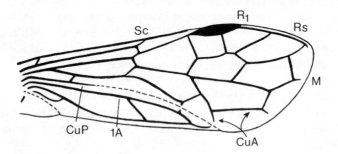

Fig. 133. Hypothetical fore-wing of the Hymenoptera. From Ross (1937).

are in fact rarely visible in fossils; and secondly, assignments to recent subgroups of the *Hymenoptera have usually been based on superficial similarities.

These reservations apply to the assignment of *Archexyela crosbyi* Riek (Fig. 134), from the Upper Triassic of Australia, and of *Liadoxyela praecox* Martynov, from the lower Lias of Central Asia, to the family Xyelidae of the Megalodontoidea. The characters in common are probably symplesiomorphies. In this connection, I should again point out that Ross at first (1937) regarded the Megalodontoidea as the relatively primitive sister-group of the

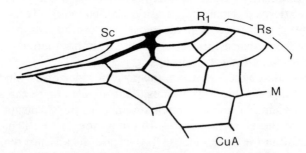

Fig. 134. Fore-wing of *Archexyela crosbyi* Riek (Hymenoptera), from the Triassic of Australia. From Riek (1955).

rest of the Orthandria, but later (1965) treated them as the sister-group of the Tenthredinoidea (Strophandria). In fact, there is probably no good reason for not assigning all the Upper Triassic and Liassic fossils to the stem-group of the Hymenoptera. Consequently, they are of no use for determining whether the *Hymenoptera had already separated into a number of subgroups by this epoch. Martynova (in Rodendorf, 1962) also assigned the genus *Anaxyela*, containing two species from the Upper Jurassic of Kazakhstan, to the Megalodontoidea, but Tillyard (1927) had previously stated that it belonged to the Siricoidea.

The situation is more promising during the Upper Jurassic (Malm) as several fossils are known that can almost definitely be assigned to the Apocrita. This shows that some of the groups that have been included in the 'Symphyta' must also have been in existence at the same time. It is impossible to say how many of these groups there were or which ones they may have been because relationships within the Symphyta are still not absolutely clear.

The genus *Pseudosirex*, from the Malm of Solnhofen, has usually been assigned to the Siricoidea. Handlirsch described no fewer than 14 species, but Carpenter (1932a) considered that these belonged to only two different species. According to Carpenter, the appearance and the wings of these fossils are so typical of the Siricoidea that they were even assigned to this group by the early authors. The compressed female ovipositor is clearly visible even in poorly preserved specimens, whilst well preserved ones show the numerous corrugations over the distal part of the wing which Carpenter considered to be characteristic of the Siricoidea. It is obviously possible that the Upper Jurassic 'Pseudosiricidae' do in fact belong to the Siricoidea. Nevertheless, these fossils do not enable us to draw any conclusions that we could not have drawn without them. They would be most useful if they could be definitely assigned to the stem-group of some recent family or group of families, but it is not known where the assignment should be made: to the Siricidae alone, the Siricidae + Xiphydridae, the Siricidae + Xiphydridae + Orussidae, or even the Siricidae + Xiphydridae + Orussidae + Cephidae.

It may be possible to decide this when the phylogenetic classification has been worked out in more detail. However, part of the answer would emerge immediately if *Parorussus extensus* Martynov, from the Malm of Karatau, could definitely be assigned to the 'Orussidea'. This classification was adopted by Martynova (in Rodendorf, 1962): this and the generic name(!) suggest that *Parorussus* belongs to the stem-group of the Orussidae. This family, whose larvae are parasitoids of wood-boring beetles, was often assigned to the Ichneumonoidea, as for example by Börner (1919), but there is no longer any doubt that it belongs to the Siricoidea. If *Parorussus*, from the Malm, really does belong to the stem-group of the Orussidae, this would give a good indication of the age of the other subgroups of the Siricoidea. Tillyard (1927) considered *Parorussus* to be a connectant form between the Orussidae and the Megalyridae (Terebrantes-Ichneumonoidea). However, this can be no more than a typological assessment because there can be no genealogical, or actual historical transitional forms between the Orussidae and the Apocrita. Consequently, the significance of *Parorussus* is still not clear.

In general, the Upper Jurassic Apocrita are primitive species, and they have all been assigned to various subgroups of the *Terebrantes. However, the characters in common may only be symplesiomorphies, and as a result some of these species may actually belong to the stem-group of the Apocrita and not to the Terebrantes at all. Unfortunately, the vital constitutive character of the Terebrantes is not visible in fossils.

These reservations particularly apply to *Ephialtites jurassicus* Meunier, from the Malm of Spain; *Mesaulacinus oviformis* Martynov, from the Malm of Karatau; and *Paraulacus*, from the Upper Jurassic of China (originally thought to be Cretaceous). These have all been assigned to the 'Ichneumonoidea', but this only means that they are primitive Terebrantes or possibly only primitive Apocrita. Handlirsch (1920–21) stated that *Ephialtites* appeared to have a rather broad base to the abdomen and that it 'might be transitional between the Symphyta and Apocrita'. Tillyard (1927) assigned it to the Megalyridae, a relict group containing a few species restricted to Australia and South America. He was probably encouraged to do this by the very long ovipositor of *Ephialtites* which is certainly not a primitive ground-plan character of the Apocrita.

Mesohelorus muchini Martynov (Fig. 135), from the Malm of Karatau, is

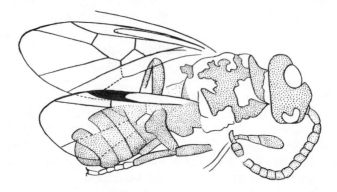

Fig. 135. *Mesohelorus muchini* Martynov (Hymenoptera–Apocrita), from the Upper Jurassic of Karatau (Kazakhstan). From Rodendorf (1962).

probably the most interesting Mesozoic fossil. In the Russian Textbook of Palaeontology (Rodendorf, 1962) it was assigned to the recent family Heloridae, but the characters in common are probably only symplesiomorphies. Tillyard (1927) considered even the assignment of *Mesohelorus* to the Proctotrupoidea to be provisional. However, the genus does have a comparatively derived venation. Tillyard (1927) thought that the venation of the

Proctotrupoidea, Chrysidoidea (= Bethyloidea), and Cynipoidea could be derived from that of *Mesohelorus*. This does not mean a great deal because these groups are not particularly closely related. Nonetheless, *Mesohelorus* does show that either Terebrantes or Aculeata with relatively derived venation must have been in existence during the Upper Jurassic. We would be justified to conclude from this that at least the sister-group relationship between the Terebrantes and Aculeata had arisen by the Upper Jurassic. Consequently, it matters very little that the first fossil Aculeata are not known until the Upper Cretaceous: *Cretavus sibiricus* Sharov (Fig. 136), from the

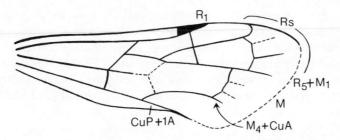

Fig. 136. Fore-wing of *Cretavus sibiricus* Sharov (Hymenoptera–Aculeata), from the Upper Cretaceous of Krasnoyarsk. From Rodendorf (1962).

Senonian of Krasnoyarsk in eastern Siberia. *Cretavus* is represented by a single incomplete fore-wing (Sharov, 1957b), and does not give any clue as to whether the *Aculeata had separated into subgroups by this period. However, this separation must have taken place before the Cretaceous, as is proved by an ant that I have seen from the Lower Cretaceous (Neocomian) of the Lebanon (unpublished work)[419].

So far a very modest amount of information about the history of the Hymenoptera during the Mesozoic has been provided by the fossils. It is probable that knowledge could be increased considerably by careful study of the associations between the parasitic Apocrita and particular groups of hosts. It seems likely that since the Middle Cretaceous considerable radiation has taken place in certain groups of the *Hymenoptera, particularly the Aculeata and Terebrantes, but so far we have only the most general ideas about this.

Revisionary notes†

412. Several comprehensive papers have been published recently: the origin and evolution of parasitism (Telenga, 1969); the evolution of instinct (Iwata, 1976); the evolution of the Hymenoptera (Malyshev, 1968); the evolution of

---

† These notes have been prepared by Dieter Schlee from Königsmann (1977, 1978a, 1978b) and are published here with the approval of Dr Eberhard Königsmann.

the ovipositor in relation to parasitism (Rasnitsyn, 1969); and a key to the superfamilies and families (Rasnitsyn, 1966). The morphology of the genitalia has been dealt with by Smith (1970). [Willi Hennig.]

See also notes 322, 323 and 326 on p. 318. [Eberhard Königsmann.]

413. Many Mesozoic Hymenoptera, including a number of amber fossils, have been described by Rasnitsyn (1975a), Evans (1973), and Townes (1973). Rasnitsyn (1972) has described the family Praeaulacidae ('Evanioidea') from the Jurassic of Karatau. [Eberhard Königsmann.]

414. Hennig's suggestion that the 'Symphyta' are a paraphyletic group has been confirmed. The 'Symphyta' without the Cephoidea are monophyletic. The Tenthredinoidea are the sister-group of the Blasticotomidae, and these two together are the sister-group of the Xyeloidea. The Megalodontoidea and Siricoidea are sister-groups. The Cephoidea are probably the sister-group of the Apocrita. See Fig. XI and Königsmann (1977). [Eberhard Königsmann.]

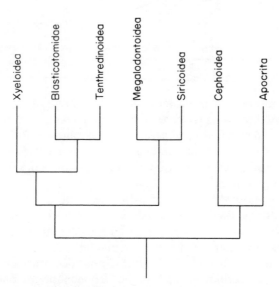

Fig. XI. Phylogenetic diagram of the 'Symphyta' and their subgroups (Hymenoptera). From Königsmann (1977).

415. The early evolution of the Apocrita has been discussed by Rasnitsyn (1975b). [Willi Hennig.]

416. The Apocrita are monophyletic, but the 'Terebrantes' are not. The phylogenetic relationships amongst the subgroups of the 'Terebrantes' are still not clear. The Chalcidoidea are probably the sister-group of the Cynipoidea; these two together are the sister-group of the Evanioidea, and these three together may be the sister-group of the Aculeata. It cannot be determined whether the Proctotrupoidea, or even the Proctotrupoidea without the Ceraphronoidea, are monophyletic. The Proctotrupidae + Heloridae + Vanhorniidae (possibly including the Roproniidae) form a single monophyletic group. The Ichneumonidae + Braconidae (= Ichneumonoidea) are probably sister-groups. The relationships amongst the other subgroups of the 'Terebrantes' are still unclear. See Fig. XII, and Königsmann (1978a).

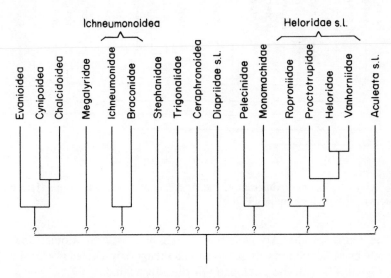

Fig. XII. Phylogenetic diagram of the 'Terebrantes' and their subgroups (Hymenoptera). From Königsmann (1978a).

The Aculeata s.lat. are monophyletic. The Chrysidoidea (= Bethyloidea) and Aculeata s.str. (= the Aculeata without Chrysidoidea) are sister-groups. The relationships of the Sclerogibbidae are very problematic. The 'Scolioidea' are not monophyletic: there appears to be no close relationship between the Scoliidae and the Tiphiidae, but there probably is between the Scoliidae and the Vespoidea; the Formicoidea may also be related to this complex. The Mutillidae (probably in a more restricted sense than usual) are probably closely related to the Sapygidae, and this complex could prove to be closest to the Tiphiidae. The subgroups excluded from the Mutillidae, possibly with the Pompiloidea, may be close to the Vespoidea complex. The Apoidea are very probably monophyletic, but the Sphecoidea are not: the Apoidea are probably most closely related to one of the subgroups of the Sphecoidea, and the Apoidea + 'Sphecoidea' would form a single monophyletic group. See Fig. XIII, and Königsmann (1978b). [Eberhard Königsmann.]

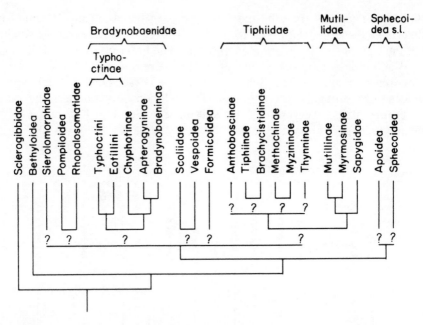

Fig. XIII. Phylogenetic diagram of the Aculeata and their subgroups (Hymenoptera). From Königsmann (1978b).

417. The wings of the Apoidea have been discussed by Louis, and the traditional classification of this group was challenged by Tkalcu (1972). Peters (1972b) has dealt with the phylogeny of the Megachilidae.

A phylogenetic tree of the subfamilies of the Braconidae has been given by Matthews (1974). See also Fischer (1972).

The classification of the Proctotrupoidea has been discussed by Kozlov (1970). [Willi Hennig.]

418. See, for example, the Upper Jurassic Praeaulacidae, which Rasnitsyn (1972) assigned to the 'Evanioidea'. [Willi Hennig.]

419. Wilson and Brown (in Wilson *et al.*, 1967) have described the first Mesozoic ant, *Sphecomyrma freyi*, from the Magothy Formation of New Jersey (Upper Cretaceous). They considered this genus to represent a new subfamily, and presented a cladogram (1967, Fig. 4) of the ant subfamilies. Wheeler and Wheeler (1972) have given a key to the subfamilies of the ants. [Willi Hennig.]

2.2.2.2..4.5. **Siphonaptera**

Nothing is known about the history of the Siphonaptera during the Mesozoic[420]. Furthermore, it is unlikely that the discovery of any fossils will answer the question of when the latest common stem-species of the *Siphonaptera lived. This is more likely to be answered by a careful study of the connections between recent groups of fleas and particular groups of hosts. All the fleas are parasites of mammals and birds. According to Holland (1964), parasitism of birds is undoubtedly secondary. This would mean that the bird parasites have arisen independently in six 'families' of fleas.

In these circumstances, a careful study of the relationships between fleas of the Marsupialia and those of the Placentalia will probably produce the most significant results. According to Wagner (1939), the Stephanocircidae are restricted to Australia and South America, where their hosts are principally Marsupialia, but also some rodents and birds. 'Their present distribution can be explained by suggesting that their ancestors were living on Marsupialia when these dispersed far to the north' (Wagner, 1939, pp. 99–100). It would probably be more correct to say that the Stephanocircidae originated in the north on the Marsupialia. Wagner pointed out that this family, partly including the Australian genus *Macropsylla*, differs in several characters from the rest of the Siphonaptera. Surely the suggestion that there is a sister-group relationship between the Stephanocircidae (possibly including *Macropsylla*) and the rest of the fleas, reflecting the sister-group relationship between the Marsupialia and Placentalia, could be a fruitful working-hypothesis? There is no hint of it in Holland's (1964) phylogenetic tree.

Additional note

420. For comments on the Lower Cretaceous 'fleas', see note 328 on p. 321. [Dieter Schlee.] For discussion of the development of the fleas, see Crowson (1970) and Rothschild (1975). [Willi Hennig.]

2.2.2.2..4.6..1.1. Trichoptera

Although various fossil 'Trichoptera' from the Mesozoic have been described, they provide hardly any information about the phylogenetic development of the group.

This is mainly because the constitutive characters of the Trichoptera and many of their monophyletic subgroups are found in the larvae and in anatomical details of the adults which are never visible in fossils. It is even impossible to separate the two sister-groups Trichoptera and Lepidoptera using characters of the venation. This can be formulated more precisely by saying that only the Lepidoptera can be definitely recognized by a derived character in the venation, the fusion of M_4 and CuA_1. It is usually impossible to decide whether fossil wings that agree formally with those of the *Trichoptera do actually belong to the Trichoptera or to the stem-group of the Amphiesmenoptera. Ross (1967) was right to make this reservation over the assignment of the Jurassic 'Necrotauliidae' to the Trichoptera. He considered that the only valid reason for regarding these fossils as Trichoptera was that the present geographical distribution of the *Trichoptera suggested that the order had already separated into a number of subgroups by the Jurassic.

The most reliable evidence that the Trichoptera were in existence during the Triassic, and far more reliable than any fossils that can be assigned to the Trichoptera, should be provided by the genus *Eoses*, from the Middle Triassic of Australia. This presupposes that its assignment to the Lepidoptera is correct, but unfortunately this is not certain (see p. 411). Ross's (1967) suggestion that the Trichoptera, and therefore the sister-group relationship between the Trichoptera and Lepidoptera, originated in the Upper Triassic is not improbable but has not yet been confirmed by the fossil record.

The Jurassic fossils would only be important if they could be assigned to particular subgroups of the *Trichoptera because this would then indicate the minimum age of these groups. Ulmer (1912) thought that some of the fossils from the Lias 'could even be assigned to recent genera' and suggested that they might belong to the Rhyacophilidae and Polycentropidae. Even Martynova (1961a) mentioned that the Mesozoic Necrotauliidae should be more closely studied because many of the 'Tertiary families' appeared to have arisen in the Mesozoic.

Martynova (in Rodendorf, 1962) assigned all the Triassic and Jurassic fossils, including the Necrotauliidae and even some species ('*Cladochorista*') from the Upper Permian of Australia, to the 'Annulipalpia'. This is one of the two divisions (Annulipalpia and Integripalpia) recognized in the recent *Trichoptera. However, the characters shared by these fossils and the *Annulipalpia are symplesiomorphies[421].

Ross (1967) has shown convincingly that the Annulipalpia and Integripalpia really are monophyletic sister-groups. The constitutive characters of the Annulipalpia are the annulate apical segment of the maxillary palp and the reduction of the supratentorium: neither character is visible in fossils. It is only the Integripalpia which have a derived (constitutive) character in the wing venation (Fig. 138, compare Fig. 137): the loss of the m–m cross-veins. Undoubted fossil Integripalpia are not known until the Cretaceous, and Ross (1967) thought that these few species could be primitive members of several different families.

Ross (1967) showed that the phylogenetic development of the Trichoptera is reflected mainly in the mode of life and morphology of the larvae[422]. Originally these must have lived in small, cool, moderately fast-flowing woodland streams. In fact, the most primitive species of several families still follow this mode of life, and as adults they are difficult to distinguish. So far as the venation is concerned, Ross did not mention any derived characters, even for monophyletic subgroups, except for the limnephilid lineage where M_4 in the male fore-wing is reduced.

In these circumstances, it is doubtful if the fossils will ever provide any information about the history of the Trichoptera during the Mesozoic. For example, they provide no support at all for the suggestion that the sister-group relationship between the Annulipalpia and Integripalpia had originated by the Jurassic, although this is more than likely.

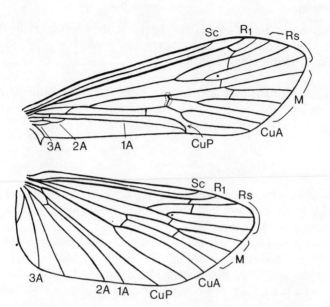

Fig. 137. Wings of *Stenopsychodes hiemalis* Tillyard
(Trichoptera–Annulipalpia, Polycentropidae; recent).
From Tillyard (1935a).

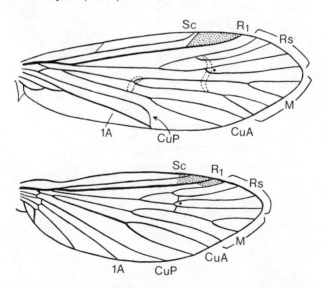

Fig. 138. Wings of *Hydrobiosella stenocera* Tillyard
(Trichoptera–Integripalpia, Rhyacophilidae; recent).
From Tillyard (1935a).

410

421. Furthermore, Sukacheva (1968) has described some caddis flies from the Lower Cretaceous of Transbaikalia, all of which she assigned to the 'Annulipalpia'. [Willi Hennig.]

422. A rather uninformative chapter on 'phylogeny' has been given by Hickin (1967) in his book on caddis larvae. Tomaszewski (1973) has discussed the adaptive evolution of caddis larvae. [Willi Hennig.]

2.2.2.2..4.6..1.2. **Lepidoptera**

The preceding remarks on the Trichoptera are also true of the Lepidoptera, with the additional comment that absolutely no undoubted fossil Lepidoptera are known from the Mesozoic[423].

What is supposed to be the oldest lepidopterous fossil was described by Tindale (1945) as *Eoses triassica* (Fig. 139), from the Middle Triassic of Mt. Crosby in Australia. Two wings were found, which Tindale stated to be a fore-wing and a hind-wing, but it is not absolutely certain that they actually

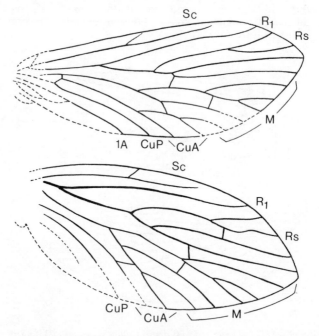

Fig. 139. Wings of *Eoses triassica* Tindale (supposedly Lepidoptera, but more probably Mecoptera), from the Triassic of Australia. From Tindale (1945). The veins indicated by broken lines, including the vitally important anal veins, are reconstructed and are not actually preserved.

belong together. However, this seems probable in view of their great similarity. One character that could support the assignment of *Eoses* to the Lepidoptera is the partial fusion of M_4 and CuA_1 in both wings. In all recent Lepidoptera, these two veins are always completely fused in both wings, and this is a derived ground-plan character which distinguishes the Lepidoptera from the Trichoptera. Evidently *Eoses* could represent the first step in this process of fusion and could therefore belong to the stem-group of the Lepidoptera.

However, the force of this conclusion is weakened by the fact that M_4 and CuA_1 are also fused in all the Diptera, which have retained only the fore-wing, and in most Mecoptera (see below). In recent Choristidae this fusion is not complete in the fore-wing (see *Chorista australis* Klug, Fig. 140): CuA_1 and M_4 are fused for a short distance and then separate again. In these

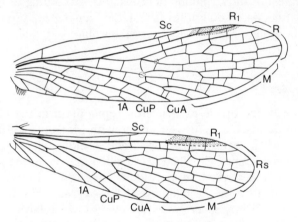

Fig. 140. Wings of *Chorista australis* Klug (Mecoptera; recent). From Tillyard (1933).

circumstances the conformation of the anal veins would be vital for an accurate appraisal of *Eoses*: looped anal veins are part of the ground-plan of the Amphiesmenoptera, a derived character that never occurs in the Antliophora (Diptera + Mecoptera). Unfortunately, the area containing the anal veins is not preserved in *Eoses*. In his illustration of the fore-wing, Tindale (1945) has only drawn a short section of the anal vein (labelled as 'PCu'). There is no trace of the apical section of a second anal vein. This may of course simply be because the relevant part of the wing has not been preserved, but it is the absence of this character that makes any reliable interpretation of *Eoses* impossible. If better preserved fossils of *Eoses* are found to have looped anal veins, this would be a good reason for assigning the genus to the Amphiesmenoptera and to the stem-group of the Lepidoptera. For the present, however, it is just as likely that *Eoses* belongs to the Mecoptera[424].

In any case, *Eoses* is only of importance for determining when the sister-group relationship between the Trichoptera and Lepidoptera originated: if it were to belong to the Lepidoptera, it could only be assigned to the stem-group of this order. I still do not think that the wing described by Kuhn (1951) as *Geisfeldiella benkerti*, from the Lias of Bamberg, will ever be interpreted.

The fossil record provides no evidence as to when the oldest subgroups of the *Lepidoptera originated. Moreover, it seems unlikely that it will ever be able to do so because none of the vital constitutive characters of the oldest monophyletic subgroups of the Lepidoptera are found in the wing venation. As with the Trichoptera, the best line of enquiry will have to be a careful analysis of the geographical distribution together with the connections with particular groups of host plants. So far the best founded suggestion is still that there is a sister-group relationship between the Zeugloptera (Micropterigidae) and the rest of the Lepidoptera (Glossata). According to Ross's (1965) phylogenetic tree, based on Forbes (1923), the Frenatae are probably another monophyletic group, and should be recognizable as fossils by their wing venation[425].

Revisionary notes

423. Direct evidence of the existence of Lepidoptera in the Mesozoic was only obtained during the 1970s. This came with MacKay's (1970) report on the remains of a caterpillar, supposedly belonging to the Ditrysia (Tineoidea), from Upper Cretaceous Canadian amber. Subsequently, wing scales have been described from Middle Cretaceous French amber and tentatively assigned to the Zeugloptera (Kühne *et al.*, 1973; Schlüter, 1974, 1975), and very recently several new discoveries of Cretaceous moths have been reported. Whalley (1977, 1978) described an alleged micropterigid, *Parasabatinca aftimacra*, from Lower Cretaceous Lebanese amber, and referred some further, less well preserved moths from the same formation to the Incurvariina. Skalski (1979a, 1979b) described two Cretaceous compression fossils: *Undopterix sukatschevae*, which he tentatively assigned to the Micropterigidae, from the Lower Cretaceous of Transbaikalia, and a leaf-mine, which he believed to belong to the Nepticulidae (i.e. to the Nannolepidoptera), from the Upper Cretaceous of Kazakhstan. He also described several moths from the Upper Cretaceous amber of northern Siberia which he considered to be similar to recent Lophocoronidae, Mnesarchaeidae and Incurvariidae.

The assignment of *Parasabatinca* to the Micropterigidae appears to be relatively well founded: it has moniliform antennae and shortened, probably two-segmented labial palps, both of which are derived features characteristic of this family (Kristensen and Nielsen, 1979). Whalley's assignment of the other Lebanese amber moth to the Incurvariina also seems to be reasonably well founded, since it possesses scales on the wing margin which are strongly dentate apically, a derived type of wing scale that is also present in the

Incurvariina. On the other hand, I cannot detect any of the undoubted autapomorphies of extant family lineages in the published accounts of the other fossils mentioned above. Skalski's Upper Cretaceous 'incurvariid' certainly belongs to the Heteroneura, but the assignment of the Canadian amber larva to the Ditrysia cannot be definitively confirmed. However, there seems little doubt that all the major lepidopteran lineages had differentiated before the Tertiary. [Niels Kristensen.]

424. Although the existence of a differentiated lepidopteran fauna in the pre-Cretaceous seems very likely, there is still no direct evidence. Hennig's suggestion that *Eoses* belongs to the Mecoptera has been vindicated, and *Eoses triassica* is a junior synonym of *Mesochorista proavita* Tillyard (see Riek, 1955). Riek (1976d) has treated the 'Paratrichoptera', a group including a number of Triassic species from the southern hemisphere, as a suborder of the Lepidoptera, but this taxon is not defined by any autapomorphies. In fact it has usually been considered as belonging to the Mecoptera. Its assignment to the Lepidoptera has been greeted with scepticism by Whalley (1978) and Skalski (1979b), and it does indeed seem to be based on very weak evidence. Riek particularly stressed that R is straight at the point where RS separates off, but this condition is neither characteristically present in lower Lepidoptera nor characteristically absent in other mecopteroid insects. Similarly, none of the other so-called 'lepidopteran' features of 'Paratrichoptera' wings are recognized apomorphies. [Niels Kristensen, Rainer Willmann.]

425. The time-honoured model of lepidopteran phylogeny is that there is a sister-group relationship between the Zeugloptera and all other Lepidoptera, but for various reasons this has been questioned during the last decade. According to Niculescu (1967, 1970), the family Eriocraniidae is the sister-group of all the other Lepidoptera, whereas according to Friese (1970) it is the superfamily Hepialoidea which has this status. Niculescu's views were based on reasoning incompatible with phylogenetic systematics, and Kristensen's (1968) criticisms of the first version of his views are equally applicable to the 1970 paper and need not be repeated here. On the other hand, Friese's hypothesis was based on an apparently sound application of the principles of phylogenetic systematics and therefore deserves closer attention. Common (1975) rejected Friese's hypothesis because it did not account for some larval and pupal apomorphies common to the Hepialoidea and other Glossata. Kristensen (1978) also rejected it and set out the evidence for a sister-group relationship between the Hepialoidea and the Mnesarchaeoidea (unquestionably a subgroup of the Glossata). A fuller examination of Friese's hypothesis, with a re-assessment of the alleged synapomorphies between the Zeugloptera and the other non-hepialoids, is now in preparation.

Very recently the view that there is a sister-group relationship between the Zeugloptera and all the other Lepidoptera has been challenged by new information on the genus *Agathiphaga* (see note 336). Both *Agathiphaga* and the Zeugloptera represent a pre-glossatan evolutionary grade, as they have articulated imaginal mandibles and non-haustellate maxillae, but *Agathiphaga* has also retained some further plesiomorphies that are not present in other Lepidoptera, such as the free M_4. The previous assignment of *Agathiphaga* to the Glossata (Hinton, 1958; Kristensen, 1967) has been based primarily on Hinton's assertion that the larva of *Agathiphaga* is 'in every aspect a typical lepidopterous larva' (when compared with the Zeugloptera). However, as Dumbleton (1952) originally pointed out, it does not have the Y-shaped adfrontal ridge that is characteristic of glossatan larvae, and certain other characters are also not present in derived (glossatan) states (Kristensen, unpublished work). Unfortunately, in some respects the larva of *Agathiphaga*, which feeds internally in coniferous seeds, is highly autapomorphic: for example, it is entirely apodous. My continuing anatomical study of *Agathiphaga* (Kristensen, in preparation) has not yet been able to provide an unambiguous answer to the question of the phylogenetic position of these moths. The 'conservative' concept of a sister-group relationship between the Zeugloptera and *Agathiphaga* + Glossata is supported by several character sets and may still be the best substantiated working hypothesis, but other characters seem to indicate that the primary division within the Lepidoptera is between *Agathiphaga* and the Zeugloptera + Glossata; nor can we entirely rule out the possibility of a sister-group relationship between *Agathiphaga* + Zeugloptera and the Glossata. It is likely that further observations on *Agathiphaga* and other primitive moths will resolve some of the discrepancies and provide definitive support for one of these phylogenetic models, but it is clear that there must have been a substantial amount of parallel evolution during the differentiation of the earliest lepidopteran lineages.

There are also different opinions of the phylogeny of the primary lineages within the Glossata. Various recent suggestions (Mutuura, 1972; Dugdale, 1974) have been authoritatively discussed by Common (1975). The conclusions to be drawn from Common's account and the recent findings by Kristensen and Nielsen (in press a, b) are in essence an elaboration of the classification proposed earlier by Hinton (1946) and Hennig (1953), and are as follows. There is a sister-group relationship between the Dacnonypha (which includes at least the Eriocraniidae; the position of the small families Lophocoronidae and Acanthopteroctetidae is still uncertain) and the rest of the Glossata (the Myoglossata). Within the Myoglossata there is a sister-group relationship between the Neopseustina (sole family Neopseustidae) and the rest of the group (the Neolepidoptera) and within the Neolepidoptera there is a sister-group relationship between the Exoporia (Mnesarchaeoidea + Hepialoidea) and the rest of the group (the Heteroneura or Frenatae). The Heteroneura are divided into two groups, the 'Monotrysia (s.str.)' and Ditrysia: the Monotrysia, which include the Nannolepidoptera and Incurvariina, are almost certainly paraphyletic in terms of the Ditrysia. The derived female genital apparatus, where the sexual opening is separated from the ovipore, is the main character used to establish that the Ditrysia are a monophyletic group. The Ditrysia include more than 95% of all the Lepidoptera and, in view of the enormous number of species, this is a remarkably homogeneous group structurally and ecologically. [Niels Kristensen.]

2.2.2.2..4.6..2.1. **Mecoptera**

This is a small group containing little more than 350 recent species, but numerous fossil species have been described from the Palaeozoic and Mesozoic[426]. The Mecoptera are more suitable than any other group for use as a paradigm of the phylogenetic development of an insect order.

Unfortunately, it is still not known for certain whether the Mecoptera are in fact a monophyletic group or whether, as Tillyard believed, the Eumecoptera are more closely related to the Diptera than are the Protomecoptera[427].

The Eumecoptera definitely appear to be a monophyletic group. It is usually stated that they have Cu_1 (= CuA) unforked. However, I think that it is most likely that the anterior branch (Cu_{1a} = CuA_1) is fused with M_4, as in the Diptera, and that the connection between CuA and M, which has usually been known as cross-vein m–cu, is actually the basal section of CuA_1 (Fig. 141). Whatever the true nature of this vein, whether actually or apparently

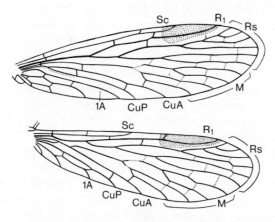

Fig. 141. Wings of *Panorpa communis* Linnaeus (Mecoptera; recent). From Grassé (in Grassé, 1951).

unforked, the simple CuA is a derived ground-plan character common to the Eumecoptera and Diptera.

A second derived character is the almost complete absence of cross-veins in the costal field, between Sc and C, and this is also common to the Eumecoptera and Diptera. In fact, if there were not some derived characters shared by the Eumecoptera and the (recent!) Protomecoptera which appear to be synapomorphies (see p. 333), there would be no alternative but to accept Tillyard's suggestion that there is a sister-group relationship between the Eumecoptera and Diptera.

The Eumecoptera include the recent families Bittacidae, Panorpidae (including the Panorpodidae of Byers, 1965), Choristidae, and Nannochoristidae. The position of the wingless Boreidae (genus *Boreus* with several Holarctic species) and Apteropanorpidae (only the Tasmanian *Apteropanorpa tasmanica* Carpenter) is uncertain[428]. According to Hinton (1958), the Boreidae are more primitive than 'all other' Mecoptera in certain characters. Riek (1954b) continued to assign *Apteropanorpa* to the Panorpidae, but it may well be asked why he did not consider it to be more closely related to the other Australian families rather than to the Panorpidae which do not occur in Australia. Both the Boreidae and the Apteropanorpidae probably belong to the Eumecoptera[429].

The two southern families Choristidae and Nannochoristidae probably form a single monophyletic group[430]. They have suppressed the connection between M and CuA which has been interpreted as the base of a vein 'M_5' that has fused with CuA: CuA touches M at one point (Choristidae) or is fused with it for a short distance. This is certainly a derived character and, taken in conjunction with the restricted distribution, indicates that the group is monophyletic. The Nannochoristidae are more derived than the Choristidae in having M and CuA fused for a longer distance and RS with only three branches: R_{2+3} is unforked. The media has only four simple branches. In view of this it is highly significant that Riek (1953a) has described two species with the same derived characters, from the Upper Permian of Australia (*Neochoristella optata* and *Nannochoristella reducta*). It would be taking scepticism too far to deny that these fossils belong to the stem-group of the Nannochoristidae, a family which still occurs in the same region. These two fossils would then prove that the more primitive Choristidae must also have been in existence as a separate group during the Upper Permian. They would also prove that the Eumecoptera, which are now found throughout the northern hemisphere and which have some even more primitive characters in their venation than the Choristidae–Nannochoristidae, must also have had their own separate ancestors living during the Upper Permian. In view of this, it really does not matter how we interpret the rest of the Upper Permian fossils from Australia. However, these are rather numerous and consequently are of considerable zoogeographical importance. The recent Nannochoristidae consist of some five species in Australia, Tasmania, and New Zealand, and a further two species in Chile and Argentina. As no recent or fossil species have ever been found in the northern hemisphere, the Nannochoristidae are one of several insect groups with a similar distribution which most probably reached southern South America by crossing Antarctica from Australia or New Zealand. However, there are very few species in South America, and none in this region belonging to the Choristidae or any other forms that could have descended from the abundant Eumecoptera known from the Permian of Australia. For this reason it is unlikely that dispersal of the Nannochoristidae took place as early as the Permian. Like other insects with this type of distribution, it is much more likely that the South American *Nannochorista* did not disperse from Australia to South America until the Mesozoic. The discovery of fossils in the Permian of Australia is important because it proves that this group really was in existence in Australia when the dispersal route across Antarctica first opened up during the Mesozoic. This proof is lacking for most groups of insects.

The next important task is to work out the relationships between the Choristidae–Nannochoristidae and the rest of the Eumecoptera, because this will enable us to deduce the age of the remaining eumecopterous groups. Unfortunately, these relationships are still not clear. Some authors have treated the Choristidae and Nannochoristidae as no more than subfamilies of the Panorpidae. It is always possible that the Panorpidae (including the Panorpodidae) are closely related to the Bittacidae. Even if this could be proved, however, the position of the Boreidae and the Apteropanorpidae would still be unclear[431].

Fossils which appear to be most closely related to the Panorpidae (+ Panorpodidae) and Bittacidae have only been found in the northern hemisphere. The Bittacidae are characterized by their slender build and by their modified tarsi: the 5th tarsal segment has a single claw and can close on the 4th like the blade of a clasp-knife. In the fore-wing the cross-vein-like basal section of M_5 has been suppressed, as in the Choristidae–Nannochoristidae: CuA is fused for a short distance with M. The wing base is petiolate. On the basis of these characters, except for the tarsi which have not been preserved, *Probittacus avitus* Martynov from the Upper Jurassic (Malm) of Karatau (Kazakhstan) can be assigned with some confidence to the Bittacidae, and probably to the stem-group of this family. However, Tillyard (1933), Martynova (in Rodendorf, 1962), and even Enderlein (1910a) have also assigned the Neorthophlebiidae to the stem-group of the Bittacidae. They differ from the Bittacidae in having M_4 forked, so that the media has five branches. This appears to be a primitive character and has been retained by the recent Choristidae. However, Tillyard (1933) considered that in the Neorthophlebiidae, as in the Bittacidae, the base of M_5 has been suppressed and CuA is fused for a short distance with M. Martynova (in Rodendorf, 1962) did not mention this character, although it is also of interest for the study of the Choristidae and Nannochoristidae. Various species have been described from the Jurassic of the northern hemisphere. These include *Protobittacus maculatus* Tillyard and *P. liassicus* Tillyard, from the lower Lias of England (Worcestershire). Martynova (in Rodendorf, 1962) also listed several 'Neorthophlebiidae' from the Upper Triassic of Issyk-Kul' (*Neorthophlebia nana* Martynova and *N. unica* Martynova), from the Urals (*Bittacopanorpa javorskii* Zalesskiy), and from Australia (*Archebittacus exilis* Riek). This means that the Bittacidae, as represented by their stem-group the 'Neorthophlebiidae', must have separated from the rest of the Panorpidae, and probably from the Boreidae too, at least as early as the Upper Triassic[432].

Martynova (in Rodendorf, 1962), Tillyard (1933), and even Enderlein (1910a) considered the Upper Triassic Orthophlebiidae to be the stem-group of the Panorpidae. They also have M_4 forked and the media consequently with five branches, a primitive character that they share with the Neorthophlebiidae. This means that the bifurcation of M_4 must have been lost independently in the Bittacidae and Panorpidae. The Orthophlebiidae are more primitive than the Neorthophlebiidae (and Bittacidae) in retaining the basal section of M_5, as in the Panorpidae, so that CuA is not fused with M.

One obstacle in the way of an accurate appraisal of the Orthophlebiidae is that the relationships of the wingless Boreidae are not known. It is usually accepted that the Boreidae are most closely related to the Panorpidae, and if this is correct then the Orthophlebiidae could well be the common stem-group of the Panorpidae (including the Panorpodidae) + Boreidae, as was suggested by Enderlein (1910a). But even if the Boreidae should prove to be more primitive than all the other Eumecoptera, as Hinton believed, it would still be impossible to disprove that the Orthophlebiidae also include the ancestors of the Boreidae. The Orthophlebiidae would then be an invalid stem-group.

The interpretation of the Permian fossils which I discussed in the last part of this chapter (p. 334) is even more problematic. The existence of the Nannochoristidae during the Upper Permian in Australia proves that this family, the Choristidae, and in addition the ancestors of several other groups of the Eumecoptera must all have been in existence at the same time. However, the fossils do not indicate how many such groups there were or which ones they may have been. These will remain unresolved questions until the phylogenetic relationships amongst the recent subgroups of the Eumecoptera have been worked out.

Nothing can be said about the history of the Protomecoptera (Meropeidae and Notiothaumidae) during the Mesozoic because no fossils have been found[433], nor is it possible to say whether they are in fact a monophyletic group. Moreover, it is not even certain whether it is really necessary to go back as far as the Lower Permian to find the common ancestors from which the Protomecoptera and Eumecoptera have descended.

Revisionary notes

426. A bibliography of recent and fossil Mecoptera has been published by Schlee and Schlee (1976). [Willi Hennig.] See also Willmann's (1978) catalogue of fossil Mecoptera. [Dieter Schlee.]

427. There is no longer any doubt that the Mecoptera are a monophyletic group (see note 344 on p. 340). [Dieter Schlee.]

428. There has been considerable dispute over the systematic position of the

Boreidae. Hinton (1958) did not think that they belonged to the Mecoptera at all, and assigned them to a separate order, the Neomecoptera (Crampton, 1930). He did not publish a dendrogram, but in his discussion he appeared to regard the Neomecoptera as the sister-group of the Siphonaptera + Diptera + Mecoptera. He justified this by reference to the primitive structure of boreid larvae, in which the cardo is not fused with the stipes, the tentorial adductors of the cardo are present, and a well developed postmentum is present. He also referred to the panoistic ovaries of adult Boreidae.

Since Hinton's paper was published, it has been confirmed that the Mecoptera, including the Boreidae, are indeed a monophyletic group. Consequently, Hinton's arguments can do no more than demonstrate that there is a sister-group relationship between the Boreidae and the rest of the Mecoptera.

Other more recent attempts to resolve the phylogenetic position of the Boreidae have placed them close to more highly derived families of the Mecoptera. Hepburn (1969b) thought that they were probably closely related to the Panorpidae, and he based this on the shift in the position of the tentorio-cardinal adductor from the tentorium to the clypeus, which he found to be restricted to the families Boreidae and Panorpidae. Mickoleit (1969), who made a study of the genital and post-genital segments of female Mecoptera, found that the Boreidae have circumapical sclerites and concluded from this that female Boreidae must already have passed through an evolutionary stage that involved a telescopic abdominal tip. This is the stage that has been reached by the Choristidae, Eomeropidae (= Notiothaumidae), Panorpodidae, Panorpidae, and Apteropanorpidae. For this reason, Mickoleit suggested that these five families and the Boreidae form a monophyletic subgroup within the Mecoptera.

Penny (1975) suggested that the Boreidae are most closely related to the Panorpodidae, and he noted the scarabaeiform larvae as a synapomorphy of these two families.

Russell (1979a, 1979b) has recently described a Nearctic genus of Boreidae, *Caurinus*, which has several primitive character states and which Russell placed at the base of the boreid phylogenetic tree. The most striking character of *Caurinus* is that the secondary ovipositor, which is characteristic of female Boreidae, appears to be primarily absent. The female of *Caurinus dectes* has free, one-segmented cerci that fit into the abdomen as in the Bittacidae, whereas all other known Boreidae have the cerci fused with the 11th abdominal segment to form a wedge-shaped probe. This means that the 'circumapical sclerites' of the Boreidae and the circumapical sclerites of the Eomeropidae (= Notiothaumidae) + Choristidae + Apteropanorpidae + Panorpodidae + Panorpidae must have arisen convergently. This is naturally of considerable importance for working out the phylogenetic position of the Boreidae. Russell was right to point out that the Boreidae must be excluded from the group of five families around the Eomeropidae because of the conformation of their cerci. [Gerhard Mickoleit.]

429. The structure of the male genitalia shows that *Apteropanorpa* is in fact closely related to the Australian family Choristidae. The Apteropanorpidae appear to be the sister-group of the Choristidae + Panorpodidae + Panorpidae. This interpretation, which conflicts with the views of Mickoleit (1978; see also note 431), will be discussed elsewhere. As a further point, it is worth noting that Riek mentioned a winged panorpid (*Austropanorpa*) from the Tertiary of Australia. However, Willmann (1977) came to the conclusion that the similarities between *Austropanorpa* and the Panorpidae are symplesiomorphies and that this genus should be excluded from the Panorpidae. This family is therefore restricted to the northern hemisphere. [Rainer Willmann.]

430. Recent work has challenged the view that the Choristidae and Nannochoristidae are closely related to each other. As I have already mentioned, there are several apomorphic characters (from the point of view of the Nannochoristidae) which the Choristidae share with the Eomeropidae (= Notiothaumidae), Apteropanorpidae, Panorpodidae, and Panorpidae, and which have certainly not arisen convergently: the displacement of the opening of the female genital chamber, the subdivision of gonocoxosternite 8, and the position of the cerci in the 11th abdominal segment. [Gerhard Mickoleit.]

431. The Apteropanorpidae appear to be most closely related to the Panorpodidae and Panorpidae. According to Mickoleit (1978), these three families have the lateral sclerites of the medigynium, which is one of the groups of sclerites surrounding the ductus receptaculi, produced into a caudally-directed, wing-shaped process on each side. Mickoleit also showed that the most primitive stage of this type of medigynium is found in the Apteropanorpidae. [Gerhard Mickoleit.]

432. It is generally believed that the highly specialized Bittacidae are a relatively recent family of the Mecoptera and are closely related to the Panorpidae, but this has recently been challenged by a substantial body of evidence which has involved the methods of argumentation as well as the facts themselves. So far as the methodology is concerned, previous authors have arranged the morphologically and chronologically appropriate fossils into a phylogenetic tree without mentioning any credible synapomorphies to support the suggested relationships. For example, Tillyard (1935a) and Handlirsch (1906–08) wanted to place the Orthophlebiidae and Neorthophlebiidae (the ancestors of the Panorpidae and Bittacidae) close to the Permochoristidae, but this presupposes two things: there must be some synapomorphies common to the Orthophlebiidae and Neorthophlebiidae, and it must be shown that the Permochoristidae belong to the stem-group of the Orthophlebiidae + Neorthophlebiidae. This phylogenetic relationship must also be supported by the presence of synapomorphies: at least one of the synapomorphies of the Orthophlebiidae + Neorthophlebiidae must be in the ground-plan of the Permochoristidae. This in turn presupposes that the Permochoristidae, Orthophlebiidae, and Neorthophlebiidae are monophyletic groups in the strict sense, but this is certainly not the case with the Permochoristidae. There is only one character mentioned by Tillyard that might be a synapomorphy of the Orthophlebiidae + Neorthophlebiidae: the media originally had six branches, but one of these has been reduced. However, no great weight can be attached to this character because even recent species of Mecoptera can have the media with a variable number of branches: Mickoleit (1967) found that two out of seven specimens of *Merope tuber* Newman had a media with five branches instead of six, and Ohm (1961) showed that in *Panorpa communis* Linnaeus an increase or decrease in the number of branches is directly correlated with body size. This suggests that reductions in the number of branches may have taken place on several occasions and independently in the course of phylogeny. Because of the high probability of convergence, the reduction of a single branch of the media cannot be used as evidence that there is a close relationship between the Orthophlebiidae and Neorthophlebiidae.

On the basis of his studies on the female post-genital segments, Mickoleit (1978) has been able to show that the Bittacidae probably occupy a relatively primitive position in the phylogenetic tree of the Mecoptera. His studies have shown that there are synapomorphies common to the Eomeropidae (= Notiothaumidae), Choristidae, Apteropanorpidae, Panorpodidae, and Panorpidae. The basal segments of the cerci are fused to the dorsum of the 11st abdominal segment. The dorsum of this segment has been extensively reduced, and the marginal areas have been retained as the so-called 'circumapical sclerites'. The situation is different in the Bittacidae, which have a well developed tergum and unattached cerci.

It follows from this that the Eomeropidae (= Notiothaumidae), Choristidae, Apteropanorpidae, Panorpodidae, and Panorpidae probably form a monophyletic subgroup of the Mecoptera from which the Bittacidae should be excluded. Embryological work also lends some support to this. The work of Ando (1973) has shown that the eggs of *Bittacus* have astonishingly little ooplasma but are rich in yolk. Embryogenesis is characterized by a very small germ band. This situation is reminiscent of hemimetabolous eggs. On the other hand, the eggs of *Panorpa* and *Panorpodes* have a rich plasma, little yolk, and a relatively large germ band. This represents a more derived egg type. Ando concluded that the peculiarities in the egg of *Bittacus* are an example of palingenesis. [Gerhard Mickoleit.]

433. Ponomarenko and Rasnitsyn (1974) have recently described several Triassic Mecoptera from the Soviet Union which they believed to be Meropeidae. However, these species (genera *Thaumatomerope, Blattomerope, Pronotiothauma*) all belong to the Eomeropidae (= Notiothaumidae), a family which is now restricted to South America. For a discussion of the relationships of these three genera, see Willmann (1978). There is another eomeropid genus (*Eomerope*) which is known from the Tertiary of the United States and the Soviet Union (Carpenter, 1972; Ponomarenko and Rasnitsyn, 1974). [Rainer Willmann.]

2.2.2.2..4.6..2.2. Diptera[434]

Amongst the Mesozoic fossils the Diptera are better represented than any other order, and so I shall be discussing their history in greater detail.

As in other insects, the wing venation is the vital character because it is almost the only feature on which the interpretation of the fossils can be based. Fig. 112 shows what I believe to be the ground-plan venation of the *Diptera, and this differs only in small details from my previous views (Hennig, 1954, Fig. 5). There are a number of fundamental features. The radial sector is dichotomously forked, and four free branches reach the wing margin. MA^+ is fused with the last branch (R_5). However, this is not confined to the Diptera: in my view it is characteristic of most of the Neoptera except for the Paurometabola (see p. 163). The main difference between Séguy (1959) and myself lies in our interpretation of what happened to MA^+. Séguy used the term R_5 or MA^+ for the vein that I consider to be the last branch of the radial sector (R_5), that is to say the result of the fusion of R_5 and MA^+. He based his interpretation on the presence of a fold which he considered to be the rudiment of MA^+ and which often extends from the radial sector almost to the wing margin.

I should repeat here what I wrote in 1954 about the relationship between wing folds and longitudinal veins. The veins were originally confined to certain longitudinal folds in the wing. It is possible for them to move away from their original folds, but this does not mean that they thereby change their identities. Consequently, it is wrong to continue to call a fold MA^+ when the vein MA^+ that originally ran along it and that has long been fused in its distal section with R_5 has already moved away from the fold together with R_5.

MP^- is the only representative of the media in the Diptera, as in most Neoptera, and it is also dichotomously forked into four branches. However, the fourth branch (M_4) is not free but is fused with the anterior branch of CuA^+ (CuA_1). This interpretation has also been challenged. Séguy considered that the media (MP^-) only has three branches and that the vein I have labelled M_4 in Fig. 112 is the anterior branch of CuA^+. This interpretation has been followed by authors such as Rodendorf who believed that the Diptera are very closely related to the Eumecoptera and that the anterior branch of CuA^+ has been reduced in the Eumecoptera.

These differences of opinion are of no importance for working out the phylogeny of the Diptera, where there is never more than one vein between M_3 and CuA^+. How this vein originated may well affect our assessment of the relationships between the Diptera and other groups but it can hardly influence our ideas about the history of the Diptera themselves.

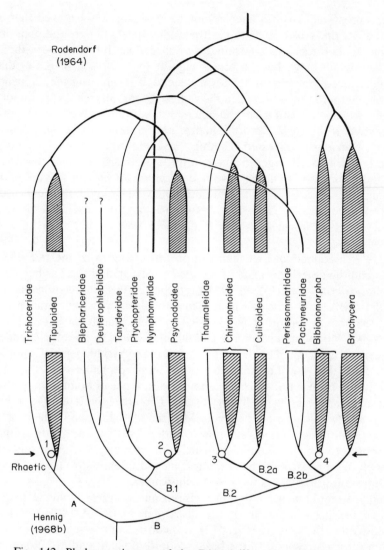

Fig. 142. Phylogenetic tree of the Diptera[435]. A comparison of the views of Rodendorf and Hennig. 1, *Architipula radiata* Rodendorf; 2, *Tanyderophryne*[436]; 3, Architendipedidae; 4, *Rhaetomyia, Protorhyphus, Protolbiogaster*, etc. (see p. 433). A, Tipulomorpha ('Polyneura'); B, 'Oligoneura'; B.1, Psychodomorpha; B.2, unnamed group; B.2a, Culicomorpha; B.2b, Bibionomorpha + Brachycera. From Hennig (1968b).

Our interpretation of the real or apparent cross-veins also depends partly on how we interpret the longitudinal veins. There are very few cross-veins in the *Diptera. There is no cross-vein between the two anal veins (1A and 2A) which are part of the ground-plan of the *Diptera. The costal and subcostal fields are also almost entirely without cross-veins: in the costal field, only the humeral cross-vein (h) has been retained. It is generally believed that Sc is forked at its tip. However, I think it possible that in the ground-plan of the Holometabola the subcosta terminates in the radius. If this were so, then Sc_1 would originally have been a cross-vein. In the *Diptera, it is Sc_2 that is usually reduced, but sometimes Sc_1 is instead. The subcosta then appears to terminate either in the costa or in the radius. These differences are important as they provide constitutive characters for certain monophyletic groups, irrespective of the correct morphological interpretation of Sc_1 and Sc_2.

I have previously suggested (Hennig, 1954, Fig. 5) that in the ground-plan of the Diptera Sc_2 diverged from R_1 after a short distance and reached the costa independently. In doing this I had *Permotipula patricia* Tillyard in mind (Hennig, 1954, Fig. 6). This interpretation may be wrong, but in any case the *Diptera never have a 'cross-vein' between R_1 and the costa, irrespective of whether it is interpreted as the free apical section of Sc_2 or as a true cross-vein.

There is a connection between the posterior branch of the radial sector (R_{4+5}) and the anterior branch of the media (M_{1+2}), and this has been called the anterior cross-vein (ta). In 1954 I interpreted this cross-vein as the basal section of MA^+, the rest of which is fused with the posterior branch of the radial sector. On the other hand, Séguy (1959) considered it to be a true cross-vein. Although this 'cross-vein' is intersected by the old channel of MA^+, which is occupied by the 'vena spuria' in the Syrphidae and Conopidae, I still think that my interpretation is the most probable one, and it has certainly not been refuted. This is another difference of opinion that is of no importance in discussing the history of the *Diptera. I think that the so-called 'posterior cross-vein' (tp) between M_2 (or $M_1 + M_2$) and M_3 is a true cross-vein, although Séguy did not consider it to be one. I think that it corresponds to the cross-vein m–m which, for example, is so important in the Trichoptera for separating the Annulipalpia and Integripalpia (see p. 408). In the Diptera it often closes the discal cell distally.

Finally, I should mention that there is a connection resembling a cross-vein between the posterior edge of the discal cell and CuA^+ (CuA_2). In the ground-plan of the Diptera (Fig. 112), this consists of two sections, and in my opinion these correspond to the basal sections of CuA_1 and M_4. Whether or not this is correct depends on how the vein between M_3 and CuA_2 is interpreted. As I mentioned above, opinions differ over the identity of this vein. As a result of changes in the venation of the *Diptera, the two sections of this cross-vein developed differently. Unfortunately, I was inconsistent in my use of the term 'basal cross-vein' (tb) in my earlier paper (Hennig, 1954), sometimes using it for the anterior section of this cross-vein and sometimes for the posterior one. In the discussion that follows I shall refer to these sections as tb_1 and tb_2, respectively.

In a number of papers, Rodendorf has described species from the Upper Triassic (Rhaetic) that can be reliably assigned to subgroups of the *Diptera, rather than simply to the *Diptera. *Pseudodiptera gallica* Laurentiaux &

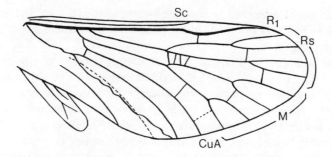

Fig. 143. *Pseudodiptera gallica* Laurentiaux & Grauvogel ('Paratrichoptera'), from the Lower Triassic (Bunter Sandstone) of the Vosges (France). From Rodendorf (1964).

Grauvogel[†] (Fig. 143) was described from the Lower Triassic (Bunter Sandstone). Martynova (in Rodendorf, 1962) assigned this species to the Paratrichoptera, which she regarded as the stem-group of the Diptera and Hymenoptera. However, everything that is known about the morphology of the recent species of these two orders indicates that there cannot have been a stem-group from which they alone have descended. The very reduced hind-wing could support the assignment of *Pseudodiptera* to the stem-group of the Diptera. However, all recent Diptera have CuP (Cu_2) very reduced, which is not the case in *Pseudodiptera*. In this character *Permotipula patricia* Tillyard, from the Upper Permian of Australia, appears to agree better with the *Diptera (Tillyard, 1929). However, *Permotipula* has two pairs of well developed wings. Other Mesozoic Paratrichoptera, such as *Pseudopolycentropus latipennis* Martynov, from the Upper Jurassic of Karatau, which is probably the most recent species, have the hind-wings rather reduced but still well developed, CuP developed normally as in *Pseudodiptera*, and the cross-veins between the anal veins still present.

I do not think that we should conclude from this that the stem-group of the Diptera, with these species as 'side-branches', survived into the Upper Jurassic alongside the *Diptera.

Two groups or 'suborders' are usually recognized in the *Diptera, the Nematocera and Brachycera. However, these are unlikely to be sister-groups. The Nematocera may well be a paraphyletic group[438], and so, like the Orthorrhapha of earlier authors, would have no place in a phylogenetic classification.

The Tipulomorpha, Psychodomorpha[439], Culicomorpha, Bibionomorpha, and Brachycera are well founded monophyletic groups. The most important problem in the phylogenetic classification of the Diptera concerns the relationship amongst these five groups.

I have recently dealt with this problem again (Hennig, 1968b), and to

[†]The generic name *Pseudodiptera* is preoccupied by *Pseudodiptera* Kaye, 1918 (Lepidoptera)[437].

facilitate future discussion I have made a comparison of my views with those published by Rodendorf (1964) in his book on the historical development of the Diptera (Fig. 142).

Nothing need be said about the six species from the Upper Triassic (Rhaetic) of Central Asia which Rodendorf assigned to his suborder 'Archidiptera': *Dictyodiptera multinervis* Rodendorf, *Paradictyodiptera trianalis* Rodendorf, *Dipterodictya tipuloides* Rodendorf, *Hyperpolyneura phryganeoides* Rodendorf, *Dyspolyneura longipennis* Rodendorf, and *Diplopolyneura mirabilis* Rodendorf. In my view they do not belong to the Diptera at all, or else Rodendorf's figures of the venation are incorrect. The wing shape and certain characters in the venation undoubtedly fit the Diptera. Other 'characters' in the venation rule out the possibility that they could be Diptera, and so I would like to believe that Rodendorf has interpreted cracks and fissures in the stone as veins, as he has certainly done with other fossils. These fossils are extremely small, ranging from barely 3 mm to just over 4 mm, and this will add to the difficulties of making a definitive interpretation of them, unless new palaeontological methods become available.

Rodendorf (1964) attempted to derive the problematic recent Nymphomyiidae (three species), and apparently the no less problematic Deuterophlebiidae too, from the Archidiptera and in particular from the 'Dictyodipteromorpha' (*Dictyodiptera, Paradictyodiptera, Dipterodictya*). He considered the Nymphomyiidae, if not the entire 'Dictyodipteromorpha', to be the sister-group of the rest of the *Diptera. I think that this is pure speculation, for which there is absolutely no evidence at present.

The Tipulomorpha (Trichoceridae, Limnobiidae, Cylindrotomidae, Tipulidae) are a monophyletic group that can easily be recognized by their wing venation. The radial sector always has only three free branches: it is generally agreed that the first of the four original free branches (R_2) terminates in R_1. In addition, the wing is long and narrow, and this has had most effect on the proximal half of the wing: as a result, many of the forks of the veins have been displaced towards the wing-tip, and so have the tips of 1A and (except in the Trichoceridae) 2A. The short discal cell is always in the distal half of the wing.

The oldest species with these characters is *Architipula radiata* Rodendorf, from the Upper Triassic (Rhaetic) of Issyk-Kul'. It is very striking that the venation of this species is not at all primitive. If my interpretation of the development of the venation in the Tipulomorpha is correct, then many recent species have a more primitive venation than *Architipula radiata*. The displacement of R_{2+3} towards R_4 described by Alexander has already taken place in *Architipula*, and R_2, the tip of which is fixed to R_1, has become recurved. This has also taken place in certain recent species. It is possible that the Upper Triassic *Architipula radiata* actually belongs to one of the subgroups of the *Tipulomorpha and that at least the sister-group relationship between the Trichoceridae and the rest of the Tipulomorpha, which is evidently well founded, had already arisen. This was also Rodendorf's view (1964: Fig. 82).

Handlirsch (1939) and Bode (1953) described many species from the Jurassic (post-Lias) which they assigned to the 'Architipulidae'. Nothing is known about the extent of intra-specific variation in the Jurassic, and so it is impossible to say how many of these fossils actually do represent distinct species. Some of them have extremely derived characters which are also found in recent species. It is very probable that the 'Architipulidae' are an invalid stem-group containing species that really belong to different sub-groups of the *Tipulomorpha. However, as long as there is no well founded phylogenetic classification of the *Tipulomorpha and no one has clearly worked out which characters of the venation can be reliably used for recognition of the various monophyletic groups, then this large number of fossils is of no particular value.

In the paper mentioned above (Hennig, 1968b), I suggested that there is probably a sister-group relationship between the Tipulomorpha ('Polyneura') and the rest of the Diptera (which can then be called the 'Oligoneura'), as earlier authors had already suggested. In developing this idea, I began with Rodendorf's (1959) views. He believed that in the ground-plan of the Diptera only the fore-wings were retained as organs of flight and that the strengthening of the anterior part of the fore-wing was associated with an intensification in the speed of the wing stroke. As CuP moved towards CuA, the posterior part of the wing was able to evade wing resistance passively. The posterior parts of the wing base became narrow, and this led to the loss of the 3rd anal vein, which was retained in the Mecoptera. Basically, the Tipulomorpha have remained at this stage of development, although in other respects they have acquired a number of derived characters (see above).

In the rest of the Diptera, which I am calling the 'Oligoneura', there is a deep indentation on the hind-margin of the wing which has created a clear division between the 'pre-arcular field' (Rodendorf's 'basiala') and the 'post-arcular field'. Correlated with this, the 2nd anal vein retained by the Tipulomorpha has been reduced.

In the Oligoneura there may be a sister-group relationship between the Psychodomorpha and the Culicomorpha + Bibionomorpha + Brachycera.

As a monophyletic group the Psychodomorpha[440] are not nearly as well founded as the Tipulomorpha. They have few constitutive characters (Hennig, 1968b), and none of these are in the wing venation. The venation of the Tanyderidae is more primitive than that of all the other Diptera, except in the reduction of the 2nd anal vein (see above): the radial sector has four branches which reach the wing margin independently; the discal cell is large and is situated at the middle of the wing; and the anal lobe is broad. It is surprising that such a primitive venation has not been found in any of the Mesozoic fossils.

None of the other families of the Psychodomorpha has a discal cell: this is certainly a derived character, but is evidently not a synapomorphy of these families. Males of the Tanyderidae and Ptychopteridae have a clasping organ on all six legs, consisting of modified bristles on the 4th and 5th tarsal segments, and this character suggests that these two families form a single monophyletic group (Ptychopteroidea). Rodendorf (1964) was wrong to show the Tipulomorpha (in my sense) and the Ptychopteridae as a single 'monophyletic' group in his phylogenetic tree. The Ptychopteridae have a very derived venation. Rodendorf (1964) treated the 'Eolimnobiidae', which are known from a single wing (*Eolimnobia geinitzi* Handlirsch) from the upper Lias of Mecklenburg, as the stem-group of the Ptychopteridae. This interpretation cannot be refuted, but neither is it supported by any evidence (Hennig, 1954).

There is some evidence, which I have discussed elsewhere (Hennig, 1968b), to indicate that the Blephariceridae and Deuterophlebiidae (the Blephariceroidea?) also belong to the Psychodomorpha. No fossils are known from these families.

The Psychodoidea form a third monophyletic group of the Psychodomorpha. They include a single family which Rodendorf (1964) divided into the Psychodidae, Nemopalpidae, and Phlebotomidae. The discal cell is absent as in the Ptychopteridae and 'Blephariceroidea'. The wing narrows uniformly towards the base so that the division between the pre- and post-arcular fields is completely or almost completely obliterated. Consequently, the wing is elongate–oval in outline, like a laurel leaf. Moreover, the origins and forks of some or all of the longitudinal veins have been displaced towards the wing base, unlike the Tipulomorpha where they have all moved towards the wing-tip. The Psychodidae also have a short subcosta, which is apparently correlated with this trend[441].

The relationships between the Blephariceroidea, Ptychopteroidea, and Psychodoidea are not clear. The Ptychopteroidea and Psychodoidea may be more closely related to each other than to the Blephariceroidea.

It is possible that the Nymphomyiidae also belong to the Psychodomorpha. However, I shall deal with this controversial family in my discussion of the Culicomorpha.

Rodendorf (1964) considered the 'Tanyderophryneidea' to be the stem-group of the Psychodidae (his Psychodidae + Nemopalpidae + Phlebotomidae). This group only includes a single species, *Tanyderophryne multinervis* Rodendorf, from the Middle Jurassic of Galkino (Karatau)[442]. Some of the details in Rodendorf's illustration of the wing of this species, such as the constriction of the anal lobe, undoubtedly point to the Psychodidae, and it would be difficult to show that it is related to any family other than the Psychodidae. On the other hand, characters such as the course of the media and its connections with the radius and the radial sector make me think that Rodendorf's drawing of the venation is not entirely accurate.

If this genus really does belong to the stem-group of the Psychodidae (or Psychodoidea), then we would have to conclude that at least the Blephariceroidea and Ptychopteroidea were also in existence as separate groups during the Rhaetic[442].

It is possible that the Culicomorpha, Bibionomorpha, and Brachycera, which I shall be discussing next, also form a single monophyletic group. However, this is no more than a tentative suggestion and as such should be regarded only as a stimulus for further detailed studies. The synapomorphic characters of these three groups may be the development of true lobe-like pulvilli and the alula, though this may only be weakly distinct. I have already put forward some ideas about the significance of the alula in the paper previously mentioned (Hennig, 1968b).

If the hypothesis that these three groups form a single monophyletic group can be substantiated, then I would make a further suggestion that there could be a sister-group relationship between the Culicomorpha and the Bibionomorpha + Brachycera.

The Culicomorpha[443] are undoubtedly a monophyletic group. Reductions in their venation are also evident, and once again a certain constriction of the wing is responsible for this. As in the Tipulomorpha and Psychodomorpha, the discal cell is absent, the result of reduction of the posterior cross-vein. Furthermore, in the ground-plan both media and radial sector have only three free branches. As far as the media is concerned, it is undoubtedly M_3 that is absent. On the other hand, there has been some dispute over the interpretation of the radial sector. A few years ago, Lindeberg (1964) put forward a new interpretation of the venation of the Culicomorpha which differed in several respects from my own (Hennig, 1954), but I am not convinced that he was correct. I think that Lindeberg has made the same mistake as Séguy in regarding each wing fold as the vestige of a longitudinal vein. Consequently, the interpretations that both Séguy and Lindeberg made of the veins actually present forced them to draw conclusions that only led to fresh confusion, by suggesting that individual veins were alternately preserved or reduced even within closely related groups. I think that they took far too little account of certain mechanical considerations: when the wings became enlarged, which is such a striking secondary development in the Simuliidae, or when the veins became concentrated towards the fore-margin of the wing, for which Russian authors coined the term 'costalization', the simultaneous movement of the tips of the veins towards the wing base produced empty areas on the wing surface. These still contained the folds in which the veins originally ran and, possibly for mechanical reasons, these became more strongly developed or even sclerotized.

I am in no way suggesting that my interpretations have never been wrong, but I think that the method used by Lindeberg and Séguy is too static and relies too heavily on formal comparisons. It takes too little account of the wings and their venation as part of a continuous historical process of functionally conditioned change. I think that the interpretation of the three branches of the radial sector in the ground-plan of the Culicomorpha is still an unresolved question. Lindeberg believed that R_4 had been reduced and that the simplest, most obvious interpretation of the other two branches was that they are R_2 and R_3. On the other hand, it is an established fact that in many Tipulomorpha, the Blephariceridae, and *Phlebotomus* (Psychodidae s.lat.)

the base of R_{2+3} has been displaced towards the wing-tip: this vein thus appears to originate from R_4. At the same time R_2 has been reduced. I have previously suggested, with some hesitation, that this has also taken place in the Culicomorpha. However, this suggestion cannot be proved, any more than Lindeberg's can.

The Culicomorpha probably contain two sister-groups, the Culicoidea and Chironomoidea[443], which have followed separate paths in their mode of life and larval morphology (Hennig, 1950).

The most characteristic feature of the venation of the Chironomoidea is the conformation of the 1st anal vein (1A), which does not reach the wing margin and which runs very close to the cubitus. In addition, the costa does not extend far beyond R_5 (in my sense); the first branch of the radial sector (either R_{2+3} or R_2: see above), which is still free in the ground-plan of the Culicomorpha, terminates in the radius (R_1). Lindeberg (1964) considered this connection to be a true, newly acquired cross-vein rather than the first branch of the radial sector.

Rodendorf (1964) assigned the Architendipedidae and Protendipedidae to the stem-group of the Chironomoidea. I cannot make much of the three genera comprising the 'Protendipedidae', from the 'Middle Jurassic' of Karatau: *Protendipes dasypterus* Rodendorf (an adult without any recognizable venation), *Eopodonomus nymphalis* Rodendorf, and *Pachyuronympha karatauensis* Rodendorf (both larvae). The 'Architendipedidae', from the Upper Triassic of Issyk-Kul', are more interesting (*Architendipes tshernovskyi* Rodendorf and *Palaeotendipes alexii* Rodendorf). On the basis of their narrow wings and venation, these species could hardly be assigned anywhere except for the Chironomidae. Nevertheless, Rodendorf's illustrations of the venation, like so many of his drawings, show certain peculiarities which are almost certainly due to his inability to distinguish cracks and fissures from elements of the venation. For example, the region of the radial sector is particularly unclear. However, it seems certain that M_3 is absent, and this would be consistent with the view that the Architendipedidae belong to the stem-group of the Culicomorpha. However, *Architendipes* has 1A running into Cu_{1b} shortly before the wing margin, and the Chironomoidea but not the Culicoidea could easily be derived from this forerunner.

I think that the Thaumaleidae also belong to the Chironomoidea. Rodendorf placed the Thaumaleidae and Perissommatidae in a single monophyletic group which he regarded as the sister-group of all the other 'Tipulomorpha': his 'Tipulomorpha' correspond to my Tipulomorpha + Psychodomorpha + Culicomorpha. I think that this view is untenable. The Thaumaleidae belong to the Chironomoidea and agree with them in larval morphology, though the presence of two pairs of spiracles is more primitive than all the other Chironomoidea, and in having the three derived venational characters listed above. On the other hand, the Perissommatidae probably belong to the Bibionomorpha (see p. 433). If the 'Architendipedidae' really do belong to the stem-group of the Chironomoidea, then their descendants include

the Thaumaleidae and all the other Chironomoidea, but not the Perissommatidae.

Rodendorf considered the 'Rhaetomyiidae' (*Rhaetomyia necopinata*
Rodendorf, from the Upper Triassic of Issyk-Kul') to be the common
stem-group of the Thaumaleidae and Perissommatidae. However, *Rhaetomyia* can only be associated with the Perissommatidae, but definitely not with
the Thaumaleidae. Like the recent Perissommatidae, *Rhaetomyia* probably
belongs to the Bibionomorpha.

A question that is still unresolved is when the *Chironomoidea separated
into subgroups. In fact, not even the relationships amongst the various
'families' of the Chironomoidea have been satisfactorily worked out yet. I
have previously suggested that there is a sister-group relationship between the
Thaumaleidae and the rest of the Chironomoidea. I still think that this is the
most probable suggestion, because it is the best founded. However, the
relationships between the Ceratopogonidae, Simuliidae, and Chironomidae
are still unclear. The Podonominae, a subfamily of the Chironomidae, are
definitely known from the Lower Cretaceous (Neocomian) of the Lebanon[444].

If there really is a sister-group relationship between the Chironomoidea
and Culicoidea, and if *Architendipes* does in fact belong to the stem-group of
the Chironomidae, then the Culicoidea must also have been in existence
during the Upper Triassic. Rodendorf (1964) thought that he could recognize
the 'Eopolyneuridea' as the stem-group of the Culicoidea: *Eopolyneura
tenuinervis* Rodendorf, *Pareopolyneura costalis* Rodendorf, and *Musidoromima crassinervis* Rodendorf, from the Upper Triassic of Issyk-Kul'. However, the venation of the Culicoidea contains only plesiomorphic characters
when compared with that of the Chironomoidea, and consequently it is
unlikely that the stem-group of the Culicoidea will ever be recognized by its
venation. Fossils where the venation agrees formally with that of the
Culicoidea could just as well belong to the stem-group of the Culicomorpha
(Culicoidea + Chironomoidea). Rodendorf stated that the Eopolyneuridea
have a media with four to six branches. However, only *Pareopolyneura* and
Musidoromima actually appear to have more than four branches, and in this
character they differ so strikingly from all known recent and fossil Diptera
that I have the gravest doubts whether they can in fact be Diptera at all. This
of course presupposes that Rodendorf's descriptions and drawings are
accurate. On the other hand, *Eopolyneura tenuinervis* does appear to belong
to the Diptera. The media has four branches. If it is assigned to the
Culicoidea, then it would show that M_3 must have been lost independently in
the Culicoidea and Chironomoidea. It would also provide evidence that my
tentative(!) interpretation of the radial sector in the Culicomorpha is correct,
and that Lindeberg's is not: *Eopolyneura* has the first branch of the radial
sector (R_{2+3}) unforked.

Such far reaching implications impress me with the need to be extremely
cautious in interpreting *Eopolyneura*. I think that it is impossible to decide
whether this species belongs to the stem-group of the Culicomorpha (or to

one of their subgroups) or to the Bibionomorpha (where it would be morphologically intermediate between the Anisopodiformia and Bibioniformia). Nor can I see why it could not belong to the stem-group of the Blephariceridae, if someone wanted to suggest this! *Eopolyneura tenuinervis* is one of those fossils from which formal 'derivations' of the most diverse recent groups can be made and which are therefore of almost no value for the study of phylogeny.

There are two sister-groups in the *Culicoidea, the Dixidea and Culicidea. Rodendorf believed that this sister-group relationship arose as early as the Jurassic. He considered the Eoptychopteridea to be the stem-group of the Culicidea. This group contains eight species in the genera *Eoptychoptera*, *Proptychoptera*, and *Palaeolimnobia*, from the Lias of Mecklenburg and Lower Saxony, and this figure includes the species described by Bode (1953) in addition to those I have previously listed (Hennig, 1964). Rodendorf also considered the Dixamimidae, containing the single species *Dixamima villosa* Rodendorf, from the Malm of Karatau, to be the stem-group of the Dixidea.

However, I still think that it is impossible to make any interpretation of the Eoptychopteridea, just as I did in 1954. The arguments that I have just applied to *Eopolyneura* can be used to exclude the Eoptychopteridea from the Culicidea, as their stem-group. In addition, some species (*Proptychoptera*) have retained a closed discal cell. If these species really do belong to the Culicidea, then both M_3 and the posterior cross-vein (which closes the discal cell distally) must have been reduced independently in the Culicoidea and Chironomoidea. Nothing in the fossils under discussion suggests that this could have happened.

I have recently discussed the interpretation of *Dixamima* (Hennig, 1966b) and concluded that whilst it might well belong to the stem-group of the Culicidea there is no evidence to show that it belongs to the Dixidae.

It is possible that the small family Nymphomyiidae, which only contains three species, also belongs to the Culicomorpha[445]. These flies are very small, and this has undoubtedly resulted in many reductions which make it extraordinarily difficult to work out their systematic position. I think that they can only be assigned to the Culicomorpha. Two characters exclude them from the Bibionomorpha: the costa extends round the entire wing, and the 9th sternite of the male postabdomen is not fused with the 'gonocoxites' (the basal segments of the gonopods) as appears to be the case in all the Bibionomorpha, or at least in the ground-plan of the Bibionomorpha.

Reductions in the male genitalia could provide evidence that the Nymphomyiidae belong to the Culicomorpha. The Culicomorpha appear to be the only group of the Diptera in which the sperm pump at the base of the ductus ejaculatorius is absent, i.e. has been reduced. Neither Tokunaga (1935) nor Ide (1965) mentioned whether the Nymphomyiidae have a sperm pump, though it is possible that they did not pay any particular attention to it because they were both unaware of its significance. The Nymphomyiidae have retained both ocelli, and this argues against any relationship with the

Culicomorpha: ocelli appear to be absent in the ground-plan of the Culico-morpha.

The wing shape of the Nymphomyiidae has several parallels in the Psychodomorpha. Ocelli are usually reduced in the Psychodomorpha too, but if the Blephariceridae really belong here then the presence of ocelli must be part of the ground-plan of the Psychodomorpha.

It is probable that the discovery of the larva will shed some valuable light on the relationships of the Nymphomyiidae. It is also important that some detailed studies should be made of certain characters in the adults: the male genitalia, to determine the fate of the 9th sternite and the sperm pump, the pretarsus, and the thorax. For example, it is still not certain whether the Nymphomyiidae have the meron of the mid coxa fused with the epimeron (an apomorphic ground-plan character of the Psychodomorpha!), or whether true pulvilli are present or not. The work of Tokunaga (1935) and Ide (1965) appears to show that lobe-like pulvilli, which may be a derived ground-plan character of the Culicomorpha + Bibionomorpha + Brachycera, are absent in the Nymphomyiidae; in their place are spinose pads like those present in the Tipulomorpha and Psychodomorpha as the precursors of true pulvilli. I think it most probable that the Nymphomyiidae belong to the Psychodomorpha, but further detailed morphological studies of the vital characters need to be made.

No one disputes that the Bibionomorpha are a monophyletic group, or that the Bibioniformia and Mycetophiliformia form a single monophyletic group. On the other hand, the position of several small families is still in dispute: Pachyneuridae (Okada, 1938), Perissommatidae (Colless, 1962), Axymyiidae, Anisopodidae, and Cramptonomyiidae. I have recently suggested (Hennig, 1968b) that there is a sister-group relationship between the Pachyneuridae (including the Axymyiidae) + Perissommatidae and all the other families of the Bibionomorpha. There are some important differences in the venation of these two groups which, circumstances permitting, could be of value for interpreting the fossils. In the first group (Pachyneuridae + Perissommatidae) the first branch of the radial sector is forked (R_2 and R_3), whereas in the second group it is not. On the other hand, the two posterior branches (R_4 and R_5) have been retained in the second group, at least in the ground-plan, whereas they have not been in the first group. The loss of one branch of the media, either the reduction of M_3 or the suppression of the bifurcation of M_{1+2}, must have taken place independently in each sister-group.

There is no space here to discuss the completely different assessment of the Pachyneuridae and Perissommatidae published by Rodendorf (see Fig. 142), nor the views of Tuomikoski (1961) and Tollet (1959) on the phylogenetic relationships of the various subgroups of the *Bibionomorpha.

Rhaetomyia necopinata Rodendorf, from the Upper Triassic of Issyk-Kul', is the only fossil from which the Pachyneuridae (including the Axymyiidae) and Perissommatidae could be derived. This wing appears to be more primitive than all the recent species in having a much larger discal cell and a three-branched media (apart from M_4). The radial sector appears to have three branches, exactly as in recent species: the anterior branch is forked (R_2 and R_3) but the posterior one is not (R_{4+5}?). This could be a synapomorphic agreement. Rodendorf too derived the Perissommatidae from *Rhaetomyia*, from which he also derived the Thaumaleidae (= Orphnephilidae) but not the Pachyneuridae. In this respect, his views are different from mine.

The problems that still exist in working out the relationships of the other families of the Bibionomorpha have considerable influence on how the fossils are interpreted. The recent species provide inescapable evidence of an astonishing amount of convergence in the venation, and comparable convergent developments must also have taken place in the fossils. As fossils are usually represented only by wings, we need to be extremely circumspect with our interpretations.

I propose using the name 'Bibionomorpha s.str.' for the other families of the Bibionomorpha, i.e. excluding the Pachyneuridae (Axymyiidae) and Perissommatidae. Of these, the Anisopodidae and Cramptonomyiidae have the most primitive venation: they are the only families to have retained a closed discal cell and a media with the complete number of branches. However, the recent Anisopodidae have only retained two branches of the radial sector, like almost all the other Bibionomorpha s.str. The Cramptonomyiidae (genera *Cramptonomyia* and *Haruka*) have these two veins connected by a 'cross-vein'. I think it possible that this 'cross-vein' is in fact the basal section of R_4. In this case, the apical section of this branch would be fused with R_{2+3}. Two species have been described from the Upper Triassic of Issyk-Kul' (*Protorhyphus turanicus* Rodendorf and *Oligophryne fungivoroides* Rodendorf) in which the radial sector still has three branches (R_{2+3}, R_4 and R_5). A further species with this character is known from the Lias (*Protorhyphus stigmaticus* Handlirsch). These three species could belong to a stem-group from which the recent Cramptonomyiidae have descended almost unchanged.

There is another Upper Triassic species (*Protolbiogaster rhaetica* Rodendorf), known only from a single wing, which hardly differs from the recent Anisopodidae and could well belong to the stem-group of this family.

All the other recent Bibionomorpha s.str., i.e. all the families except for the Cramptonomyiidae and Anisopodidae, have the discal cell open (posterior cross-vein reduced), one branch of the media reduced (M_3 reduced or M_{1+2} unforked?) and the radial sector with only two branches.

Rodendorf (1964) described an amazingly large number of fossils with these characters from the Upper Triassic of Issyk-Kul': 28 species in the genera *Rhaetofungivora*, *Rhaetofungivorella*, *Rhaetofungivorodes*, *Archipleciofungivora*, *Archihesperinus*, *Protallactoneura*, *Archipleciomima*, and *Palaeoplecia*.

It is impossible to believe that Rodendorf has illustrated the venation of all these species correctly. Such a task may not even be possible with the techniques currently available. However, if only a few of his drawings are accurate, they show that there must have been species in existence during the Upper Triassic which agree with very derived recent species of the Bibiono-morpha in having the radial sector reduced to two branches, or even to one.

Furthermore, Rodendorf's figures give the impression that another feature that is very characteristic of many recent Mycetophiliformia had not only begun to develop in some of the Upper Triassic species but had already progressed a fair way towards its modern state: the fork of CuA_2 and M_4 has been displaced towards the wing base, whilst the main stem of the media has become very pale and short, and the 'cross-vein' connecting M_4 and the media has become directed obliquely.

This suggests that the *Bibionomorpha s.str., and even the *Mycetophili-formia, had separated into a considerable number of subgroups by the Upper Triassic. Attempts to reach more precise conclusions have failed for two reasons: the vital characters have not been adequately described, probably because they are not visible in the fossils; and, more importantly, the relationships amongst the recent subgroups of the Mycetophiliformia have not been adequately worked out.

It is still not known whether the Bibionomorpha and Brachycera are a monophyletic group of higher rank. Brundin pointed out that the enlarge-ment of the 2nd pleurotergite could be a synapomorphy of these two groups (Hennig, 1968b). However, further studies of this problem are urgently required.

The Brachycera are undoubtedly a monophyletic group. The earliest fossils that can be assigned here are from the Lower Jurassic. However, the group must have been in existence during the Upper Triassic because all the possible sister-groups of the Brachycera have been reliably recorded from this epoch.

Unfortunately, very few changes took place in the venation when the *Brachycera first separated into subgroups. Significant derived characters in the venation are only found in comparatively subordinate groups, and as a result it is very difficult to interpret the oldest fossils.

Even the division of the Brachycera into monophyletic subgroups is still not settled. For example, it is not certain whether the families placed in the 'Homoeodactyla' (= Tabaniformia of Hennig, 1954) actually form a monophyletic group. They have three similar lobe-like appendages on the pretarsus. There is no doubt that the lateral pulvilli did not first arise in the Brachycera, and they are therefore a primitive character in the group. On the other hand, I am not sure how the median lobe should be interpreted. I have previously considered it to be a primitive character, but Ross (1965) thought that the 'lobe-like empodium' was a derived character of the Homoeodactyla. If he was correct, then this would be an indication that the group is probably monophyletic, as no other derived characters are known. However, there is no generally agreed answer to this. A careful comparative study of the pretarsus and its appendages needs to be made. The available studies, which were listed by Crampton (1942), are all defective, mainly because they take too little account of the presumed relationships amongst the families of the Diptera. As a result they do not clarify what has actually taken place within the individual monophyletic groups or what the recognizable changes in the pretarsus can contribute to the unresolved questions of brachyceran classification[446].

All the groups that could be considered as the possible sister-group of the Brachycera have been recorded from the Upper Triassic, and so we should infer that the Brachycera too must have been in existence at that time. No Triassic fossils have been found, but the Jurassic fossils can be assigned fairly confidently to various subgroups of the Homoeodactyla (see Hennig, 1967a).

It is possible, but still uncertain, that the Brachycera exclusive of the Homoeodactyla are a monophyletic group, which could be called the Heterodactyla. There are only two possible derived characters in this group: the bristle-like[447] 'empodium', and the reduction of the antennal flagellum to four segments. It is doubtful whether the bristle-like empodium really is a derived character: the Clinocerinae (Empididae) and possibly other groups have a lobe-like empodium, and I do not know how to interpret this. As far as I know, no Brachycera have an antennal flagellum consisting of more than four segments. The first segment (the '3rd antennal segment') is large, and the remaining three segments form the 'arista' or antennal style. Some Homoeodactyla also have a reduced number of flagellar segments: for example, the Acroceridae have only one segment. However, this reduction appears to have taken place independently in different families. Whether or not the reduction of the flagellum to four segments is a derived ground-plan character of my Heterodactyla still needs to be confirmed.

The number of antennal segments may be of importance for working out the systematic position of the Empidiformia (Empididae + Dolichopodidae).

In an earlier paper (Hennig, 1954), I suggested that the Empidiformia are the sister-group of the Cyclorrhapha, and Rodendorf's (1964) phylogenetic tree showed that he too shared this view. However, in their ground-plan (and in almost all their species) the Cyclorrhapha have four flagellar segments, which consist of a '3rd antennal segment' and a three-segmented 'arista'. The Empidiformia, on the other hand, have three flagellar segments, consisting of a '3rd antennal segment' and a two-segmented style or arista, as in the Asiliformia of my earlier paper (Hennig, 1954). The Empidiformia agree with the Cyclorrhapha in having M_3 absent and the anal cell closed and rather short. I cannot decide whether the antennal structure is a synapomorphic character of the Asiliformia and Empidiformia or whether the venational characters are synapomorphies of the Empidiformia and Cyclorrhapha. At present I am inclined to believe that the first of these suggestions is most likely to be correct. However, further detailed studies are urgently needed[448].

Usachev (1968) has recently described *Protempis antennata*, from the Upper Jurassic of Karatau. This could belong to the stem-group of the Empidiformia.

No derived ground-plan characters are known for the Asiliformia[449], and as a result it is difficult to interpret the older fossils. Only a few species have been described from the Mesozoic. *Protomphrale martynovi* Rodendorf, from the Upper Jurassic of Karatau, may in fact belong to the stem-group of the Scenopinidae (= Omphralidae), which is one of the more derived families of the Asiliformia. Rodendorf himself (1964) considered the position of *Palaeophora ancestrix* Rodendorf, also from the Upper Jurassic of Karatau, to be uncertain. Rodendorf (in Rodendorf, 1968) has also described some undoubted Nemestrinidae from the Upper Jurassic.

There can be no doubt that the Cyclorrhapha are a monophyletic group. However, like the other subgroups of the Brachycera, their derived ground-plan characters are not found in the venation.

The only Mesozoic fossils that can be assigned to the Cyclorrhapha[450] are from Canadian amber of the Upper Cretaceous. McAlpine and Martin (1966) described two species, *Sciadophora bostoni* and *Priophora canadambra*, which they assigned to the Sciadoceridae, in the sense of typological systematics. However, these species have characters which show clearly that they are very primitive Phoridae and belong to different layers of the stem-group of this family[451]. In fact, this was correctly shown by McAlpine and Martin in their phylogenetic tree (1966, Fig. 23). If this phylogenetic tree is correct, and there is little doubt of this, then all the families of the Phoroidea (Platypezidae, Ironomyiidae, Sciadoceridae, Phoridae) must have been in existence as separate groups during the Upper Cretaceous.

None of the other subgroups of the Cyclorrhapha[452] occur as fossils before the Tertiary. This is an extremely large group and cannot have arisen any later than the Jurassic, as the sister-group either of the Empidiformia or of the Asiliformia + Empidiformia. It would be a particularly fascinating task to investigate how far the principal radiation of this group has followed that of the angiosperms and the placental mammals during and after the Cretaceous. I hope that an intensive investigation of the inclusions in Upper and Lower Cretaceous fossil resins will contribute to a solution of this problem.

Revisionary notes

434. A recent monograph on the Diptera has been published by Hennig (1973). Griffiths (1972) has proposed a revised classification of the Cyclorrhapha, and Hennig himself has published a number of papers dealing with amber fossils and classification (Hennig, 1964–76).

In his revision of Volume 1 of *Die Larvenformen der Dipteren*, on which he was working at the time of his death, Hennig gave his final views on the phylogeny of the major subgroups of the Diptera and in particular on the classification of the Nematocera. This volume is being prepared for publication (Hennig, in preparation) and consequently any thorough revision of this chapter seems superfluous: reference should be made to this new book. [Dieter Schlee.]

435. Hennig's later phylogenetic tree (Hennig, 1973, Fig. 3) indicates a far greater number of unresolved questions than does the present Fig. 142. [Dieter Schlee.]

436. According to Rodendorf (1964), *Tanyderophryne* is from the Middle Jurassic, not from the Rhaetic (Triassic). [Willi Hennig.]

437. The name *Pseudodiptera* has been replaced by *Laurentiptera* Martynova and Willmann (in Willmann, 1978). [Rainer Willmann.]

438. An investigation needs to be made of the following characters to determine whether the Nematocera are monophyletic: the pretarsus, especially the origin of the pulvilli; the ventral closure of the head in the larva and the displacement of the prementum from the upper surface; the origin of the prothoracic spiracular horns in the larvae (a synapomorphy or the result of convergence). [Willi Hennig.]

439. Hennig (1973) has subsequently shown that the Psychodomorpha are unlikely to be a monophyletic group, because there is only a single character that will support this: the fusion of the meron of the mid-legs with the epimeron of the metathorax. The phylogenetic classification of the recent and fossil Psychodidae and their subgroups has been discussed by Hennig (1972b). [Dieter Schlee.]

440. See the preceding note. The Psychodomorpha = Blephariceridae + Deuterophlebiidae (= Blephariceroidea); Nymphomyiidae; Tanyderidae + Ptychopteridae (= Ptychopteroidea); Psychodidae. A new monograph of the Blephariceridae, including a Hennigian analysis of the phylogeny of the family, has been published by Zwick (1977). [Dieter Schlee.]

441. For further discussion, see Hennig (1972b). [Dieter Schlee.]

442. *Tanyderophryne* is a Middle Jurassic fossil (see note 436). [Willi Hennig.]

443. The Culicomorpha = Dixidae + Chaoboridae + Culicidae (= Culicoidea); Thaumaleidae + Simuliidae + Ceratopogonidae + Chironomidae (= Chironomoidea). [Dieter Schlee.]

444. Schlee (1975) argued that it is still not certain that the Podonominae are a monophyletic group. [Dieter Schlee.]
   According to Saether (1977), the female genitalia indicate that the Ceratopogonidae alone are the sister-group of the Simuliidae, and not the Ceratopogonidae + Chironomidae as was suggested by Hennig (1973). [Adrian Pont.]

445. Because of the pupal characters, Hennig (1973) has moved the Nymphomyiidae from the Culicomorpha and provisionally assigned them to the Psychodomorpha. [Dieter Schlee.]

446. The significance of two further characters (the absence of macrochaetae and the apical position of the posterior larval spiracles) is also unclear (Hennig, 1973). [Dieter Schlee.]

447. If the broad empodium is considered to be a synapomorphy of the Homoeodactyla (see two paragraphs above), then the presence of a bristle-like empodium cannot be used to prove that the 'Heterodactyla' are a monophyletic group. [Dieter Schlee.]

448. Hennig (1971b) discussed this problem again in some detail in his analysis of the Lower Cretaceous Empidiformia ('Microphorinae'), from Lebanese amber, and of the Cyclorrhapha fossils. See also Hennig (1970, 1972c). [Dieter Schlee.]

449. Hennig (1972c) referred to a groove in the 8th abdominal segment of the larva, which he considered to be an apomorphy of the Asiliformia. [Dieter Schlee.]

450. Hennig (1971b) described a member of the Cyclorrhapha from Lower Cretaceous Lebanese amber, but he was unable to assign it to any subgroup. [Dieter Schlee.]

451. The problem of the stem-group of the Sciadoceridae + Phoridae was discussed by Hennig (1971a). [Dieter Schlee.]

452. Phylogenetic investigations of the Cyclorrhapha have been published by Griffiths (1972), based mainly on the hypopygium, and by Hennig (1976), who also included a discussion of Griffiths' paper. For a list of Hennig's papers on acalyptrate fossils and phylogeny, see the references in this book (pp. 459–460) and the bibliography of Hennig's publications (Anon., 1978). The phylogeny of the Piophilidae (including the 'Neottiophilidae' and 'Thyreophoridae') has been dealt with by McAlpine (1977). [Dieter Schlee.]
McAlpine (1970) has described some Upper Cretaceous compression fossils, from the Edmonton Formation of Alberta, as *Cretaphormia fowleri*, which he believed to be puparia of the Calliphoridae. [Willi Hennig.]

CHAPTER FOUR

Concluding Remarks and Prospects for the Future

In the preceding chapter I have shown that many of the prerequisites for a really satisfactory account of insect phylogeny are still lacking. The most obvious shortcoming is the almost complete absence of fossils from before the Lower Carboniferous, when a considerable number of insect groups must have arisen according to my phylogenetic trees (Figs. 21, 34, 40, 64, and 95). However, an increased number of Lower Carboniferous and Devonian fossils would only be of value if the relationships of certain groups had already been worked out. Particularly important problems are whether the Palaeoptera (Ephemeroptera and Odonata) really are a monophyletic group and, perhaps even more so, what are the relationships of the Plecoptera. These are not at all clear at present[453].

Upper Carboniferous fossils have been found only in the northern hemisphere[454]. Some belong to problematic groups that appear to have become extinct subsequently, but the rest belong to the Palaeoptera and Paurometabola. No Paraneoptera or Holometabola have been recorded from the Carboniferous, although both groups must have arisen during this epoch. Even the occurrence of Plecoptera in the Carboniferous of the northern hemisphere is still doubtful[455].

One of the most serious obstacles in the way of interpreting the fossils is that it is usually only the wings that are preserved, and as often as not the constitutive characters of individual monophyletic groups are not present in the wings. It is important that in the future zoologists should make use of the greatest possible number of organs and characters to work out clearly the constitutive characters of all ranks of monophyletic groups amongst the recent insects. In many cases, everything necessary for this is already available in the morphological and systematic literature: it is just that it has not been sufficiently heeded. In this field the authors of comprehensive accounts in zoological textbooks have a great responsibility. So often these authors seem to regard their main task as the description of the 'typical structural plan' of each insect group. Deviations from this are mentioned only in passing, if at all. I think that it is just as important to work out the *ground-plan* of each group. For example, little can be made of the statement that 'pseudotracheae' are 'usually' present in the labella of the dipterous proboscis, but are sometimes absent. What is really needed is a statement of whether the pseudotracheae are part, or are not yet part, of the ground-plan of the dipterous proboscis (which they do actually appear to be).

It is not always possible to give such a direct answer to questions of this kind. In such cases I would expect to find a clear statement that this is an unresolved question that can only be answered by a fresh and very specific investigation. I regard this as one of the most important and urgent tasks for future textbooks.

By the nature of things the study of fossils is inevitably the study of wings, and so it is vitally important to work out which features in the wings will provide the constitutive characters of the various monophyletic groups. Nachtigall (1968) has recently given a popular and brilliant account of modern biophysical investigations of the insect wing, and I believe that his astonishing results have opened a most promising field of study. However, I must admit that for the phylogeneticist the investigations carried out so far have raised more questions than they have answered. Nevertheless, I am convinced that the results to be obtained from the interplay of functional–morphological, physiological, and biophysical studies will be of considerable value for phylogenetic research and the appraisal of fossils.

I have shown that it is important to distinguish clearly between the Lower Permian and the Upper Permian. There are numbers of monophyletic groups which are represented by several of their subgroups in the Upper Permian but which are only represented in the Lower Permian by fossils belonging to their stem-groups.

On the whole, I would say that the problems arising from the study of insect phylogeny during the Palaeozoic are concerned with the origin of the monophyletic groups that modern classifications usually treat as subclasses, superorders, and orders, and occasionally as suborders. The study of the problems that are still outstanding does not demand a detailed knowledge of the development of characters in each group. It is only necessary to recognize and analyse what were the most primitive character states at any particular time.

The situation is different in the Mesozoic. The largest 'orders' are represented by fossils that belong to very subordinate groups, sometimes even to those that are ranked as families in most classifications, and these are reliably known from the Jurassic and in many cases even from the Triassic. A very detailed knowledge of the individual orders is needed for the study of insect phylogeny during the Mesozoic, and such a task can only be undertaken by specialists. However, specialists usually only work on single 'families' or groups of families. There are very few who have a comprehensive knowledge of any of the large orders and of the unresolved phylogenetic–systematic questions involved. New and fundamental morphological investigations are often needed to deal with these questions and, because these take up a great deal of time, they are usually given as subjects for graduate theses, at least in Germany. Progress is therefore slow, simply because the outstanding problems and the methods for dealing with them are not recognized with sufficient clarity. Several of my illustrations (Figs. 115, 121, 122, 126, 128, and 142) show very clearly how widely the conclusions of different authors on the phylogenetic–systematic structure of a particular group can diverge, even when published at almost the same time.

These differences of opinion sometimes arise because there are not enough morphological facts available to allow a clear decision to be made. Very often, however, they result from a methodologically incorrect interpretation of the morphological results.

These problems can be overcome if we take each group which can be definitely recognized as monophyletic, which would mean most of the so-called 'orders', and first set ourselves the task of recognizing the oldest sister-group relationship in each. It would mean a great deal in the *Diptera, for example, if we knew whether the Tipulomorpha really do have a sister-group relationship with the rest of the Diptera; or, in the *Hymen-optera, whether there is a sister-group relationship between the Tenthre-dinoidea and the rest of the Hymenoptera. Using the available fossils it would be possible in most cases to work out the minimum age of these sister-group relationships. In order to make a more complete study of insect phylogeny during the Mesozoic and to extract as much information as possible from the available fossils, the sister-group relationships of much more subordinate categories also need to be worked out. The elucidation of these relationships should be our next task.

These goals need to be recognized and consciously followed. If this is not done, then the steady increase in the number of descriptions of Mesozoic insects will not shed any light on the problems of classification and phylogeny. On the contrary, it will only lead to an increasingly chaotic array of different opinions, of the kind that have already become disturbingly evident in the Auchenorrhyncha (p. 272) and Planipennia (p. 298).

Additional notes

453. See note 166 on p. 172. [Peter Zwick.]

454. See note 66 on p. 57. [Adrian Pont.]

455. See note 172 on p. 173. [Peter Zwick.]

Bibliography

Abdullah, M. (1974). 'World Entomophaga Abdullah, a new suborder of Coleoptera including Stepsiptera (Insecta)', *Zool. Beitr. (n. F.)*, **20**, 177–211.

Abdullah, M. (1975). 'The higher classification of the insect order Coleoptera including fossil records and a classified directory of the Coleopterists and Coleoptera collections of the world', *Zool. Beitr. (n. F.)*, **21**, 363–461.

Abel, O. (1922). *Lebensbilder aus der Tierwelt der Vorzeit*, Fischer, Jena.

Abul-Nasr, S. (1954). 'Origin and development of the reproductive system in insects', *Bull. Soc. Fouad 1. Ent.*, **38**, 33–45.

Achtelig, M. (1967). 'Ueber die Anatomie des Kopfes von *Raphidia flavipes* Stein und die Verwandtschaftsbeziehungen der Raphidiidae zu den Megaloptera', *Zool. Jb. (Anat.)*, **84**, 249–312.

Achtelig, M. (1975). 'Die Abdomenbasis der Neuropteroidea (Insecta, Holometabola). Eine vergleichend anatomische Untersuchung des Skeletts und der Muskulatur', *Zoomorphologie*, **82**, 201–242.

Achtelig, M. (1976). 'Indizien zur Monophylie der Raphidioptera und Megaloptera (Insecta, Holometabola)', *Verh. dt. zool. Ges.*, **1976**, 233.

Achtelig, M. (1977). 'Skelett und Muskulatur des Abdomens weiblicher Raphidioptera (Insecta, Neuropteroidea)', *Zool. Jb. (Anat.)*, **98**, 137–164.

Achtelig, M. (1978). 'Entwicklung und Morphologie der inneren und äusseren weiblichen Genitalorgane der Kamelhalsfliegen (Neuropteroidea: Raphidioptera)', *Entomologica germ.*, **4**, 140–163.

Achtelig, M., and Kristensen, N. P. (1973). 'A re-examination of the relationships of the Raphidioptera (Insecta)', *Z. zool. Syst. EvolForsch.*, **11**, 268–274.

Acker, T. S. (1960). 'The comparative morphology of the male terminalia of Neuroptera (Insecta)', *Microentomology*, **24**, 27–83.

Adams, P. A. (1958). 'The relationship of the Protoperlaria and the Endopterygota', *Psyche, Camb.*, **65**, 115–127.

Ahmad, M. (1950). 'The phylogeny of termite genera based on imago-worker mandibles', *Bull. Am. Mus. nat. Hist.*, **95**, 37–86.

Ali, H. A. (1969). 'The higher classification of the Carabidae and the significance of internal characters (Coleoptera)', *Bull. Soc. ent. Égypte*, **51** [1967], 211–231.

Ander, K. (1939a). 'Vergleichend-anatomische und phylogenetische Studien über die Ensifera (Saltatoria)', *Opusc. ent.*, **Suppl. 2**, 1–306.

Ander, K. (1939b). 'Systematische Einteilung und Phylogenie der Ensifera (Saltatoria) auf Grund von vergleichend-anatomischen Untersuchungen', *Proc. 7th Int. Congr. Ent.*, **2**, 621–627.

Anderson, H. M. (1974). 'A brief review of the flora of the Molteno "Formation" (Triassic), South Africa', *Palaeont. afr.*, **17**, 1–10.

Ando, R. (1973). 'Old oocytes and newly laid eggs of scorpion-flies and hanging-flies (Mecoptera: Panorpidae and Bittacidae)', *Sci. Rep. Tokyo Kyoiku Daig. (B)*, **15**, No. 230, 163–187.

Anonymous (1978). 'In memoriam: Willi Hennig (*20.4.1913, †5.11.1976). Verzeichnis seiner Veröffentlichungen', *Beitr. Ent.*, **28**, 169–177.

Arnold, J. W. (1963). 'A note on the pterostigma in insects', *Can. Ent.*, **95**, 13–16.

Arora, G. L. (1956). 'The relationship of the Symphyta (Hymenoptera) to other orders of insects on the basis of adult external morphology', *Res. Bull. Panjab Univ. (Zool.)*, **90**, 85–119.

446

Asahina, S. (1954). *A morphological study of a relic dragonfly* Epiophlebia superstes *Sélys (Odonata, Anisozygoptera)*, Jap. Soc. Prom. Sci., Tokyo.

Aspöck, H., and Aspöck, U. (1971). 'Raphidioptera (Kamelhalsfliegen)', *Handb. Zool., Berl.*, **4** (2), 2/25, 1–50, De Gruyter, Berlin and New York.

Atkins, M. D. (1958). 'On the phylogeny and biogeography of the family Cupedidae (Coleoptera)', *Can. Ent.*, **90**, 532–537.

Aubert, J. (1950). 'L'origine et l'évolution des insectes', *Bull. Soc. vaud. Sci. nat.*, **64**, 461–477.

Ax, P. (1977a). 'Willi Hennig 20.4.1913 bis 5.11.1976', *Verh. dt. zool. Ges.*, **1977**, 346–347.

Ax, P. (1977b). 'Professor Dr. Dr. h. c. Willi Hennig†', *Zoomorphologie*, **86**, 1–2.

Baccetti, B., Dallai, R., and Rosati, F. (1970). 'The spermatozoon of Arthropoda. VII. Plecoptera and Trichoptera', *J. Ultrastruct. Res.*, **31**, 212–228.

Bachmayer, F., and Vasicek, W. (1967). 'Insektenreste aus dem Perm von Zöbing bei Krems in Niederösterreich', *Annln naturh. Mus. Wien*, **71**, 13–18.

Baehr, M. (1975). 'Skelett und Muskulatur des Thorax von *Priacma serrata* Leconte (Coleoptera, Cupedidae)', *Z. Morph. Tiere*, **81**, 55–101.

Baehr, M. (1976). 'Das Prothorakalskelett von *Atractocerus* (Lymexylonidae) und seine Bedeutung für die Phylogenie der Coleopteren, besonders der Polyphagen (Insecta: Coleoptera)', *Zoomorphologie*, **85**, 39–58.

Baehr, M. (1979). 'Vergleichende Untersuchungen am Skelett und an der Coxalmuskulatur des Prothorax der Coleoptera. Ein Beitrag zur Klärung der phylogenetischen Beziehungen der Adephaga (Coleoptera, Insecta)', *Zoologica, Stuttg.*, **44**, Hft 130, 1–76.

Balachowsky, A. (1942). 'Essai sur la classification des Cochenilles', *Annls Éc. natn. Agric. Grignon (3)*, **3**, 34–48.

Barlet, J. (1972). 'Sur le thorax des certains Myxophaga Crowson', *Bull. Inst. r. Sci. nat. Belg. (Ent.)*, **48** (14), 1–6.

Barlet, J. (1974). 'À propos du thorax d'un Torridincolide (Coleoptera)', *Bull. Annls Soc. R. ent. Belg.*, **110**, 287–289.

Beardsley, J. W. (1969). 'A new fossil scale insect (Homoptera: Coccoidea) from Canadian amber', *Psyche, Camb.*, **76**, 270–279.

Beier, M. (1968a). 'Mantodea (Fangheuschrecken)', *Handb. Zool., Berl.*, **4** (2), 2/12, 1–47, De Gruyter, Berlin.

Beier, M. (1968b). 'Phasmida (Stab- oder Gespenstheuschrecken)', *Handb. Zool., Berl.*, **4** (2), 2/10, 1–56, De Gruyter, Berlin.

Beier, M. (1972). 'Saltatoria (Grillen und Heuschrecken)'. *Handb. Zool., Berl.*, **4** (2), 2/9, 1–217, De Gruyter, Berlin and New York.

Beier, M. (1974). 'Blattariae (Schaben)', *Handb. Zool., Berl.*, **4** (2), 2/13, 1–127, De Gruyter, Berlin and New York.

Bekker, E. G. (1958). 'On the problem of the origin and development of the wing in insects. V. A contribution to the knowledge of the ontogeny and phylogeny of the organs of flight in Orthoptera s. Saltatoria', *Ént. Obozr.*, **37**, 775–784. [In Russian; English translation in *Ent. Rev., Wash.*, **37**, 671–677.]

Bekker, E. G. (1966). *Theory of the morphological evolution of the insects*, edited by E. S. Smirnov, University of Moscow. [In Russian.]

Bekker-Migdisova, E. E. (1947). 'A transitional form between the Permian Prosbolidae and the recent Cicadidae', *Dokl. Akad. Nauk SSSR*, **55**, 445–448. [In Russian.]

Bekker-Migdisova, E. E. (1959). 'Some new species of the Sternorrhyncha from the Permian and Mesozoic of the USSR', *Mater. Osnov. Paleont.*, **3**, 109–116. [In Russian.]

Bekker-Migdisova, E. E. (1960a). 'Palaeozoic Homoptera from the USSR and problems in the phylogeny of the order', *Paleont. Zh.*, **1960** (3), 28–42. [In Russian.]

Bekker-Migdisova, E. E. (1960b). 'New Permian Homoptera from the European USSR', *Trudỹ paleont. Inst.*, **76**, 1–112. [In Russian.]

Bekker-Migdisova, E. E. (1961). 'Die Archescytinidae als vermutliche Vorfahren der Blattläuse', *Proc. 11th Int. Congr. Ent.*, **1**, 298–301.

Bekker-Migdisova, E. E. (1971). 'On the evolution of the Homoptera-Psylomorpha [*sic*]', *Proc. 13th Int. Congr. Ent.*, **1**, 231.

Bekker-Migdisova, E. E. (1973). 'The classification of the Psyllomorpha and their position in the order Homoptera', *Dokl. ezheg. Chten. N. A. Kholodkovskogo*, **24** [1971], 90–118. [In Russian.]

Bell, R. T. (1964). 'Does *Gehringia* belong to the Isochaeta? (Coleoptera, Carabidae)', *Coleopts Bull.*, **18**, 59–61.

Bell, R. T. (1966). '*Trachypachus* and the origin of the Hydradephaga (Coleoptera)', *Coleopts Bull.*, **20**, 107–112.

Bell, R. T. (1967). 'Coxal cavities and the classification of the Adephaga', *Ann. ent. Soc. Am.*, **60**, 101–107.

Benson, R. B. (1938). On the classification of sawflies (Hymenoptera, Symphyta)', *Trans. R. ent. Soc. Lond.*, **87**, 353–384.

Bils, W. (1976). 'Das Abdomenende weiblicher, terrestrisch lebender Adephaga (Coleoptera) und seine Bedeutung für die Phylogenie', *Zoomorphologie*, **84**, 113–193.

Birket-Smith, S. J. R. (1974). 'On the abdominal morphology of Thysanura (Archaeognatha and Thysanura s.str.)', *Entomologica scand.*, **Suppl. 6**, 1–67.

Bitsch, J. (1966). 'L'évolution des structures céphaliques chez les larves de Coléoptères', *Annls Soc. ent. Fr. (n.s.)*, **2**, 255–324.

Bitsch, J. (1974). 'Fonction et ultrastructures des vésicules exertiles de l'abdomen des Machilides', *Pedobiologia*, **14**, 142–143.

Bitsch, J., and Ramond, S. (1970). 'The prothoracic skeleton and musculature of *Embia ramburi* R.-K. (Emb., Embiidae). Comparison with the prothoracic structure of the other Polyneoptera and the Apterygota', *Zool. Jb. (Anat.)*, **87**, 63–93.

Bock, W. J., and Wahlert, G. von (1965). 'Adaptation and the form-function complex', *Evolution, Lancaster, Pa.*, **19**, 269–299.

Bode, A. (1953). 'Die Insektenfauna des ostniedersächsischen oberen Lias', *Palaeontographica*, **103** (A), 1–375.

Boettger, C. R. (1958). 'Die systematische Stellung der Apterygota', *Proc. 10th Int. Congr. Ent.*, **1**, 509–516.

Bolton, H. (1921–22). 'A monograph of the fossil insects of the British Coal Measures. Parts I + II', *Palaeontogr. Soc. [Monogr.]*, 1–156.

Bolton, H. (1925). 'Insects from the coal measures of Commentry', *Fossil insects*, **2**, 1–56, British Museum, London.

Bolton, H. (1930). 'Fossil insects of the South Wales coalfield', *Q. Jl geol. Soc. Lond.*, **86**, 9–49.

Bolton, H. (1934). 'New forms from the insect fauna of the British coal measures', *Q. Jl geol. Soc. Lond.*, **90**, 277–304.

Bonhag, P. F. (1958). 'Ovarian structure and vitellogenesis in insects', *A. Rev. Ent.*, **3**, 137–160.

Borkhsenius, N. S. (1958). 'On the evolution and phylogenic interrelations of Coccoidea (Insecta, Homoptera)', *Zool. Zh.*, **37**, 765–780. [In Russian.]

Börner, C. (1904). 'Zur Systematik der Hexapoden', *Zool. Anz.*, **27**, 511–533.

Börner, C. (1909). 'Neue Homologien zwischen Crustaceen und Hexapoden. Die Beissmandibel der Insekten und ihre phylogenetische Bedeutung. Archi- und Metapterygota', *Zool. Anz.*, **34**, 100–125.

Börner, C. (1919). 'Stammesgeschichte der Hautflügler', *Biol. Zbl.*, **39**, 145–186.

Börner, C. (1934). 'Ueber System und Stammesgeschichte der Schnabelkerfe', *Ent. Beih. Berl.-Dahlem*, **1**, 138–144.

448

Boudreaux, H. B. (1979). *Arthropod phylogeny with special reference to insects*, Wiley–Interscience, New York.

Bradbury, J. P., and Kirkland, D. W. (1967). 'Upper Jurassic Hemiptera from the Todilto Formation, northern New Mexico', *Progm a. Mtgs Geol. Soc. Am.*, **1966**, 24.

Bradley, J. C. (1947). 'The classification of Coleoptera', *Coleopts Bull.*, **1**, 75–84.

Bradley, J. C. (1958). 'The phylogeny of the Hymenoptera', *Proc. 10th Int. Congr. Ent.*, **1**, 265–269.

Branson, C. C. (1948). 'Bibliographic index of Permian invertebrates', *Mem. geol. Soc. Am.*, **26**, 1–1049. [Insecta: 895–984.]

Brinkman, R. (1954). *Abriss der Geologie begründet durch Emanuel Kayser*, 7th ed., vol. 2, *Historische Geologie*, Enke, Stuttgart.

Britton, E. B. (1966). 'On the larva of *Sphaerius* and the systematic position of the Sphaeriidae', *Aust. J. Zool.*, **14**, 1193–1198.

Brock, J. P. (1971). 'A contribution towards an understanding of the morphology and phylogeny of the ditrysian Lepidoptera', *J. nat. Hist.*, **5**, 29–102.

Brodskiy, A. K. (1970). 'Organisation of the flight system of *Ephemera vulgata* L. (Ephemeroptera)', *Ént. Obozr.*, **49**, 307–315. [In Russian, with English summary; English translation in *Ent. Rev., Wash.*, **49**, 184–188.]

Brodskiy, A. K. (1974). 'Evolution of the flight apparatus of Ephemeroptera', *Ént. Obozr.*, **53**, 291–303. [In Russian; English translation in *Ent. Rev., Wash.*, **53** (2), 35–43.]

Brown, R. W. (1941). 'The comb of a wasp nest from the Upper Cretaceous of Utah', *Am. J. Sci.*, **239**, 54–56.

Brown, R. W. (1957). 'Cockroach egg case from the Eocene of Wyoming', *J. Wash. Acad. Sci.*, **47**, 340–342.

Brues, C. F., Melander, A. L., and Carpenter, F. M. (1954). 'Classification of insects. Keys to the living and extinct families of insects, and to the living families of other terrestrial arthropods', *Bull. Mus. comp. Zool. Harv.*, **108**, 1–917.

Brundin, L. (1976). 'A Neocomian chironomid and Podonominae–Aphroteniinae (Diptera) in the light of phylogenetics and biogeography', *Zoologica Scr.*, **5**, 139–160.

Bubnoff, S. von (1941). *Einführung in die Erdgeschichte*, vol. 1, Borntraeger [?], Berlin.

Bubnoff, S. von (1949). *Einführung in die Erdgeschichte*, 2nd ed., vols. 1 and 2, Mitteldeutsche Druckerei und Verlagsanstalt, Halle.

Buchner, P. E. C. (1953). *Endosymbiose der Tiere mit pflanzlichen Mikroorganismen*, Birkhäuser, Basle and Stuttgart. [Revised English version, 1965, *Endosymbiosis of animals with plant microorganisms*, Interscience, New York.]

Bulanova-Zakhvatkina, E. M. (1974). 'A new genus of mite (Acariformes, Oribatei) from the Upper Cretaceous of Taymyr', *Paleont. Zh.*, **1974** (2), 141–144. [In Russian; English translation in *Paleont. J.*, **8**, 247–250.]

Burmeister, E. G. (1976). 'Der Ovipositor der Hydradephaga (Coleoptera) und seine phylogenetische Bedeutung unter besonderer Berücksichtigung der Dytiscidae', *Zoomorphologie*, **85**, 165–257.

Byers, G. W. (1954). 'Notes on North American Mecoptera', *Ann. ent. Soc. Am.*, **47**, 484–510.

Byers, G. W. (1965). 'Families and genera of Mecoptera'. *Proc. 12th Int. Congr. Ent.*, 123.

Byers, G. W. (1977). 'In memoriam Willi Hennig (1913–1976)', *J. Kans. ent. Soc.*, **50**, 272–274.

Cabrera, A. (1928). 'Un segundo ortóptero del Triásico argentino', *Eos, Madr.*, **4**, 371–373.

Callahan, P. S. (1975). 'Insect antennae with special reference to the mechanism of scent detection and the evolution of the sensilla', *Int. J. Insect Morph. Embryol.*, **4**, 381–430.

Calvert, P. P. (1929). 'The significance of odonate larvae for insect phylogeny', *Proc. 4th Int. Congr. Ent.*, **2**, 919–925.

Caroll, E. J. (1962). 'Mesozoic fossil insects from Koonwarra, South Gippsland, Victoria', *Aust. J. Sci.*, **25**, 264–265.

Carpenter, F. M. (1928). 'A new protodonatan from the Grand Canyon', *Psyche, Camb.*, **35**, 186–190.

Carpenter, F. M. (1930a). 'Um blattide permiano do Brasil', *Boln Serv. geol. mineral. Brasil*, **50**, 1–9.

Carpenter, F. M. (1930b). 'A review of our present knowledge of the geological history of insects', *Psyche, Camb.*, **37**, 15–34.

Carpenter, F. M. (1930c). 'The Lower Permian insects of Kansas. 1. Introduction and the order Mecoptera', *Bull. Mus. comp. Zool. Harv.*, **70**, 69–101.

Carpenter, F. M. (1930d). 'The Lower Permian insects of Kansas. Part 3. The Protohymenoptera', *Psyche, Camb.*, **37**, 343–374.

Carpenter, F. M. (1931a). 'The Lower Permian insects of Kansas. Part 2. The orders Paleodictyoptera, Protodonata and Odonata', *Am. J. Sci. (5)*, **21**, 97–139.

Carpenter, F. M. (1931b). 'The Lower Permian insects of Kansas. Part 4. The order Hemiptera and additions to the Paleodictyoptera and Protohymenoptera', *Am. J. Sci. (5)*, **22**, 113–130.

Carpenter, F. M. (1932a). 'Jurassic insects from Solenhofen in the Carnegie Museum and the Museum of Comparative Zoology', *Ann. Carneg. Mus.*, **21**, 97–129.

Carpenter, F. M. (1932b). 'The Lower Permian insects of Kansas. Part 5. Psocoptera and additions to the Homoptera', *Am. J. Sci. (5)*, **24**, 1–22.

Carpenter, F. M. (1933). 'The Lower Permian insects of Kansas. Part 6. Delopteridae, Protelytroptera, Plectoptera, and a new collection of Protodonata, Odonata, Megasecoptera, Homoptera, and Psocoptera', *Proc. Am. Acad. Arts Sci.*, **68**, 411–504.

Carpenter, F. M. (1934a). 'A new megasecopteron from the Carboniferous of Kansas', *Kans. Univ. Sci. Bull.*, **21**, 365–367.

Carpenter, F. M. (1934b). 'Carboniferous insects from Pennsylvania in the Carnegie Museum and the Museum of Comparative Zoology', *Ann. Carneg. Mus.*, **22**, 323–342.

Carpenter, F. M. (1935). 'The Lower Permian insects of Kansas. Part 7. The order Protoperlaria', *Proc. Am. Acad. Arts Sci.*, **70**, 101–146.

Carpenter, F. M. (1938). 'Two Carboniferous insects from the vicinity of Mazon Creek, Illinois', *Am. J. Sci. (5)*, **36**, 445–452.

Carpenter, F. M. (1939). 'The Lower Permian insects of Kansas. Part 8. Additional Megasecoptera, Protodonata, Odonata, Homoptera, Psocoptera, Protelytroptera, Plectoptera, and Protoperlaria', *Proc. Am. Acad. Arts Sci.*, **73**, 29–70.

Carpenter, F. M. (1943a). 'The Lower Permian insects of Kansas. Part 9. The orders Neuroptera, Raphidiodea, Caloneurodea, and Protorthoptera (Probnisidae), with additional Protodonata and Megasecoptera', *Proc. Am. Acad. Arts Sci.*, **75**, 55–84.

Carpenter, F. M. (1943b). 'Studies on Carboniferous insects from Commentry, France. Part I. Introduction and families Protagriidae, Meganeuridae, and Campylopteridae', *Bull. geol. Soc. Am.*, **54**, 527–554.

Carpenter, F. M. (1945). 'Lower Permian insects from Oklahoma. Part I. Introduction and the orders Megasecoptera, Protodonata, and Odonata', *Proc. Am. Acad. Arts Sci.*, **76**, 25–54.

Carpenter, F. M. (1947). 'Early insect life', *Psyche, Camb.*, **54**, 65–85.

Carpenter, F. M. (1948a). 'The supposed nymphs of the Palaeodictyoptera', *Psyche, Camb.*, **55**, 41–47.

Carpenter, F. M. (1948b). 'A Permian insect from Texas', *Psyche, Camb.*, **55**, 101–103.

Carpenter, F. M. (1950). 'The Lower Permian insects of Kansas. Part 10. The order Protorthoptera: the family Liomopteridae and its relatives', *Proc. Am. Acad. Arts Sci.*, **78**, 185–219.

Carpenter, F. M. (1951). 'Studies on Carboniferous insects from Commentry, France. Part II. The Megasecoptera', *J. Paleont.*, **25**, 336–355.

Carpenter, F. M. (1953). 'The geological history and evolution of insects', *Am. Scient.*, **41**, 256–270.

Carpenter, F. M. (1954a). 'The geological history and evolution of insects', *Rep. Smithson. Instn*, **1953**, 339–350. [Reprinted almost without change from Carpenter, 1953.]

Carpenter, F. M. (1954b). 'Fossil orders', in Brues, C. F., Melander, A. L., and Carpenter, F. M., 'Classification of insects. Keys to the living and extinct families of insects, and to the living families of other terrestrial arthropods', *Bull. Mus. comp. Zool. Harv.*, **108**, 777–827.

Carpenter, F. M. (1961a). 'A Triassic odonate from Argentina', *Psyche, Camb.*, **67**, 71–75.

Carpenter, F. M. (1961b). 'Studies on North American Carboniferous insects. 1. The Protodonata', *Psyche, Camb.*, **67**, 98–110.

Carpenter, F. M. (1962a). 'A Permian megasecopteron from Texas', *Psyche, Camb.*, **69**, 37–41.

Carpenter, F. M. (1962b). 'Studies on Carboniferous insects of Commentry, France. Part III. The Caloneurodea', *Psyche, Camb.*, **68**, 145–153.

Carpenter, F. M. (1963). 'Studies on North American Carboniferous insects. 2. The genus *Brodioptera* from the Maritime Provinces, Canada', *Psyche, Camb.*, **70**, 59–63.

Carpenter, F. M. (1964). 'Studies on North American Carboniferous insects. 3. A spilapterid from the vicinity of Mazon Creek, Illinois (Palaeodictyoptera)', *Psyche, Camb.*, **71**, 117–124.

Carpenter, F. M. (1965). 'Studies on North American Carboniferous insects. 4. The genera *Metropator, Eubleptus, Hapaloptera* and *Hadentomum*', *Psyche, Camb.*, **72**, 175–190.

Carpenter, F. M. (1966). 'The Lower Permian insects of Kansas. Part 11. The orders Protorthoptera and Orthoptera', *Psyche, Camb.*, **73**, 46–88.

Carpenter, F. M. (1967). 'Cretaceous insects from Labrador. 2. A new family of snake-flies (Neuroptera: Alloraphidiidae)', *Psyche, Camb.*, **74**, 270–275.

Carpenter, F. M. (1970). 'Fossil insects from Antarctica', *Psyche, Camb.*, **76**, 418–425.

Carpenter, F. M. (1972). 'The affinities of *Eomerope* and *Dinopanorpa* (Mecoptera)', *Psyche, Camb.*, **79**, 79–87.

Carpenter, F. M., Folsom, J. W., Essig, E. O., Kinsey, A. C., Brues, C. T., Boesel, M. W., and Ewing, H. E. (1937). 'Insects and arachnids from Canadian amber', *Univ. Toronto Stud. geol. Ser.*, **40**, 7–62.

Carpenter, F. M., and Kukalová, J. (1964). 'The structure of the Protelytroptera, with descriptions of a new genus from Permian strata of Moravia', *Psyche, Camb.*, **71**, 183–197.

Carpenter, F. M., and Miller, A. K. (1937). 'A Permian insect from Coahuila, Mexico', *Am. J. Sci. (5)*, **34**, 125–127.

Carpenter, F. M., and Richardson, E. S. (1972). 'Additional insects in Pennsylvanian concretions from Illinois', *Psyche, Camb.*, **78**, 267–295.

Carpentier, F., and Barlet, J. (1959). 'The first leg segments in the Crustacea Malacostraca and the insects', *Smithson. misc. Collns*, **137**, 99–115.

Carpentier, F., and Carpentier, M. (1949). 'Observations sur la morphologie des

Méganeurides (Insectes, Protodonates) du Stéphanien de Commentry, France', *C. r. 13e Congr. Int. Zool.*, 553–554.

Carpentier, F., and Lejeune-Carpentier, M. (1949). 'Conformation de l'abdomen d'un insecte Protodonate du Stéphanien de Commentry (Allier, France)', *Annls Soc. géol. Belg.*, **72**, B317–B326.

Casey, R., and Rawson, P. F. (1973). 'The boreal Lower Cretaceous', *Geol. J.*, Special Issue, **5**, 1–448, Seel House, Liverpool.

Chadwick, L. E. (1959). 'Spinasternal musculature in certain insect orders', *Smithson. misc. Collns*, **137**, 117–156.

Chen, S., and T'an, C. C. (1973). 'A new family of Coleoptera from the Lower Cretaceous of Kansu', *Acta ent. sin.*, **16**, 169–178. [In Chinese and English.]

Chen, S. H. (1946). 'Evolution of the insect larva', *Trans. R. ent. Soc. Lond.*, **97**, 381–404.

Chernova, O. A. (1961). 'On taxonomical position and geological age of the genus *Ephemeropsis* Eichwald (Ephemeroptera, Hexagenitidae)', *Ént. Obozr.*, **40**, 858–869. [In Russian, with English summary; English translation in *Ent. Rev., Wash.*, **40**, 485–493.]

Chernova, O. A. (1970). 'On the classification of the fossil and recent Ephemeroptera', *Ént. Obozr.*, **49**, 124–145. [In Russian, with English summary; English translation in *Ent. Rev., Wash.*, **49**, 71–81.]

Chernova, O. A. (1971). 'A may-fly from fossil resin of Cretaceous deposits in polar Siberia (Ephemeroptera, Leptophlebiidae)', *Ént. Obozr.*, **50**, 612–618. [In Russian, with English summary; English translation in *Ent. Rev., Wash.*, **50**, 346–349.]

China, W. E. (1933). 'A new family of Hemiptera–Heteroptera with notes on the phylogeny of the suborder', *Ann. Mag. nat. Hist.* (10), **12**, 180–196.

China, W. E. (1955). 'The evolution of the water-bugs', *Bull. natn. Inst. Sci. India*, **7**, 91–103.

China, W. E. (1962). 'South American Peloridiidae (Hemiptera–Homoptera: Coleorrhyncha)', *Trans. R. ent. Soc. Lond.*, **114**, 131–161.

Clay, T. R. (1970). 'The Amblycera (Phthiraptera: Insecta)', *Bull. Br. Mus. nat. Hist.* (Ent.), **25**, 75–98.

Cobben, R. H. (1965). 'Das aero-mikropylare System der Homoptereneier und Evolutionstrends bei Zikadeneiern (Homoptera, Auchenorrhyncha)', *Zool. Beitr.* (n. F.), **11**, 13–69.

Cobben, R. H. (1968). 'Evolutionary trends in Heteroptera. Part I. Eggs, architecture of the shell, gross embryology and eclosion', *Meded. Lab. ent., Wageningen*, **151**, 1–475.

Cobben, R. H. (1978). 'Evolutionary trends in Heteroptera. Part II. Mouthpart-structures and feeding strategies', *Meded. LandbHoogesch. Wageningen*, **289**, 1–407.

Cockerell, T. D. A. (1925). 'Fossils in the Ondai Sair formation, Mongolia', *Bull. Am. Mus. nat. Hist.*, **51**, 129–144.

Cockerell, T. D. A. (1927a). 'The Carboniferous insects of Maryland', *Ann. Mag. nat. Hist.* (9), **19**, 385–416.

Cockerell, T. D. A. (1927b). 'New light on the giant fossil may-fly of Mongolia', *Am. Mus. Novit.*, **244**, 1–4.

Colbert, E. H. (1953). 'Explosive evolution', *Evolution, Lancaster, Pa.*, **7**, 89–90.

Colless, D. H. (1962). 'A new Australian genus and family of Diptera (Nematocera: Perissommatidae)', *Aust. J. Zool.*, **10**, 519–535.

Common, I. F. B. (1973). 'A new family of Dacnonypha (Lepidoptera) based on three new species from southern Australia, with notes on the Agathiphagidae', *J. Aust. ent. Soc.*, **12**, 11–23.

Common, I. F. B. (1975). 'Evolution and classification of the Lepidoptera', *A. Rev. Ent.*, **20**, 183–203.

Condal, L. F. (1951). 'Nuevos hallazgos en el jurásico superior del Montsech', *Notas Comun. Inst. geol. min. Esp.*, **23**, 45–62.

Crampton, G. C. (1923). 'A comparison of the terminal abdominal structures of an adult female of the primitive termite *Mastotermes darwiniensis* with those of the roach *Periplaneta americana'*, *Bull. Brooklyn ent. Soc.*, **18**, 85–93.

Crampton, G. C. (1924). 'The phylogeny and classification of insects', *J. Ent. Zool.*, **16**, 33–47.

Crampton, G. C. (1930). 'The wings of the remarkable archaic mecopteron *Notiothauma reedi* McLachlan with remarks on their protoblattoid affinities', *Psyche, Camb.*, **37**, 83–103.

Crampton, G. C. (1931). 'A claim for priority in dividing pterygotan insects into two sections on the basis of the position of the wings in repose, with remarks on the relationships of the insect orders', *Ent. News*, **42**, 130–136.

Crampton, G. C. (1938). 'The interrelationships and lines of descent of living insects', *Psyche, Camb.*, **45**, 165–181.

Crampton, G. C. (1942). 'The external morphology of the Diptera, *in* Guide to the insects of Connecticut. VI. The Diptera or true flies of Connecticut', *Bull. Conn. St. geol. nat. Hist. Surv.*, **64**, 10–165.

Crowson, R. A. (1955). *The natural classification of the families of Coleoptera*, Lloyd, London. [2nd ed.: Crowson, 1968.]

Crowson, R. A. (1960). 'The phylogeny of Coleoptera', *A. Rev. Ent.*, **5**, 111–134.

Crowson, R. A. (1962). 'Observations on the beetle family Cupedidae, with descriptions of two new fossil forms and a key to the recent genera', *Ann. Mag. nat. Hist. (13)*, **5**, 147–157.

Crowson, R. A. (1968 [not 1967]). *The natural classification of the families of Coleoptera*, reprint [with addenda and corrigenda], Classey, Hampton.

Crowson, R. A. (1970). *Classification and biology*, Heinemann, London.

CSIRO (1970). *The insects of Australia*, Melbourne University Press, Melbourne.

CSIRO (1974). *The insects of Australia, Supplement 1974*, Melbourne University Press, Melbourne.

Dahl, E. (1969). 'The insects as arthropods', *Opusc. ent.*, **34**, 1–9.

Daiber, M. (1913). 'Myriapoda', in Lang, A. (ed.), *Handbuch der Morphologie der wirbellosen Tiere*, vol. 4, *Arthropoda*, pp. 373–414, Fischer, Jena.

Davey, K. G. (1960). 'The evolution of the spermatophores in insects', *Proc. R. ent. Soc. Lond.*, **35**, 107–113.

Davies, L. (1965). 'On spermatophores in Simuliidae', *Proc. R. ent. Soc. Lond. (A)*, **40**, 30–34.

Davis, C. (1940a). 'Family classification of the order Embioptera', *Ann. ent. Soc. Am.*, **33**, 677–682.

Davis, C. (1940b). 'Taxonomic notes on the order Embioptera. XX. The distribution and comparative morphology of the order Embioptera', *Proc. Linn. Soc. N.S.W.*, **65**, 533–542.

Davis, D. R. (1975). 'Systematics and zoogeography of the family Neopseustidae with the proposal of a new superfamily (Lepidoptera: Neopseustoidea)', *Smithson. Contr. Zool.*, **210**, 1–45.

Debaizieux, P. (1935). 'Organes scolopidiaux des pattes d'insectes. I. Lépidoptères et Trichoptères', *Cellule*, **44**, 271–314.

Dehm, R. (1963). 'Das Ursachen- und Zeitproblem in der Stammesgeschichte', *Naturw. Rdsch. Stutt.*, **16**, 127–134.

Delamare Deboutteville, C., and Massoud, Z. (1967). 'Un groupe panchronique: les collemboles. Essai critique sur *Rhyniella praecursor'*, *Annls Soc. ent. Fr. (n.s.)*, **3**, 625–630.

Delamare Deboutteville, C., and Massoud, Z. (1968). 'Révision de *Protentomobrya walkeri* Folsom, Collembole du Crétacé, et remarques sur sa position systématique', *Revue Ecol. Biol. Sol*, **5**, 619–630.

Demoulin, G. (1953). 'À propos d'*Hexagenites weyenberghi* Scudder, Ephéméroptère du Jurassique supérieur de Solenhofen', *Bull. Inst. r. Sci. nat. Belg.*, **29** (25), 1–8.

Demoulin, G. (1954a). 'Les Ephéméroptères jurassiques du Sinkiang', *Bull. Annls Soc. r. ent. Belg.*, **90**, 322–326.

Demoulin, G. (1954b). 'Quelques remarques sur les Archodonates', *Bull. Annls Soc. r. ent. Belg.*, **90**, 327–337.

Demoulin, G. (1954c). 'Essai sur quelques Ephéméroptères fossiles adultes', *Vol. jubil. Victor van Straelen*, **1**, 547–574, Brussels.

Demoulin, G. (1955a). 'Recherches sur les Ephémères du Jurassique bavarois', *Bull. Annls Soc. r. ent. Belg.*, **91**, 33.

Demoulin, G. (1955b). 'Contribution à l'étude morphologique, systématique et phylogénique des Ephéméroptères jurassiques d'Europe centrale. I', *Bull. Inst. r. Sci. nat. Belg.*, **31** (39), 1–14.

Demoulin, G. (1955c). 'Contribution à l'étude morphologique, systématique et phylogénique des Ephéméroptères jurassiques d'Europe centrale, II', *Bull. Inst. r. Sci. nat. Belg.*, **31** (55), 1–10.

Demoulin, G. (1955d). 'Contribution à l'étude des Ephéméroptères jurassiques d'Europe centrale. III. Phylogénie et zoogéographie', *Mém. Soc. r. ent. Belg.*, **27**, 176–183.

Demoulin, G. (1956a). 'Nouvelles recherches sur *Triplosoba pulchella* Brongniart (Insectes Ephéméroptères)', *Bull. Inst. r. Sci. nat. Belg.*, **32** (14), 1–8.

Demoulin, G. (1956b). 'Le mystère des *Ephemeropsis* (Ephéméroptères jurassicocrétacés d'Asie paléarctique)', *Bull. Inst. r. Sci. nat. Belg.*, **32** (53), 1–8.

Demoulin, G. (1958a). 'Nouvelles recherches sur *Patteiskya bouckaerti* Laurentiaux (insecte paléodictyoptère)', *Bull. Annls Soc. r. ent. Belg.*, **94**, 357–364.

Demoulin, G. (1958b). 'Nouveau schéma de classification des Archodonates et des Ephéméroptères', *Bull. Inst. r. Sci. nat. Belg.*, **34** (27), 1–19.

Demoulin, G. (1958c). 'Nouvelles observations sur l'aile de *Lithoptilus boulei* (Meunier) (insecte paléodictyoptère)', *Bull. Inst. r. Sci. nat. Belg.*, **34** (39), 1–5.

Demoulin, G. (1960). 'Quelques remarques sur un insecte fossile abracadabrant: *Lycocercus goldenbergi* (Brongniart 1885)', *Bull. Inst. r. Sci. nat. Belg.*, **36** (44), 1–4.

Demoulin, G. (1966). 'Remarques sur les Ephéméroptères Misthodotidae d'Europe et sur leurs rapports avec autres Ephéméroptères permiens', *Bull. Inst. r. Sci. nat. Belg.*, **42** (15), 1–5.

Demoulin, G. (1967a). 'Contribution à l'étude morphologique, systématique et phylogénique des Ephéméroptères jurassiques d'Europe centrale. IV. Hexagenitidae et Paedephemeridae', *Bull. Inst. r. Sci. nat. Belg.*, **43** (21), 1–9.

Demoulin, G. (1967b). 'Redescription de l'holotype ♂ imago de *Chromarcys magnifica* Navas et discussion des affinités phylétiques du genre *Chromarcys* Navas (Ephemeroptera, Chromarcyinae)', *Bull. Inst. r. Sci. nat. Belg.*, **43** (31), 1–10.

Demoulin, G. (1969a). 'Sur les rapports phylétiques des Aenigmephemeridae avec les autres familles des Siphlonuroidea (Ephemeroptera)', *Bull. Inst. r. Sci. nat. Belg.*, **45** (13), 1–5.

Demoulin, G. (1969b). 'Sur la position systématique et phylogénétique des Rallidentinae (Ephemeroptera)', *Bull. Inst. r. Sci. nat. Belg.*, **45** (15), 1–5.

Demoulin, G. (1969c). 'Remarques critiques sur la position systématique des Baetiscidae et des Prosopistomatidae (Ephemeroptera)', *Bull. Inst. r. Sci. nat. Belg.*, **45** (17), 1–8.

Demoulin, G. (1969d). 'Sur l'origine et les tendances évolutives des Baetidae et des Siphlaenigmatidae (Ephemeroptera)', *Bull. Inst. r. Sci. nat. Belg.*, **45** (18), 1–8.

454

Demoulin, G. (1969e). 'Quelques remarques sur certains Ephemeroptera triasiques et jurassiques', *Bull. Inst. r. Sci. nat. Belg.*, **45** (42), 1–10.

Demoulin, G. (1970a). 'Remarques critiques sur des larves "Ephemeromorphes" du Permien', *Bull. Inst. r. Sci. nat. Belg.*, **46** (3), 1–10.

Demoulin, G. (1970b). 'Contribution à l'étude morphologique, systématique et phylogénique des Ephéméroptères jurassiques d'Europe centrale. V. Hexageniti-dae = Paedephemeridae (syn.nov.)', *Bull. Inst. r. Sci. nat. Belg.*, **46** (4), 1–8.

Demoulin, G. (1971). 'Contribution à l'étude morphologique, systématique et phylogénétique des Ephéméroptères jurassiques. VI. L'aile postérieure des *Hexagenites* Scudder et les rapports Hexagenitidae–Chromarcyidae–Oligoneuriidae', *Bull. Inst. r. Sci. nat. Belg.*, **47** (29), 1–10.

Dietrich, H. G. (1975). 'Zur Entstehung und Erhaltung von Bernstein-Lagerstätten—1: Allgemeine Aspekte', *Neues Jb. Geol. Paläont. Abh.*, **149**, 39–72.

Dietrich, H. G. (1976). 'Zur Entstehung und Erhaltung von Bernstein-Lagerstätten—2: Bernstein-Lagerstätten im Libanon', *Neues Jb. Geol. Paläont. Abh.*, **152**, 222–279.

Dirsh, V. M. (1961). 'A preliminary revision of the families and subfamilies of Acridoidea', *Bull. Br. Mus. nat. Hist.* (Ent.), **10**, 350–419.

Dirsh, V. M. (1975). *Classification of the Acridomorphoid insects*, Classey, Faringdon.

Dodds, B. (1949). 'Mid-Triassic Blattoidea from the Mount Crosby insect bed', *Pap. Dep. Geol. Univ. Qd*, **3** (10), 1–11.

Dohle, W. (1965). 'Ueber die Stellung der Diplopoden im System', *Verh. dt. zool. Ges.*, **1964**, 597–606.

Dolin, V. G. (1975). 'A contribution to the systematics of Mesozoic click beetles (Coleoptera, Elateridae)', *Paleont. Zh.*, **1975** (4), 51–62. [In Russian; English translation in *Paleont. J.*, **9**, 474–486.]

Donovan, D. T. (1964). 'Cephalopod phylogeny and classification', *Biol. Rev.*, **39**, 259–287.

Dorf, E. (1967). 'Cretaceous insects from Labrador. 1. Geologic occurrence', *Psyche, Camb.*, **74**, 267–269.

Downes, J. A. (1958). 'The feeding habits of biting flies and their significance in classification', *A. Rev. Ent.*, **3**, 249–266.

Downes, J. A. (1968). 'Notes on the organs and processes of sperm-transfer in the lower Diptera', *Can. Ent.*, **100**, 608–617.

Drevermann, F. (1930). 'Permische Insekten mit erhaltener Farbe', *Natur. Mus., Frankf.*, **60**, 507–513.

Duffield, R. M., Blum, M. S., Wallace, J. B., Lloyd, H. A., and Regnier, F. E. (1977). 'Chemistry of the defensive secretion of the caddisfly *Pycnopsyche scabripennis* (Trichoptera: Limnephilidae)', *J. chem. Ecol.*, **3**, 649–656.

Dugdale, J. S. (1974). 'Female genital configuration in the classification of Lepidoptera', *N.Z. Jl Zool.*, **1**, 127–146.

Dumbleton, L. J. (1952). 'A new genus of seed-infesting micropterygid moths', *Pacif. Sci.*, **6**, 17–29.

Dunbar, C. O., and Tillyard, R. J. (1924). 'Kansas Permian insects. Part I. The geologic occurrence and the environment of the insects [by C. O. Dunbar]; with a description of a new palaeodictyopterid [by R. J. Tillyard]', *Am. J. Sci. (5)*, **7**, 171–209.

DuPorte, E. M. (1957). 'The comparative morphology of the insect head', *A. Rev. Ent.*, **2**, 55–70.

DuPorte, E. M. (1958a). 'The origin and evolution of the pupa', *Can. Ent.*, **90**, 436–439.

DuPorte, E. M. (1958b). 'Are there preoral structures in insects?', *Proc. 10th Int. Congr. Ent.*, **1**, 479–485.

DuPorte, E. M. (1960). 'Evolution of cranial structure in adult Coleoptera', *Can. J. Zool.*, **38**, 655–675.

DuPorte, E. M. (1962). 'The morphology of the insect mandible', *Can. J. Zool.*, **40**, 1229–1232.

Dupuis, C. (1979). 'Permanence et actualité de la systématique: la "systématique phylogénétique" de W. Hennig (historique, discussion, choix de références)', *Cah. Nat. (n.s.)*, **34** [1978], 1–69.

Edmunds, G. F. (1962). 'The principles applied in determining the hierarchic level in the higher categories of Ephemeroptera', *Syst. Zool.*, **11**, 22–31.

Edmunds, G. F. (1965). 'The classification of Ephemeroptera in relation to the evolutionary grade of nymphal and adult stages', *Proc. 12th Int. Congr. Ent.*, 112.

Edmunds, G. F. (1972). 'Biogeography and evolution of Ephemeroptera', *A. Rev. Ent.*, **17**, 21–42.

Edmunds, G. F., and Traver, J. R. (1954). 'The flight mechanics and evolution of the wings of Ephemeroptera, with notes on the archetype insect wing', *J. Wash. Acad. Sci.*, **44**, 390–399.

Edwards, J. G. (1953). 'The morphology of the male terminalia of beetles belonging to the genus *Priacma* (Cupesidae)', *Bull. Inst. r. Sci. nat. Belg.*, **29** (28), 1–8.

Ellenberger, F., Ellenberger, P., Laurentiaux, D., and Ricour, J. (1953). 'Note préliminaire sur la faune et un niveau insectifère des lentilles de grès et schistes noires des gypses de la Vanoise (Trias supérieur)', *Bull. Soc. géol. Fr.* (6), **2**, 269–274.

Emden, F. I. van (1934). 'Sind Polyphaga-Larven mit selbständigem Tarsus bekannt? (Col.)', *Stettin. ent. Ztg*, **95**, 61–64.

Emden, F. I. van (1957). 'The taxonomic significance of the characters of immature insects', *A. Rev. Ent.*, **2**, 91–106.

Emerson, A. E. (1961). 'Vestigial characters of termites and processes of regressive evolution', *Evolution, Lancaster, Pa.*, **15**, 115–131.

Emerson, A. E. (1965). 'A review of the Mastotermitidae (Isoptera), including a new fossil genus from Brazil', *Am. Mus. Novit.*, **2236**, 1–46.

Enderlein, G. (1909). 'Zur Kenntnis frühjurassischer Copeognathen und Coniopterygiden und über das Schicksal der Archipsylliden', *Zool. Anz.*, **34**, 770–776.

Enderlein, G. (1910a). 'Ueber die Phylogenie und die Klassifikation der Mecopteren unter Berücksichtigung der fossilen Formen', *Zool. Anz.*, **35**, 385–399.

Enderlein, G. (1910b). '*Panisopelma quadrigibbiceps*, eine neue Psyllidengattung aus Argentinien', *Zool. Anz.*, **36**, 280–281.

Enderlein, G. (1912). 'Embiidinen', *Collections Zoologiques du Baron Edm. de Sélys Longchamps*, **3**, 1–121. Hayez, Bruxelles.

Esaki, T. (1949). 'The occurrence of the Mesozoic insect *Chresmoda* in the Far East', *Insecta matsum.*, **17**, 4–5.

Evans, H. E. (1969). 'Three new Cretaceous aculeate wasps (Hymenoptera)', *Psyche, Camb.*, **76**, 251–261.

Evans, H. E. (1973). 'Cretaceous aculeate wasps from Taimyr, Siberia (Hymenoptera)', *Psyche, Camb.*, **80**, 166–178.

Evans, J. W. (1943). 'Upper Permian Homoptera from New South Wales', *Rec. Aust. Mus.*, **21**, 180–198.

Evans, J. W. (1948). 'Some observations on the classification of the Membracidae and the ancestry, phylogeny and distribution of the Jassoidea', *Trans. R. ent. Soc. Lond.*, **99**, 497–575.

Evans, J. W. (1950). 'A re-examination of an Upper Permian insect, *Paraknightia magnifica* Ev.', *Rec. Aust. Mus.*, **22**, 246–250.

Evans, J. W. (1956). 'Palaeozoic and Mesozoic Hemiptera', *Aust. J. Zool.*, **4**, 165–258.

456

Evans, J. W. (1957). 'Some aspects of the morphology and inter-relationships of extinct and recent Homoptera', *Trans. R. ent. Soc. Lond.*, **109**, 275–294.

Evans, J. W. (1958). 'Upper Permian Homoptera from the Belmont Beds', *Rec. Aust. Mus.*, **24**, 109–114.

Evans, J. W. (1963). 'The phylogeny of the Homoptera', *A. Rev. Ent.*, **8**, 77–94.

Evans, J. W. (1964). 'The periods of origin and diversification of the superfamilies of the Homoptera–Auchenorrhyncha as determined by a study of the wings of Palaeozoic and Mesozoic fossils', *Proc. Linn. Soc. Lond.*, **175**, 171–181.

Evans, J. W. (1971). 'Some Upper Triassic Hemiptera from Mount Crosby, Queensland', *Mem. Qd Mus.*, **16**, 145–151.

Evans, J. W. (1972). 'A new species of Peloridiidae (Homoptera, Coleorrhyncha) from north Queensland', *Proc. R. Soc. Qd*, **83** [1971], 83–88.

Ewing, H. E. (1928). 'The legs and leg-bearing segments of some primitive arthropod groups, with notes on leg-segmentation in the Arachnida', *Smithson. misc. Collns*, **80** (11), 1–41.

Fahlander, K. (1938). 'Beiträge zur Anatomie und systematischen Einteilung der Chilopoden', *Zool. Bidr. Upps.*, **17**, 1–148.

Fennah, R. G. (1961). 'The occurrence of a cixiine fulgoroid in the Weald Clay', *Ann. Mag. nat. Hist. (13)*, **4**, 161–163.

Fischer, F. C. J. (1965). 'Zwei Jahrhunderte Systematik der Köcherfliegen (Trichoptera)', *Proc. 12th Int. Congr. Ent.*, 124–125.

Fischer, M. (1972). 'Hymenoptera, Braconidae (Opiinae I)', *Tierreich*, **91**, 1–620.

Fleury, E. (1936). 'Sur quelques insectes du Stéphanien portugais', *C.r. 12e Congr. Int. Zool.*, 1453–1457.

Flower, J. W. (1964). 'On the origin of flight in insects', *J. Insect Physiol.*, **10**, 81–88.

Forbes, W. T. M. (1923). 'The Lepidoptera of New York and neighboring states', *Mem. Cornell Univ. agric. Exp. Stn*, **68**, 1–729.

Forbes, W. T. M. (1926). 'The wing folding patterns of the Coleoptera', *Jl N.Y. ent. Soc.*, **34**, 91–140.

Forbes, W. T. M. (1928). 'The Protocoleoptera', *Psyche, Camb.*, **35**, 32–35.

Forbes, W. T. M. (1943). 'Origin of wings and venational types in insects', *Am. Midl. Nat.*, **29**, 381–405.

Fossa-Manzini, E. (1941). 'Noticias sobre hallazgos de insectos fósiles en la América del Sur', *Notas Mus. La Plata*, **6**, 101–140.

François, J. (1969). 'Anatomie et morphologie céphalique des protoures (Insecta Apterygota)', *Mém. Mus. natn. Hist. nat. Paris (A)*, **59**, 1–144.

Fraser, F. C. (1954). 'The origin and descent of the order Odonata based on the evidence of persistent archaic characters', *Proc. R. ent. Soc. Lond. (B)*, **23**, 89–94.

Friese, G. (1970). 'Zur Phylogenie der älteren Teilgruppen der Lepidopteren', *Ber. Wanderversamm. dt. Ent.*, **10**, 203–222.

Fujiyama, I. (1973). 'Mesozoic insect fauna of East Asia. Part I. Introduction and Upper Triassic faunas'. *Bull. natn. Sci. Mus. Tokyo*, **16**, 331–386.

Fujiyama, I. (1974). 'A Liassic cockroach from Toyora, Japan', *Bull. natn. Sci. Mus. Tokyo*, **17**, 311–314.

Gagné, R. J. (1973). 'Cecidomyiidae from Mexican Tertiary amber (Diptera)', *Proc. ent. Soc. Wash.*, **75**, 169–171.

Gangwere, S. K. (1965). 'The phylogenetic development of food selection in Orthoptera (sens. lat.)', *Proc. 12th Int. Congr. Ent.*, 333–334.

Gardiner, B. G. (1961). 'New Rhaetic and Lias beetles', *Palaeontology*, **4**, 87–89.

George, V. P. (1969). 'Record of an apparently micro-fossil species (Insecta: Corrodentia) from the Cretaceous limestone clay, Seminar Hills, Nagpur (India)', *Bull. Ent. ent. Soc. India*, **10**, 1–3.

Ghauri, M. S. K. (1962). *The morphology and taxonomy of male scale insects (Homoptera: Coccoidea)*, British Museum (Natural History), London.

Gibbs, D. G. (1966). 'Goblet cells in the midgut epithelium of some trichopterous larvae', *Proc. R. ent. Soc. Lond. (A)*, **40**, 81–82.

Giles, E. T. (1963). 'The comparative external morphology and the affinities of the Dermaptera', *Trans. R. ent. Soc. Lond.*, **115**, 95–164.

Giles, E. T. (1974). 'The relationship between the Hemimerina and the other Dermaptera: a case for reinstating the Hemimerina within the Dermaptera, based upon a numerical procedure', *Trans. R. ent. Soc. Lond.*, **126**, 189–206.

Gilyarov, M. S. (1957). 'Evolution of the postembryonic development and the types of insect larvae', *Zool. Zh.*, **36**, 1683–1697. [In Russian, with English summary.]

Gilyarov, M. S. (1960). 'Evolution of the insemination type in insects as the result of the transition from aquatic to terrestrial life in the course of phylogenesis', in Hrdý, I. (ed.), *The ontogeny of insects. Acta symposii de evolutione insectorum Praha 1959*, pp. 50–55, Czechoslovak Academy of Sciences, Prague.

Gilyarov, M. S. (1969). 'Trends in the evolution of passively dispersing insects and the feed-back control in phylogenesis', *Z. zool. Syst. EvolForsch.*, **7**, 1–18.

Goel, S. C., and Schaefer, C. W. (1970). 'The structure of the pulvillus and its taxonomic value in the land Heteroptera (Hemiptera)', *Ann. ent. Soc. Am.*, **63**, 307–313.

Gouin, F. (1968). 'Morphologie, Histologie und Entwicklungsgeschichte der Insekten und der Myriapoden. IV. Die Strukturen des Kopfes', *Fortschr. Zool.*, **19**, 194–282.

Grabau, A. W. (1923). 'Cretaceous fossils from Shantung', *Bull. geol. Surv. China*, **5** (2), 164–181.

Grandi, M. (1947a). 'Contributi allo studio degli "Efemeroidei" italiani. VIII. Gli scleriti ascellari (pseudopteralia) degli efemeroidei, loro morfologia e miologia comparate', *Boll. Ist. Ent. Univ. Bologna*, **16**, 85–114.

Grandi, M. (1947b). 'Gli scleriti ascellari degli odonati, loro morfologia e miologia comparate', *Boll. Ist. Ent. Univ. Bologna*, **16**, 254–278.

Grassé, P. P. (ed.) (1949). *Traité de Zoologie. Anatomie, systématique, biologie*, vol. 9, *Insectes. Paléontologie, géonémie, aptérygotes, ephéméroptères, odonatoptères, blattoptéroïdes, orthoptéroïdes, dermaptéroïdes, coléoptères*, Masson, Paris.

Grassé, P. P. (ed.) (1951). *Traité de Zoologie. Anatomie, systématique, biologie*, vol. 10, *Insectes supérieurs et Hémiptéroïdes*, Masson, Paris.

Grassé, P. P. (ed.) (1973). *Traité de Zoologie. Anatomie, systématique, biologie*, vol. 8, Part 1, *Insectes. Tête, aile, vol*, Masson, Paris.

Grassé, P. P., and Noirot, C. (1959). 'L'évolution de la symbiose chez les Isoptères', *Experientia*, **15**, 365–372.

Grauvogel, L., and Laurentiaux, D. (1952). 'Un Protodonate du Trias des Vosges', *Annls. Paléont.*, **38**, 121–129.

Grell, K. G. (1942). 'Der Genitalapparat von *Panorpa communis* L.', *Zool. Jb. (Anat.)*, **67**, 513–588.

Griffiths, G. C. D. (1972). 'The phylogenetic classification of Diptera Cyclorrhapha with special reference to the structure of the male postabdomen', *Series Ent.*, **8**, 1–340.

Gross, W. (1964). 'Polyphyletische Stämme im System der Wirbeltiere', *Zool. Anz.*, **173**, 1–22.

Günther, K. (1962). 'Systematik und Stammesgeschichte der Tiere', *Fortschr. Zool.*, **14**, 268–547.

Günther, K., and Herter, K. (1974). 'Dermaptera (Ohrwürmer)', *Handb. Zool., Berl.*, **4** (2), 2/11, 1–158, De Gruyter, Berlin and New York.

Günther, K. K. (1974). 'Stabläuse, Psocoptera', *Tierwelt Dtl.*, **61**, 1–314.

Gupta, A. P. (1963a). 'A consideration of the systematic position of the Saldidae and the Mesoveliidae', *Proc. ent. Soc. Wash.*, **65**, 31–38.

Gupta, A. P. (1963b). 'Comparative morphology of the Saldidae and Mesoveliidae', *Tijdschr. Ent.*, **106**, 169–196.

458

Gupta, A. P. (ed.) (1979). *Arthropod phylogeny*, Van Nostrand Reinhold, New York and London.

Guthörl, P. (1934). 'Die Arthropoden aus dem Carbon und Perm des Saar–Nahe–Pfalz-Gebietes', *Abh. preuss. geol. Landesanst. (n. F.)*, **164**, 1–219. [Insects: 48–178.]

Guthörl, P. (1963). 'Zur Arthropoden-Fauna des Karbons und Perms. 17. *Saaromioptera jordani* n.g. n.sp. (Ins., Miomoptera) aus dem Stefan A des Saarkarbons', *Mitt. bayer. St. Paläont. Hist. Geol.*, **3**, 21–26.

Guthörl, P. (1965). 'Zur Arthropoden-Fauna des Karbons und Perms. 21. *Protereisma rossenrayensis* n.sp., ein Ephemeropteren-Fund (Insecta) aus dem niederrheinischen Zechstein', *Palaeont. Z.*, **39**, 229–233.

Hamilton, K. G. A. (1971a). 'The insect wing, part I. Origin and development of wings from notal lobes', *J. Kans. ent. Soc.*, **44**, 421–433.

Hamilton, K. G. A. (1971b). 'A remarkable fossil homopteran from Canadian Cretaceous amber representing a new family', *Can. Ent.*, **103**, 943–946.

Hamilton, K. G. A. (1972a). 'The insect wing, part II. Vein homology and the archetypal insect wing', *J. Kans. ent. Soc.*, **45**, 54–58.

Hamilton, K. G. A. (1972b). 'The insect wing, part III. Venation of the orders', *J. Kans. ent. Soc.*, **45**, 145–162.

Hamilton, K. G. A. (1972c). 'The insect wing, part IV. Venational trends and phylogeny of the winged orders', *J. Kans. ent. Soc.*, **45**, 295–308.

Handlirsch, A. (1906–08). *Die fossilen Insekten und die Phylogenie der rezenten Formen. Ein Handbuch für Paläontologen und Zoologen*, Engelmann, Leipzig.

Handlirsch, A. (1919). 'Revision der palaeozoischen Insekten', *Denkschr. Akad. Wiss. Wien*, **96**, 511–592.

Handlirsch, A. (1920–21). 'Palaeontologie', in Schröder, C. (ed.), *Handb. Ent.*, vol. 3, pp. 117–306, Fischer, Jena.

Handlirsch, A. (1922). 'Insecta palaeozoica', *Fossilium Cat. (I)*, **16**, 1–230, Junk, Berlin.

Handlirsch, A. (1937). 'Neue Untersuchungen über die fossilen Insekten. I', *Ann. naturhist. Mus. Wien*, **48**, 1–140.

Handlirsch, A. (1939). 'Neue Untersuchungen über die fossilen Insekten. II', *Ann. naturhist. Mus. Wien*, **49**, 1–240.

Handschin, E. (1958). 'Die systematische Stellung der Collembolen', *Proc. 10th Int. Congr. Ent.*, **1**, 499–508.

Haub, F. (1973). 'Das Cibarium der Mallophagen. Untersuchungen zur morphologischen Differenzierung', *Zool. Jb. (Anat.)*, **90**, 483–525.

Haudour, J., *et al.* (1960). [This paper was cited by Hennig in the original text, but it has not been possible to locate it.]

Haughton, S. H. (1924). 'The fauna and stratigraphy of the Stormberg series in South and Central Africa', *Ann. S. Afr. Mus.*, **12**, 323–497.

Haupt, H. (1935). 'Unterordnung Gleichflügler, Homoptera', *Tierwelt Mitteleur.*, **4** (3), 115–262.

Haupt, H. (1940). 'Die ältesten geflügelten Insekten und ihre Beziehungen zur Fauna der Jetztzeit', *Z. Naturw.*, **94**, 60–121.

Haupt, H. (1944). 'Die Beziehungen der permo-carbonischen zur rezenten Insektenwelt und die sich daraus ergebenden Lehren', *Nova Acta Acad. Caesar. Leop. Carol. (n. F.)*, **13**, 463–472.

Haupt, H. (1949). 'Rekonstruktionen permokarbonischer Insekten', *Beitr. tax. Zool.*, **1**, 23–43.

Haupt, H. (1951). '*Eugereon freygangi* n.sp.', *Hallesches Jb. mitteldt. Erdgesch.*, **1**, 182–183.

Haupt, H. (1952). 'Insektenfunde aus den Goldlauterer Schichten des Thüringer Waldes', *Hallesches Jb. mitteldt. Erdgesch.*, **1**, 241–258.

Heie, O. E. (1967). 'Studies on fossil aphids (Homoptera: Aphidoidea)', *Spolia zool. Mus. haun.*, **26**, 1–274.

Heming, B. S. (1971). 'Functional morphology of the thysanopteran pretarsus', *Can. J. Zool.*, **49**, 91–108.

Hennig, W. (1948). *Die Larvenformen der Dipteren. Eine Uebersicht über die bisher bekannten Jugendstadien der zweiflügeligen Insekten*, vol. 1, Akademie-Verlag, Berlin.

Hennig, W. (1950): *Die Larvenformen der Dipteren. Eine Uebersicht über die bisher bekannten Jugendstadien der zweiflügeligen Insekten*, vol. 2, Akademie-Verlag, Berlin.

Hennig, W. (1953). 'Kritische Bemerkungen zum phylogenetischen System der Insekten', *Beitr. Ent.*, **3**, Sonderheft, 1–85.

Hennig, W. (1954). 'Flügelgeäder und System der Dipteren unter Berücksichtigung der aus dem Mesozoikum beschriebenen Fossilien', *Beitr. Ent.*, **4**, 245–388.

Hennig, W. (1964). 'Die Dipteren-Familie Sciadoceridae im Baltischen Bernstein (Diptera: Cyclorrhapha Aschiza)', *Stuttg. Beitr. Naturk.*, **127**, 1–10.

Hennig, W. (1965a). 'Die Acalyptratae des Baltischen Bernsteins und ihre Bedeutung für die Erforschung der phylogenetischen Entwicklung dieser Dipteren-Gruppe', *Stuttg. Beitr. Naturk.*, **145**, 1–215.

Hennig, W. (1965b). 'Phylogenetic systematics', *A. Rev. Ent.*, **10**, 97–116.

Hennig, W. (1966a). '*Fannia scalaris* Fabricius, eine rezente Art im Baltischen Bernstein? (Diptera: Muscidae)', *Stuttg. Beitr. Naturk.*, **150**, 1–12.

Hennig, W. (1966b). 'Dixidae aus dem Baltischen Bernstein, mit Bemerkungen über einige andere fossile Arten aus der Gruppe Culicoidea (Diptera Nematocera)', *Stuttg. Beitr. Naturk.*, **153**, 1–16.

Hennig, W. (1966c). 'Conopidae im Baltischen Bernstein (Diptera: Cyclorrhapha)', *Stuttg. Beitr. Naturk.*, **154**, 1–24.

Hennig, W. (1966d). 'Einige Bemerkungen über die Typen der von Giebel 1862 angeblich aus dem Bernstein beschriebenen Insektenarten', *Stuttg. Beitr. Naturk.*, **162**, 1–7.

Hennig, W. (1966e). 'Spinnenparasiten der Familie Acroceridae im Baltischen Bernstein', *Stuttg. Beitr. Naturk.*, **165**, 1–21.

Hennig, W. (1966f). 'Bombyliidae im Kopal und im Baltischen Bernstein (Diptera: Brachycera)', *Stuttg. Beitr. Naturk.*, **166**, 1–20.

Hennig, W. (1966g). *Phylogenetic systematics*, translated by D. D. Davis and R. Zangerl, University of Illinois Press, Urbana.

Hennig, W. (1967a). 'Die sogenannten "niederen Brachycera" im Baltischen Bernstein (Diptera: Fam. Xylophagidae, Xylomyidae, Rhagionidae, Tabanidae)', *Stuttg. Beitr. Naturk.*, **174**, 1–51.

Hennig, W. (1967b). 'Neue Acalyptratae aus dem Baltischen Bernstein (Diptera: Cyclorrhapha)', *Stuttg. Beitr. Naturk.*, **175**, 1–27.

Hennig, W. (1967c). 'Therevidae aus dem Baltischen Bernstein mit einigen Bemerkungen über Asilidae und Bombyliidae (Diptera Brachycera)', *Stuttg. Beitr. Naturk.*, **176**, 1–14.

Hennig, W. (1968a). 'Ein weiterer Vertreter der Familie Acroceridae im Baltischen Bernstein (Diptera: Brachycera)', *Stuttg. Beitr. Naturk.*, **185**, 1–6.

Hennig, W. (1968b). 'Kritische Bemerkungen über den Bau der Flügelwurzel bei den Dipteren und die Frage nach der Monophylie der Nematocera', *Stuttg. Beitr. Naturk.*, **193**, 1–23.

Hennig, W. (1969a). 'Neue Uebersicht über die aus dem Baltischen Bernstein bekannten Acalyptratae', *Stuttg. Beitr. Naturk.*, **209**, 1–42.

Hennig, W. (1969b). 'Bernsteinfossilien', *Naturwissenschaft Med.*, **6**, 10–24.

Hennig, W. (1969c). 'Kritische Betrachtungen über die phylogenetische Bedeutung von Bernsteinfossilien; die Gattungen *Proplatypygus* (Diptera: Bombyliidae) und *Palaeopsylla* (Siphonaptera)', *Memorie Soc. ent. ital.*, **48**, 57–67.

Hennig, W. (1969d). 'Die Stammesgeschichte der Insekten', *Senckenberg-Büch.*, **49**, 1–436.

Hennig, W. (1970). 'Insektenfossilien aus der unteren Kreide. II. Empididae (Diptera, Brachycera)', *Stuttg. Beitr. Naturk.*, **214**, 1–12.

Hennig, W. (1971a). 'Zur Situation der biologischen Systematik', *Erlanger Forsch. (B)*, **4**, 7–15.

Hennig, W. (1971b), 'Insektenfossilien aus der unteren Kreide. III. Empidiformia ("Microphorinae") aus der unteren Kreide und aus dem Baltischen Bernstein; ein Vertreter der Cyclorrhapha aus der unteren Kreide', *Stuttg. Beitr. Naturk.*, **232**, 1–28.

Hennig, W. (1971c). 'Die Familien Pseudopomyzidae und Milichiidae im Baltischen Bernstein', *Stuttg. Beitr. Naturk.*, **233**, 1–16.

Hennig, W. (1972a). 'Beiträge zur Kenntnis der rezenten und fossilen Carnidae, mit besonderer Berücksichtigung einer neuen Gattung aus Chile (Diptera: Cyclorrhapha)', *Stuttg. Beitr. Naturk.*, **240**, 1–20.

Hennig, W. (1972b). 'Insektenfossilien aus der unteren Kreide. IV. Psychodidae (Phlebotominae), mit einer kritischen Uebersicht über das phylogenetische System der Familie und die bisher beschriebenen Fossilien (Diptera)', *Stuttg. Beitr. Naturk.*, **241**, 1–69.

Hennig, W. (1972c). 'Eine neue Art der Rhagionidengattung *Litoleptis* aus Chile, mit Bemerkungen über Fühlerbildung und Verwandtschaftsbeziehungen einiger Brachycerenfamilien (Diptera: Brachycera)', *Stuttg. Beitr. Naturk.*, **242**, 1–18.

Hennig, W. (1973). 'Diptera (Zweiflügler)', *Handb. Zool., Berl.*, **4** (2), 2/31, 1–337, De Gruyter, Berlin and New York.

Hennig, W. (1974). 'Kritische Bemerkungen zur Frage "cladistic analysis or cladistic classification?"', *Z. zool. Syst. EvolForsch.*, **12**, 279–294. [English translation: Hennig, 1975.]

Hennig, W. (1975). '"Cladistic analysis or cladistic classification?": A reply to Ernst Mayr', *Syst. Zool.*, **24**, 244–256. [English translation of Hennig, 1974.]

Hennig, W. (1976). 'Das Hypopygium von *Lonchoptera lutea* Panzer und die phylogenetischen Verwandtschaftsbeziehungen der Cyclorrhapha (Diptera)', *Stuttg. Beitr. Naturk. (A)*, **283**, 1–63.

Hennig, W. (in preparation). '*Die Larvenformen der Dipteren. Eine Uebersicht über die bisher bekannten Jugendstadien der zweiflügeligen Insekten*, vol. 1, revised edition, Akademie-Verlag, Berlin.

Hennig, W., and Schlee, D. (1978). 'Abriss der phylogenetischen Systematik', *Stuttg. Beitr. Naturk. (A)*, **319**, 1–11.

Henning, B. (1974). 'Morphologie und Histologie der Tarsen von *Tettigonia viridissima* L. (Orthoptera, Ensifera)', *Z. Morph. Tiere*, **79**, 323–342.

Henriksen, K. L. (1932). 'The manner of moulting in Arthropoda', *Notul. Ent.*, **11**, 103–127.

Hepburn, H. R. (1969a). 'The proventriculus of Mecoptera', *J. Georgia ent. Soc.*, **4**, 159–167.

Hepburn, H. R. (1969b). 'The skeleto-muscular system of Mecoptera: the head', *Kans. Univ. Sci. Bull.*, **48**, 721–765.

Hepburn, H. R. (1970). 'The skeleto-muscular system of Mecoptera: the thorax', *Kans. Univ. Sci. Bull.*, **48**, 801–844.

Heslop-Harrison, G. (1951). 'Preliminary notes on the ancestry, family relations, evolution and speciation of the homopterous Psyllidae [part]', *Ann. Mag. nat. Hist. (12)*, **4**, 1057–1072.

Heslop-Harrison, G. (1952a). 'Preliminary notes on the ancestry, family relations, evolution and speciation of the homopterous Psyllidae [conclusion]', *Ann. Mag. nat. Hist. (12)*, **5**, 679–696.

Heslop-Harrison, G. (1952b). 'The probable origin, phylogeny, and evolution of the class Insecta, with special reference to the classification of the Hemiptera–Homoptera (Psyllidae)', *Proc. Leeds phil. lit. Soc. (Sci. Sect.)*, **6**, 54–58.

Heslop-Harrison, G. (1956). 'The age and origin of the Hemiptera with special reference to the sub-order Homoptera', *Proc. Univ. Durham phil. Soc. (12)*, **15**, 150–169.

Heslop-Harrison, G. (1957). 'The age and origin of the Hemiptera with special reference to the suborder Homoptera. Part II. The assessment, integration, and use of the evidence from fossil and modern Homoptera in phylogenetic deduction', *Proc. Univ. Durham phil. Soc. (A)*, **13**, 41–53.

Heslop-Harrison, G. (1958). 'On the origin and function of the pupal stadia in holometabolous Insecta', *Proc. Univ. Durham phil. Soc. (A)*, **13**, 59–79.

Heslop-Harrison, G. (1961). 'The Berlese theory of the insect metamorphosis: the views of Hinton and Poulton', *Proc. Univ. Durham phil. Soc. (A)*, **13**, 174–192.

Heymer, A. (1973). 'Verhaltensstudien an Prachtlibellen. Beiträge zur Ethologie und Evolution der Calopterygidae Sélys, 1850 (Odonata; Zygoptera)', *Z. Tierpsychol.*, Beiheft **11**, 1–100.

Heymons, R. (1915). 'Die Vielfüssler, Insekten und Spinnenkerfer', in *Brehms Tierleben. Allgemeine Kunde des Tierreichs*, 4th ed., vol. 2, Bibliographisches Institut, Leipzig and Vienna.

Hickin, N. E. (1967). *Caddis larvae. Larvae of the British Trichoptera*, Hutchinson, London.

Hinton, H. E. (1946). 'On the homology and nomenclature of the setae of the lepidopterous larvae, with some notes on the phylogeny of the Lepidoptera', *Trans. R. ent. Soc. Lond.*, **97**, 1–37.

Hinton, H. E. (1948). 'On the origin and function of the pupal stage', *Trans. R. ent. Soc. Lond.*, **99**, 395–409.

Hinton, H. E. (1949). 'On the function, origin and classification of pupae', *Trans. Proc. S. Lond. ent. nat. Hist. Soc.*, **1947–48**, 111–154.

Hinton, H. E. (1958). 'The phylogeny of the panorpoid orders', *A. Rev. Ent.*, **3**, 181–206.

Hinton, H. E. (1963). 'The origin and function of the pupal stage', *Proc. R. ent. Soc. Lond. (A)*, **38**, 77–85.

Hinton, H. E. (1971). 'Some neglected phases in metamorphosis', *Proc. R. ent. Soc. Lond. (C)*, **35**, 55–64.

Hinton, H. E. (1980). *Biology of insect eggs*, 3 vols., Pergamon Press, Oxford.

Hirst, S., and Maulik, S. (1926). 'On some arthropod remains from the Rhynie Chert (Old Red Sandstone)', *Geol. Mag.*, **63**, 69–71.

Hlavac, T. F. (1972). 'The prothorax of Coleoptera: origin, major features of variation', *Psyche, Camb.*, **79**, 123–149.

Hlavac, T. F. (1975). 'The prothorax of Coleoptera (except Bostrichiformia–Cucujiformia)', *Bull. Mus. comp. Zool. Harv.*, **147**, 137–183.

Hoffmann, C. (1964). 'Bau und Vorkommen von proprioreceptiven Sinnesorganen bie den Arthropoden', *Ergebn. Biol.*, **27**, 1–38.

Holdsworth, R. (1940). 'Histology of the wing pads of the early instars of *Pteronarcys proteus* Newman', *Psyche, Camb.*, **47**, 112–120.

Holdsworth, R. (1941). 'The wing development of *Pteronarcys proteus* Newman', *Morphology*, **70**, 431–461.

Holland, G. P. (1964). 'Evolution, classification and host relationship of Siphonaptera', *A. Rev. Ent.*, **9**, 123–146.

Holland, T. H. de P., Cotten, G., Krihnan, M. S., Jacob, K., Karanth, K. R., Seth, N. N., Agrawal, S. K., Khan, M. H., and Haque, A. F. M. M. (1957). 'India, Pakistan, Nepal, Bhutan', *Lexique Strat. Int., Asie*, **3** (8), 1–404.

Holmgren, N. (1909). 'Termitenstudien. 1. Anatomische Untersuchungen', *K. svenska VetenskAkad. Handl.*, **44** (3), 1–215.

462

Holmgren, N. (1911), 'Termitenstudien. 2. Systematik der Termiten. Die Familien Mastotermitidae, Protermitidae und Mesotermitidae', *K. svenska VetenskAkad. Handl.*, **46** (6), 1–88.

Hong, Y. Ch. (1979). 'On Eocene *Philolimnias* gen.nov. (Ephemeroptera, Insecta) in amber from Fushun coalfield, Liaoning province', *Scientia sin.*, **22**, 331–339.

House, M. R., Richardson, J. B., Chaloner, W. G., Allen, J. R. L., Holland, C. H., and Westoll, T. S. (1977). 'A correlation of Devonian rocks of the British Isles', *Spec. Rep. geol. Soc. Lond.*, **8**, 1–110.

Hoyt, C. P. (1952). 'The evolution of the mouth parts of adult Diptera', *Microentomology*, **17**, 61–125.

Huber, I. (1974). 'Taxonomic and ontogenetic studies of cockroaches (Blattaria)', *Kans. Univ. Sci. Bull.*, **50**, 233–332.

Hughes, N. F. (1961). 'Fossil evidence and angiosperm ancestry', *Sci. Progr., Lond.*, **49**, 84–102.

Hughes-Schrader, S. (1975). 'Male meiosis in camel-flies (Raphidioptera; Neuropteroidea)', *Chromosoma*, **51**, 99–110.

Hurd, P. D., Smith, R. F., and Usinger, R. L. (1958). 'Cretaceous and Tertiary insects in Arctic and Mexican amber', *Proc. 10th Int. Congr. Ent.*, **1**, 851.

Ide, F. P. (1965). 'A fly of the archaic family Nymphomyiidae (Diptera) from North America', *Can. Ent.*, **97**, 496–507.

Illies, J. (1963). 'Revision der südamerikanischen Gripopterygidae', *Mitt. Schweiz. ent. Ges.*, **36**, 145–248.

Illies, J. (1965). 'Phylogeny and zoogeography of the Plecoptera', *A. Rev. Ent.*, **10**, 117–140.

Illies, J. (1967). 'Die Gattung *Megaleuctra* (Plecopt., Ins.)', *Z. Morph. Oekol. Tiere*, **60**, 124–134.

Illies, J. (1968). 'Ephemeroptera (Eintagsfliegen)', *Handb. Zool., Berl.*, **4** (2), 2/5, 1–63, De Gruyter, Berlin.

Illies, J. (1974). *Introduction to zoogeography*, translated by W. D. Williams, Macmillan, London and Basingstoke.

Imamura, S. (1974). 'New discovery and its meaning of a Blattodea from the Lower Jurassic Nishinakayama Formation, Toyora Group at Ishimachi, Yamaguchi Prefecture, Japan', *Bull. Suzugamine Wom. Coll. nat. Sci.*, **18**, 7–12. [In Japanese.]

Imms, A. D. (1945). 'The phylogeny of insects', *Tijdschr. Ent.*, **88**, 63–66.

Insects of Australia—see CSIRO.

Iwata, K. (1976). *Evolution of instinct. Comparative ethology of Hymenoptera*, translated by A. Gopal, edited by K. Krombein and V. S. Kothekar, Amerind Publishing Co., New Delhi.

Jacot-Guillarmot, C. F. (1970). 'Catalogue of the Thysanoptera of the world (Part 1)', *Ann. Cape prov. Mus.*, **7**, 1–216.

Jacot-Guillarmot, C. F. (1971). 'Catalogue of the Thysanoptera of the world (Part 2)', *Ann. Cape prov. Mus.*, **7**, 217–515.

Jander, U. (1966). 'Untersuchungen zur Stammesgeschichte von Putzbewegungen von Tracheaten', *Z. Tierpsychol.*, **23**, 799–844.

Janetschek, H. (1957). 'Ueber die mögliche phyletische Reversion eines Merkmals bei Felsenspringern mit einigen Bemerkungen über die Natur der Styli bei den Thysanuren', *Broteria (Ser. trim)*, **26**, 1–22.

Janetschek, H. (1970). 'Protura (Beintastler)', *Handb. Zool., Berl.*, **4** (2), 2/3, 1–72, De Gruyter, Berlin.

Jeannel, R. (1945). 'Sur la position systématique des Strepsiptères', *Revue fr. Ent.*, **11**, 111–118.

Jeannel, R. (1950). 'Origine et évolution des insectes', *Proc. 8th Int. Congr. Ent.*, 80–86.

Joleaud, L. (1939). *Atlas de paléobiogéographie*, Lechevalier, Paris.

Jordan, K. H. C. (1972). 'Heteroptera (Wanzen)', *Handb. Zool., Berl.*, **4** (2), 2/20, 1–113, De Gruyter, Berlin and New York.

Jurzitza, G. (1969). 'Ein Diskussionsbeitrag zur Bedeutung des Pterostigmas der Libellen', *Faun.-Oekol. Mitt.*, **3**, 257–259.

Kaltenbach, A. (1968). 'Embiodea (Spinnfüsser)', *Handb. Zool., Berl.*, **4** (2), 2/8, 1–29, De Gruyter, Berlin.

Kaltenbach, A. (1978). 'Mecoptera (Schnabelhafte, Schnabelfliegen)', *Handb. Zool., Berl.*, **4** (2), 2/28, 1–111, De Gruyter, Berlin & New York.

Kalugina, N. S. (1976). 'Non-biting midges of the subfamily Diamesinae (Diptera, Chironomidae) from the Upper Cretaceous of the Taymyr', *Paleont. Zh.*, **1976** (1), 87–93. [In Russian; English translation in *Paleont. J.*, **10**, 78–83.]

Kéler, S. von (1955). *Entomologisches Wörterbuch*, Akademie-Verlag, Berlin.

Kéler, S. von (1956). *Entomologisches Wörterbuch*, 2nd ed., Akademie-Verlag, Berlin.

Kéler, S. von (1957). 'Ueber die Deszendenz und die Differenzierung der Mallophagen', *Z. Parasitenk.*, **18**, 55–160.

Kéler, S. von (1969). 'Mallophaga (Federlinge und Haarlinge)', *Handb. Zool., Berl.*, **4** (2), 2/17, 1–72, De Gruyter, Berlin.

Kevan, P. G., Chaloner, W. G., and Savile, D. B. O. (1975). 'Interrelationships of early terrestrial arthropods and plants', *Palaeontology*, **18**, 391–417.

Kiauta, B. (1968). 'Considerations on the evolution of the chromosome complement in Odonata', *Genetica*, **38**, 430–468.

Killington, F. J. (1936). *A monograph of the British Neuroptera*, vol. 1, Ray Society, London.

Kinzelbach, R. (1967). 'Zur Kopfmorphologie der Fächerflügler (Strepsiptera, Insecta)', *Zool. Jb. (Anat.)*, **84**, 559–684.

Kinzelbach, R. (1970). 'Eine fossile Eintagsfliege aus dem Perm des Saar-Nahe-Pfalz-Gebietes (Ephemeroptera: Permoplectoptera: ?Misthodotidae Tillyard, 1932)', *Mainz. Wiss. Arch.*, **9**, 318–322.

Kinzelbach, R. (1971a). 'Strepsiptera (Fächerflügler)', *Handb. Zool., Berl.*, **4** (2), 2/24, 1–73, De Gruyter, Berlin and New York.

Kinzelbach, R. (1971b). 'Morphologische Befunde an Fächerflüglern und ihre phylogenetische Bedeutung (Insecta: Strepsiptera)', *Zoologica, Stuttg.*, **41**, Hft 119, 1–256.

Kinzelbach, R. (1978). 'Insecta. Fächerflügler (Strepsiptera)', *Tierwelt Dtl.*, **65**, 1–166.

Klausnitzer, B. (1975). 'Probleme der Abgrenzung von Unterordnungen bei den Coleoptera', *Ent. Abh. Mus. Tierk. Dresden*, **40**, 269–275.

Klemm, N. (1966). 'Die Morphologie des Kopfes von *Rhyacophila* Pict. (Trichoptera)', *Zool. Jb. (Anat.)*, **83**, 1–51.

Klingstedt, H. (1937). 'Chromosome behaviour and phylogeny in the Neuroptera', *Nature, Lond.*, **139**, 468–469.

Knight, O. LeM. (1950). 'Fossil insect beds of Belmont, N.S.W.', *Rec. Aust. Mus.*, **22**, 251–253.

Kölbel, H. (1969). 'Entwicklung und heutiger Stand der Grossgliederung der Erdgeschichte', *Leopoldina (3)*, **14** [1968], 113–120.

Königsmann, E. (1960). 'Zur Phylogenie der Parametabola unter besonderer Berücksichtigung der Phthiraptera', *Beitr. Ent.*, **10**, 705–744.

Königsmann, E. (1976). 'Das phylogenetische System der Hymenoptera. Teil 1: Einführung, Grundplanmerkmale, Schwestergruppe und Fossilfunde', *Dt. ent. Z. (n. F.)*, **23**, 253–279.

Königsmann, E. (1977). 'Das phylogenetische System der Hymenoptera. Teil 2: "Symphyta"', *Dt. ent. Z. (n. F.)*, **24**, 1–40.

464

Königsmann, E. (1978a). 'Das phylogenetische System der Hymenoptera. Teil 3: "Terebrantes" (Unterordnung Apocrita)', *Dt. Ent. Z. (n. F.)*, **25**, 1–55.

Königsmann, E. (1978b). 'Das phylogenetische System der Hymenoptera. Teil 4: Aculeata (Unterordnung Apocrita)', *Dt. ent. Z. (n. F.)*, **25**, 365–435.

Kononova, E. L. (1975). 'A new family of aphids (Homoptera, Aphidinea) from the Upper Cretaceous of the Taymyr', *Ént. Obozr.*, **54**, 795–807. [In Russian; English translation in *Ent. Rev., Wash.*, **54** (4), 60–68.]

Kononova, E. L. (1976). 'Extinct aphid families (Homoptera, Aphidinea) of the late Cretaceous', *Paleont. Zh.*, **1976** (3), 117–126. [In Russian; English translation in *Paleont. J.*, **10**, 352–360.]

Koss, R. W., and Edmunds, G. F. (1974). 'Ephemeroptera eggs and their contribution to phylogenetic studies of the order', *Zool. J. Linn. Soc.*, **55**, 267–349.

Kossmat, F. (1936). *Paläogeographie und Tektonik*, Berlin.

Koteja, J. (1974). 'On the phylogeny and classification of the scale insects (Homoptera, Coccinea) (discussion based on the morphology of the mouthparts)', *Acta zool. cracov.*, **19**, 267–325.

Kovalev, V. G. (1974). 'A new genus of the family Empididae (Diptera) and its phylogenetic relationships', *Paleont. Zh.*, **1974** (2), 84–94. [In Russian; English translation in *Paleont. J.*, **8**, 196–204.]

Kozlov, M. A. (1970). 'Suprageneric groupings of the Proctotrupoidea (Hymenoptera)', *Ént. Obozr.*, **49**, 203–226. [In Russian; English translation in *Ent. Rev., Wash.*, **49**, 115–127.]

Kristensen, N. P. (1967). 'Erection of a new family in the lepidopterous suborder Dacnonypha', *Ent. Meddr*, **35**, 341–345.

Kristensen, N. P. (1968). 'The anatomy of the head and the alimentary canal of adult Eriocraniidae (Lep., Dacnonypha)', *Ent. Meddr*, **36**, 239–315.

Kristensen, N. P. (1971). 'The systematic position of the Zeugloptera in the light of recent anatomical investigations', *Proc. 13th Int. Congr. Ent.*, **1**, 261.

Kristensen, N. P. (1972). 'Sommerfuglenes stilling i insektsystemet', *Lepidoptera, Kbh. (n.s.)*, **2**, 61–67.

Kristensen, N. P. (1975). 'The phylogeny of hexapod "orders". A critical review of recent accounts', *Z. zool. Syst. EvolForsch.*, **13**, 1–44.

Kristensen, N. P. (1978). 'A new familia of Hepialoidea from South America, with remarks on the phylogeny of the suborder Exoporia (Lepidoptera)', *Entomologica germ.*, **4**, 272–294.

Kristensen, N. P., and Nielsen, E. S. (1979). 'A new subfamily of micropterigid moths from South America. A contribution to the morphology and phylogeny of the Micropterigidae with a generic catalogue of the family (Lepidoptera: Zeugloptera)', *Steenstrupia*, **5**, 69–147.

Kristensen, N. P., and Nielsen, E. S. (in press a). 'The ventral diaphragm in primitive (non-ditrysian) Lepidoptera. A morphological and phylogenetic study', *Z. zool. Syst. EvolForsch.*

Kristensen, N. P., and Nielsen, E. S. (in press b). 'Intrinsic proboscis musculature in non-ditrysian Lepidoptera–Glossata: structure and phylogenetic significance', *Entomologica scand.*

Kryzhanovskiy, O. L. (1976). 'An attempt at a revised classification of the family Carabidae (Coleoptera)', *Ént. Obozr.*, **55**, 80–91. [In Russian, with English summary; English translation in *Ent. Rev., Wash.*, **55** (1), 56–64.]

Kuhn, O. (1937). 'Insekten aus dem Buntsandstein von Thüringen', *Beitr. Geol. Thür.*, **4**, 190–193.

Kuhn, O. (1938). 'Drei neue Insekten aus dem Mesozoikum von Bayern', *Paläont. Z.*, **20**, 318–320.

Kuhn, O. (1951). 'Ein vermutlicher Schmetterling, *Geisfeldiella benkerti* n.g. n.sp. aus dem Lias Nordfrankens', *Neues Jb. Geol. Paläont. Mh.*, **1951**, 58–61.

Kuhn, O. (1952). 'Neue Crustacea Decapoda und Insecta aus dem Lias von Nordfranken', *Palaeontographica (A)*, **101**, 153–166.

Kuhn, O. (1961). 'Die Tier- und Pflanzenwelt des Solnhofer Schiefers mit vollständigem Arten- und Schriftenverzeichnis', *Geologica bav.*, **48**, 1–68.

Kuhn, O. (1966). 'Die Tierwelt des Solnhofener Schiefers', *Neue Brehm Büch.*, vol. 318, 2nd ed., Ziemsen, Wittenberg.

Kühne, W. G. (1978). 'Willi Hennig 1913–1976: Die Schaffung einer Wissenschaftstheorie', *Entomologica germ.*, **4**, 374–376.

Kühne, W. G., Kubig, L., and Schlüter, T. (1973). 'Eine Micropterygide (Lepidoptera: Homoneura) aus mittelcretazischem Harz Westfrankreichs', *Mitt. dt. ent. Ges.*, **32**, 61–64.

Kuhn-Schnyder, E. (1965?). 'Das Leben im Strom der Zeit', in Meyer, R. W. (ed.), *Das Zeitproblem im 20. Jahrhundert*, Francke, Berne and Munich.

Kukalová, J. (1958a). 'On the systematic position of *Ampeliptera limburgica* Pruvost, 1927 (Insecta Protorthoptera)', *Vest. ústřed. Úst. geol.*, **33**, 377–379.

Kukalová, J. (1958b). 'Paoliidae Handlirsch (Insecta–Protorthoptera) aus dem oberschlesischen Steinkohlenbecken', *Geologie*, **7**, 935–959.

Kukalová, J. (1958c). 'On Czechoslovakian Spilapteridae Handlirsch (Insecta–Palaeodictyoptera)', *Acta Univ. Carol. (Geol.)*, **3**, 231–240.

Kukalová, J. (1958d). 'New Palaeodictyoptera of the Carboniferous and Permian of Czechoslovakia', *Sb. st. geol. Úst. čsl. Repub.*, **25** (*pal. odd.*), 239–251.

Kukalová, J. (1959). 'On the family Blattinopsidae Bolton, 1925 (Insecta, Protorthoptera)', *Rozpr. čsl. Akad. Věd.*, **69**, 1–27.

Kukalová, J. (1961). 'Palaeozoische Insekten in der Tschechoslowakei', *Proc. 11th Int. Congr. Ent.*, **1**, 292–294.

Kukalová, J. (1965). 'The ecology of Permian insects', *Proc. 12th Int. Congr. Ent.*, 132.

Kukalová, J. (1966). 'Protelytroptera from the Upper Permian of Australia, with a discussion of the Protocoleoptera and Paracoleoptera', *Psyche, Camb.*, **73**, 89–111.

Kukalová, J. (1969a). 'Revisional study of the order Palaeodictyoptera in the Upper Carboniferous shales of Commentry, France, Part I', *Psyche, Camb.*, **76**, 163–215.

Kukalová, J. (1969b). 'Revisional study of the order Palaeodictyoptera in the Upper Carboniferous shales of Commentry, France, Part II', *Psyche, Camb.*, **76**, 439–486.

Kukalová, J. (1970). 'Revisional study of the order Palaeodictyoptera in the Upper Carboniferous shales of Commentry, France, Part III', *Psyche, Camb.*, **77**, 1–44.

Kukalová-Peck, J. (1972a). 'The structure of *Dunbaria* (Palaeodictyoptera)', *Psyche, Camb.*, **78** [1971], 306–318.

Kukalová-Peck, J. (1972b). 'Unusual structures in the Paleozoic insect orders Megasecoptera and Palaeodictyoptera with a description of a new family', *Psyche, Camb.*, **79**, 243–268.

Kukalová-Peck, J. (1974). 'Pteralia of the Paleozoic insect orders Palaeodictyoptera, Megasecoptera and Diaphanopterodea (Paleoptera)', *Psyche, Camb.*, **81**, 416–430.

Kukalová-Peck, J. (1975). 'Megasecoptera from the Lower Permian of Moravia', *Psyche, Camb.*, **82**, 1–19.

Kukalová-Peck, J. (1978). 'Origin and evolution of insect wings and their relation to metamorphosis, as documented by the fossil record', *J. Morph.*, **156**, 53–126.

Kukalová-Peck, J., and Peck, S. B. (1976). 'Adult and immature Calvertiellidae (Insecta: Palaeodictyoptera) from the Upper Palaeozoic of New Mexico and Czechoslovakia', *Psyche, Camb.*, **83**, 79–93.

Kurtén, B. (1968). *The age of the dinosaurs*, Weidenfeld and Nicolson, London.

Kuschel, G. (1959). 'Un Curculionido del Cretaceo superior primer insecto fosil de Chile', *Investnes zool. chil.*, **5**, 49–54.

Lameere, A. (1917). 'Revision sommaire des insectes fossiles du Stephanien de Commentry', *Bull. Mus. Hist. nat., Paris*, **23**, 141–200.

466

Lameere, A. (1935–1936). *Précis de Zoologie*, vol. 4, Université de Bruxelles, Brussels.

Landa, V. (1948). 'Contributions to the anatomy of ephemerid larvae. I. Topography and anatomy of tracheal system', *Věst. čsl. zeměd. Mus.*, **12**, 25–82.

Landa, V. (1959). 'Problems of internal anatomy of Ephemeroptera and their relation to the phylogeny of their order', *Proc. 15th Int. Congr. Zool.*, 113–115.

Langer, H., and Schneider, L. (1970). 'Zur Struktur und Funktion offener Rhabdome in Facettaugen', *Verh. dt. zool. Ges.*, **1969**, 494–503.

Larsén, O. (1945a). 'Das Meron der Insekten', *K. fysiogr. Sällsk. Lund Förh.*, **15**, 96–104.

Larsén, O. (1945b). 'Die hintere Region der Insektenhüfte', *K. fysiogr. Sällsk. Lund Förh.*, **15**, 105–116.

Larsson, S. G. (1978). 'Baltic Amber—a palaeobiological study', *Entomonograph*, **1**, 1–192, Scandinavian Science Press, Klampenborg.

Laurentiaux, D. (1947). 'Note préliminaire sur une faune d'insectes de l'Assisse de Tongshan (bassin houiller de Kaiping, China)', *C. r. somm. Séanc. Soc. géol. Fr.*, **1947** (5–6), 92–94.

Laurentiaux, D. (1949). 'Ocorrênicia de Blatidios do género *Eneriblatta* Teix. no estefaniano de Saint-Eloy-Les-Mines (França)', *Bolm Soc. geol. Port.*, **8**, 121–127.

Laurentiaux, D. (1950). 'Les insectes des bassins houillers du Gard et de la Loire', *Annls Paléont.*, **36**, 63–84.

Laurentiaux, D. (1951). 'La problème des blattes paléozoiques à ovipositeur externe', *Annls Paléont.*, **37**, 185–196.

Laurentiaux, D. (1952). 'Découverte d'un Homoptère Prosboloide dans le Namurien belge', *Publs Ass. Étude Paléont. Stratigr. houill.*, **14**, 1–16.

Laurentiaux, D. (1958). '*Patteiskya bouckaerti* nov.gen. et sp., Insekt aus dem Namur des Ruhrkarbons', *Neues Jb. Geol. Paläont. Mh.*, **1958**, 302–306.

Laurentiaux, D. (1960a). 'Le réproduction chez les blattes carbonifères. Essai d'explication du panchronisme des Blattaires et classification sous-ordinal', *C. r. hebd. Séanc. Acad. Sci., Paris*, **250**, 1700–1702.

Laurentiaux, D. (1960b). 'Le pliement de l'aile métathoracique des blattes et son évolution', *C. r. hebd. Séanc. Acad. Sci., Paris*, **250**, 1884–1886.

Laurentiaux, D. (1967). 'Révision du genre *Archimylacris* Scudder (Insectes Blattaires Archimylacridae) et description d'un genre nouveau westphalien de sa descendance', *Annls Univ. ARERS*, **5**, 59–64.

Laurentiaux, D., and Laurentiaux-Vieira, F. (1960). 'À propos du gigantisme des insectes carbonifères', *C. r. somm. Séanc. Soc. géol. Fr.*, **1960** (5), 96.

Laurentiaux, D., and Teixeira, C. (1948). 'Découverte d'un *Blattinopsis* dans le carbonifère du Bas-Douro (Portugal)', *Bolm Soc. geol. Port.*, **7**, 165–166.

Laurentiaux, D., and Teixeira, C. (1950). 'Novos blatídios fósseis das bacias de Valongo (Portugal) e de Saint-Eloy-Les-Mines (França)', *Comunções Servs geol. Port.*, **31**, 299–308.

Laurentiaux, D., and Teixeira, C. (1958a). 'Um novo género de insecto Palaedictyóptero do carbónico continental do Baixa-Douro (Portugal)', *Bolm Soc. geol. Port.*, **12**, 41–49.

Laurentiaux, D., and Teixeira, C. (1958b). 'Occorrência do género *Metoedischia* Mart. (Ins. Protorthopt. Saltat.) no Estefanio do Baixo-Douro (Portugal)', *Revta Fac. Ciênc. Univ. Lisb. (2 C)*, **6**, 211–218.

Laurentiaux-Vieira, F., Ricour, J., and Laurentiaux, D. (1953). 'Un Protodonate du Trias de la Dent de Villard (Savoie)', *Bull. Soc. géol. Fr. (6)*, **2**, 319–324.

Lauterbach, G. (1954). 'Begattung und Larvengeburt bei den Strepsipteren. Zugleich ein Beitrag zur Anatomie der *Stylops*-Weibchen', *Z. Parasitenk.*, **16**, 255–297.

Lauterbach, K. E. (1972a). 'Ueber die sogenannte Ganzbein-Mandibel der Tracheata, insbesondere der Myriapoda', *Zool. Anz.*, **188**, 145–154.

Lauterbach, K. E. (1972b). 'Die morphologischen Grundlagen für die Entstehung der Entognathie bei den apterygoten Insekten in phylogenetischer Sicht', *Zool. Beitr. (n. F.)*, **18**, 25–69.

Lauterbach, K. E. (1972c). 'Beschreibung zweier neuer europäischer Inocelliiden (Insecta Raphidioptera), zugleich ein Beitrag zur vergleichenden Morphologie und Phylogenie der Kamelhalsfliegen', *Bonn. zool. Beitr.*, **23**, 219–252.

Lauterbach, K. E. (1973). 'Schlüsselereignisse in der Evolution der Stammgruppe der Euarthropoda', *Zool. Beitr. (n.F.)*, **19**, 251–299.

Lemche, H. (1940). 'The origin of winged insects', *Vidensk. Meddr dansk naturh. Foren.*, **104**, 127–168.

Leston, D., Pendergrast, J. G., and Southwood, T. R. E. (1954). 'Classification of the terrestrial Heteroptera (Geocorisae)', *Nature, Lond.*, **174**, 91–92.

Lewis, S. (1970). 'Fossil caddisfly cases (Trichoptera) from the Cobb's Creek site (Cretaceous) near New Ulm, Minnesota', *Ann. ent. Soc. Am.*, **63**, 1779–1780.

Lin, Q. B. [C. P.] (1965). 'Two insects from the lower part of Jurassic, Inner Mongolia', *Acta palaeont. sin.*, **13**, 363–368. [In Chinese and English.]

Lin, Q. B. (1976). 'The Jurassic fossil insects from Western Liaoning', *Acta palaeont. sin.*, **15**, 97–116. [In Chinese, with English summary.]

Lin, Q. B. (1978a). 'On the fossil Blattoidea of China', *Acta ent. sin.*, **21**, 335–342. [In Chinese and English.]

Lin, Q. B. (1978b). 'Upper Permian and Triassic fossil insects of Guizhou', *Acta palaeont. sin.*, **17**, 313–317. [In Chinese, with English summary.]

Linck, O. (1949). 'Fossile Bohrgänge (*Anobichnidium simile* n.g. n.sp.) an einem Keuperholz', *Neues Jb. Miner. Geol. Paläont. Mh. (B)*, **1949**, 180–185.

Lindeberg, B. (1964). 'Nomenclature of the wing-venation of the Chironomidae and of some other families of the Nematocerous Diptera', *Annls zool. fenn.*, **1**, 147–152.

Lindroth, C. H. (1969). 'The ground beetles (Carabidae, excl. Cicindelinae) of Canada and Alaska, Part 1', *Opusc. ent.*, **Suppl. 35**, i–xlviii.

Lombardo, C. A. (1973). 'On the presence of two coxal sense organs in Pterygota insects', *Monitore zool. ital. (n.s.)*, **7**, 243–246.

Lorenz, R. E. (1961). 'Biologie und Morphologie vom *Micropteryx calthella* (L.)', *Dt. ent. Z. (n.F.)*, **8**, 1–23.

Louis, J. (1972). 'Études sur les ailes des Hyménoptères. VII. Situation de l'aile des abeilles dans le schéma général d'évolution de l'aile animale', *Apidologie*, **3**, 233–245.

Ludwig, H. W. (1968). 'Zahl, Vorkommen und Verbreitung der Anoplura', *Z. Parasitenk.*, **31**, 254–265.

Lull, R. S. (1953). 'Triassic life of the Connecticut Valley', *Bull. Conn. St. geol. nat. Hist. Surv.*, **81**, 1–336. [Insecta: 34–37.]

Machatschke, J. W. (1962). 'Bemerkungen zum System der Coleoptera', *Ber. 9. Wanderversamm. dt. ent. Ges.*, **1961**, 121–137.

Mackauer, M. (1965). 'Parasitological data as an aid in aphid classification', *Can. Ent.*, **97**, 1016–1024.

MacKay, M. R. (1970). 'Lepidoptera in Cretaceous amber', *Science, N.Y.*, **167**, 379–380.

Mägdefrau, K. (1959). *Vegetationsbilder der Vorzeit*, 3rd ed., Fischer, Jena.

Malicky, H. (1973). 'Trichoptera (Köcherfliegen)', *Handb. Zool., Berl.*, **4** (2), 2/29, 1–114. De Gruyter, Berlin and New York.

Malyshev, S. I. (1968). *Genesis of the Hymenoptera and the phases of their evolution*, translated by B. Haigh, edited by O. W. Richards and B. P. Uvarov, Methuen, London.

Manton, S. M. (1960). 'Concerning head development in the Arthropods', *Biol. Rev.*, **35**, 265–282.

468

Manton, S. M. (1973). 'Arthropod phylogeny—a modern synthesis', *J. Zool., Lond.*, **171**, 111–130.

Manton, S. M. (1977). *The Arthropoda: habits, functional morphology and evolution*, Oxford University Press, London.

Marcus, H. (1958). 'Ueber die Atmungsorgane bei Tracheaten', *Z. wiss. Zool. (A)*, **160**, 165–212.

Mariammal, N., and Sundara Rajulu, G. (1975). 'A serological investigation on the phylogenetic relationship of arthropod classes', *Z. zool. Syst. EvolForsch.*, **13**, 91–97.

Martynov, A. V. (1924a). 'Sur l'interprétation de la nervuration et de la trachéation des ailes des Odonates et des Agnathes', *Russk. ent. Obozr.*, **18**, 145–174. [In Russian, with French summary; English translation by F. M. Carpenter in 1930, *Psyche, Camb.*, **37**, 245–280.]

Martynov, A. V. (1924b). 'L'évolution de deux formes d'ailes différentes chez les insectes', *Russk. zool. Zh.*, **4**, 155–185. [In Russian, with French summary.]

Martynov, A. V. (1925a). 'Ueber zwei Grundtypen der Flügel bei den Insekten und ihre Evolution', *Z. Morph. Oekol. Tiere*, **4**, 465–501.

Martynov, A. V. (1925b). 'To the knowledge of fossil insects from Jurassic beds in Turkestan. 2. Raphidioptera (continued), Orthoptera (s.l.), Odonata, Neuroptera', *Izv. Akad. Nauk SSSR (6)*, **19**, 569–597.

Martynov, A. V. (1926). 'To the knowledge of fossil insects from Jurassic beds in Turkestan. 5. On some interesting Coleoptera', *Ezheg. russk. paleont. Obshch.*, **5**, 1–38. [In Russian, with English summary.]

Martynov, A. V. (1928). 'Permian fossil insects of North-East Europe', *Trudŷ geol. Muz.*, **4**, 1–118.

Martynov, A. V. (1929). 'Permian entomofauna of North Russia and its relation to that of Kansas', *Proc. 4th Int. Congr. Ent.*, **2**, 595–599.

Martynov, A. V. (1930). 'New Permian insects from Tikhie Gory, Kazan province. Pt. I. Palaeoptera', *Trudŷ geol. Muz.*, **6**, 69–86.

Martynov, A. V. (1931). 'New Permian insects from Tikhie Gory, Kazan province. II. Neoptera (excluding Miomoptera)', *Trudŷ geol. Muz.*, **8**, 149–212. [In Russian, with English summary.]

Martynov, A. V. (1932). 'New Permian Palaeoptera with the discussion of some problems of their evolution', *Trudŷ paleont. Inst.*, **1**, 1–44.

Martynov, A. V. (1933). 'On the Permian family Archescytinidae (Homoptera) and its relationships', *Izv. Akad. Nauk SSSR (otd. mat. est. Nauk) (7)*, **1933**, 883–894.

Martynov, A. V. (1935). 'Note on fossil insects from Mesozoic deposits in the Chelyabinsk district', *Trudŷ paleont. Inst.*, **4**, 37–48. [In Russian and English.]

Martynov, A. V. (1937). 'Liassic insects from Shurab and Kisyl-Kiya', *Trudŷ paleont. Inst.*, **7** (1), 1–232. [In Russian, with English summary.]

Martynov, A. V. (1938a). 'Review of localities of fossil insects in USSR', *Trudŷ paleont. Inst.*, **7** (3), 6–28. [In Russian, with English summary.]

Martynov, A. V. (1938b). 'Studies on the geological history and phylogeny of the orders of insects (Pterygota)', *Trudŷ paleont. Inst.*, **7** (4), 1–149. [In Russian, with French summary.]

Martynov, A. V. (1940). 'Permian fossil insects from Tshekarda', *Trudŷ paleont. Inst.*, **11** (1), 1–62. [In Russian, with English summary.]

Martynova, O. M. (1948). 'Materials for the evolution of the Mecoptera', *Trudŷ paleont. Inst.*, **14** (1), 1–76. [In Russian.]

Martynova, O. M. (1952). 'Permian lacewings from the USSR', *Trudŷ paleont. Inst.*, **40**, 187–237. [In Russian.]

Martynova, O. M. (1957). 'Phylogeny of the orders of insects with complete metamorphosis', *Tez. Dokl. 2. Sov. Vses. Ent. Obsh., Tiflis 1957*, **1**, 113–114. [In Russian.]

Martynova, O. M. (1958). 'New insects from the Permian and Mesozoic of the USSR', *Mater. Osnov. Paleont.*, **2**, 69–94. [In Russian.]

Martynova, O. M. (1959). 'The phylogenetic relationships of the insects of the mecopteroid complex', *Trudȳ Inst. Morf. Zhivot.*, **27**, 221–230. [In Russian.]

Martynova, O. M. (1961a). 'Palaeoentomology', *A. Rev. Ent.*, **6**, 285–294.

Martynova, O. M. (1961b). 'Die Kamelhalsfliegen aus dem Perm und Karbon', *Proc. 11th Int. Congr. Ent.*, **1**, 302–304.

Martynova, O. M. (1961c). 'Recent and fossil snakeflies (Insecta, Raphidioptera)', *Paleont. Zh.*, **1961** (3), 73–83. [In Russian.]

Mason, J. B. (1969). 'The tympanal organ of Acridomorpha (Orthoptera)', *Eos, Madr.*, **44** [1968], 267–355.

Massoud, Z. (1967). 'Contribution à l'étude de *Rhyniella praecursor* Hirst et Maulik, 1926, Collembole fossile du Dévonien', *Revue Ecol. Biol. Sol*, **4**, 497–505.

Matsuda, R. (1956). 'Morphology of the thoracic exoskeleton and musculature of a mayfly *Siphlonurus columbianus* McDunough (Siphlonuridae, Ephemeroptera), a contribution to the subcoxal theory of the insect thorax', *J. Kans. ent. Soc.*, **29**, 92–113.

Matsuda, R. (1958). 'On the origin of the external genitalia of insects', *Ann. ent. Soc. Am.*, **51**, 84–94.

Matsuda, R. (1960). 'Morphology of the pleurosternal region of the pterothorax in insects', *Ann. ent. Soc. Am.*, **53**, 712–731.

Matsuda, R. (1965a). 'Evolution of the head musculature in insects', *Proc. 12th Int. Congr. Ent.*, 141.

Matsuda, R. (1965b). 'Morphology and evolution of the insect head', *Mem. Am. ent. Inst.*, **4**, 1–334.

Matsuda, R. (1970). 'Morphology and evolution of the insect thorax', *Mem. ent. Soc. Can.*, **76**, 1–431.

Matsuda, R. (1976). *Morphology and evolution of the insect abdomen with special reference to developmental patterns and their bearings upon systematics*, Pergamon Press, Oxford.

Matthews, R. W. (1974). 'Biology of Braconidae', *A. Rev. Ent.*, **19**, 15–32.

Mayr, E. (1974). 'Cladistic analysis or cladistic classification?', *Z. zool. Syst. EvolForsch.*, **12**, 94–128.

McAlester, A. L. (1970). 'Animal extinctions, oxygen consumption, and atmospheric history', *J. Paleont.*, **44**, 405–409.

McAlpine, J. F. (1970). 'First record of Calypterate flies in the Mesozoic era (Diptera: Calliphoridae)', *Can. Ent.*, **102**, 342–346.

McAlpine, J. F. (1973). 'A fossil ironomyiid fly from Canadian amber (Diptera: Ironomyiidae)', *Can. Ent.*, **105**, 105–111.

McAlpine, J. F. (1977). 'A revised classification of the Piophilidae, including "Neottiophilidae" and "Thyreophoridae" (Diptera: Schizophora)', *Mem. ent. Soc. Can.*, **103**, 1–66.

McAlpine, J. F., and Martin, J. E. H. (1966). 'Systematics of Sciadoceridae and relatives with descriptions of two new genera and species from Canadian amber and erection of family Ironomyiidae (Diptera: Phoroidea)', *Can. Ent.*, **98**, 527–544.

McAlpine, J. F., and Martin, J. E. H. (1969a). 'Canadian amber: a palaeontological treasure-chest', *Can. Ent.*, **101**, 819–838.

McAlpine, J. F., and Martin, J. E. H. (1969b). 'Canadian amber', *Beaver*, **Summer 1969**, 28–37.

McCafferty, W. P., and Edmunds, G. F. (1976). 'Redefinition of the family Palingeniidae and its implications for the higher classification of Ephemeroptera', *Ann. ent. Soc. Am.*, **69**, 486–490.

McKittrick, F. A. (1964). 'Evolutionary studies of cockroaches', *Mem. Cornell Univ. agric. Exper. Stn*, **389**, 1–197.

McKittrick, F. A., and Mackerras, M. J. (1965). 'Phyletic relationships within the Blattidae', *Ann. ent. Soc. Am.*, **58**, 224–229.

McLellan, I. D. (1977). 'New alpine and southern Plecoptera from New Zealand, and a new classification of the Gripopterygidae', *N.Z. Jl Zool.*, **4**, 119–147.

Meinander, M. (1972). 'A revision of the family Coniopterygidae (Planipennia)', *Acta zool. fenn.*, **136**, 1–357.

Meinander, M. (1975). 'Fossil Coniopterygidae (Neuroptera)', *Notul. ent.*, **55**, 53–57.

Mickoleit, G. (1966). 'Zur Kenntnis einer neuen Spezialhomologie (Syanapomorphie) der Panorpoidea', *Zool. Jb. (Anat.)*, **83**, 483–496.

Mickoleit, G. (1967). 'Das Thoraxskelett von *Merope tuber* Newman (Protomecoptera), *Zool. Jb. (Anat.)*, **84**, 313–342.

Mickoleit, G. (1969). 'Vergleichend-anatomische Untersuchungen an der pterothorakalen Pleurotergalmuskulatur der Neuropteria und Mecopteria (Insecta, Holometabola)', *Z. Morph. Tiere*, **64**, 151–178.

Mickoleit, G. (1971). 'Das Exoskelett von *Notiothauma reedi* McLachlan, ein Beitrag zur Morphologie und Phylogenie der Mecoptera (Insecta)', *Z. Morph. Tiere*, **69**, 318–362.

Mickoleit, G. (1973). 'Ueber den Ovipositor der Neuropteroidea und Coleoptera und seine phylogenetische Bedeutung (Insecta, Holometabola)', *Z. Morph. Tiere*, **74**, 37–64.

Mickoleit, G. (1974). 'Ueber die Spermatophore von *Boreus westwoodi* Hagen (Insecta, Mecoptera)', *Z. Morph. Tiere*, **77**, 271–284.

Mickoleit, G. (1975). 'Die Genital- und Postgenitalsegmente der Mecoptera-Weibchen (Insecta, Holometabola). I. Das Exoskelett', *Z. Morph. Tiere*, **80**, 97–135.

Mickoleit, G. (1976). 'Die Genital- und Postgenitalsegmente der Mecoptera-Weibchen (Insecta, Holometabola). II. Das Dach der Genitalkammer', *Zoomorphologie*, **85**, 133–156.

Mickoleit, G. (1978). 'Die phylogenetischen Beziehungen der Schnabelfliegen-Familien auf Grund morphologischer Ausprägungen der weiblichen Genital- und Postgenitalsegmente (Mecoptera)', *Entomologica germ.*, **4**, 258–271.

Mierzejewski, P. (1976a). 'Scanning electron microscope studies on the fossilization of Baltic amber spiders (preliminary note)', *Annls med. Sect. Pol. Acad. Sci.*, **21**, 81–82.

Mierzejewski, P. (1976b). 'On application of scanning electron microscope to the study of organic inclusions from the Baltic amber', *Roczn. pol. Tow. geol.*, **46**, 291–295.

Moulins, M. (1968). 'Contribution à la connaissance anatomique des Plécoptères: la région céphalique de la larve de *Nemoura cinerea* (Nemouridae)', *Annls Soc. ent. Fr. (n.s.)*, **4**, 91–143.

Mound, L. A., Heming, B. S., and Palmer, J. M. (1980). 'Phylogenetic relationships between the families of recent Thysanoptera (Insecta)', *Zool. J. Linn. Soc.*, **69**, 111–141.

Müller, A. H. (1975). 'Zur Entomofauna des Permokarbon: 1. Mylacridae (Blattodea) aus dem Unterrotliegenden (Unterperm, Autun) von Thüringen', *Z. geol. Wiss.*, **3**, 621–641.

Müller, A. H. (1977a). 'Zur Entomofauna des Permokarbon: 2. Ueber einige Blattinopsidae (Protorthoptera) aus dem Unterrotliegenden (Unteres Autun) von Mitteleuropa', *Z. geol. Wiss.*, **5**, 1029–1051.

Müller, A. H. (1977b). 'Ueber interessante Insektenreste aus dem mitteleuropäischen Unterperm (Rotliegendes) mit allgemeinen Bemerkungen Zur Morphologie, Oekologie und Phylogenetik', *Biol. Rdsch.*, **15**, 41–58.

Müller, H. J. (1962). 'Neuere Vorstellungen über Verbreitung und Phylogenie der Endosymbiosen der Zikaden', *Z. Morph. Oekol. Tiere*, **51**, 190–210.

Mutuura, A. (1972). 'Morphology of the female terminalia in Lepidoptera and its taxonomic significance', *Can. Ent.*, **104**, 1055–1071.

Myers, J. G., and China, W. E. (1929). 'The systematic position of the Peloridiidae as elucidated by a further study of the external anatomy of *Hemiodoecus leai* China (Hemiptera, Peloridiidae)', *Ann. Mag. nat. Hist. (10)*, **3**, 282–294.

Nachtigall, W. (1968). *Gläserne Schwingen. Aus einer Werkstatt biophysikalischer Forschung*, Moos Verlag, Munich.

Nakata, S., and Maa, T. C. (1974). 'A review of the parasitic earwigs (Dermaptera: Arixeniina; Hemimerina)', *Pacif. Insects*, **16**, 307–374.

New, T. R. (1974). 'Psocoptera', *Handbk Ident. Br. Insects*, **1** (7), 1–102.

Niculescu, E. V. (1967). 'Origina si evolutia lepidopterelor', *Studii Cerc. Biol. (Ser. Zool.)*, **19**, 83–88.

Niculescu, E. V. (1970). 'Aperçu critique sur la systématique et la phylogénie des Lépidoptères', *Bull. Soc. ent. Mulhouse*, **1970**, 1–16.

Nielsen, A. (1961). 'Some thoughts on arthropod phylogeny', *Proc. 11th Int. Congr. Ent.*, **1**, 13–14.

North, F. J. (1931). 'Insect life in coal forests, with special reference to South Wales', *Trans. Cardiff Nat. Soc.*, **62**, 16–44.

Northrop, S. A. (1928). 'Beetles from the Fox Hills Cretaceous strata of South Dakota', *Am. J. Sci. (5)*, **15**, 28–38.

Novák, V. (1955). 'The metamorphosis in insects and its origin and evolution from the point of view of the facts about the metamorphosis hormones', *Roč. čsl. Spol. ent.*, **52**, 31–48.

Oeser, R. (1961). 'Vergleichend-morphologische Untersuchungen über den Ovipositor der Hymenopteren', *Mitt. zool. Mus. Berl.*, **37**, 1–119.

Ohm, P. (1961). 'Beziehungen zwischen Körpergrösse und Flügeläderung bei *Panorpa* (Mecoptera)', *Zool. Anz.*, **166**, 1–8.

Okada, I. (1938). 'Die Phryneiden und Pachyneuriden Japans', *J. Fac. Agric. Hokkaido imp. Univ.*, **42**, 221–238.

Oliveira, E. de (1930). 'Insectos permianos do Estado do Paraná', *Anais Acad. bras. Cienc.*, **2**, 215–218.

Ossianilsson, F. (1949). 'Insect drummers. A study on the morphology and function of the sound-producing organs of Swedish Homoptera Auchenorrhyncha, with notes on their sound-production', *Opusc. ent.*, **Suppl. 10**, 1–146.

Paclt, J. (1972). 'Grundsätzliches zur Chorologie und Systematik der Felsenspringer', *Zool. Anz.*, **188**, 422–429.

Panov, A. A., and Davydova, E. D. (1976). 'Medial neurosecretory cells in the brain of Mecoptera and Neuropteroidea (Insecta)', *Zool. Anz.*, **197**, 187–206.

Parsons, M. C. (1964). 'The origin and development of the hemipteran cranium', *Can. J. Zool.*, **42**, 409–432.

Parsons, M. C. (1974). 'The morphology and possible origin of hemipteran loral lobes', *Can. J. Zool.*, **52**, 189–202.

Paulus, H. F. (1972a). 'Die Feinstruktur der Stirnaugen einiger Collembolen (Insecta, Entognatha) und ihre Bedeutung für die Stammesgeschichte der Insekten', *Z. zool. Syst. EvolForsch.*, **10**, 81–122.

Paulus, H. F. (1972b). 'Die Feinstruktur der Stirnaugen einiger Collembolen (Insecta, Entognatha) und ihre Bedeutung für die Stammesgeschichte der Mandibulaten', *Verh. dt. zool. Ges.*, **66**, 56–60.

Paulus, H. F. (1974). 'Die phylogenetische Bedeutung der Ommatidien der apterygoten Insekten (Collembola, Archaeognatha, Zygentoma)', *Pedobiologia*, **14**, 123–133.

Péneau, J. (1930). 'Description d'un insecte fossile du Stéphanien de l'ouest de la France', *Bull. Soc. géol. Fr. (4)*, **30**, 251–252.

Penny, N. D. (1975). 'Evolution of the extant Mecoptera', *J. Kans. ent. Soc.*, **48**, 331–350.

Pérez, C. (1910). 'Signification phylétique de la nymphe chez les insectes métaboles', *Bull. scient. Fr. Belg.*, **44**, 221–234.

Peters, D. S. (1972a). 'Das Problem konvergent entstandener Strukturen in der anagenetischen und genealogischen Systematik', *Z. zool. Syst. EvolForsch.*, **10**, 161–173.

Peters, D. S. (1972b). 'Ueber die Stellung von *Aspidosmia* Brauns 1926 nebst allgemeinen Erörterungen der phylogenetischen Systematik der Megachilidae (Insecta, Hymenoptera, Apoidea)', *Apidologie*, **3**, 167–186.

Peterson, B. V. (1975). 'A new Cretaceous bibionid from Canadian amber (Diptera: Bibionidae)', *Can. Ent.*, **107**, 711–715.

Peyrimhoff, P. de (1934). 'Les Coléoptères remontent ils au Permien?', *Bull. Soc. ent. Fr.*, **39**, 39–44.

Pfau, H. K. (1971). 'Struktur und Funktion des sekundären Kopulationsapparates der Odonaten (Insecta, Palaeoptera), ihre Wandlung in der Stammesgeschichte und Bedeutung für die adaptive Entfaltung der Ordnung', *Z. Morph. Tiere*, **70**, 281–371.

Philpott, A. (1924). 'The tibial strigil of the Lepidoptera', *Trans. Proc. N.Z. Inst.*, **55**, 215–224.

Pierce, W. D. (1908). 'A preliminary review of the classification of the order Strepsiptera', *Proc. ent. Soc. Wash.*, **9**, 75–85.

Pierce, W. D. (1964). 'The Strepsiptera are a true order, unrelated to Coleoptera', *Ann. ent. Soc. Am.*, **57**, 603–605.

Ping, C. (1928). 'Study of the Cretaceous fossil insects of China', *Palaeont. sin. (B)*, **13** (1), 1–47.

Ping, C. (1931). 'On a blattoid insect in the Fushun amber', *Bull. geol. Soc. China*, **11**, 245.

Ping, C. (1935). 'On our fossil insects from Sinkiang', *Chin. J. Zool.*, **1**, 107–115.

Pinto, I. (1972a). 'Permian insects from the Parana basin, South Brazil. I. Mecoptera', *Revta bras. Geociênc.*, **2**, 105–116.

Pinto, I. (1972b). 'A new Insecta, *Archangelskyblatta vishniakovae* Pinto gen. nov., sp.nov., a Permian blattoid from Patagonia, Argentina', *Ameghiniana*, **9**, 79–82.

Pistor, D. (1955). 'Die Sprungmuskulatur der Collembolen', *Zool. Jb. (Syst.)*, **83**, 511–540.

Pistor, D., and Schaller, F. (1955). 'Die Sprungmuskulatur der Collembolen', *Verh. dt. zool. Ges.*, **1954**, 230–234.

Piveteau, J. (1953). *Traité de Paléontologie*, vol. 3, Masson, Paris. [Insects, by D. Laurentiaux: pp. 397–527.]

Poinar, G. O. (1977). 'Fossil nematodes from Mexican amber', *Nematologica*, **23**, 232–238.

Pollock, J. N. (1972). 'The evolution of sperm transfer mechanisms in the Diptera', *J. Ent. (A)*, **47**, 29–35.

Pomerantsev, B. I. (1932). 'On the morphology and anatomy of the genitalia of *Culicoides* (Diptera, Nematocera), *Parazit. Sb.*, **3**, 183–214. [In Russian, with German summary.]

Ponomarenko, A. G. (1969a). 'The historical development of the Coleoptera–Archostemata', *Trudÿ paleont. Inst.*, **125**, 1–239. [In Russian.]

Ponomarenko, A. G. (1969b). 'Cretaceous insects from Labrador. 4. A new family of beetles (Coleoptera, Archostemata)', *Psyche, Camb.*, **76**, 306–310.

Ponomarenko, A. G. (1976). 'A new insect from the Cretaceous of Transbaikalia, a possible parasite of pterosaurians', *Paleont. Zh.*, **1976** (3), 102–106. [In Russian; English translation in *Paleont. J.*, **10**, 339–343.]

Ponomarenko, A. G., and Rasnitsyn, A. P. (1974). 'New Mesozoic and Coenozoic Protomecoptera', *Paleont. Zh.*, **1974** (4), 59–73. [In Russian; English translation in *Paleont. J.*, **8**, 493–507.]

Popham, E. J. (1961). 'On the systematic position of *Hemimerus* Walker—a case for ordinal status', *Proc. R. ent. Soc. Lond. (B)*, **30**, 19–25.

Popham, E. J. (1962). 'Prognathism and brain structure in *Forficula auricularia* L. and *Dysdercus intermedius* Dist.', *Zool. Anz.*, **168**, 36–43.

Popham, E. J. (1965). 'Towards a natural classification of the Dermaptera', *Proc. 12th Int. Congr. Ent.*, 114–115.

Popham, E. J. (1973). 'Is *Hemimerus* an earwig?', *Entomologist*, **106**, 193–195.

Popov, Yu. A. (1970). 'Notes on the classification of the recent Naucoridae (Heteroptera, Nepomorpha)', *Bull. Acad. pol. Sci. Ser. biol.*, **18**, 93–98.

Popov, Yu. A. (1971). 'Historical development of Hemiptera of the infraorder Nepomorpha (Heteroptera)', *Trudȳ paleont. Inst.*, **129**, 1–230. [In Russian.]

Poyarkov'', E. F. (1914a). 'A new theory concerning the nymph of the Insecta Holometabola', *Hor. Soc. ent. ross.*, **41**, 1–51. [In Russian, with French summary.]

Poyarkov'', E. F. (1914b). 'Essai d'une théorie de la nymphe des insectes holométaboles', *Archs Zool. exp. gén.*, **54**, 221–265.

Priesner, H. (1924). 'Bernstein-Thysanopteren', *Ent. Mitt.*, **13**, 130–151.

Priesner, H. (1968). 'Thysanoptera (Physopoda, Blasenfüsser)', *Handb. Zool., Berl.*, **4** (2), 2/19, 1–32, De Gruyter, Berlin.

Princis, K. (1960). 'Zur Systematik der Blattarien', *Eos, Madr.*, **36**, 427–449.

Pringle, J. W. S. (1957). *Insect flight*, Cambridge University Press, Cambridge.

Pruvost, P. (1919). 'La faune continentale du terrain houiller du Nord de la France', *Mém. Serv. Carte géol. dét. Fr.*, 1–584, Imprimerie Nationale, Paris. [Insects: pp. 93–321.]

Pruvost, P. (1927). 'Sur une aile d'insecte fossile trouvée au sondage de Gulpen', *Jversl. geol. Bur. ned. Mijngeb. Heerlen*, **1926**, 76–77.

Pruvost, P. (1930). 'La faune continentale du terrain houiller de la Belgique', *Mém. Mus. Hist. nat. Belg.*, **44**, 103–280. [Insects: pp. 142–167.]

Pruvost, P. (1934). 'Description d'un insecte fossile des couches de la Lukuga (Kivu)', *Mém. Inst. géol. Univ. Louvain*, **9** (4), 1–8.

Quadri, A. H. (1940). 'On the development of the genitalia and their ducts of orthopteroid insects', *Trans. R. ent. Soc. Lond.*, **90**, 121–175.

Quentin, D. St. (1969). 'Odonatenmerkmale im Geäder anderer Insekten', *Ent. Abh. Mus. Tierk. Dresden*, **36**, 193–199.

Quentin, D. St., and Beier, M. (1968). 'Odonata (Libellen), *Handb. Zool., Berl.*, **4** (2), 2/6, 1–39, De Gruyter, Berlin.

Ragge, D. R. (1955a). *The wing-venation of Orthoptera Saltatoria, with notes on dictyopteran wing-venation*, British Museum (Natural History), London.

Ragge, D. R. (1955b). 'The wing-venation of the order Phasmida', *Trans. R. ent. Soc. Lond.*, **106**, 375–392.

Rähle, W. (1970). 'Untersuchungen an Kopf und Prothorax von *Embia ramburi* Rimsky-Korsakow 1906 (Embioptera, Embiidae)', *Zool. Jb. (Anat.)*, **87**, 248–330.

Rainey, R. C. (ed.) (1976). 'Insect flight', *Symp. R. ent. Soc. Lond.*, **7**, 1–287, Blackwell, Oxford.

Rasnitsyn, A. P. (1966). 'A key to the superfamilies and families of the Hymenoptera', *Ént. Obozr.*, **45**, 599–611. [In Russian; English translation in *Ent. Rev., Wash.*, **45**, 340–347.]

Rasnitsyn, A. P. (1968). 'On the evolution of the function of the ovipositor in relation to the origin of parasitism in the Hymenoptera', *Ént. Obozr.*, **47**, 61–70. [In Russian; English translation in *Ent. Rev., Wash.*, **47**, 35–40.]

Rasnitsyn, A. P. (1969). 'The origin and evolution of the lower Hymenoptera', *Trudȳ paleont. Inst.*, **123**, 1–196. [In Russian. English translation: Rasnitsyn, 1979.]

Rasnitsyn, A. P. (1972). 'Hymenopterous insects of the Praeaulacidae from the Late Jurassic of Karatau', *Paleont. Zh.*, **1972** (1), 70–87. [In Russian; English translation in *Paleont. J.*, **6**, 62–77.]

Rasnitsyn, A. P. (1975a). 'The higher Hymenoptera of the Mesozoic', *Trudȳ paleont. Inst.*, **147**, 1–134. [In Russian.]

Rasnitsyn, A. P. (1975b). 'Early evolution of higher Hymenoptera (Apocrita)', *Zool. Zh.*, **54**, 848–860. [In Russian, with English summary.]

Rasnitsyn, A. P. (1979). *Origin and evolution of Lower Hymenoptera*, translated by Indira Nair, edited by V. S. Kothekar, Amerind Publishing Co., New Delhi. [English translation of Rasnitsyn, 1969.]

Rau, P. (1941). 'Cockroaches: the forerunners of termites (Orthoptera: Blattidae; Isoptera)', *Ent. News*, **52**, 256–259.

Ravoux, P. (1948). 'Observations sur l'anamorphose de *Scutigerella immaculata* Newport', *Archs Zool. exp. gén. (notes et revues)*, **85** (4), 189–198.

Reichardt, H. (1973a). 'A critical study of the suborder Myxophaga, with a taxonomic revision of the Brazilian Torridincolidae and Hydroscaphidae (Coleoptera)', *Archos Zool. S. Paulo*, **24**, 73–162.

Reichardt, H. (1973b). 'More on Myxophaga: On the morphology of *Scaphydra angra* (Reichardt, 1971) (Coleoptera)', *Revta bras. Ent.*, **17**, 109–110.

Reichardt, H. (1974). 'Relationships between Hydroscaphidae and Torridincolidae, based on larvae and pupae, with the description of the immature stages of *Scaphydra angra* (Coleoptera, Myxophaga)', *Revta bras. Ent.*, **18**, 117–122.

Reichardt, H. (1976). 'On the new world beetles of the family Hydroscaphidae', *Papéis avuls. Zool. S. Paulo*, **30**, 1–24.

Reichardt, H., and Vanin, S. A. (1977). 'The torridincolid genus *Ytu* (Coleoptera, Myxophaga)', *Papéis avuls. Zool. S. Paulo*, **31**, 119–140.

Rice, H. M. A. (1969). 'An antlion (Neuroptera) and a stonefly (Plecoptera) of Cretaceous age from Labrador, Newfoundland', *Geol. Surv. Pap. Can.*, **1968** (65), 1–11.

Richards, A. G. (1965). 'The proventriculus of adult Mecoptera and Siphonaptera', *Ent. News*, **76**, 253–256.

Richards, W. R. (1966). 'Systematics of fossil aphids from Canadian amber (Homoptera: Aphididae)', *Can. Ent.*, **98**, 746–760.

Richardson, E. S. (1956). 'Pennsylvanian invertebrates of the Mazon Creek area, Illinois', *Fieldiana, Geol.*, **12**, 15–56.

Richardson, J. B. (1967). 'Some British Lower Devonian spore assemblages and their stratigraphic significance', *Rev. Palaeobot. Palynol.*, **1**, 111–129.

Ricker, W. E. (1935). 'New Canadian Perlids (Part II)', *Can. Ent.*, **67**, 256–264.

Rieger, C. (1976). 'Skelett und Muskulatur des Kopfes und Prothorax von *Ochterus marginatus* Latreille. Beitrag zur Klärung der phylogenetischen Verwandtschaftsbeziehungen der Ochteridae (Insecta, Heteroptera)', *Zoomorphologie*, **83**, 109–191.

Riek, E. F. (1950). 'A fossil mecopteron from Triassic beds at Brookvale, N.S.W.', *Rec. Aust. Mus.*, **22**, 254–256.

Riek, E. F. (1953a). 'Fossil mecopteroid insects from the Upper Permian of New South Wales', *Rec. Aust. Mus.*, **23**, 55–87.

Riek, E. F. (1953b). 'Further Triassic insects from Brookvale, N.S.W. (order Orthoptera Saltatoria, Protorthoptera, Perlaria)', *Rec. Aust. Mus.*, **23**, 161–168.

Riek, E. F. (1954a). 'A second specimen of the dragonfly *Aeschnidiopsis flindersiensis* (Woodward 1884) from the Queensland Cretaceous', *Proc. Linn. Soc. N.S.W.*, **79**, 61–64.

Riek, E. F. (1954b). 'The Australian Mecoptera or scorpion-flies', *Aust. J. Zool.*, **2**, 143–168.

Riek, E. F. (1955). 'Fossil insects from the Triassic beds at Mt. Crosby, Queensland', *Aust. J. Zool.*, **3**, 654–691.

Riek, E. F. (1956). 'A re-examination of the mecopteroid and orthopteroid fossils (Insecta) from the Triassic beds at Denmark Hill, Queensland, with descriptions of further specimens', *Aust. J. Zool.*, **4**, 98–110.

Riek, E. F. (1962). 'Fossil insects from the Triassic at Hobart, Tasmania', *Pap. Proc. R. Soc. Tasm.*, **96**, 39–40.

Riek, E. F. (1967a). 'A fossil cockroach (Blattodea: Poroblattinidae) from the Mt. Nicholas coal measures, Tasmania', *J. Aust. ent. Soc.*, **6**, 73.

Riek, E. F. (1967b). 'Structures of unknown, possibly stridulatory, function on the wings and body of Neuroptera; with an appendix on other endopterygote orders', *Aust. J. Zool.*, **15**, 337–348.

Riek, E. F. (1968a). '*Robinjohnia tillyardi* Martynova, a mecopteron from the Upper Permian of Belmont, New South Wales', *Rec. Aust. Mus.*, **27**, 299–302.

Riek, E. F. (1968b). 'Undescribed fossil insects from the Upper Permian of Belmont, New South Wales (with an appendix listing the described species)', *Rec. Aust. Mus.*, **27**, 303–310.

Riek, E. F. (1968c). 'On the occurrence of fossil insects in the Mesozoic rocks of Western Australia', *Rec. Aust. Mus.*, **27**, 311–312.

Riek, E. F. (1970). 'Lower Cretaceous fleas', *Nature, Lond.*, **227**, 746–747.

Riek, E. F. (1971). 'Origin of the Australian insect fauna', *Proc. Pap. 2nd Gondwana Symp.*, 593–598.

Riek, E. F. (1973a). 'Fossil insects from the Upper Permian of Natal, South Africa', *Ann. Natal Mus.*, **21**, 513–532.

Riek, E. F. (1973b). 'A Carboniferous insect', *Nature, Lond.*, **244**, 455–456.

Riek, E. F. (1974a). 'A fossil insect from the Dwyka Series of Rhodesia', *Palaeont. afr.*, **17**, 15–18.

Riek, E. F. (1974b). 'Upper Triassic insects from the Molteno "Formation", South Africa', *Palaeont. afr.*, **17**, 19–31.

Riek, E. F. (1974c). 'An unusual immature insect from the Upper Permian of Natal', *Ann. Natal Mus.*, **22**, 271–274.

Riek, E. F. (1976a). 'New Upper Permian insects from Natal, South Africa', *Ann. Natal Mus.*, **22**, 755–790.

Riek, E. F. (1976b). 'Fossil insects from the Middle Ecca (Lower Permian) of Southern Africa', *Palaeont. afr.*, **19**, 145–148.

Riek, E. F. (1976c). 'An entomobryid collembolan (Hexapoda: Collembola) from the Lower Permian of southern Africa', *Palaeont. afr.*, **19**, 141–143.

Riek, E. F. (1976d). 'A new collection of insects from the Upper Triassic of South Africa', *Ann. Natal Mus.*, **22**, 791–820.

Roberts, H. R. (1941). 'A comparative study of the subfamilies of the Acrididae (Orthoptera) primarily on the basis of their phallic structures', *Proc. Acad. nat. Sci. Philad.*, **93**, 201–246.

Robinson, G. S., and Tuck, K. R. (1976). 'The kauri moth *Agathiphaga*—A preliminary report', unpublished (cyclostyled) report, British Museum (Natural History), London.

Rodendorf, B. B. (1939). 'A new protelytropteron from the Permian of the Urals', *Dokl. Akad. Nauk SSSR (n.s.)*, **23**, 506–508.

Rodendorf, B. B. (1944). 'A new family of Coleoptera from the Permian of the Urals', *Dokl. Akad. Nauk SSSR (n.s.)*, **44**, 252–253.

Rodendorf, B. B. (1947). 'The fauna of two-winged insects of Karatau and their importance for the understanding of the evolution of this order', *Dokl. Akad. Nauk SSSR (n.s.)*, **55**, 757–760. [In Russian.]

Rodendorf, B. B. (1956). 'Palaeozoic insects from southern Siberia', *Ént. Obozr.*, **35**, 611–619. [In Russian, with French summary.]

Rodendorf, B. B. (1957a). 'Basic trends in the historical development of the Diptera', *Tez. Dokl. 2. Sov. Vses. Ent. Obsh., Tiflis 1957*, **1**, 115–119. [In Russian.]

Rodendorf, B. B. (1957b). 'Palaeoentomological investigations in the USSR', *Trudȳ paleont. Inst.*, **66**, 1–102. [In Russian.]

476

Rodendorf, B. B. (1958). 'Les insectes paléozoïques du Sud de la Sibérie', *Proc. 10th Int. Congr. Ent.*, **1**, 853–859.

Rodendorf, B. B. (1959). 'Die Bewegungsorgane der Zweiflügler-Insekten und ihre Entwicklung I–III', *Wiss. Z. Humboldt-Univ. Berl.*, **8**, 1–119, 269–308, 435–454.

Rodendorf, B. B. (1961a). 'Die Paläoentomologie in der UdSSR', *Proc. 11th Int. Congr. Ent.*, **1**, 313–318.

Rodendorf, B. B. (1961b). 'The description of the first winged insect from the Devonian beds of the Timan', *Ént. Obozr.*, **40**, 485–489. [In Russian, with English summary; English translation in *Ent. Rev., Wash.*, **40**, 260–262.]

Rodendorf, B. B. (1961c). 'The oldest infraorders of Diptera from the Triassic of Middle Asia', *Paleont. Zh.*, **1961** (2), 90–100. [In Russian.]

Rodendorf, B. B. (ed.) (1962). 'Arthropoda—Tracheata and Chelicerata', in *Textbook of Palaeontology. A Textbook for the Palaeontologists and Geologists of the USSR*, vol. 9, pp. 1–561, Academy of Sciences of the USSR, Moscow.

Rodendorf, B. B. (1964). 'The historical development of the Diptera', *Trudȳ paleont. Inst.*, **100**, 1–311. [In Russian. English translation: Rodendorf, 1974.]

Rodendorf, B. B. (ed.) (1968). *The Jurassic insects of Karatau*, Academy of Sciences of the USSR, Moscow. [In Russian.]

Rodendorf, B. B. (1969a). 'Paläontologie', *Handb. Zool., Berl.*, **4** (2), 1/3, 1–27, De Gruyter, Berlin.

Rodendorf, B. B. (1969b). 'Phylogenie', *Handb. Zool., Berl.*, **4** (2), 1/4, 1–28, De Gruyter, Berlin.

Rodendorf, B. B. (1970). 'The second finding of the remains of winged insects', *Ént. Obozr.*, **49**, 835–837. [In Russian, with English summary; English translation in *Ent. Rev., Wash.*, **49**, 508–509.]

Rodendorf, B. B. (1972). 'The Devonian eopterids are not insects but Crustacea Eumalacostraca', *Ént. Obozr.*, **51**, 96–7. [In Russian; English translation in *Ent. Rev., Wash.*, **51**, 58–59.]

Rodendorf, B. B. (1974). *The historical development of the Diptera*, translated by J. E. Moore and I. Thiele, edited by B. Hocking, H. Oldroyd and G. E. Ball, University of Alberta, Edmonton. [English translation of Rodendorf, 1964.]

Rodendorf, B. B., Bekker-Migdisova, E. E., Martynova, O. M., and Sharov, A. G. (1961). 'Palaeozoic insects from the Kuznetsk basin', *Trudȳ paleont. Inst.*, **85**, 1–705. [In Russian.]

Rodendorf, B. B., and Rasnitsyn, A. P. (1980). 'The historical development of the class Insecta', *Trudȳ paleont. Inst.*, **175**, 1–269. [In Russian.]

Rodendorf, B. B., and Zherikhin, V. V. (1974). 'Palaeontology and nature conservation', *Priroda, Mosk.*, **1974** (5), 82–91. [In Russian.]

Roonwal, M. L. (1975). 'Phylogeny and status of termite families Stylotermitidae and Indotermitidae with three-segmented tarsi, and the evolution of tarsal segmentation in the Isoptera', *Biol. Zbl.*, **94**, 27–43.

Ross, E. S. (1970). 'Biosystematics of the Embioptera', *A. Rev. Ent.*, **15**, 157–172.

Ross, H. H. (1936). 'The ancestry and wing venation of the Hymenoptera', *Ann. ent. Soc. Am.*, **29**, 99–111.

Ross, H. H. (1937). 'A generic classification of the Nearctic sawflies', *Illinois biol. Monogr.*, **15** (2), 1–175.

Ross, H. H. (1955). 'The evolution of the insect orders', *Ent. News*, **66**, 197–208.

Ross, H. H. (1965). *A Textbook of Entomology*, 3rd ed., Wiley, New York.

Ross, H. H. (1967). 'The evolution and past dispersal of Trichoptera', *A. Rev. Ent.*, **12**, 169–206.

Roth, L. M. (1970). 'Evolution and taxonomic significance of reproduction in Blattaria', *A. Rev. Ent.*, **15**, 75–96.

Rothschild, M. (1975). 'Recent advances in our knowledge of the order Siphonaptera', *A. Rev. Ent.*, **20**, 241–259.

Rupprecht, R. (1976). 'Struktur und Funktion der Bauchblase und des Hammers von Plecopteren', *Zool. Jb. (Anat.)*, **95**, 9–80.

Rusek, J. (1974). 'Zur Morphologie und Phylogenesis der Abdominalbeine der Protura', *Pedobiologia*, **14**, 134–137.

Russell, L. K. (1979a). 'A new genus and a new species of Boreidae from Oregon (Mecoptera)', *Proc. ent. Soc. Wash.*, **81**, 22–31.

Russell, L. K. (1979b). *A study of the armored boreid* Caurinus dectes (*Mecoptera*), thesis submitted to Oregon State University.

Sabrosky, C. W. (1953). 'How many insects are there?', *Syst. Zool.*, **2**, 31–36.

Saether, O. A. (1977). 'Female genitalia in Chironomidae and other Nematocera: morphology, phylogeny, keys', *Bull. Fish. Res. Bd Can.*, **197**, 1–209.

Schaefer, C. W. (1975). 'The mayfly subimago: a possible explanation', *Ann. ent. Soc. Am.*, **68**, 183.

Schaller, F. (1970). 'Collembola (Springschwänze)', *Handb. Zool., Berl.*, **4** (2), 2/1, 1–72, De Gruyter, Berlin.

Schawaller, W. (1978). 'Neue Pseudoskorpione aus dem Baltischen Bernstein der Stuttgarter Bernsteinsammlung (Arachnida: Pseudoscorpionidea)', *Stuttg. Beitr. Naturk. (B)*, **42**, 1–22.

Schilder, M. (1949). 'Zahl und Verbreitung der Coleoptera', *Biol. Zbl.*, **68**, 385–397.

Schlee, D. (1969a). 'Sperma-Uebertragung in ihrer Bedeutung für das phyogenetische System der Sternorrhyncha. Phylogenetische Studien an Hemiptera I. Psylliformes (Psyllina + Aleyrodina) als monophyletische Gruppe', *Z. Morph. Tiere*, **64**, 95–138.

Schlee, D. (1969b). 'Die Verwandtschaftsbeziehungen innerhalb der Sternorrhyncha aufgrund synapomorpher Merkmale. Phylogenetische Studien an Hemiptera. II: Aphidiformes als monophyletische Gruppe', *Stuttg. Beitr. Naturk.*, **199**, 1–19.

Schlee, D. (1969c). 'Bau und Funktion des Aedeagus bei Psyllina und deren Bedeutung für systematische Untersuchungen. Phylogenetische Studien an Hemiptera. III. Entkräftung eines Arguments gegen die Monophylie der Sternorrhyncha', *Z. Morph. Tiere*, **64**, 139–150.

Schlee, D. (1969d). 'Morphologie und Symbiose; ihre Beweiskraft für die Verwandtschaftsbeziehungen der Coleorrhyncha. Phylogenetische Studien an Hemiptera. IV: Heteropteroidea (Heteroptera + Coleorrhyncha) als monophyletische Gruppe', *Stuttg. Beitr. Naturk.*, **210**, 1–27.

Schlee, D. (1969e). 'Der Flügel von *Sphaeraspis*, prinzipiell identisch mit Aphidina-Flügeln. Phylogenetische Studien an Hemiptera. V: Synapomorphe Flügelmerkmale bei Aphidina und Coccina', *Stuttg. Beitr. Naturk.*, **211**, 1–11.

Schlee, D. (1970). 'Insektenfossilien aus der unteren Kreide—1. Verwandtschaftsforschung an fossilen und rezenten Aleyrodina (Insecta, Hemiptera)', *Stuttg. Beitr. Naturk.*, **213**, 1–72.

Schlee, D. (1971). 'Die Rekonstruktion der Phylogenese mit Hennig's Prinzip', *Aufs. Reden senckenb. naturf. Ges.*, **20**, 1–62.

Schlee, D. (1972). 'Bernstein aus dem Libanon', *Kosmos, Stuttg.*, **68**, 460–463.

Schlee, D. (1973). 'Harzkonservierte Vogelfedern aus der untersten Kreide', *J. Orn., Lpz.*, **114**, 207–219.

Schlee, D. (1975). 'Das Problem der Podonominae-Monophylie; Fossiliendiagnose und Chironomidae-Phylogenetik (Diptera)', *Entomologica germ.*, **1**, 316–351.

Schlee, D. (1977a). 'Florale und extraflorale Nektarien sowie Insektenkot als Nahrungsquelle für Chironomidae-Imagines (und andere Diptera)', *Stuttg. Beitr. Naturk. (A)*, **300**, 1–16.

Schlee, D. (1977b). 'Willi Hennig', *Jh. Ges. Naturk. Württ.*, **132**, 196–197.

Schlee, D. (1978a). 'Anmerkungen zur phylogenetischen Systematik: Stellungnahme zu einigen Missverständnissen', *Stuttg. Beitr. Naturk. (A)*, **320**, 1–14.

Schlee, D. (1978b). 'In memoriam Willi Hennig 1913–1976. Eine biographische Skizze', *Entomologica germ.*, **4**, 377–391.

478

Schlee, D., and Dietrich, H. G. (1970). 'Insektführender Bernstein aus der Unter-kreide des Libanon', *Neues Jb. Geol. Paläont. Mh.*, **1970**, 40–50.
Schlee, D., and Glöckner, W. (1978). 'Bernstein. Bernsteine und Bernstein-Fossilien', *Stuttg. Beitr. Naturk. (C)*, **8**, 1–72.
Schlee, H. B., and Schlee, D. (1976). 'Bibliographie der rezenten und fossilen Mecoptera', *Stuttg. Beitr. Naturk. (A)*, **282**, 1–76.
Schliephake, G. (1975). 'Beitrag zur Phylogenetischen Systematik bei Thysanoptera (Insecta)', *Beitr. Ent.*, **25**, 5–13.
Schlüter, T. (1974). 'Kritisches zum Nachweis von Schmetterlingsschuppen aus einem fossilen Harz der mittleren Kreide Nordwestfrankreichs', *Ent. Z., Frankf. a. M.*, **84**, 253–256.
Schlüter, T. (1975). 'Nachweis verschiedener Insecta-Ordines in einem mittelcretazi-schen Harz Nordwestfrankreichs', *Entomologica germ.*, **1**, 151–161.
Schlüter, T. (1978). 'Zur Systematik und Palökologie harzkonservierter Arthropoda einer Taphozönose aus dem Cenomanium von NW-Frankreich', *Berl. geowiss. Abh. (A)*, **9**, 1–150.
Schmidt, W. (1962). 'Neue Insekten aus dem rheinisch-westfälischen Oberkarbon', *Fortschr. Geol. Rheinld Westf.*, **3**, 819–860.
Schmidt, W. (1963). 'Die Evolution der ältesten Insekten', *Natur Mus., Frankf.*, **93**, 449–461.
Schuchert, C. (1926). 'Review of the late Paleozoic formations and faunas, with special reference to the ice-age of middle Permian time', *Bull. geol. Soc. Am.*, **39**, 769–886.
Schuchert, C., and Dunbar, C. O. (1945). *A Textbook of Geology*, vol. 2, *Historical Geology*, 4th ed., Wiley, New York.
Schwarzbach, M. (1939). 'Der älteste Insektenflügel. Bemerkungen zu einem ober-schlesischen Funde', *Jber. geol. Verein. Oberschles.*, **1939**, 28–30.
Schwarzbach, M. (1950). *Das Klima der Vorzeit. Eine Einführung in die Paläoklimato-logie*, Enke, Stuttgart.
Scourfield, D. J. (1940). 'The oldest known fossil insect (*Rhyniella praecursor* Hirst & Maulik)—further details from additional specimens', *Proc. Linn. Soc. Lond.*, **152**, 113–131.
Scudder, G. G. E. (1961). 'The functional morphology and interpretation of the insect ovipositor', *Can. Ent.*, **93**, 267–272.
Scudder, G. G. E. (1964). 'Further problems in the interpretation and homology of the insect ovipositor', *Can. Ent.*, **96**, 405–417.
Scudder, G. G. E. (1971). 'Comparative morphology of insect genitalia', *A. Rev. Ent.*, **16**, 379–406.
Seeger, W. (1975). 'Funktionsmorphologie an Spezialbildungen der Fühlergeissel von Psocoptera und anderen Paraneoptera (Insecta); Psocodea als monophyletische Gruppe', *Z. Morph. Tiere*, **81**, 137–159.
Seeger, W. (1979). 'Spezialmerkmale an Eihüllen und Embryonen von Psocoptera im Vergleich zu anderen Paraneoptera (Insecta); Psocoptera als monophyletische Gruppe', *Stuttg. Beitr. Naturk.*, **329**, 1–57.
Séguy, E. (1959). 'Introduction à l'étude morphologique de l'aile des insectes', *Mém. Mus. natn. Hist. nat., Paris (n.s.) (A)*, **21**, 1–248.
Shaposhnikov, G. K. (1971). 'The principal trends and modes of evolution in aphids', *Proc. 13th Int. Congr. Ent.*, **1**, 196–197.
Sharov, A. G. (1948). 'Triassic Thysanura from the Ural foreland', *Dokl. Akad. Nauk SSSR (n.s.)*, **61**, 517–519. [In Russian.]
Sharov, A. G. (1953). 'The discovery of Permian larvae of alderflies (Megaloptera) from Kargala', *Dokl. Akad. Nauk SSSR (n.s.)*, **89**, 731–732. [In Russian.]
Sharov, A. G. (1957a). 'Types of insect metamorphosis and their relationship', *Ént. Obozr.*, **36**, 569–576. [In Russian, with English summary.]

Sharov, A. G. (1957b). 'The first record of a Cretaceous Aculeata', *Dokl. Akad. Nauk SSSR (n.s.)*, **112**, 943–944. [In Russian.]

Sharov, A. G. (1957c). 'Comparative ontogenetic method and its application in systematics and phylogeny (on the example of the insects)', *Zool. Zh.*, **36**, 64–84. [In Russian, with English summary.]

Sharov, A. G. (1957d). 'The distinctive Palaeozoic wingless insects of the new order Monura (Insecta, Apterygota)', *Dokl. Akad. Nauk SSSR (n.s.)*, **115**, 795–798. [In Russian.]

Sharov, A. G. (1959a). 'Evolution as the process of ontogeny alteration', *Proc. 15th Int. Congr. Zool.*, 105–108.

Sharov, A. G. (1959b). 'The classification of the primarily wingless insects', *Trudӯ Inst. Morf. Zhivot.*, **27**, 175–186. [In Russian.]

Sharov, A. G. (1961a). 'On the system of the orthopterous insects', *Proc. 11th Int. Congr. Ent.*, **1**, 295–296.

Sharov, A. G. (1961b). 'The origin of the order Plecoptera', *Proc. 11th Int. Congr. Ent.*, **1**, 296–298.

Sharov, A. G. (1965). 'Evolution and taxonomy', *Z. zool. Syst. EvolForsch.*, **3**, 349–358.

Sharov, A. G. (1966a). 'The position of the orders Glosselytrodea and Caloneurodea in the insect classification', *Paleont. Zh.*, **1966** (3), 84–93. [In Russian.]

Sharov, A. G. (1966b). *Basic arthropodan stock with special reference to insects*, Pergamon Press, Oxford.

Sharov, A. G. (1967). 'Basic steps in the evolution of the Orthopteroidea', *Dokl. ezheg. Chten. N. A. Kholodkovskogo*, **17**, 3–16. [In Russian.]

Sharov, A. G. (1968). 'The phylogeny of the orthopteroid insects', *Trudӯ paleont. Inst.*, **118**, 1–217. [In Russian. English translation: Sharov, 1971.]

Sharov, A. G. (1971). *Phylogeny of the Orthopteroidea*, translated by J. Salkind, edited by O. Theodor, Israel Program for Scientific Translation, Jerusalem. [English translation of Sharov, 1968.]

Sharov, A. G. (1972). 'On the phylogenetic relationships of the order of the thrips (Thysanoptera)', *Ént. Obozr.*, **51**, 854–858. [In Russian; English translation in *Ent. Rev., Wash.*, **51**, 506–508.]

Shimer, H. W. (1934). 'Correlation chart of geologic formations of North America', *Bull. geol. Soc. Am.*, **45**, 909–936.

Short, J. R. T. (1952). 'The morphology of the head of larval Hymenoptera with special reference to the head of Ichneumonoidea, including a classification of the final instar larvae of the Braconidae', *Trans. R. ent. Soc. Lond.*, **103**, 27–84.

Shvanvich, B. N. (1943). 'Subdivision of Insecta Pterygota into subordinate groups', *Nature, Lond.*, **152**, 727–728.

Shvanvich, B. N. (1946). 'On the interrelationships of the orders of Insecta Pterygota as dependent on the origin of flight', *Zool. Zh.*, **25**, 529–542. [In Russian, with English summary.]

Shvanvich, B. N. (1958). 'Alary system as a basis of the system of pterygote insects', *Proc. 10th Int. Congr. Ent.*, **1**, 605–610.

Siewing, R. (1960). 'Zum Problem der Arthropoden', *Z. wiss. Zool.*, **164**, 238–270.

Simon, R. (1971). 'Neue Arthropodenfunde aus dem Stephan der Halleschen Mulde', *Ber. dt. Ges. geol. Wiss. (A)*, **16**, 53–62.

Skalski, A. W. (1979a). 'A new Lower Cretaceous representative of the family Micropterigidae (Lepidoptera) from Transbaikalia', *Paleont. Zh.*, **1979**, 190–197. [In Russian.] [Original journal not seen.]

Skalski, A. W. (1979b). 'Records of oldest Lepidoptera', *Nota lepid.*, **2**, 61–66.

Smart, J. (1951). 'The wing-venation of the American cockroach *Periplaneta americana* Linn.', *Proc. zool. Soc. Lond.*, **121**, 501–509.

Smart, J. (1963). 'Explosive evolution and the phylogeny of insects', *Proc. Linn. Soc. Lond.*, **174**, 125–126.

Smith, E. L. (1969). 'Evolutionary morphology of external insect genitalia. 1. Origin and relationships to other appendages', *Ann. ent. Soc. Am.*, **62**, 1051–1079.

Smith, E. L. (1970). 'Evolutionary morphology of the external insect genitalia. 2. Hymenoptera', *Ann. ent. Soc. Am.*, **63**, 1–27.

Smithers, C. N. (1965). 'A bibliography of the Psocoptera (Insecta)', *Aust. Zool.*, **13**, 137–209.

Smithers, C. N. (1967). 'A catalogue of the Psocoptera of the world', *Aust. Zool.*, **14**, 1–145.

Smithers, C. N. (1972). 'The classification and phylogeny of the Psocoptera', *Mem. Aust. Mus.*, **14**, 1–349.

Snodgrass, R. E. (1935). *Principles of insect morphology*, McGraw-Hill, New York and London.

Snodgrass, R. E. (1937). 'The male genitalia of orthopteroid insects', *Smithson. misc. Collns*, **96** (5), 1–107.

Snodgrass, R. E. (1938). 'Evolution of the Annelida, Onychophora and Arthropoda', *Smithson. misc. Collns*, **97** (6), 1–159.

Snodgrass, R. E. (1952). *A textbook of arthropodan anatomy*, Comstock, New York.

Snodgrass, R. E. (1954). 'Insect metamorphosis', *Smithson. misc. Collns*, **122** (9), 1–124.

Snodgrass, R. E. (1957). 'A revised interpretation of the external reproductive organs of male insects', *Smithson. misc. Collns*, **135** (6), 1–60.

Snodgrass, R. E. (1958a). 'The insect tentorium and its antecedents', *Proc. 10th Int. Congr. Ent.*, **1**, 487.

Snodgrass, R. E. (1958b). 'Evolution of arthropod mechanisms', *Smithson. misc. Collns*, **138** (2), 1–77.

Snodgrass, R. E. (1960). 'Facts and theories concerning the insect head', *Smithson. misc. Collns*, **142** (1), 1–61.

Snodgrass, R. E. (1961). 'Insect metamorphosis and retromorphosis', *Trans. Am. ent. Soc.*, **87**, 273–280.

Spooner, C. S. (1938). 'The phylogeny of the Hemiptera based on a study of the head capsule', *Illinois biol. Monogr.*, **16** (3), 1–102.

Staesche, K. (1963). 'Uebersicht über die Fauna des deutschen Rotliegenden (Unteres Perm) B. Insekten', *Stuttg. Beitr. Naturk.*, **110**, 1–6.

Stannard, L. J. (1956). 'A note on the relationship of the hemipteroid insects', *Syst. Zool.*, **5**, 94–95.

Stark, B. P., and Gaufin, A. R. (1976). 'The Nearctic genera of Perlidae (Plecoptera)', *Misc. Publs ent. Soc. Am.*, **10**, 1–80.

Steffan, A. W. (1964). 'Torridincolidae, coleopterorum nova familia e regione aethiopica', *Ent. Z., Frankf. a. M.*, **74**, 193–200.

Steffan, A. W. (1968). 'Elektraphididae, aphidinorum nova familia e sucino baltico (Insecta: Homoptera: Phylloxeroidea)', *Zool. Jb. (Syst.)*, **95**, 1–15.

Steffan, A. W. (1978). 'Dem Gedenken an Professor Dr. phil. Dr. rer. nat. h. c. Willi Hennig, Mitglied des wissenschaftlichen Beirates der Zeitschrift für wissenschaftliche Entomologie, Entomologica germanica', *Entomologica germ.*, **4**, cover page before page 193.

Steiner, P. (1930). 'Studien an *Panorpa communis*', *Z. Morph. Oekol. Tiere*, **17**, 1–67.

Steinmann, H. (1973). 'A zoogeographical checklist of World Dermaptera', *Folia ent. hung. (n.s.)*, **26**, 145–154.

Steinmann, H. (1975). 'Suprageneric classification of Dermaptera', *Acta zool. hung.*, **21**, 195–220.

Strassen—see Zur Strassen.

Streng, R. (1973). 'Die Erzeugung eines chitinigen Kokonfadens aus peritrophischer

Membran bei der Larve von *Rhynchaenus fagi* L. (Coleoptera, Curculionidae)', *Z. Morph. Tiere*, **75**, 137–164.

Strümpel, H. (1972). 'Beitrag zur Phylogenie der Membracidae Rafinesque', *Zool. Jb. (Syst.)*, **99**, 313–407.

Štys, P. (1969). 'Revision of fossil and pseudofossil Enicocephalidae (Heteroptera)', *Acta ent. bohemoslovaca*, **66**, 352–365.

Sukacheva, I. D. (1968). 'Mesozoic caddis-flies (Trichoptera) from Transbaikalia', *Paleont. Zh.*, **1968** (2), 59–75. [In Russian; English translation in *Paleont. J.*, **2**, 202–216.]

Sukacheva, I. D. (1976). 'Caddis-flies of the suborder Permotrichoptera', *Paleont. Zh.*, **1976** (2), 94–105. [In Russian; English translation in *Paleont. J.*, **10**, 198–209.]

Suomalainen, E. (1966). 'Achiasmatische Oogenese bei Trichopteren', *Chromosoma*, **18**, 201–207.

Symmons, S. (1952). 'Comparative anatomy of the mallophagan head', *Trans. zool. Soc. Lond.*, **27**, 349–436.

Szelegiewicz, H. (1971). 'Autapomorphic wing-characters in the recent subgroups of the Sternorrhyncha and their significance for the classification of the Palaeozoic members of this group', *Annls zool. Warsz.*, **29**, 15–81. [In Polish, with German summary.]

Szelegiewicz, H., and Popov, Yu. A. (1978). 'Revision der fossilen "Permaphidopsidae" aus dem Perm der UdSSR (Hemiptera: Sternorrhyncha)', *Entomologica germ.*, **4**, 234–241.

Tannert, W. (1958). 'Die Flügelgelenkung bei Odonaten', *Dt. ent. Z. (n.F.)*, **5**, 394–455.

Tannert, W. (1961). 'Die Flügelgelenkung der Odonata, ein Beispiel phylogenetischer Entwicklungstendenzen', *Proc. 11th Int. Congr. Ent.*, **1**, 328–332.

Tasch, P. (1971). 'Permian insect bed—Ohio Range, Antarctica', *Trans. Kans. Acad. Sci.*, **74**, 34–37.

Tasch, P. (1973). 'Jurassic beetle from southern Victoria Land, Antarctica', *J. Paleont.*, **47**, 590–591.

Tasch, P., and Riek, E. F. (1969). 'A Permian insect wing from the Antarctic Sentinel Mountains', *Science, N.Y.*, **164**, 1529–1530.

Tasch, P., and Zimmermann, J. R. (1959). 'New Permian insects discovered in Kansas and Oklahoma', *Science, N.Y.*, **130**, 1656.

Teixeira, C. (1939). 'Insectos de Estefanio do Douro-Litoral', *Publções Mus. Lab. miner. geol. Univ. Porto*, **9**, 1–20.

Teixeira, C. (1941). 'Nouveaux insectes du Stéphanien portugais', *Bolm Soc. geol. Port.*, **1**, 13–31.

Teixeira, C. (1942a). 'Insectos do carbónico alentejano', *Ciencias*, **7**, 331–335.

Teixeira, C. (1942b). 'Sôbre a fauna do estefanio norte de Portugal. Novos elementos para o seu estudo', *Ciencias*, **7**, 548–553.

Teixeira, C. (1942c). 'O carbónico das margens do Douro e seus caracteres paleontológicos', *Bolm Soc. port. Ciênc. nat.*, **13**, Suppl. 3, 527–541.

Teixeira, C. (1944). 'Sur l'étrange forme des ailes de *Stephanomylacris duriensis* Teix.', *Bolm Ass. Filos. nat.*, **2**, 17–18.

Teixeira, C. (1947). 'Nota sôbre um Blatídio fóssil do Retiano de Coimbra', *Bolm Soc. geol. Port.*, **6**, 243–244.

Teixeira, C. (1948). 'Nota sôbre um insecto fóssil do Autuniano do Buçaco', *Comunções Servs geol. Port.*, **28**, 107–109.

Telenga, N. A. (1969). *Origin and evolution of parasitism of Hymenoptera Parasitica and development of their fauna in the USSR*, translated by E. V. Zverezomb-Zubovskii, Israel Program for Scientific Translations, Jerusalem.

Termier, H., and Termier, G. (1960). *Atlas de paléogéographie*, Masson, Paris.

Teskey, H. J. (1971). 'A new soldier fly from Canadian amber (Diptera: Stratiomyidae)', *Can. Ent.*, **103**, 1659–1661.

Théodoridès, J. (1952). 'Les Coléoptères fossiles', *Annls Soc. ent. Fr.*, **121** [1951], 23–48.

Theron, J. G. (1958). 'Comparative studies on the morphology of male scale insects', *Annale Univ. Stellenbosch (A)*, **34**, 1–71.

Thompson, D. B. (1966). 'The occurrence of an insect wing and branchiopods (*Euestheria*) in the lower Keuper marls at Styal, Cheshire', *Mercian Geol.*, **1** [1965], 237–245.

Tiegs, O. W., and Manton, S. M. (1958). 'The evolution of the Arthropoda', *Biol. Rev.*, **33**, 255–337.

Tillyard, R. J. (1916). 'Studies in Australian Neuroptera. No. iv', *Proc. Linn. Soc. N.S.W.*, **41**, 269–332.

Tillyard, R. J. (1917a). 'Mesozoic insects of Queensland. No. 1. Planipennia, Trichoptera, and the new order Protomecoptera', *Proc. Linn. Soc. N.S.W.*, **42**, 175–200.

Tillyard, R. J. (1917b). 'Mesozoic insects of Queensland. No. 2. The fossil dragonfly *Aeschnidiopsis (Aeshna) flindersiensis* Woodward, from the Rolling Downs (Cretaceous) series', *Proc. Linn. Soc. N.S.W.*, **42**, 676–692.

Tillyard, R. J. (1918a). 'Mesozoic insects of Queensland. No. 3. Odonata and Protodonata', *Proc. Linn. Soc. N.S.W.*, **43**, 417–436.

Tillyard, R. J. (1918b). 'Mesozoic insects of Queensland. No. 4. Hemiptera Heteroptera: the family Dunstaniidae. With a note on the origin of the Heteroptera', *Proc. Linn. Soc. N.S.W.*, **43**, 568–592.

Tillyard, R. J. (1919a). 'The panorpoid complex, Part 3. The wing venation', *Proc. Linn. Soc. N.S.W.*, **44**, 533–718.

Tillyard, R. J. (1919b). 'Mesozoic insects of Queensland. No. 5. Mecoptera, the new order Paratrichoptera, and additions to the Planipennia', *Proc. Linn. Soc. N.S.W.*, **44**, 194–212.

Tillyard, R. J. (1919c). 'Mesozoic insects of Queensland. No. 6. Blattoidea', *Proc. Linn. Soc. N.S.W.*, **44**, 358–382.

Tillyard, R. J. (1919d). 'A fossil insect wing belonging to the new order Paramecoptera, ancestral to the Trichoptera and Lepidoptera, from the upper coal-measures of Newcastle, N.S.W.', *Proc. Linn. Soc. N.S.W.*, **44**, 231–256.

Tillyard, R. J. (1920). 'Mesozoic insects of Queensland. No. 7. Hemiptera Homoptera; with a note on the phylogeny of the suborder', *Proc. Linn. Soc. N.S.W.*, **44**, 857–896.

Tillyard, R. J. (1921a). 'Two fossil insect wings in the collection of Mr. John Mitchell, from the Upper Permian of Newcastle, N. S. Wales, belonging to the order Hemiptera', *Proc. Linn. Soc. N.S.W.*, **46**, 413–422.

Tillyard, R. J. (1921b). 'Mesozoic insects of Queensland. No. 8. Hemiptera Homoptera (continued). The genus *Mesogereon*; with a discussion of its relationship with the Jurassic Palaeontinidae', *Proc. Linn. Soc. N.S.W.*, **46**, 270–284.

Tillyard, R. J. (1922a). 'Some new Permian insects from Belmont N.S.W. in the collection of Mr. John Mitchell', *Proc. Linn. Soc. N.S.W.*, **47**, 279–292.

Tillyard, R. J. (1922b). 'Mesozoic insects of Queensland. No. 9. Orthoptera, and additions to the Protorthoptera, Odonata, Hemiptera and Planipennia', *Proc. Linn. Soc. N.S.W.*, **47**, 447–470.

Tillyard, R. J. (1923a). 'The Lower Permian insects of Kansas. Preliminary announcement', *Ent. News*, **34**, 292–295. [See also Dunbar and Tillyard, 1924.]

Tillyard, R. J. (1923b). 'Mesozoic insects of Queensland. No. 10. Summary of the Upper Triassic insect fauna of Ipswich, Q. (with an appendix describing new Hemiptera and Planipennia)', *Proc. Linn. Soc. N.S.W.*, **48**, 481–498.

Tillyard, R. J. (1923c). 'Mesozoic insects of Queensland', *Publs geol. Surv. Qd*, **273**, 1–75.

Tillyard, R. J. (1924). 'Kansas Permian insects. Part III. The new order Pro-
tohymenoptera', *Am. J. Sci. (5)*, **8**, 111–122.
Tillyard, R. J. (1925a). 'The British Liassic dragonflies (Odonata)', *Fossil Insects*, **1**,
1–38, British Museum, London.
Tillyard, R. J. (1925b). 'A new fossil insect wing from Triassic beds near Deewhy,
N.S.W.', *Proc. Linn. Soc. N.S.W.*, **50**, 374–377.
Tillyard, R. J. (1925c). 'Kansas Permian insects. No. IV. The order Palaeodictyop-
tera', *Am. J. Sci. (5)*, **9**, 328–335.
Tillyard, R. J. (1925d). 'Kansas Permian insects. No. V. The orders Protodonata and
Odonata', *Am. J. Sci. (5)*, **10**, 41–73.
Tillyard, R. J. (1926a). 'Fossil insects in relation to living forms', *Nature, Lond.*, **117**,
828–830.
Tillyard, R. J. (1926b). 'Upper Permian insects of New South Wales, Part I.
Introduction and the order Hemiptera. Part II. The orders Mecoptera, Paramecop-
tera and Neuroptera', *Proc. Linn. Soc. N.S.W.*, **51**, 1–30, 265–282.
Tillyard, R. J. (1926c). 'The Rhaetic "crane-flies" from South America not Diptera
but Homoptera', *Am. J. Sci. (5)*, **11**, 265–272.
Tillyard, R. J. (1926d). 'Kansas Permian insects. Part VI. Additions to the orders
Protohymenoptera and Odonata', *Am. J. Sci. (5)*, **11**, 58–73.
Tillyard, R. J. (1926e). 'Kansas Permian insects. Part VII. The order Mecoptera', *Am.
J. Sci. (5)*, **11**, 133–164.
Tillyard, R. J. (1926f). 'Kansas Permian insects. Part VIII. The order Copeognatha',
Am. J. Sci. (5), **11**, 315–349.
Tillyard, R. J. (1926g). 'Kansas Permian insects. Part IX. The order Hemiptera', *Am.
J. Sci. (5)*, **11**, 381–395.
Tillyard, R. J. (1926h). *The insects of Australia and New Zealand*, Angus and
Robertson, Sydney.
Tillyard, R. J. (1927). 'The ancestry of the order Hymenoptera', *Trans. ent. Soc.
Lond.*, **75**, 307–318.
Tillyard, R. J. (1928a). 'A Permian fossil damselfly wing from the Falkland Islands',
Trans. ent. Soc. Lond., **76**, 55–63.
Tillyard, R. J. (1928b). 'Some remarks on the Devonian fossil insects from the Rhynie
Chert beds, Old Red Sandstone', *Trans. ent. Soc. Lond.*, **76**, 65–71.
Tillyard, R. J. (1928c). 'Kansas Permian insects. Part X. The new order Protoperlaria:
a study of the typical genus *Lemmatophora* Sellards', *Am. J. Sci. (5)*, **16**, 185–
220.
Tillyard, R. J. (1928d). 'Kansas Permian insects. Part XI. Order Protoperlaria, family
Lemmatophoridae (continued)', *Am. J. Sci. (5)*, **16**, 313–348.
Tillyard, R. J. (1928e). 'Kansas Permian insects. Part XII. The family Delopteridae,
with a discussion of its ordinal position', *Am. J. Sci. (5)*, **16**, 469–484.
Tillyard, R. J. (1929). 'Permian Diptera from Warner's Bay, N.S.W.', *Nature, Lond.*,
123, 778–779.
Tillyard, R. J. (1931a). 'The evolution of the Class Insecta', *Pap. Proc. R. Soc. Tasm.*,
1930, 1–89. [Very detailed review in *Bernstein-Forsch.*, **3**, 181–184.]
Tillyard, R. J. (1931b). 'Kansas Permian insects. Part XIII. The new order Protelyt-
roptera, with a discussion of its relationships', *Am. J. Sci. (5)*, **21**, 232–266.
Tillyard, R. J. (1932a). 'The evolution of the class Insecta', *Am. J. Sci. (5)*, **23**,
529–539. [Reply to Petrunkevitch, Raymond and Carpenter.]
Tillyard, R. J. (1932b). 'Kansas Permian insects. Part XIV. The order Neuroptera',
Am. J. Sci. (5), **23**, 1–30.
Tillyard, R. J. (1932c). 'Kansas Permian insects. Part XV. The order Plectoptera',
Am. J. Sci. (5), **23**, 97–134, 237–272.
Tillyard, R. J. (1933). 'The panorpoid complex in the British Rhaetic and Lias', *Fossil
insects*, **3**, 1–79, British Museum, London.

Tillyard, R. J. (1935a). 'The evolution of the scorpion-flies and their derivatives (order Mecoptera)', *Ann. ent. Soc. Am.*, **28**, 1–45.

Tillyard, R. J. (1935b). 'Upper Permian insects of New South Wales. Part V. The order Perlaria or stoneflies', *Proc. Linn. Soc. N.S.W.*, **60**, 385–391.

Tillyard, R. J. (1936a). 'Are termites descended from true cockroaches?', *Nature, Lond.*, **137**, 655.

Tillyard, R. J. (1936b). 'A new Upper Triassic fossil insect bed in Queensland', *Nature, Lond.*, **138**, 719–720.

Tillyard, R. J. (1936c). 'Kansas Permian insects. Part XVI. The order Plectoptera (continued): the family Doteridae, with a note on the affinities of the order Protohymenoptera', *Am. J. Sci. (5)*, **32**, 435–453.

Tillyard, R. J. (1937a). 'The ancestors of the Diptera', *Nature, Lond.*, **139**, 66–67.

Tillyard, R. J. (1937b). 'Kansas Permian insects. Part XVII. The order Megasecoptera and additions to the Palaeodictyoptera, Odonata, Protoperlaria, Copeognatha and Neuroptera', *Am. J. Sci. (5)*, **33**, 81–110.

Tillyard, R. J. (1937c). 'Kansas Permian insects. Part XVIII. The order Embiaria', *Am. J. Sci. (5)*, **33**, 241–251.

Tillyard, R. J. (1937d). 'Kansas Permian insects. Part XIX. The order Protoperlaria (continued). The family Probnisidae', *Am. J. Sci. (5)*, **33**, 401–425.

Tillyard, R. J. (1937e). 'Kansas Permian insects. Part XX. The cockroaches, or order Blattaria I, II', *Am. J. Sci. (5)*, **34**, 169–202, 249–276.

Tillyard, R. J., and Fraser, F. C. (1938–1940). 'A reclassification of the order Odonata based on some new interpretation of the venation of the dragonfly wing', *Aust. Zool.*, **9**, 125–169 [1938] and 195–221 [1940].

Timmermann, G. (1957). 'Stellung und Gliederung der Regenpfeifervögel (Ordnung Charadriiformes) nach Massgabe des mallophagologischen Befundes', *1st Symp. Host Specif. Parasit. Vertebr., Neuchâtel*, 159–172.

Tindale, N. B. (1945). 'Triassic insects of Queensland, I. *Eoses*, a probable lepidopterous insect from the Triassic beds of Mt. Crosby, Queensland', *Proc. R. Soc. Qd*, **56**, 37–46.

Tkalcu, B. (1972). 'Arguments contre l'interprétation traditionelle de la phylogénie des abeilles (Hymenoptera, Apoidea)', *Bull. Soc. ent. Mulhouse*, **1972**, 17–28.

Tokunaga, M. (1935). 'A morphological study of a nymphomyiid fly', *Philipp. J. Sci.*, **56**, 127–214.

Tollet, R. (1959). 'Note systématique sur les Corynoscelidae fam. nov. (Diptera) du globe et description d'un Corynoscelidae nouveau de l'hémisphère austral', *Bull. Annls Soc. r. ent. Belg.*, **95**, 132–153.

Tomaszewski, C. (1973). 'Studies on the adaptive evolution of the larvae of Trichoptera', *Acta zool. cracov.*, **18**, 311–398.

Townes, H. K. (1973). 'Three tryphonine ichneumonids from Cretaceous amber (Hymenoptera)', *Proc. ent. Soc. Wash.*, **75**, 282–287.

Tuomikoski, R. (1961). 'Zur Systematik der Bibionomorpha, 1. Anisopodidae und Protorhyphidae', *Suom. hyönt. Aikak.*, **27**, 65–69.

Tuxen, S. L. (ed.) (1956). *Taxonomist's glossary of genitalia in insects*, Munksgaard, Copenhagen.

Tuxen, S. L. (1958). 'Relationships of Protura', *Proc. 10th Int. Congr. Ent.*, **1**, 493–497.

Tuxen, S. L. (1959). 'The phylogenetic significance of ontogeny in entognathous apterygotes', *Smithson. misc. Collns*, **157**, 379–416.

Tuxen, S. L. (1960). 'Ontogenie und Phylogenie bezogen auf die apterygoten Insekten', *Zool. Anz.*, **164**, 359–363.

Tuxen, S. L. (1963). 'Phylogenetic trends in the Protura, as shown by relationship between recent genera', *Z. zool. Syst. EvolForsch.*, **1**, 277–310.

Tuxen, S. L. (1964). 'The Protura. A revision of the species of the world with keys for determination', *Actual. scient. ind.*, **1311**, 1–360.

Tuxen, S. L. (1969). 'Nomenclature and homology of genitalia in insects', *Memorie Soc. ent. ital.*, **48** (1B), 6–16.

Tuxen, S. L. (1972). 'Filogenesi degli atterigoti', *Atti Congr. Nat. ital.*, **9**, 193–205.

Ulmer, G. (1912). 'Die Trichopteren des Baltischen Bernsteins', *Schr. phys.-ökon. Ges. Königsb.*, **10**, 1–380.

Ulrich, W. (1943). 'Die Mengeidae (Mengenillini) und die Phylogenie der Strepsipteren', *Z. Parasitenk.*, **13**, 62–101.

Usachev, D. A. (1968). 'New Jurassic Asilomorpha (Diptera) in Karatau', *Ént. Obozr.*, **47**, 617–628. [In Russian, with English summary; English translation in *Ent. Rev., Wash.*, **47**, 378–384.]

Verhoeff, K. W. (1904). 'Ueber vergleichende Morphologie des Kopfes niederer Insekten mit besonderer Berücksichtigung der Dermaptera und Thysanuren, nebst biologisch-physiologischen Beiträgen', *Nova Acta Acad. Caesar. Leop. Carol.*, **84** (1), 1–144.

Verhoeff, K. W. (1926). 'Diplopoda', *Bronn's Kl. Ordn. Tierreichs*, 5/2, 2, 1–288.

Vishnyakova, V. N. (1971). 'The structure of the abdominal appendages in fossil cockroaches (Insecta: Blattodae [sic])', *Proc. 13th Int. Congr. Ent.*, **1**, 315 [In Russian.]

Vishnyakova, V. N. (1975). 'Psocoptera from fossil late Cretaceous insect-bearing resins of the Taymyr', *Ent. Obozr.*, **54**, 92–106. [In Russian; English translation in *Ent. Rev., Wash.*, **54** (1), 63–75.]

Wagner, J. (1939). 'Aphaniptera', *Bronn's Kl. Ordn. Tierreichs*, 5/3, 13f, 1–114.

Walker, M. V. (1938). 'Evidence of Triassic insects in the Petrified Forest National Monument, Arizona', *Proc. U.S. natn. Mus.*, **85**, 137–141.

Wallace, F. L., and Fox, R. C. (1975). 'A comparative morphological study of the hind wing venation of the order Coleoptera, part I', *Proc. ent. Soc. Wash.*, **77**, 329–354.

Wallis, F. S. (1939a). 'New blattoid insects from the South Wales coalfield', *Geol. Mag.*, **76**, 23–35.

Wallis, F. S. (1939b). 'The fossil insects of the Bristol coalfield', *Proc. Bristol Nat. Soc. (4)*, **8**, 429–434.

Waterlot, G. (1934a). 'Paléozoologie du bassin de la Sarre', *Annls Soc. géol. N.*, **59**, 201–213.

Waterlot, G. (1934b). 'Bassin houiller de la Sarre et de la Lorraine. II. Faune fossil', *Etud. Gîtes minér. Fr.*, 1–317.

Watson, M. C. (1956). 'The utilization of mandibular armature in taxonomic studies of anisopterous species', *Trans. Am. ent. Soc.*, **81**, 155–205.

Weber, H. (1924). 'Das Thorakalskelett der Lepidopteren. Ein Beitrag zur vergleichenden Morphologie des Insekten-Thorax', *Z. Anat. EntwGesch., Leipzig*, **73**, 277–331.

Weber, H. (1930). *Biologie der Hemipteren. Eine Naturgeschichte der Schnabelkerfe*, Springer, Berlin.

Weber, H. (1933). *Lehrbuch der Entomologie*, Fischer, Jena.

Weber, H. (1949). *Grundriss der Insektenkunde*, 2nd ed., Fischer, Jena.

Weber, H. (1974). *Grundriss der Insektenkunde*, 5th ed., revised by H. Weidner, Fischer, Stuttgart.

Weesner, F. (1960). 'Evolution and biology of the termites', *A. Rev. Ent.*, **5**, 153–170.

Weidner, H. (1966). 'Betrachtungen zur Evolution der Termiten', *Dt. ent. Z. (n.F.)*, **13**, 323–350.

Weidner, H. (1969). 'Die Ordnung Zoraptera oder Bodenläuse', *Ent. Z., Frankf. a. M.*, **79**, 29–51.

486

Weidner, H. (1970a). 'Isoptera (Termiten)', *Handb. Zool., Berl.*, **4** (2), 2/14, 1–147, De Gruyter, Berlin.

Weidner, H. (1970b). 'Zoraptera (Bodenläuse)', *Handb. Zool., Berl.*, **4** (2), 2/15, 1–12, De Gruyter, Berlin.

Weidner, H. (1972). 'Copeognatha (Psocodea)', *Handb. Zool., Berl.*, **4** (2), 2/16, 1–94, De Gruyter, Berlin and New York.

Wenk, P. (1965). 'Ueber die Biologie blutsaugender Simuliiden (Diptera). II. Schwärmverhalten, Geschlechterfindung und Kopulation', *Z. Morph. Oekol. Tiere*, **55**, 671–713.

Weyda, F. (1974). 'Coxal vesicles of Machilidae', *Pedobiologia*, **14**, 138–141.

Whalley, P. E. S. (1977). 'Lower Cretaceous Lepidoptera', *Nature, Lond.*, **266**, 526.

Whalley, P. E. S. (1978). 'New taxa of fossil and recent Micropterigidae with a discussion of their evolution and a comment on the evolution of the Lepidoptera (Insecta)', *Ann. Transv. Mus.*, **31**, 71–86.

Wheeler, G. C., and Wheeler, J. (1972). 'The subfamilies of Formicidae', *Proc. ent. Soc. Wash.*, **74**, 35–45.

White, M. J. D. (1954). *Animal cytology and evolution*, 2nd ed., Cambridge University Press, Cambridge.

White, M. J. D. (1957). 'Cytogenetics and systematic entomology', *A. Rev. Ent.*, **2**, 71–90.

Wieland, G. R. (1925). 'Rhaetic crane flies from South America', *Am. J. Sci. (5)*, **9**, 21–28.

Wigglesworth, V. B. (1963). 'Origin of wings in insects', *Nature, Lond.*, **197**, 97–98.

Wigglesworth, V. B. *et al.* (1963). 'The origin of flight in insects. (The scientific proceedings of the meeting of the Royal Entomological Society held on 3rd July, 1963.)', *Proc. R. ent. Soc. Lond. (C)*, **28**, 23–32.

Wille, A. (1960). 'The phylogeny and relationships between the insect orders', *Revta Biol. trop.*, **8**, 93–122.

Willmann, R. (1977). 'Zur systematischen Stellung von *Austropanorpa* (Insecta, Mecoptera) aus dem Alttertiär Australiens', *Palaeont. Z.*, **51**, 12–18.

Willmann, R. (1978). 'Mecoptera (Insecta, Holometabola)', *Fossilium Cat. (I)*, **124**, 1–139, Junk, The Hague.

Willmann, R. (1979). 'Ueber das Exoskelett von *Austromerope poultoni* Killington (Mecoptera: Meropeidae), ein Beitrag zur Phylogenie der Schnabelfliegen', *Z. zool. Syst. EvolForsch.*, **17**, 296–309.

Wilson, E. O., Carpenter, F. M., and Brown, W. L. (1967). 'The first Mesozoic ants, with the description of a new subfamily', *Psyche, Camb.*, **74**, 1–19.

Withycombe, C. L. (1924). 'Some aspects of the biology and morphology of the Neuroptera. With special reference to immature stages and their possible phylogenetic significance', *Trans. ent. Soc. Lond.*, **1924**, 303–412.

Wittig, G. (1955). 'Untersuchungen am Thorax von *Perla abdominalis* Burm. (Larve und Imago) unter besonderer Berücksichtigung des peripheren Nervensystems und der Sinnesorgane', *Zool. Jb. (Anat.)*, **74**, 491–570.

Wong, S. K. (1970). 'The study of internal genital systems and the classification of the Psocoptera', *N.Z. Ent.*, **4** (4), 66–71.

Wootton, R. J. (1963). 'Actinoscytinidae (Hemiptera-Heteroptera) from the Upper Triassic of Queensland', *Ann. Mag. nat. Hist. (13)*, **6**, 249–255.

Wootton, R. J. (1965). 'Evidence for tracheal capture in early Heteroptera', *Proc. 12th Int. Congr. Ent.*, 65–67.

Wygodzinsky, P. (1961). 'On a surviving representative of the Lepidotrichidae (Thysanura)', *Ann. ent. Soc. Am.*, **54**, 621–627.

Yasuda, T. (1962). 'On the larva and pupa of *Neomicropteryx nipponensis* Issiki, with its biological notes (Lepidoptera, Micropterygidae)', *Kontyû*, **30**, 130–136.

Yoshimoto, C. M. (1975). 'Cretaceous chalcidoid fossils from Canadian amber', *Can. Ent.*, **107**, 499–528.

Zalesskiy, M. D. (1932). 'Observations sur les nouveaux insectes permiens de l'Europe orientale', *Bull. Soc. géol. Fr. (5)*, **2**, 183–210.

Zalesskiy, Yu. M. (1938). 'Nouveaux insectes permiens de l'ordre des Embiodea', *Annls Soc. géol. N.*, **63**, 62–81.

Zalesskiy, Yu. M. (1939). 'Studies on the Permian insects from the basin of the River Sylva and problems in the evolution of the class Insecta. III. Some new species of the Protohymenoptera, Homoptera, Hemipsocoptera, Psocoptera, Protoperlaria, Isoptera and Protoblattoidea', *Problemy̆ Paleont.*, **5**, 33–91. [In Russian, with French summary.]

Zalesskiy, Yu. M. (1946). 'A new representative of the Permian Neuroptera', *Dokl. Akad. Nauk SSSR (n.s.)*, **51**, 543–544.

Zalesskiy, Yu. M. (1953). 'New localities for Cretaceous insects in the Volga basin, Kazakhstan and Transbaikalia', *Dokl. Akad. Nauk SSSR (n.s.)*, **89**, 163–166. [In Russian.]

Zalesskiy, Yu. M. (1958). 'Morpho-functional causes of wing-folding in Palaeoptera', *Zool. Zh.*, **37**, 845–854. [In Russian, with English summary.]

Zeuner, F. E. (1934). 'Phylogenesis of the stridulating organ of locusts', *Nature, Lond.*, **134**, 460–461.

Zeuner, F. E. (1936). 'The subfamilies of the Tettigoniidae (Orthoptera)', *Proc. R. ent. Soc. Lond. (B)*, **5**, 103.

Zeuner, F. E. (1939). *Fossil Orthoptera Ensifera*, British Museum (Natural History), London.

Zeuner, F. E. (1940a). 'Biology and evolution of fossil insects', *Proc. Geol. Ass.*, **51**, 44–48.

Zeuner, F. E. (1940b). 'Saltatoria Ensifera fossilia', *Fossilium Cat. (I)*, **90**, 1–108, Feller, Neubrandenburg.

Zeuner, F. E. (1955). 'A fossil blattoid from the Permian of Rhodesia', *Ann. Mag. nat. Hist. (12)*, **8**, 685–688.

Zeuner, F. E. (1959a). 'A new Liassic dragonfly from Gloucestershire', *Palaeontology*, **1**, 406–407.

Zeuner, F. E. (1959b). 'Jurassic beetles from Grahamland, Antarctica', *Palaeontology*, **1**, 407–409.

Zeuner, F. E. (1961). 'A Triassic insect fauna from the Molteno Beds of South Africa', *Proc. 11th Int. Congr. Ent.*, **1**, 304–306.

Zeuner, F. E. (1962). 'Fossil insects from the lower Lias of Charmouth (Dorset)', *Bull. Br. Mus. nat. Hist. (Geol.)*, **7**, 155–171.

Zherikhin, V. V. (1978). 'The evolution and changes in Cretaceous and Coenozoic faunistic complexes (Tracheata and Chelicerata)', *Trudy̆ paleont. Inst.*, **165**, 1–198. [In Russian.]

Zherikhin, V. V., and Sukacheva, I. D. (1973). 'On the Cretaceous insect-bearing "ambers" (resins) of northern Siberia', *Dokl. ezheg. Chten. N. A. Kholodkovskogo*, **24** [1971], 3–48. [In Russian.]

Zimmermann, J. R. (1959). 'A new Permian insect horizon', *Proc. ent. Soc. Wash.*, **61**, 259.

Zur Strassen, R. (1973). 'Insektenfossilien aus der unteren Kreide—5: Fossile Fransenflügler aus mesozoischem Bernstein des Libanon (Insecta: Thysanoptera)', *Stuttg. Beitr. Naturk. (A)*, **256**, 1–51.

Zwick, P. (1969). *Das phylogenetische System der Plecopteren als Ergebnis vergleichend-anatomischer Untersuchungen*, Dissertation, Univ. Kiel (Kurzfassung [précis]), Kiel.

Zwick, P. (1973). 'Insecta: Plecoptera. Phylogenetisches System und Katalog', *Tierreich*, **94**, 1–465.

488

Zwick, P. (1977). 'Australian Blephariceridae (Diptera)', *Aust. J. Zool.*, **Suppl. 46**, 1–121.

Zwick, P. (1980a). 'Plecoptera (Steinfliegen)', *Handb. Zool., Berl.*, **4** (2), 2/7, 1–115, De Gruyter, Berlin and New York.

Zwick, P. (1980b, in press). 'Plecoptera', in Keast, A. (ed.), *Ecological biogeography in Australia, Monographiae biol.*

Reference added to proof

Schlee, D. (1980). *Bernstein-Raritäten. Farben, Strukturen, Fossilien, Handwerk*, Staatliches Museum für Naturkunde, Stuttgart.

Author index

Subject index

An asterisk denotes a phylogenetic tree or map; two asterisks denote morphological figures. Principal references are in bold type.

510

514